Photonics Metamaterials: Processing and Applications

Photonics Metamaterials: Processing and Applications

Guest Editors

Feng Cheng
Yao Wang

MDPI

Basel • Beijing • Wuhan • Barcelona • Belgrade • Novi Sad • Cluj • Manchester

Guest Editors

Feng Cheng
Department of Electrical and Computer Engineering
Northeastern University
Boston
USA

Yao Wang
Vaziri Laboratory
The Rockefeller University
New York
USA

Editorial Office
MDPI AG
Grosspeteranlage 5
4052 Basel, Switzerland

This is a reprint of the Special Issue, published open access by the journal *Photonics* (ISSN 2304-6732), freely accessible at: https://www.mdpi.com/journal/photonics/special_issues/281I583MX8.

For citation purposes, cite each article independently as indicated on the article page online and as indicated below:

Lastname, A.A.; Lastname, B.B. Article Title. *Journal Name* **Year**, *Volume Number*, Page Range.

ISBN 978-3-7258-7106-3 (Hbk)
ISBN 978-3-7258-7107-0 (PDF)
https://doi.org/10.3390/books978-3-7258-7107-0

Contents

About the Editors ix

Francesco Dell'Olio
Metasurface-Enabled Microphotonic Biosensors via BIC Modes
Reprinted from: *Photonics* **2025**, *12*, 48, https://doi.org/10.3390/photonics12010048 . . . . **1**

Forouzan Habibighahfarokhi, Olga Sergaeva, Luca Carletti, Paolo Franceschini, Andrea Tognazzi, Andrea Locatelli, et al.
Nonlinear Dielectric Metasurfaces for Terahertz Applications
Reprinted from: *Photonics* **2025**, *12*, 370, https://doi.org/10.3390/photonics12040370 . . . . **18**

Grazia Giuseppina Politano
Localized Effects in Graphene Oxide Systems: A Pathway to Hyperbolic Metamaterials
Reprinted from: *Photonics* **2025**, *12*, 121, https://doi.org/10.3390/photonics12020121 . . . . **37**

Zongliang He, Dong Fang and Yougen Yi
Design of a Tunable Metamaterial Absorption Device with an Absorption Band Covering the Mid-Infrared Atmospheric Window
Reprinted from: *Photonics* **2025**, *12*, 148, https://doi.org/10.3390/photonics12020148 . . . . **48**

Aijun Zhu, Mengyi Zhang, Weigang Hou, Lei Cheng, Cong Hu and Chuanpei Xu
A High-Sensitivity Graphene Metasurface and Four-Frequency Switch Application Based on Plasmon-Induced Transparency Effects
Reprinted from: *Photonics* **2025**, *12*, 218, https://doi.org/10.3390/photonics12030218 . . . . **64**

Chandra M. Adhikaril
Optoplasmonics of Single-Walled Carbon Nanotube Thin Films
Reprinted from: *Photonics* **2025**, *12*, 298, https://doi.org/10.3390/photonics12040298 . . . . **77**

Asad Khan, Jinling Zhang, Muhammad Ishfaq, Ibrar Ahmad, Shahbaz Khan and Kamlesh Kumar Soothar
Compact Reflective Metasurface: Production of Broadband Vortex Beams in Millimeter Waves
Reprinted from: *Photonics* **2025**, *12*, 305, https://doi.org/10.3390/photonics12040305 . . . . **91**

Thanh Son Pham, Bui Xuan Khuyen, Vu Dinh Lam, Liangyao Chen and Youngpak Lee
Wide-Angle, Polarization-Independent Broadband Metamaterial Absorber by Using Plasmonic Metasurface-Based Split-Circular Structure
Reprinted from: *Photonics* **2025**, *12*, 334, https://doi.org/10.3390/photonics12040334 . . . . **104**

Jorge Parra
Polarization-Insensitive Silicon Grating Couplers via Subwavelength Metamaterials and Metaheuristic Optimization
Reprinted from: *Photonics* **2025**, *12*, 428, https://doi.org/10.3390/photonics12050428 . . . . **118**

Hongfu Liu, Jijun Li, Hua Yang, Junqiao Wang, Boxun Li, Han Zhang and Yougen Yi
TiN-Only Metasurface Absorber for Solar Energy Harvesting
Reprinted from: *Photonics* **2025**, *12*, 443, https://doi.org/10.3390/photonics12050443 . . . . **137**

Lei Zhu, Shujie Wang, Yun Wang, Liang Dong, Hailong Li, Yiya Wang and Xumin Ding
A Dual-Band Tunable Electromagnetically Induced Transparency (EIT) Metamaterial Based on Vanadium Dioxide
Reprinted from: *Photonics* **2025**, *12*, 463, https://doi.org/10.3390/photonics12050463 . . . . **152**

Sayed El. Soliman, Israa Abood, Naglaa Abdel All and Chii-Chang Chen
Topological Rainbow Trapping with Expanded Bandwidth in Valley Photonic Crystals
Reprinted from: *Photonics* **2025**, *12*, 487, https://doi.org/10.3390/photonics12050487 . . . . **165**

Oleg Kameshkov, Vasily Gerasimov, Boris Goldenberg and Vladimir Nazmov
Spectral Tuning and Angular–Gap Interrogation of Terahertz Spoof Surface Plasmon Resonances Excited on Rectangular Subwavelength Grating Using Attenuated Total Reflection in Otto Configuration
Reprinted from: *Photonics* **2025**, *12*, 651, https://doi.org/10.3390/photonics12070651 . . . . **180**

Maxim Durach
Beyond Green's Functions: Inverse Helmholtz and "Om'ॐ-Potential Methods for Macroscopic Electromagnetism in Isotropy-Broken Media
Reprinted from: *Photonics* **2025**, *12*, 660, https://doi.org/10.3390/photonics12070660 . . . . **200**

Siqiang Zhao, Daoye Zheng, Yunche Zhu, Shuyan Zou and Yu-Sheng Lin
Flexible Color Filter Using Lithium Niobate Metamaterial with Ultrahigh Purity and Brightness Characteristics
Reprinted from: *Photonics* **2025**, *12*, 768, https://doi.org/10.3390/photonics12080768 . . . . **210**

Lin Jiang, Qi Hu and Yijun Feng
Reconfigurable High-Efficiency Power Dividers Using Waveguide Epsilon-Near-Zero Media for On-Demand Splitting
Reprinted from: *Photonics* **2025**, *12*, 897, https://doi.org/10.3390/photonics12090897 . . . . **222**

Kele Chen, Zhengning Wang, Meizhang Guan, Shubo Cheng, Hongyu Ma, Zao Yi and Boxun Li
Tunable Ultra-Wideband VO_2–Graphene Hybrid Metasurface Terahertz Absorption Devices Based on Dual Regulation
Reprinted from: *Photonics* **2025**, *12*, 987, https://doi.org/10.3390/photonics12100987 . . . . **233**

Ioannis S. Fosteris and George S. Kliros
Analytical Design of Optically Transparent, Wideband, and Tunable Microwave Absorber Based on Graphene Spiral Resonator Metasurface
Reprinted from: *Photonics* **2025**, *12*, 1006, https://doi.org/10.3390/photonics12101006 . . . . . . **251**

Xingyi Li, Peixuan Wu, Yuanyuan Xing, Peng Shi, Xinjian Yao and Yaoguang Ma
Design Methodology of a VIS Hybrid Refractive–Metalens System with a Wide FOV
Reprinted from: *Photonics* **2025**, *12*, 1023, https://doi.org/10.3390/photonics12101023 . . . . . . **267**

Ying Yao, Hongming Fei, Xin Liu, Mingda Zhang, Pengqi Dong, Junjun Ren and Han Lin
Mach–Zehnder Interferometer Electro-Optic Modulator Based on Thin-Film Lithium Niobate Valley Photonic Crystal
Reprinted from: *Photonics* **2026**, *13*, 33, https://doi.org/10.3390/photonics13010033 . . . . **279**

Kao-Peng Min, Yu-Ting Gao, Cheng-Fu Yang, Walter Water and Chi-Ting Ho
Effect of Mo Layer Thickness on Bandwidth Tunability and Absorption Properties of Planar Ultra-Wideband Optical Absorbers
Reprinted from: *Photonics* **2026**, *13*, 86, https://doi.org/10.3390/photonics13010086 . . . . **292**

Eric Amoateng, Ellis Mubarak Sani, Kingsford Sarkodie Obeng Kwakye and Alexandros Pitilakis
Analysis and Design of a Hybrid Graphene/Vanadium-Dioxide Terahertz Metasurface with Independently Reconfigurable Reflection Phase and Magnitude
Reprinted from: *Photonics* **2026**, *13*, 195, https://doi.org/10.3390/photonics13020195 . . . . **311**

Sy Khiem Nguyen, Ba Thong Trinh, Dayeon Kim, Netrapal Singh, Young Kyu Hwang, Vu Dinh Lam and Ilsun Yoon
Quasi-Bound States in the Continuum in PDMS-Supported Silicon Metasurfaces
Reprinted from: *Photonics* **2026**, *13*, 226, https://doi.org/10.3390/photonics13030226 . . . . **332**

About the Editors

Feng Cheng

Feng Cheng earned his Ph.D. from the Department of Electrical and Computer Engineering at Northeastern University, Boston. His academic research mainly focuses on metamaterials, nanophotonics, and the all-optical magnetic switching phenomenon, with an emphasis on understanding and engineering light-matter interactions at the nanoscale. He has extensive experience in computational electromagnetic modeling and ultrafast laser experiments, integrating theoretical analysis with experimental validation to explore nanoscale optical phenomena. His work contributes to the development of advanced functional optical materials and the next-generation all-optical data storage technologies.

Yao Wang

Yao Wang earned his Ph.D. in Bioengineering from Northeastern University, Boston. His academic research mainly focuses on bioimaging, instrumentation, and image processing, with an emphasis on developing automated laser surgery systems and illumination techniques for neuroscience applications. He has extensive experience in high-resolution volumetric two-photon imaging, femtosecond laser experiments, and computational image processing. He integrates hardware development with novel algorithms to improve neuronal imaging, and contributes to the development of high-throughput microscope platforms and large-scale optogenetics.

photonics

MDPI

Review

Metasurface-Enabled Microphotonic Biosensors via BIC Modes

Francesco Dell'Olio

Micro Nano Sensor Group, Polytechnic University of Bari, 70126 Bari, Italy; francesco.dellolio@poliba.it

Abstract: Photonic biosensors based on bound states in the continuum (BIC) resonant modes exhibit a transformative potential for high-sensitivity, label-free detection across various diagnostic applications. BIC-enabled metasurfaces, utilizing dielectric, plasmonic, and hybrid structures, achieve ultra-high Q-factors and amplify target molecule interactions on functionalized sensor surfaces. These unique properties result in increased refractive index sensitivity and low detection limits, essential for monitoring biomolecules in clinical diagnostics, environmental analysis, and food safety. Recent advancements in BIC-enabled metasurfaces have demonstrated ultra-low detection limits in the zeptomolar range, making these devices highly promising for real-world applications. This review paper critically discusses the design principles of BIC-based biosensors, emphasizing key factors such as material selection, structural asymmetry, and functionalization strategies that enhance both sensitivity and specificity. Additionally, recent advancements in fabrication techniques that enable precise BIC control with scalable approaches for practical biosensing applications are examined. Case studies demonstrate the effectiveness of BIC metasurfaces for real-time, low-concentration detection, highlighting their versatility and adaptability. Finally, the review discusses future challenges and opportunities, such as integration with microfluidics for point-of-care testing and multiplexed sensing, underscoring the potential of BIC-based platforms to revolutionize the field of biosensing.

Keywords: nanophotonics; metasurface; biosensing; label-free detection; point-of-care diagnostics

1. Introduction

Biosensors are analytical devices designed to detect biological molecules, which integrate a biorecognition element with a signal transducer. They hold immense potential across healthcare, biodefence, environmental monitoring, and food safety, driven by the ongoing demand for rapid, accurate, and cost-effective diagnostics [1]. Current research focuses on achieving three key goals: (i) integration into point-of-care testing (PoCT) systems comprising low-cost disposable biochips and simple reader devices; (ii) ultra-low limits of detection (LoDs) that extend down to the zeptomolar (zM) range; and (iii) ultra-wide detection ranges spanning, for example, from zM to picomolar (pM) concentrations. These objectives are pursued alongside standard requirements such as selectivity, specificity, stability, and reproducibility. Among biosensors, optical chip-scale platforms exhibit unique advantages in addressing these challenges. Their label-free detection capabilities, high sensitivity, and compatibility with CMOS fabrication processes make them ideal for miniaturized, portable PoCT applications. Silicon photonic biosensors [2,3], such as those developed by SiPhox and Genalyte, exemplify this trend. SiPhox leverages silicon photonic ring resonators to measure biomarkers, such as high-sensitivity C-reactive protein, with compact reader instruments, while Genalyte's Maverick system integrates multiplexed detection with cloud connectivity for advanced diagnostics. These platforms demonstrate the feasibility of high-performance biosensors within disposable and user-friendly systems. The development of chip-scale optical biosensors offers a promising pathway toward fulfilling the ambitious goals of modern biosensing. By enabling precise and scalable integration of biosensing functions, these devices pave the way for transformative advances in diagnostics.

Metasurfaces—thin layers composed of subwavelength nanostructures arranged according to periodic or pseudo-periodic grids—have demonstrated remarkable potential in controlling the behavior of electromagnetic waves, enabling precise manipulation of light properties such as phase, amplitude, and polarization [4–12]. By tailoring the geometry, size, and arrangement of these nanostructures, metasurfaces can achieve effects that traditionally require much thicker optical components, thus paving the way for highly compact, lightweight, and versatile optical devices. Among the various resonant phenomena that can be engineered within metasurfaces, BICs have emerged as particularly promising because of their unique ability to provide high-Q resonances. This feature makes BICs highly suitable for applications requiring extreme sensitivity and strong light–matter interaction, such as biosensing [13–16].

BICs are distinctive, resonant states that remain perfectly confined within a radiation continuum, thus achieving theoretically infinite quality factors [17–19]. Unlike typical resonant modes, which tend to couple with external radiation and lose energy, BICs avoid such coupling by virtue of symmetry constraints or destructive interference mechanisms. This enables them to retain energy without radiation leakage, resulting in confined resonances that are exceptionally sharp and robust against radiative decay. As illustrated in Figure 1, BICs reside in the radiation continuum without coupling to radiative modes, producing enhanced field localization that is beneficial for applications requiring intense light–matter interactions. This intrinsic property has spurred substantial interest in BIC-based metasurfaces across multiple fields, from lasing and nonlinear optics to sensing and imaging.

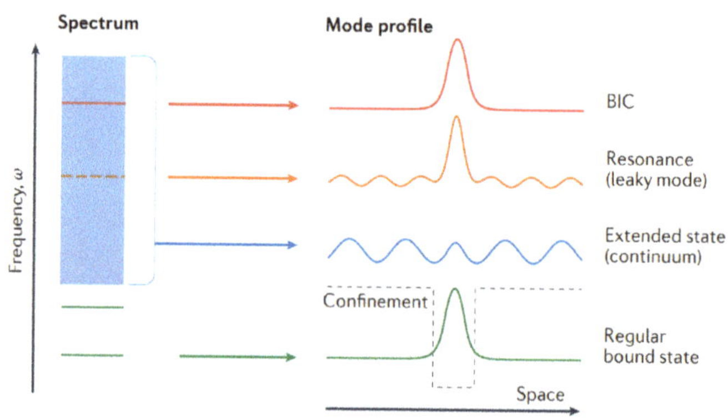

Figure 1. Illustrative concept of BICs. Within an open system, the frequency spectrum includes a continuum of extended states (blue) and discrete bound states (green) that do not radiate. Structures with specific symmetry and interference properties enable BICs (red), which reside within the continuum but remain localized without energy loss. Reprinted with permission from [17].

The high-Q resonances achievable on BIC-based metasurfaces are particularly advantageous for biosensing applications, where a sensor's performance is often determined by its sensitivity to changes in the refractive index near the sensing surface. High-Q BIC resonances amplify the interaction between incident light and analytes within the functionalized sensing region, significantly enhancing the sensor's response to minute changes in refractive index [20,21]. This sensitivity is critical in detecting low concentrations of biomolecules—such as proteins, DNA, and other biomarkers—that might be indicative of disease states. Additionally, the high-Q nature of BICs reduces the linewidth of the resonance, leading to improved signal-to-noise ratios that are essential for high-precision biosensing.

Similarly to other chip-scale technologies, a further advantage of BIC-based biosensors is their capability for label-free detection. Traditional biosensing methods often require fluorescent or colorimetric labels to visualize molecular interactions, which can add complexity and limit the application scope. BIC-based sensors, however, achieve strong field localization within the active sensing layer, enabling the direct detection of biomolecules without the need for additional labels. This label-free approach not only simplifies the sensing process but also enables real-time monitoring of biological interactions, an important feature in diagnostic and environmental sensing applications.

BIC-based metasurfaces exploit a very simple excitation scheme, which is normal-to-surface. As shown in Figure 2, the metasurface is typically excited by a beam that is normal to the surface, and reflection and transmission spectra are observed. The metasurface supports a BIC-wave that is delocalized in a wide area, thus enhancing light–matter interaction.

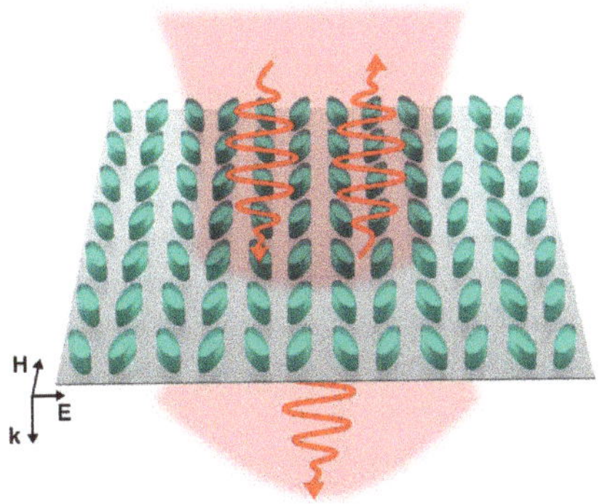

Figure 2. Light scattered by the metasurface. The metasurface is top-illuminated. Reflection and transmission can be observed. The metasurface is a periodic 2D array of dielectric metaunits, each consisting of a pair of tilted silicon nanobars. Each metaunit has a mirror symmetry across the major vertical axis and consists of elliptical dielectric nanoresonators with major and minor axes of approximately 280 nm and 100 nm, respectively. These dimensions are fine-tuned to achieve the desired resonance wavelength. The all-dielectric metasurface was fabricated by nanopatterning a thin layer (100 nm) of amorphous silicon, which was deposited on a fused-silica substrate. Reprinted with permission from [22].

The versatility of BIC-based metasurfaces is enhanced by the range of material choices available, including dielectric, plasmonic, and hybrid metal–dielectric materials. Dielectric metasurfaces, for instance, offer low-loss, high-Q resonances due to minimal intrinsic absorption, making them ideal for applications where high sensitivity and stability are paramount. Plasmonic metasurfaces, which utilize metals such as gold or silver, provide strong field enhancement near the metal surface, making them well-suited for applications requiring intense local field effects. However, their intrinsic losses can limit the achievable Q-factors. Hybrid metal–dielectric BIC metasurfaces represent a promising compromise, combining the strong field localization of plasmonic structures with the low-loss advantages of dielectric materials. This flexibility allows BIC-based sensors to be tailored to specific application requirements, optimizing parameters such as sensitivity, robustness, and compatibility with lab-on-chip platforms [23–26].

This review explores the principles and applications of BIC-based metasurfaces for biosensing. The focus is on metasurfaces that operate in the visible and infrared spectral regions, where BIC resonances are particularly relevant for optical biosensing applications. Several metasurfaces for biosensing that support BIC have been demonstrated in the THz regime, achieving an LoD as low as a few tens of aM [27–29]. Section 2 delves into the mechanisms behind BIC resonances in metasurfaces, providing an overview of symmetry-protected, Friedrich–Wintgen, and Fabry–Perot BICs. Section 3 examines the key design parameters that are essential for maximizing sensitivity and specificity in BIC-based biosensors, including refractive index sensitivity, material choice, and functionalization techniques. Section 4 discusses the fabrication techniques and challenges associated with producing high-Q BIC metasurfaces. Section 5 highlights applications of BIC-based biosensors in clinical diagnostics and environmental monitoring, emphasizing performance benchmarks. Finally, Section 6 outlines potential future developments and challenges for BIC-based biosensing technologies.

2. Principles of BIC Resonance in Metasurfaces

The BIC phenomenon occurs due to specific conditions of symmetry and interference that prevent coupling with radiative modes, allowing for high field confinement within the metasurface structure [30–33]. BICs, originally introduced in quantum mechanics, have become a fundamental concept in nanophotonics and are utilized in various applications, including enhanced biosensing, which is the topic of this review paper.

BICs can be categorized into three primary types based on their mechanisms of confinement: symmetry-protected BICs [34–36], Friedrich–Wintgen (or accidental) BICs [37–39], and Fabry–Perot BICs [40–42]. Symmetry-protected BICs arise in systems where the confined mode's symmetry prevents it from coupling with the radiative continuum. In metasurfaces, this form of BIC typically occurs in symmetric structures where the symmetry of the mode is incompatible with the surrounding radiation modes. This symmetry mismatch results in a non-radiating state that remains perfectly confined. When the symmetry of these structures is intentionally broken, the symmetry-protected BIC transitions to a quasi-BIC (qBIC), introducing a controlled level of radiative leakage. This slight asymmetry enables the creation of ultra-high-Q resonances that retain much of the confinement characteristics of a pure BIC, making them suitable for experimental applications where complete isolation is not feasible. Friedrich–Wintgen BICs, named after the physicists who first identified this mechanism, rely on destructive interference between two or more resonant modes, resulting in a non-radiating state within the continuum. This type of BIC does not depend on the system's symmetry but instead occurs due to phase-matching conditions that allow multiple resonances to cancel each other's radiative components. Fabry–Perot BICs, on the other hand, emerge in periodic or multilayer structures where phase-matching conditions lead to constructive interference within the structure, effectively confining light due to balanced internal and external phase conditions. These BICs are particularly relevant in multilayer and waveguide structures, where they exploit internal reflection mechanisms to achieve confinement.

The ability to confine light within a metasurface structure using BICs has profound implications for enhancing light–matter interactions. This high-Q resonance produces a stronger electromagnetic field within the region of confinement, which is particularly advantageous for applications like biosensing that rely on detecting minute changes in refractive index. By focusing light within a small volume, BICs amplify the interaction between light and biomolecules, increasing sensitivity to low concentrations of analytes. The enhancement of the electromagnetic field within the metasurface also enables label-free detection, as it eliminates the need for external tags or markers to visualize biomolecular interactions.

BICs can be realized in dielectric, plasmonic, and metal–dielectric hybrid metasurfaces, each offering distinct advantages based on their material properties and resonant behaviors.

Dielectric metasurfaces, typically composed of materials like silicon or titanium dioxide, are particularly well-suited for supporting high-Q BICs due to their low intrinsic optical losses. At the state-of-the-art, the highest Q-factor in such BIC-supporting metasurfaces is close to 500,000 [43]. Dielectric materials facilitate the generation of Mie resonances, which allow for strong confinement of electromagnetic fields within the nanostructure while minimizing dissipative losses. This characteristic makes dielectric metasurfaces ideal for applications demanding high sensitivity and stability, such as biosensors designed to detect minute changes in refractive index or low concentrations of analytes. The ability to achieve ultra-narrow resonance linewidths further enhances their precision and signal-to-noise ratio, making them a cornerstone in the development of advanced BIC-based sensing platforms.

Plasmonic metasurfaces, in contrast, rely on metallic materials such as gold, silver, or aluminum to harness surface plasmon resonance (SPR). These metasurfaces exhibit exceptionally strong near-field enhancement due to the collective oscillation of free electrons at the metal–dielectric interface when illuminated by light at specific wavelengths. While plasmonic metasurfaces inherently suffer from higher intrinsic losses compared with their dielectric counterparts, the intense field localization near the metallic nanostructures can significantly amplify light–matter interactions. This property is especially advantageous for applications that prioritize strong surface sensitivity, such as label-free biosensing and surface-enhanced Raman spectroscopy. For instance, plasmonic metasurfaces have been shown to detect low concentrations of biomolecules by leveraging their capacity to amplify the electromagnetic field within the sensing region, albeit at the expense of broader resonance linewidths and reduced Q-factors.

Metal–dielectric hybrid metasurfaces bridge the gap between these two paradigms by combining the high-field localization of plasmonic materials with the low-loss characteristics of dielectric components. These hybrid designs enable the realization of quasi-BICs that offer a balance between field confinement and radiative leakage. By integrating plasmonic nanostructures with dielectric layers, hybrid metasurfaces can achieve enhanced near-field interactions while maintaining relatively narrow resonance linewidths compared to purely plasmonic systems. This synergy is particularly valuable for biosensing applications where both high sensitivity and robustness are required. For example, hybrid metasurfaces can be tailored to operate at specific wavelengths, optimizing the trade-off between sensitivity and Q-factor to meet the needs of diverse sensing scenarios.

The choice between dielectric, plasmonic, or hybrid metasurfaces depends on the specific requirements of the application. Dielectric metasurfaces excel in scenarios where minimizing energy loss and achieving high-Q resonances are critical, making them suitable for high-precision sensing. Plasmonic metasurfaces are preferred for applications demanding extreme near-field enhancement and strong surface interactions, despite their higher losses. Hybrid metasurfaces offer a versatile alternative, leveraging the strengths of both approaches to deliver tailored solutions for demanding biosensing environments. This versatility highlights the importance of material and structural optimization in designing metasurfaces that maximize the advantages of BICs across a broad range of applications.

As already mentioned, structural asymmetry plays a critical role in tuning the Q-factor and resonance properties of BICs, particularly in the generation of quasi-BICs. Introducing a slight asymmetry in a BIC-supporting structure—such as altering the shape, refractive index, or periodicity of the nanostructures—results in a quasi-BIC with finite but high Q-factors. This controlled asymmetry allows for selective coupling with radiative modes, enabling external access to the confined mode without fully compromising its resonant characteristics. For biosensing applications, the ability to fine-tune the Q-factor by adjusting the degree of asymmetry is essential for optimizing sensitivity to refractive index changes. Quasi-BICs created through asymmetry adjustments offer a practical approach for enhancing the detection of biomolecular interactions within the functionalization layer of the sensor. By designing the asymmetry to achieve a specific resonance linewidth, quasi-BICs provide high field localization with a controlled level of radiation, balancing confinement with accessibility.

Figure 3 illustrates the impact of introducing asymmetry in metasurfaces designed to support BICs, taking as an example the metasurface discussed in [44]. The figure demonstrates how varying the asymmetry parameter d_x, from $d_x = 0$ nm (symmetric configuration) to $d_x = 120$ nm (asymmetric configuration), influences the transition from pure BICs to quasi-BICs (qBICs). In the symmetric case, the metasurface supports BICs that are perfectly confined without radiative losses, leading to theoretically infinite Q-factors. As asymmetry increases, controlled radiation leakage is introduced, transforming the BICs into qBICs with finite yet high Q-factors. This tunable balance between field confinement and radiative coupling in qBICs is particularly advantageous for biosensing applications, as it allows for high sensitivity while maintaining practical detection capabilities.

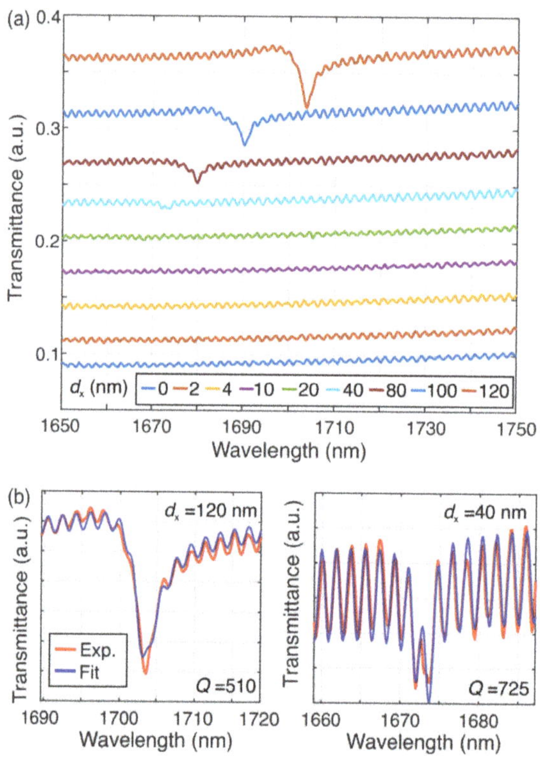

Figure 3. Illustration of the transition from BICs to qBICs in a metasurface as the asymmetry parameter d_x varies. For $d_x = 0$ nm (symmetric case), the metasurface supports perfectly confined BICs with infinite Q-factors. As d_x increases to 120 nm, qBICs emerge with controlled leakage and finite Q-factors, balancing field confinement and radiative coupling. This tunable behavior highlights the potential of such metasurfaces for applications requiring both sensitivity and accessibility, such as biosensing. (**a**) Experimentally measured transmittance spectra. (**b**) Spectra for $d_x = 40$ nm and $d_x = 120$ nm with the relevant fit. Reprinted with permission from [44].

3. Design Considerations for BIC-Based Biosensors

The design of BIC-based biosensors requires careful optimization of some parameters of the metasurface, such as refractive index sensitivity, field confinement where the target–receptor molecular interaction takes place, Q-factor of the BIC mode, and figure of merit (FOM), defined as the sensitivity-to-linewidth ratio. These parameters collectively determine the sensor's performance in detecting small refractive index changes, which is essential for biosensing. In BIC-based sensors, the refractive index sensitivity typically ranges between 100 and 500 nm/RIU, with high-Q resonances yielding experimental values above 1000 [45]. The FOM, expressed in RIU^{-1}, reaches values exceeding 100 RIU^{-1} for optimized BIC structures. This high FOM, together with the field confinement close to the sensor surface exposed to the target molecules, is crucial for achieving precise detection, as it balances sensitivity with spectral resolution. The measured Q-factor, Q_{tot}, has a significant impact on the LoD, as described by the following equation:

$$\mathrm{LoD} = \frac{\lambda_0}{S Q_{tot} \left(\frac{Q_{tot}}{Q_R} \right)} \sqrt{3\sigma}, \tag{1}$$

where λ_0 is the resonance wavelength, S is the sensitivity (nm/RIU), Q_{tot} is the measured total quality factor, and Q_R is the resonant Q-factor, i.e., the quality factor without losses. The term σ represents the noise variance. This equation highlights that the LoD decreases as Q_{tot} increases, underscoring the critical role of achieving high-Q resonances to resolve minute shifts in the resonance wavelength. For a more detailed discussion, refer to Ref. [46].

The selection of materials and geometric optimization play pivotal roles in maximizing the Q-factor and enhancing the refractive index sensitivity. Dielectric materials are preferred due to their low intrinsic losses and capability to support high-Q resonances with minimal energy dissipation. Conversely, hybrid metal–dielectric structures, which incorporate plasmonic materials like gold or silver, offer enhanced near-field interactions, albeit at the cost of slightly reduced Q-factors, typically in the order of some hundreds [47]. These hybrid designs are useful in applications where field enhancement near the sensor surface is prioritized, even if higher radiative losses are introduced.

Specificity in biosensing applications relies on functionalization strategies that ensure selective binding of target analytes. Functionalization methods often involve biochemical receptors such as antibodies, aptamers, or molecularly imprinted polymers (MIPs) tailored to interact specifically with the target molecule. The strong field confinement provided by BICs enhances the interaction between the target analyte and the functionalized surface, improving signal response and enabling low-concentration detection without the need for labeling agents.

Case studies of BIC-based biosensors demonstrate their effectiveness in real-time monitoring and low-concentration analyte detection. They will be discussed in Section 5.

Simulation Techniques for BIC-Based Metasurfaces

The design and optimization of metasurfaces supporting BICs heavily rely on advanced numerical simulation techniques. These methods allow for the detailed exploration of electromagnetic behavior, facilitating the tailoring of metasurface properties to achieve high sensitivity, Q-factors, and field confinement. Key simulation methods include rigorous coupled-wave analysis (RCWA), finite-difference time-domain (FDTD), finite element method (FEM), and eigenmode solvers.

RCWA [48–52] is a frequency-domain method particularly well-suited for analyzing periodic structures such as metasurfaces. By expanding the electromagnetic fields into Fourier series, RCWA provides a computationally efficient way to study the diffraction behavior and resonance characteristics of periodic structures. This method is highly effective for analyzing the optical response of BIC metasurfaces, particularly their reflectance, transmittance, and resonance sharpness. RCWA is frequently used in the initial stages of design to evaluate the influence of lattice parameters, feature dimensions, and material properties on the emergence of BICs and quasi-BICs (qBICs).

The FDTD method is widely employed for simulating the dynamic interaction of light with metasurfaces. Solving Maxwell's equations in the time domain, FDTD enables the detailed analysis of transient and steady-state field distributions. This method is particularly useful for visualizing field localization near BIC modes and optimizing the geometric parameters for maximum sensitivity and Q-factor. FDTD simulations are also critical for studying the effect of symmetry-breaking perturbations on the transition from BIC to qBIC modes.

The FEM is another widely used approach, particularly for metasurfaces with complex geometries or inhomogeneous material properties. FEM operates in the frequency domain, dividing the computational space into finite elements and solving Maxwell's equations locally. This method is ideal for detailed studies of near-field distributions, coupling effects, and the impact of material anisotropy on the BIC properties.

Eigenmode solvers are crucial for the direct identification and analysis of BIC modes in metasurfaces. By solving for the eigenfrequencies and eigenfields, these solvers help determine the conditions under which BICs occur, quantify Q-factors, and predict how structural parameters affect radiative losses and resonance tunability. This approach provides a robust framework for understanding the transition from pure BICs to qBICs under controlled symmetry-breaking.

The simulation process for BIC metasurfaces often involves a combination of these methods. RCWA provides rapid initial analyses of periodic structures, while FDTD and FEM offer more detailed insights into electromagnetic field behavior and resonance optimization. Eigenmode solvers complement these techniques by providing direct access to mode characteristics and guiding the design toward specific application requirements.

Environmental and mechanical considerations, such as temperature variations or substrate deformation, can also influence the performance of BIC-based sensors. Multiphysics simulation platforms that integrate optical, thermal, and mechanical modeling enable the prediction of sensor behavior under real-world conditions, ensuring robust designs for practical applications.

4. Fabrication Techniques, Functionalization, and Microfluidic Integration Challenges

The fabrication of BIC-based metasurfaces relies on precise nanofabrication techniques to achieve the high structural fidelity necessary for supporting high-Q resonances. In fact, the Q-factor is limited by the fabrication imperfections and the finite size of the metasurface, in addition to the light absorption by the non-transparent media where the electromagnetic field is confined and the angular spread collimated beam utilized for the metasurface excitation.

Commonly employed methods for metasurfaces fabrication include electron beam lithography (E-beam lithography), focused ion beam (FIB) milling, and nanoimprint lithography (NIL). Each technique offers unique advantages and faces distinct challenges when applied to the production of BIC metasurfaces for biosensing applications. This section explores these fabrication techniques, addressing the challenges in achieving reproducible, high-Q BICs, and the importance of material selection in maintaining BIC quality over large sensor areas.

E-beam lithography is widely used for the fabrication of BIC metasurfaces due to its high resolution and precise control over nanoscale features. This technique allows for the direct writing of patterns on a substrate by using a focused electron beam to expose a resist layer, which is subsequently developed to reveal the desired nanostructures. E-beam lithography is particularly effective for creating the fine features and complex geometries required for BIC formation, such as symmetry-breaking elements that enable quasi-BICs (qBICs) with controlled radiative losses. However, E-beam lithography is limited by its low throughput and high cost, making it less suitable for large-scale production. Achieving reproducibility over large areas is challenging with E-beam lithography, as variations in exposure dose and development conditions can lead to inconsistencies in feature dimensions, which directly affect the Q-factor of BICs.

FIB milling is another technique employed in the fabrication of BIC metasurfaces. FIB uses a focused beam of ions, typically gallium, to sputter material from the surface, allowing for direct patterning of nanostructures without the need for a resist layer. FIB milling offers flexibility in pattern design and is well-suited for rapid prototyping and the fabrication of complex structures. However, like E-beam lithography, FIB is limited in scalability and throughput. Additionally, the high energy of ion milling can introduce damage to the material, potentially reducing the quality of the BIC resonance due to structural imperfections and material degradation.

Nanoimprint lithography (NIL) has emerged as a highly versatile and promising method for the large-scale fabrication of metasurfaces based on bound states in the continuum (BIC), particularly for applications requiring extensive biosensing areas [53–56]. This technique employs a pre-patterned mold to replicate nanoscale structures onto a substrate. By applying heat and pressure, the resist layer conforms to the mold's intricate patterns, enabling the creation of highly precise and complex geometries. NIL offers significant advantages, including high throughput and reduced costs compared to techniques like electron-beam lithography. Recent research has demonstrated that NIL can achieve nanoscale features with accuracies comparable to those produced by electron-beam lithography while offering greater scalability. However, NIL also faces technical challenges. The precision of the process is critically dependent on the quality of the mold and the uniform distribution of pressure during the imprinting phase. Minor imperfections in the mold or uneven pressure can lead to inconsistencies in feature dimensions and shapes, which may degrade the quality of BIC resonances by introducing scattering losses and lowering the Q-factor. Furthermore, the materials used in NIL must withstand the mechanical and thermal stresses of the process, which narrows the range of suitable materials for fabricating BIC metasurfaces.

The choice of fabrication technique has a profound impact on the performance of BIC-based biosensors. While E-beam lithography and FIB milling provide high precision, their scalability is limited, making NIL a more promising option for producing large-area BIC metasurfaces at scale. However, achieving consistent high-Q performance across large areas remains a challenge that requires continued research into material stability, process optimization, and mold fabrication for NIL.

Surface functionalization is a critical step in transforming a fabricated metasurface into a functional biosensor. One commonly employed protocol utilized for all-dielectric silicon or silicon nitride metasurfaces involves silanization followed by the application of a crosslinker to immobilize biorecognition elements. The surface is functionalized and incubated with aptamers or antibodies specific to the target analyte, ensuring covalent binding. This approach enables the precise detection of biomolecular interactions.

Integration with microfluidic systems is equally essential for real-time sensing applications. PDMS-based microfluidic chambers are typically plasma-bonded to the metasurface. The chamber features inlet and outlet ports for controlled fluid injection and allows the metasurface to interact with analytes in a liquid environment. This approach facilitates precise monitoring of molecular binding events, crucial for biosensing applications. The chamber should be optimized to ensure compatibility with high-Q metasurface resonance conditions, especially minimizing signal perturbations caused by flow-induced stress.

To provide a clear summary of the key processes involved, Figure 4 illustrates the fabrication, functionalization, and microfluidic integration steps for BIC-based metasurfaces. This schematic offers an overview of the methodologies described in this section, highlighting the critical stages required to manufacture these biosensors for practical applications.

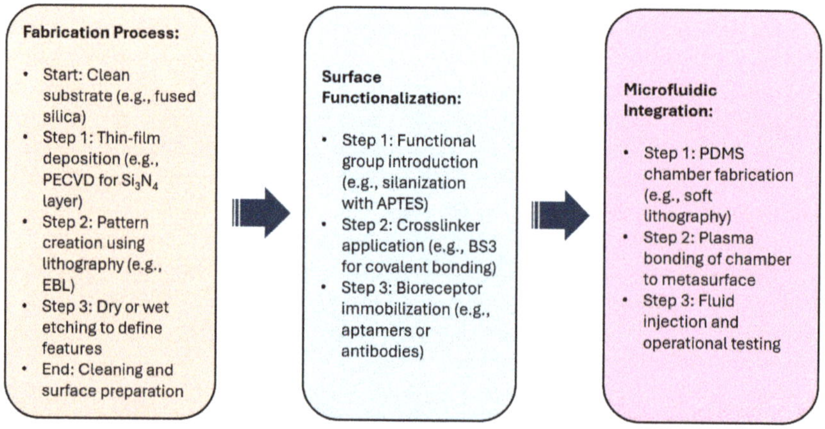

Figure 4. Overview of the fabrication, functionalization, and integration processes for BIC-based metasurfaces for biosensing.

5. Applications of BIC-Based Biosensors

As widely discussed in this paper, BIC-based biosensors leverage the unique properties of high-Q resonances and enhanced light–matter interactions to achieve exceptional performance metrics. These sensors have demonstrated transformative potential across various application domains, including clinical diagnostics, environmental monitoring, and food safety. This section provides an in-depth discussion of the performance parameters critical for biosensor applications, followed by an analysis of experimental results from BIC-based biosensors.

5.1. Key Performance Metrics for Biosensors

The performance of a biosensor is determined by several critical parameters, which define its suitability for specific applications. Among these, the LoD and dynamic range are the most fundamental metrics, but sensitivity, specificity, and time of response also play vital roles.

The LoD represents the smallest concentration of analyte that a biosensor can reliably detect. For BIC-based sensors, LoD is significantly enhanced by their high-Q resonances, which amplify the interaction between the analyte and the functionalized surface.

Dynamic range refers to the range of analyte concentrations over which the sensor provides accurate and linear detection. A wide dynamic range is essential for biosensors used in clinical and environmental monitoring, where analyte concentrations can vary significantly. BIC-based biosensors typically exhibit dynamic ranges spanning several orders of magnitude, ensuring their applicability across diverse scenarios.

Sensitivity quantifies the sensor's ability to detect small changes in analyte concentration. In BIC-based biosensors, sensitivity is often expressed as the shift in resonance wavelength per unit change in refractive index (nm/RIU).

Specificity refers to the ability of the biosensor to distinguish the target analyte from other substances in complex samples. BIC-based sensors achieve high specificity through advanced functionalization strategies, such as the use of molecularly imprinted polymers, aptamers, or antibodies, which selectively bind to the target analyte. The strong field confinement provided by BICs further enhances specificity by increasing binding affinity.

The time required for a biosensor to detect and quantify an analyte is critical in applications requiring real-time monitoring. BIC-based sensors generally offer rapid response times due to their highly localized field interactions, enabling quick and efficient analyte detection. Reported response times range from a few seconds to minutes, depending on the target analyte and functionalization method.

5.2. Experimental Applications of BIC-Based Biosensors

The unique properties of BIC-based biosensors have enabled their use in a variety of real-world applications. Experimental evidence highlights their effectiveness in clinical diagnostics, environmental monitoring, and food safety. Table 1 summarizes the experimental performance metrics of some highly representative recent BIC-based biosensors, highlighting their LoD, sensitivity, and Q-factors across various applications. Figure 5 provides a graphical summary of their application domains and performance metrics. The results presented in Table 1 highlight and confirm the trade-offs and advantages of different approaches for BIC-based biosensors. For instance, dielectric metasurfaces exhibit higher Q-factors due to their low intrinsic optical losses. On the other hand, hybrid systems combine the strengths of plasmonic and dielectric structures. These systems achieve significant near-field enhancement due to plasmonic effects while maintaining relatively low losses. Although their Q-factors are slightly lower than purely dielectric structures, hybrid metasurfaces offer advantages in detecting biomolecules at extremely low concentrations, as evidenced by LoDs in the zeptomolar range.

BIC-based biosensors have shown exceptional potential in detecting biomarkers at ultra-low concentrations, facilitating early diagnosis of diseases. For instance, Clabassi et al. [57] utilized a hybrid BIC sensor to detect transactive response DNA-binding protein 43 (TDP-43), a biomarker for neurodegenerative diseases, especially amyotrophic lateral sclerosis, achieving an LoD of 100 zM and a detection range ranging from 100 fM to 100 zM.

Zito et al. [58] demonstrated the detection of the transforming growth factor-beta (TGP-beta) at 10 fM in saliva, showcasing the applicability of BIC sensors in non-invasive diagnostics. The result is particularly interesting due to the use of MIP as a biorecognition element and the clinical relevance of the target biomarker, a cytokine, which is involved in the progression of oral cancer.

In environmental and food safety applications, BIC-based sensors have been employed to detect pathogens, toxins, allergens, pollutants, and contaminants with high precision. Schiattarella et al. [59] reported the detection of ochratoxin A in food products with an LoD of 2.3 pg/mL (=5.7 pM), highlighting their capability for ensuring compliance with safety standards. Ochratoxin A has nephrotoxic, immunotoxic, carcinogenic, and genotoxic effects. Thus, its detection, also at the level of traces, is crucial in many application contexts. Sensors based on the same operating principle or similar ones are also being explored for real-time monitoring of heavy metals and pesticides in water and air samples.

Table 1. Comparative performance of BIC-based biosensors in experimental studies.

Study	Target Analyte	LoD
Zito et al., 2024 [58]	TGF-beta	10 fM
Clabassi et al., 2024 [57]	TDP-43	100 zM
Schiattarella et al., 2022 [59]	Ochratoxin A	6 pM

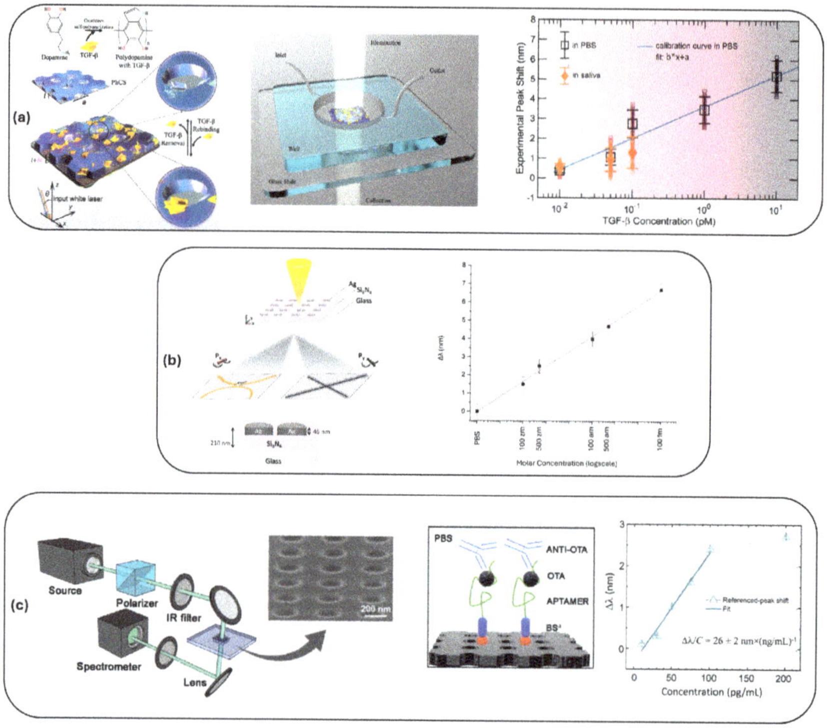

Figure 5. Graphical summary of experimental achievements and application areas for BIC-based biosensors, including clinical diagnostics, environmental monitoring, and food safety. (**a**) Scheme of an MIP–BIC biosensor for TGF-beta detection, showing the interrogation scheme and dose-response curve (LoD = 10 fM [58]). (**b**) Nanostructure based on dimers for hybrid symmetry-protected BIC generation, consisting of a square array of fully silver-filled nanoholes within a silicon nitride waveguide slab. The dose-response curve is provided (LoD = 100 zM [57]). (**c**) Sensing setup with an SEM image of the metasurface and a schematic of surface functionalization. A spectral shift vs. ochratoxin A concentration is shown (LoD = 6 pM [59]). Reprinted with permission from [57–59].

6. Future Directions, Challenges, and Conclusions

BIC-based biosensors hold immense promise for applications requiring high sensitivity, low detection limits, and precise specificity, particularly in fields such as clinical diagnostics, environmental monitoring, and food safety. Despite these advantages, the broader application and commercialization of BIC-based biosensors require addressing several technological challenges and leveraging emerging trends to enhance their functionality and adaptability. Key areas for future research include the development of hybrid BIC sensors, improving scalability, integrating with microfluidics, and advancing functionalization and multiplexing capabilities. Also, integration with hyperspectral imaging has already proven its promise with the experimental demonstration of biomolecular detection information at the extremely low molecular counts of approximately three molecules per μm^2 [22].

One of the most promising advancements in BIC-based biosensors is the integration of hybrid structures that combine BIC resonances with advanced materials, such as MIPs. MIPs offer selective binding sites that serve as a "molecular fingerprint" for target analytes, thereby enhancing the sensor's specificity. By incorporating MIPs into BIC structures, sensors can achieve greater selectivity and stability, making them especially suited for applications where high specificity is required. Additionally, materials like graphene and transition metal dichalcogenides offer unique properties that can be harnessed to further customize BIC sensors. Graphene's tunable electrical and optical properties, for instance, can facilitate dynamic control of BIC resonances, enabling real-time adjustments in sensing applications and further broadening the capabilities of BIC-based sensors.

Scalability remains a significant challenge in the development of BIC-based biosensors, especially for large-scale production. While high-precision fabrication techniques such as electron beam lithography and focused ion beam milling are effective for research purposes, their low throughput and high costs limit their feasibility for mass production. Nanoimprint lithography has emerged as a promising alternative, offering the ability to replicate high-resolution features over large areas with reduced costs. However, achieving consistent high-Q performance across large surfaces using NIL remains challenging, as even minor variations in mold quality or pressure can impact the uniformity of the BIC structures. Addressing these issues is crucial for enabling large-scale production of BIC sensors that maintain high sensitivity and reliable performance, paving the way for widespread use in commercial and industrial applications.

The integration of BIC-based biosensors with microfluidic systems presents another frontier for innovation, particularly for point-of-care testing in medical and environmental applications. Microfluidic platforms allow precise control over small fluid volumes, enabling efficient sample handling and on-site testing. By combining BIC sensors with microfluidics, it is possible to develop portable diagnostic devices capable of rapid, real-time analysis in decentralized settings, thus reducing the dependency on centralized laboratories. However, integrating BIC structures with microfluidic channels involves significant technical challenges, including alignment, material compatibility, and maintaining sensor stability over extended use. Successfully addressing these integration challenges could expand the accessibility and utility of BIC-based sensors in healthcare, offering faster diagnostics and improved patient outcomes.

Further advancements in functionalization methods could also enhance the specificity and applicability of BIC biosensors. Current functionalization techniques, such as using antibodies, aptamers, or molecularly imprinted polymers, provide a degree of selectivity that is crucial for accurate detection. However, new strategies involving multi-layered functionalization or combinatorial libraries could push the boundaries of selectivity even further, enabling BIC sensors to accurately distinguish between structurally similar biomolecules. This would be particularly beneficial for clinical diagnostics, where high specificity is required to detect disease markers in complex biological samples. Additionally, advances in multiplexing could enable BIC-based sensors to detect multiple analytes simultaneously, which is invaluable in medical diagnostics where profiling multiple biomarkers can provide a comprehensive health assessment.

The miniaturization of BIC-based sensor systems is another area with transformative potential. By integrating compact light sources, detectors, and readout electronics with BIC structures, it is feasible to create portable devices for field diagnostics and point-of-care applications. Such miniaturized systems could allow BIC biosensors to be deployed in resource-limited or remote settings, providing rapid and accurate testing capabilities where laboratory facilities are unavailable. The portability and ease of use of such devices would not only broaden the applicability of BIC sensors but also help to address healthcare disparities by improving access to diagnostics in underserved regions.

In conclusion, BIC-based biosensors have demonstrated clear advantages over traditional sensing technologies, offering higher sensitivity, lower detection limits, and superior specificity through enhanced light confinement and selective functionalization. These attributes position BIC-based sensors as a transformative tool in biosensing, with potential applications across diagnostics, environmental monitoring, and food safety. Nonetheless, challenges in scalability, material integration, and device miniaturization remain. Addressing these challenges will require a collaborative effort across disciplines, combining advancements in nanofabrication, materials science, and microfluidics to fully realize the potential of BIC technology.

Looking forward, BIC-based biosensors are poised to play a significant role in next-generation biosensing and diagnostic applications. The ongoing exploration of hybrid materials, improved fabrication techniques, and integration strategies will likely expand the scope and functionality of BIC sensors, making them increasingly competitive in the biosensor industry. As these advancements continue, BIC technology is expected to drive innovation in diagnostics and environmental sensing, offering more sensitive, specific, and accessible biosensing solutions that meet the growing demands of modern healthcare and environmental safety.

As already mentioned, the two key performance parameters of a biosensor are the LoD and the dynamic range. Significant improvements in these metrics can be achieved by enhancing the Q-factor through innovative metasurface designs and, more critically, through the optimization of fabrication techniques to reduce imperfections that limit the Q-factor.

Additionally, improving the quality and specificity of biological receptors—as well as functionalization strategies—can enhance the sensitivity and ensure reliable detection of low-concentration analytes. Advanced functionalization methods, including multi-layered biointerfaces, could further amplify performance by increasing selectivity.

Another key direction is extending the dynamic range of the sensor, ensuring operation across a broader range of analyte concentrations. This could be achieved by introducing hybrid sensor designs that combine BIC resonances with mechanisms for controlled field localization, thereby maintaining high sensitivity even at elevated analyte levels.

The stability and reusability of BIC-based biosensors are key factors for their practical deployment, particularly in applications such as point-of-care diagnostics. While extensive stability tests are beyond the scope of this review, several referenced studies suggest promising durability and operational robustness. For instance, Schiattarella et al. demonstrated consistent performance over multiple sensing cycles [59], while Romano et al. highlighted the mechanical stability of microfluidic systems integrated with BIC metasurfaces [45]. These findings indicate the potential for the reliable long-term use of BIC-based biosensors under typical operational conditions.

Funding: F.D. acknowledges the support of the project PRIN-2022 ALPHA "ALl-dielectric resonant metasurfaces enhancing PHoton emission phenomenA" (CUP:D53D23001060006) funded by the Italian Ministry of University and Research. F.D. acknowledges the support of the European Union under the Italian National Recovery and Resilience Plan (NRRP) of Next Generation EU, partnership on "Telecommunications of the Future" (PE00000001—program "RESTART").

Conflicts of Interest: The author declares no conflicts of interest.

References

1. Pasquarelli, A. *Biosensors and Biochips*, 1st ed.; Learning Materials in Biosciences; Springer International Publishing: Berlin/Heidelberg, Germany, 2021. [CrossRef]
2. Preston, K.J.; Chen, G.; Chapman, C.A.; Gundavarapu, S.; Aljohani, E.; Paredes Arroyo, A.; Lin, C.C.; Vinitsky, A.; Gonzales, A.; Yin, Z.; et al. At-home measurement of blood biomarkers on the SiPhox Home silicon photonics platform. In *Proceedings of the Frontiers in Biological Detection: From Nanosensors to Systems XVI*; Miller, B.L., Weiss, S.M., Danielli, A., Eds.; SPIE: Bellingham, WA, USA, 2024; p. 6. [CrossRef]
3. Steiner, D.J.; Bryan, M.R.; Miller, B.L. Photonic biosensing at the point-of-care. In *Biophotonics and Biosensing*; Elsevier: Amsterdam, The Netherlands, 2024; pp. 243–268. [CrossRef]
4. Yu, N.; Capasso, F. Flat optics with designer metasurfaces. *Nat. Mater.* **2014**, *13*, 139–150. [CrossRef] [PubMed]

5. Kildishev, A.V.; Boltasseva, A.; Shalaev, V.M. Planar Photonics with Metasurfaces. *Science* **2013**, *339*, 1232009. [CrossRef] [PubMed]
6. Khorasaninejad, M.; Capasso, F. Metalenses at visible wavelengths: Diffraction-limited focusing and subwavelength resolution imaging. *Science* **2016**, *352*, 1190–1194. [CrossRef]
7. Genevet, P.; Capasso, F.; Aieta, F.; Khorasaninejad, M.; Devlin, R. Recent advances in planar optics: From plasmonic to dielectric metasurfaces. *Optica* **2017**, *4*, 139. [CrossRef]
8. Schulz, S.A.; Oulton, R.F.; Kenney, M.; Alù, A.; Staude, I.; Bashiri, A.; Fedorova, Z.; Kolkowski, R.; Koenderink, A.F.; Xiao, X.; et al. Roadmap on photonic metasurfaces. *Appl. Phys. Lett.* **2024**, *124*, 260701. [CrossRef]
9. Kuznetsov, A.I.; Brongersma, M.L.; Yao, J.; Chen, M.K.; Levy, U.; Tsai, D.P.; Zheludev, N.I.; Faraon, A.; Arbabi, A.; Yu, N.; et al. Roadmap for Optical Metasurfaces. *ACS Photonics* **2024**, *11*, 816–865. [CrossRef]
10. Yuan, L.; Zhao, Y.; Toma, A.; Aglieri, V.; Gerislioglu, B.; Yuan, Y.; Lou, M.; Ogundare, A.; Alabastri, A.; Nordlander, P.; et al. A Quasi-Bound States in the Continuum Dielectric Metasurface-Based Antenna–Reactor Photocatalyst. *Nano Lett.* **2023**, *24*, 172–179. [CrossRef]
11. Yu, S.; Xu, M.; Pu, M.; Tang, X.; Zheng, Y.; Guo, Y.; Zhang, F.; Li, X.; Ma, X.; Luo, X. Dynamic nonlocal metasurface for multifunctional integration via phase-change materials. *Nanophotonics* **2024**, *13*, 4317–4325. [CrossRef] [PubMed]
12. Chen, Y.; Liu, Z.; Li, Y.; Hu, Z.; Wu, J.; Wang, J. Adjustable converter of bound state in the continuum basing on metal-graphene hybrid metasurfaces. *Opt. Express* **2022**, *30*, 23828. [CrossRef] [PubMed]
13. Zhang, S.; Wong, C.L.; Zeng, S.; Bi, R.; Tai, K.; Dholakia, K.; Olivo, M. Metasurfaces for biomedical applications: Imaging and sensing from a nanophotonics perspective. *Nanophotonics* **2020**, *10*, 259–293. [CrossRef]
14. Tseng, M.L.; Jahani, Y.; Leitis, A.; Altug, H. Dielectric Metasurfaces Enabling Advanced Optical Biosensors. *ACS Photonics* **2020**, *8*, 47–60. [CrossRef]
15. Chung, T.; Wang, H.; Cai, H. Dielectric metasurfaces for next-generation optical biosensing: A comparison with plasmonic sensing. *Nanotechnology* **2023**, *34*, 402001. [CrossRef] [PubMed]
16. Altug, H.; Oh, S.H.; Maier, S.A.; Homola, J. Advances and applications of nanophotonic biosensors. *Nat. Nanotechnol.* **2022**, *17*, 5–16. [CrossRef] [PubMed]
17. Hsu, C.W.; Zhen, B.; Stone, A.D.; Joannopoulos, J.D.; Soljačić, M. Bound states in the continuum. *Nat. Rev. Mater.* **2016**, *1*, 16048. [CrossRef]
18. Liu, Z.; Wang, K.; Zhang, Y.; Mao, D. High-Q quasi-bound states in the continuum for nonlinear metasurfaces. *Phys. Rev. Lett.* **2019**, *123*, 253901. [CrossRef]
19. Koshelev, K.; Bogdanov, A.; Kivshar, Y. Meta-optics and bound states in the continuum. *Sci. Bull. 2019*, *64*, 836–842. [CrossRef] [PubMed]
20. Tan, Z.; Su, M.; Song, Y. Metamaterials-tuned light-matter resonance for ultra-sensitivity biological diagnosis. *Matter* **2024**, *7*, 2657–2659. [CrossRef]
21. Khan, S.A.; Khan, N.Z.; Xie, Y.; Abbas, M.T.; Rauf, M.; Mehmood, I.; Runowski, M.; Agathopoulos, S.; Zhu, J. Optical Sensing by Metamaterials and Metasurfaces: From Physics to Biomolecule Detection. *Adv. Opt. Mater.* **2022**, *10*, 2200500. [CrossRef]
22. Yesilkoy, F.; Arvelo, E.R.; Jahani, Y.; Liu, M.; Tittl, A.; Cevher, V.; Kivshar, Y.; Altug, H. Ultrasensitive hyperspectral imaging and biodetection enabled by dielectric metasurfaces. *Nat. Photonics* **2019**, *13*, 390–396. [CrossRef]
23. Algorri, J.F.; Dell'Olio, F.; Roldán-Varona, P.; Rodríguez-Cobo, L.; López-Higuera, J.M.; Sánchez-Pena, J.M.; Zografopoulos, D.C. Strongly resonant silicon slot metasurfaces with symmetry-protected bound states in the continuum. *Opt. Express* **2021**, *29*, 10374. [CrossRef]
24. Algorri, J.F.; Dell'Olio, F.; Roldán-Varona, P.; Rodríguez-Cobo, L.; López-Higuera, J.M.; Sánchez-Pena, J.M.; Dmitriev, V.; Zografopoulos, D.C. Analogue of electromagnetically induced transparency in square slotted silicon metasurfaces supporting bound states in the continuum. *Opt. Express* **2022**, *30*, 4615. [CrossRef]
25. Algorri, J.; Dmitriev, V.; Hernández-Figueroa, H.; Rodríguez-Cobo, L.; Dell'Olio, F.; Cusano, A.; López-Higuera, J.; Zografopoulos, D. Polarization-independent hollow nanocuboid metasurfaces with robust quasi-bound states in the continuum. *Opt. Mater.* **2024**, *147*, 114631. [CrossRef]
26. Zhou, Y.; Guo, Z.; Zhao, X.; Wang, F.; Yu, Z.; Chen, Y.; Liu, Z.; Zhang, S.; Sun, S.; Wu, X. Dual-Quasi Bound States in the Continuum Enabled Plasmonic Metasurfaces. *Adv. Opt. Mater.* **2022**, *10*, 2200965. [CrossRef]
27. Lyu, J.; Huang, L.; Chen, L.; Zhu, Y.; Zhuang, S. Review on the terahertz metasensor: From featureless refractive index sensing to molecular identification. *Photonics Res.* **2024**, *12*, 194. [CrossRef]
28. Wang, R.; Xu, L.; Huang, L.; Zhang, X.; Ruan, H.; Yang, X.; Lou, J.; Chang, C.; Du, X. Ultrasensitive Terahertz Biodetection Enabled by Quasi-BIC-Based Metasensors. *Small* **2023**, *19*, 2301165. [CrossRef] [PubMed]
29. Zhang, W.; Lin, J.; Yuan, Z.; Lin, Y.; Shang, W.; Chin, L.K.; Zhang, M. Terahertz Metamaterials for Biosensing Applications: A Review. *Biosensors* **2023**, *14*, 3. [CrossRef] [PubMed]
30. Stillinger, F.H.; Herrick, D.R. Bound states in the continuum. *Phys. Rev. A* **1975**, *11*, 446–454. [CrossRef]

31. Marinica, D.C.; Borisov, A.G.; Shabanov, S.V. Bound States in the Continuum in Photonics. *Phys. Rev. Lett.* **2008**, *100*, 183902. [CrossRef] [PubMed]
32. Kang, M.; Liu, T.; Chan, C.T.; Xiao, M. Applications of bound states in the continuum in photonics. *Nat. Rev. Phys.* **2023**, *5*, 659–678. [CrossRef]
33. Plotnik, Y.; Peleg, O.; Dreisow, F.; Heinrich, M.; Nolte, S.; Szameit, A.; Segev, M. Experimental Observation of Optical Bound States in the Continuum. *Phys. Rev. Lett.* **2011**, *107*, 183901. [CrossRef] [PubMed]
34. Li, S.; Zhou, C.; Liu, T.; Xiao, S. Symmetry-protected bound states in the continuum supported by all-dielectric metasurfaces. *Phys. Rev. A* **2019**, *100*, 063803. [CrossRef]
35. Huang, L.; Li, S.; Zhou, C.; Zhong, H.; You, S.; Li, L.; Cheng, Y.; Miroshnichenko, A.E. Realizing Ultrahigh-Q Resonances Through Harnessing Symmetry-Protected Bound States in the Continuum. *Adv. Funct. Mater.* **2024**, *34*, 2309982. [CrossRef]
36. Sadrieva, Z.F.; Belyakov, M.A.; Balezin, M.A.; Kapitanova, P.V.; Nenasheva, E.A.; Sadreev, A.F.; Bogdanov, A.A. Experimental observation of a symmetry-protected bound state in the continuum in a chain of dielectric disks. *Phys. Rev. A* **2019**, *99*, 053804. [CrossRef]
37. Neale, S.; Muljarov, E.A. Accidental and symmetry-protected bound states in the continuum in a photonic-crystal slab: A resonant-state expansion study. *Phys. Rev. B* **2021**, *103*, 155112. [CrossRef]
38. Sidorenko, M.; Sergaeva, O.; Sadrieva, Z.; Roques-Carmes, C.; Muraev, P.; Maksimov, D.; Bogdanov, A. Observation of an Accidental Bound State in the Continuum in a Chain of Dielectric Disks. *Phys. Rev. Appl.* **2021**, *15*, 034041. [CrossRef]
39. Amrani, M.; Khattou, S.; El Boudouti, E.H.; Talbi, A.; Akjouj, A.; Dobrzynski, L.; Djafari-Rouhani, B. Friedrich-Wintgen bound states in the continuum and induced resonances in a loop laterally coupled to a waveguide. *Phys. Rev. B* **2022**, *106*, 125414. [CrossRef]
40. Luo, M.; Wu, F. Wavy optical grating: Wideband reflector and Fabry-Pérot bound states in the continuum. *Phys. Rev. A* **2022**, *106*, 063514. [CrossRef]
41. Nabol, S.V.; Pankin, P.S.; Maksimov, D.N.; Timofeev, I.V. Fabry-Perot bound states in the continuum in an anisotropic photonic crystal. *Phys. Rev. B* **2022**, *106*, 245403. [CrossRef]
42. Li, X.; Maqbool, E.; Han, Z. Narrowband mid-infrared thermal emitters based on the Fabry-Perot type of bound states in the continuum. *Optics Express* **2023**, *31*, 20338. [CrossRef] [PubMed]
43. Jin, J.; Yin, X.; Ni, L.; Soljačić, M.; Zhen, B.; Peng, C. Topologically enabled ultrahigh-Q guided resonances robust to out-of-plane scattering. *Nature* **2019**, *574*, 501–504. [CrossRef] [PubMed]
44. Algorri, J.F.; Dell'Olio, F.; Ding, Y.; Labbé, F.; Dmitriev, V. Experimental demonstration of a silicon-slot quasi-bound state in the continuum in near-infrared all-dielectric metasurfaces. *Opt. Laser Technol.* **2023**, *161*, 109199. [CrossRef]
45. Romano, S.; Zito, G.; Torino, S.; Calafiore, G.; Penzo, E.; Coppola, G.; Cabrini, S.; Rendina, I.; Mocella, V. Label-free sensing of ultralow-weight molecules with all-dielectric metasurfaces supporting bound states in the continuum. *Photonics Res.* **2018**, *6*, 726. [CrossRef]
46. Conteduca, D.; Arruda, G.S.; Barth, I.; Wang, Y.; Krauss, T.F.; Martins, E.R. Beyond Q: The Importance of the Resonance Amplitude for Photonic Sensors. *ACS Photonics* **2022**, *9*, 1757–1763. [CrossRef]
47. Luo, M.; Zhou, Y.; Zhao, X.; Guo, Z.; Li, Y.; Wang, Q.; Liu, J.; Luo, W.; Shi, Y.; Liu, A.Q.; et al. High-Sensitivity Optical Sensors Empowered by Quasi-Bound States in the Continuum in a Hybrid Metal–Dielectric Metasurface. *ACS Nano* **2024**, *18*, 6477–6486. [CrossRef] [PubMed]
48. Lalanne, P.; Morris, G.M. Highly improved convergence of the coupled-wave method for TM polarization. *JOSA A* **1996**, *13*, 779. [CrossRef]
49. Lalanne, P. Convergence performance of the coupled-wave and the differential methods for thin gratings. *JOSA A* **1997**, *14*, 1583. [CrossRef]
50. Lalanne, P. Improved formulation of the coupled-wave method for two-dimensional gratings. *JOSA A* **1997**, *14*, 1592. [CrossRef]
51. Lalanne, P.; Jurek, M.P. Computation of the near-field pattern with the coupled-wave method for transverse magnetic polarization. *J. Mod. Opt.* **1998**, *45*, 1357–1374. [CrossRef]
52. Silberstein, E.; Lalanne, P.; Hugonin, J.P.; Cao, Q. Use of grating theories in integrated optics. *JOSA A* **2001**, *18*, 2865. [CrossRef]
53. Yao, Y.; Liu, H.; Wang, Y.; Li, Y.; Song, B.; Wang, R.P.; Povinelli, M.L.; Wu, W. Nanoimprint-defined, large-area meta-surfaces for unidirectional optical transmission with superior extinction in the visible-to-infrared range. *Opt. Express* **2016**, *24*, 15362. [CrossRef] [PubMed]
54. Choi, S.; Zuo, J.; Das, N.; Yao, Y.; Wang, C. Scalable Nanoimprint Manufacturing of Functional Multilayer Metasurface Devices. *Adv. Funct. Mater.* **2024**, *34*, 2404852. [CrossRef]
55. Oh, D.K.; Lee, T.; Ko, B.; Badloe, T.; Ok, J.G.; Rho, J. Nanoimprint lithography for high-throughput fabrication of metasurfaces. *Front. Optoelectron.* **2021**, *14*, 229–251. [CrossRef]

56. Einck, V.J.; Torfeh, M.; McClung, A.; Jung, D.E.; Mansouree, M.; Arbabi, A.; Watkins, J.J. Scalable Nanoimprint Lithography Process for Manufacturing Visible Metasurfaces Composed of High Aspect Ratio TiO_2 Meta-Atoms. *ACS Photonics* **2021**, *8*, 2400–2409. [CrossRef]
57. Clabassi, E.; Balestra, G.; Siciliano, G.; Polimeno, L.; Tarantini, I.; Primiceri, E.; Tobaldi, D.M.; Cuscunà, M.; Quaranta, F.; Passaseo, A.; et al. Hybrid plasmonic Bound State in the Continuum entering the zeptomolar biodetection range. *arXiv* **2024**. [CrossRef]
58. Zito, G.; Siciliano, G.; Seifalinezhad, A.; Miranda, B.; Lanzio, V.; Schwartzberg, A.; Gigli, G.; Turco, A.; Rendina, I.; Mocella, V.; et al. Molecularly Imprinted Polymer Sensor Empowered by Bound States in the Continuum for Selective Trace-Detection of TGF-beta. *Adv. Sci.* **2024**, *11*, 2401843. [CrossRef] [PubMed]
59. Schiattarella, C.; Sanità, G.; Guilcapi Alulema, B.; Lanzio, V.; Cabrini, S.; Lamberti, A.; Rendina, I.; Mocella, V.; Zito, G.; Romano, S. High-Q photonic aptasensor based on avoided crossing bound states in the continuum and trace detection of ochratoxin A. *Biosens. Bioelectron. X* **2022**, *12*, 100262. [CrossRef]

photonics

MDPI

Review

Nonlinear Dielectric Metasurfaces for Terahertz Applications

Forouzan Habibighahfarokhi [1], Olga Sergaeva [1], Luca Carletti [1,2], Paolo Franceschini [1,2], Andrea Tognazzi [2,3], Andrea Locatelli [1,2], Unai Arregui Leon [4], Giuseppe Della Valle [4], Costantino De Angelis [1,2] and Davide Rocco [1,2,*]

1 Department of Information Engineering, University of Brescia, Via Branze 38, 25123 Brescia, Italy; forouzan.habibighahfarokhi@unibs.it (F.H.); olga.sergaeva@unibs.it (O.S.); luca.carletti@unibs.it (L.C.); paolo.franceschini@unibs.it (P.F.); andrea.locatelli@unibs.it (A.L.); costantino.deangelis@unibs.it (C.D.A.)
2 National Institute of Optics, Consiglio Nazionale delle Ricerche, Via Branze 45, 25123 Brescia, Italy
3 Department of Engineering, University of Palermo, Viale Delle Scienze, 90128 Palermo, Italy
4 Department of Physics, Politecnico di Milano, Piazza Leonardo da Vinci 32, 20133 Milano, Italy; unai.arregui@polimi.it (U.A.L.); giuseppe.dellavalle@polimi.it (G.D.V.)
* Correspondence: davide.rocco@unibs.it

Abstract: The terahertz (THz) region of the electromagnetic spectrum, spanning from 0.1 to 30 THz, represents a prospering area in photonics, with transformative applications in imaging, communications, and material analysis. However, the development of efficient and compact THz sources has long been hampered by intrinsic material limitations, inefficient conversion processes, and complex phase-matching requirements. Recent breakthroughs in nonlinear optical mechanisms, resonant metasurface engineering, and advances in the fabrication processes for materials such as lithium niobate (LN) and aluminum gallium arsenide (AlGaAs) have paved the way for innovative THz generation techniques. This review article explores the latest theoretical advances, together with key experimental results and outlines perspectives for future developments.

Keywords: optical rectification; resonant dielectric metasurfaces; localized surface phonon-polaritons

1. Introduction

The terahertz spectrum bridges the gap between microwave and infrared (IR) frequencies, enabling unique functionalities that leverage the best of both domains [1–3]. Its non-ionizing radiation is safe for biological tissues, making it an attractive tool for medical imaging and diagnostics. Unlike X-rays or higher-frequency radiation, THz photons do not carry enough energy to ionize atoms, thereby avoiding DNA damage and other harmful effects. Moreover, its ability to penetrate non-metallic and non-conductive materials such as polymers and textiles has spurred interest in security screening applications. For example, THz scanners can reveal concealed objects under clothing or inside packages without exposing individuals to hazardous radiation [4]. Beyond imaging, the THz range is sensitive to molecular vibrations, enabling high-resolution spectroscopic fingerprinting for material characterization. Many molecules and solids have characteristic absorptive or emissive features in the THz band, allowing the identification of substances—from pharmaceutical compounds to explosives—by their spectral "fingerprints" [5,6]. Despite these benefits, the THz spectrum has historically been underutilized due to the lack of efficient sources and detectors. Electronics struggle to reach THz frequencies and conventional photonics typically operate at much higher frequencies, leaving a gap in available technologies.

However, recent advances in nonlinear optics, particularly optical rectification and difference frequency generation from metasurface-based devices, have begun to address these challenges. Such nonlinear techniques bypass traditional limitations by using ultrafast optical pulses and engineered nanostructures to directly generate THz waves, mitigating issues of materials and phase-matching requirements that previously constrained THz sources efficiency [7–9]. Metasurfaces offer several concrete advantages for THz generation, detection, and related applications due to their ability to engineer electromagnetic waves at sub-wavelength scales. One of the main advantages of metasurfaces is their ultrathin and compact dimensions which allow to reduce and miniaturize the geometrical dimension of the final platform. Apart from the geometrical form factor, metasurfaces can achieve dynamic and reconfigurable control of the emitted THz radiation. In particular, optimized metamaterials and metasurfaces have been suggested to be beneficial for the development of various novel optical THz components [10]. Polarization conversion, active phase modulation, engineered absorption, and nonlinear effects have been recently demonstrated in THz meta-structures [11–15]. Moreover, engineered metasurfaces have been shown to significantly enhance THz emission from photoconductive antennas [16]. Notably, nanostructuring of a nonlinear crystal into resonant meta-atoms can dramatically enhance the local field intensity, boosting frequency conversion efficiency. For instance, in [17] an optimized 160 nm-thick GaAs metasurface emits THz radiation comparable to that of a 650 μm GaAs crystal, despite its significantly smaller material volume. Several papers have recently presented the potential of nonlinear metasurface for THz generation such as reported in [18–21]; however, the use of fully dielectric material for realizing the metasurface is a new concept which we address in detail in this manuscript.

This review is organized as follows: Section 2 presents a concise introduction to the prospective applications and implications of the THz frequency range; Section 3 addresses THz generation in dielectric slabs, while Section 4 illustrates the potential offered by dielectric metasurfaces for efficient nonlinear THz generation. Crystalline defect-free condition is assumed in the reported results. Finally, Section 5 proposes some innovative solutions and applications for the previously described thin surfaces. This structured approach allows readers to first understand the broad context and applications of THz photonics, then delve into specific mechanisms of THz generation, and finally explore forward-looking concepts and conclusions.

2. Overview

Terahertz radiation, often referred to as the bridge between electronics and photonics, has gained significant attention in recent years because of its unique properties and the vast potential for applications in sensing, wireless communication, healthcare, spectroscopy, security, quality control, and imaging, as described in Figure 1 [22–25]. Exploring and harnessing this frequency range necessitates a multidisciplinary strategy that encourages close collaboration between high-frequency semiconductor technology for radio frequencies (RF) electronics and alternative photonic-based solutions. In practice, this means engineers working on microwave/millimeter-wave devices must team up with optical scientists to develop new hybrid approaches that operate at THz frequencies. One of the main fields in which this effort is already yielding promising results is data transfer. With the growing demand for faster data rates and enhanced connectivity, the THz spectrum is gaining attention as a candidate for next-generation wireless communication systems. THz waves promise ultra-high bandwidths that could support data rates exceeding 100 Gbps, enabling technologies such as 6G and beyond [26–29]. In this context, research efforts are focused on overcoming challenges related to signal attenuation and hardware development, including efficient THz sources and detectors [30,31]. Recent studies have deeply highlighted the potential of the THz range for communication applications [32,33]. In addition to this research topic, we also underline that THz imaging and spectroscopy have introduced a significant breakthrough in material characterization through non-destructive testing and biomedical imaging [34–39]. Indeed, the ability of THz waves to probe materials without causing damage is particularly promising in recognizing defects in semiconductors, investigating pharmaceutical compounds, and detecting concealed objects. Furthermore, their interaction with molecules and biological tissues makes them ideal for medical diagnostics [40–42]. Moreover, it is relevant to stress that the penetration capabilities of THz waves through fabrics and plastics make them suitable for security screening [4,43,44]. As previously mentioned, differently from X-rays, THz radiation is non-ionizing, guaranteeing safety for routine processes in airports and other high-security environments. Additionally, THz sensing enables for the detection of explosives and hazardous substances through their spectral fingerprints [45,46], while in industry, THz systems are being integrated into quality control processes in many advanced material manufacturing [47,48]. For instance, THz quality control scanners can monitor plastic welds or inspect food products and pharmaceutical tablets for uniformity without physical contact or damage, illustrating the breadth of THz applicability [49,50]. Several comprehensive reviews on metasurfaces operating in the THz range were recently published [51–54]. Here, we intend to briefly highlight a variety of possible THz metasurfaces designs and applications. For instance, in Ref. [55], a Silicon–Silica metasurface with a graphene layer on a teflon substrate demonstrated high-Q resonances under THz excitation whereas in Ref. [56] a high-Q Silicon–Silica metasurface on gold was suggested for sensing applications. Additionally, a liquid crystal–silica–graphene metasurface THz absorber was developed in [57], while a ceramic microsphere metasurface designed as a THz reflector was fabricated in [58]. Let us also mention that in [59], a silicon cubes-based metasurface was designed and fabricated to function as a magnetic reflector in the THz.

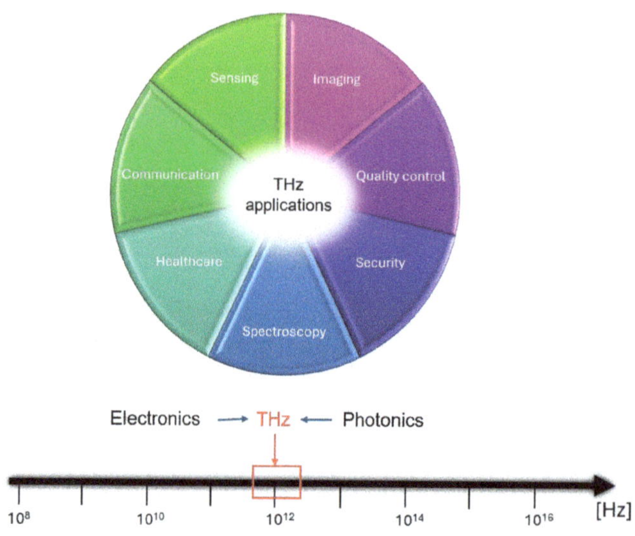

Figure 1. Terahertz technology applications.

Despite its promise, the development of efficient, compact, and reliable THz sources remains a critical challenge, particularly due to the inherent limitations of conventional materials and techniques. This difficulty in creating THz emitters and detectors has historically been known as the THz technology gap. Indeed, for THz transmitters, achieving modulated THz generation with a high repetition rate across different spectral regions is essential. Although certain spectral gaps remain where suitable THz sources are lacking, various mechanisms now enable THz generation. The most widely explored methods are THz-Quantum Cascade Lasers (THz-QCLs), photoconductive antennas (PCAs), and nonlinear optical rectification process (NORP) in bulk crystals. These approaches, which are particularly well-suited for spectroscopy and imaging applications, rely on large generation volumes and typically operate at frequencies below 5 THz [60–66]. In brief, THz-QCLs are semiconductor lasers with cascading quantum well transitions that can be lasing at THz frequencies (often around 3–4 THz), though they usually require cryogenic cooling to function. PCAs, on the other hand, use an ultrafast laser (with pulse energy of the order of 10 nJ [67]) to excite carriers in a biased photoconductive material, producing broadband THz pulses; they are common in time-domain THz spectroscopy but their output power and bandwidth are limited by, among others, carrier mobility and lifetime, phonon absorption or the applicable bias voltage [68]. Regarding NORP, difference frequency generation in the crystals (with zinc telluride and lithium niobate being the most popular) occurs between the different spectral components that comprise the optical probe resulting in a spectrally limited THz pulse (0.1–3 THz) [69]. Although widely used for time-domain THz spectroscopy, this approach requires laser pulses with pulse energy higher than 1 μJ (i.e., much larger compared to PCAs) [67].

More recently, to achieve the highest performance per unit volume using nanoscale resonant platforms, THz generation through optical rectification in plasmonic Split-Ring Resonators (SRRs) has been demonstrated. In this method, nanoscale SRRs nonlinearly generate THz signals by exciting magnetic dipole modes in the infrared regime at the pump wavelengths [70–74]. By confining the optical excitation to the surface of metallic ring structures, these plasmonic schemes significantly enhance local electromagnetic fields, which in turn strengthens the nonlinear polarization and THz emission. Such resonant plasmonic approaches have succeeded in creating THz bursts in sub-wavelength volumes, albeit often with trade-offs in efficiency due to metal losses. Importantly, a novel approach for generating structured single-cycle THz wavepackets using engineered nonlinear plasmonic metasurfaces has been reported in [75] where the generation of propagating spatiotemporal quadrupole and few-cycle THz pulses with engineered angular dispersion is demonstrated. In the report, the THz output was not a simple Gaussian beam but had a complex field pattern—a quadrupole-like spatial distribution—and only a few oscillation cycles in time, achieved by carefully tailoring a plasmonic metasurface. This spatiotemporally structured THz emission showcases the level of control possible with designed nonlinear surfaces.

A promising alternative to plasmonic THz emitters is represented by all-dielectric THz metasurfaces, which have already attracted significant attention in nonlinear optics as key components for second- and third-harmonic generation [76–78]. Such platforms may outperform plasmonic metasurfaces in nonlinear generation efficiency because of the significantly lower optical losses, as they avoid the ohmic losses inherent in metallic structures. In dielectrics, incident light does not induce the same kind of free-carrier currents that dissipate energy as heat, so more of the input optical energy can be converted into the desired THz radiation. Moreover, in high refractive index dielectrics, the bulk nonlinear contribution is mostly the dominant one, opposite to metallic structures where surface currents are responsible for the harmonic generation. This difference arises because many non-centrosymmetric dielectric materials (like certain semiconductors or ferroelectrics) possess strong second-order nonlinearities throughout their volume, whereas centrosymmetric metals have no bulk second-order response and produce even-order nonlinearities only at surfaces, where inversion symmetry is broken.

Conventional phase matching in nonlinear optics relies on satisfying strict momentum conservation conditions over a macroscopic propagation length via birefringence or periodically poled structures. In contrast, metasurfaces exploit localized, resonant nonlinear interactions at the nanoscale [55,79], where phase accumulation is governed by sub-wavelength unit cell engineering rather than bulk coherence lengths. In other words, resonant structures with sub-wavelength scale mode volumes do not require strict phase matching, as their conversion efficiency depends on the modal overlap between fundamental modes and higher harmonics. Of course, the overall small volume limits the total efficiency of the harmonic process involved. To circumvent this drawback, hybrid nonlinear integrated photonic devices consisting of phase gradient metasurfaces patterned on top of a nonlinear waveguide have been recently proposed [80]. In this context, the metasurfaces can provide additional momentum to compensate for the phase mismatch between the fundamental and the harmonic waves. Distributed phase matching in nonlinear metasurfaces is another mechanism that contributes to efficient THz generation via constructive interference across a metasurface without requiring strict bulk phase matching [81]. Also, high conversion efficiency can be achieved with high-Q quasi-bound states in the continuum (quasi-BICs) and other (Mie, Fano) resonances in metasurfaces [82,83]. Furthermore, adjusting the geometric parameters of the metasurfaces enables additional functionalities, such as polarization control, dynamic modulation, beam shaping, and steering of the emitted THz waves (usually achieved by adding an extra metasurface [51–53,79,84]). Notably, the ENZ materials (e.g., ITO) reduce phase-matching constraints due to near-zero permittivity. Recent advances in ENZ-metasurfaces [79,85] further bridge the gap between material limitations and phase-matching requirements. By patterning ENZ materials into sub-wavelength resonators, these platforms combine the field enhancement of ENZ modes with the momentum engineering of metasurfaces, achieving efficient THz generation without bulk constraints. Notably, they surpass LN in conversion efficiency per unit thickness while enabling dynamic tuning [85].

For all the abovementioned reasons, this review explores the latest advances in THz generation methods, with a focus on nonlinear optical mechanisms, in resonant dielectric metasurfaces, with a specific concentration on materials such as LN and AlGaAs. The discussed innovations address key challenges such as limited efficiency, scalability, and phase-matching constraints, paving the way for next-generation photonic devices. By synthesizing findings from experimental and theoretical studies, this article provides a comprehensive evaluation of the current state of the field, highlighting the critical role of material science and nanotechnology in driving breakthroughs in THz technologies. In addition, future perspectives are examined, including the integration of dielectric thin-film platforms and metasurface-based systems, which hold promise for transforming scientific and industrial applications. Integrating ultrathin nonlinear films with resonant metasurfaces could lead to hybrid devices that combine the advantages of each, potentially enabling on-chip THz emitters or sensors. This brief overview delves into the key challenges and motivation behind recent research efforts in this rapidly evolving domain.

3. THz Generation in Dielectric Slabs

In this section, we present recent results about THz generation through optical rectification in lithium niobate (LN) slabs. The investigation of LN's material properties dates back to the 1960s. Its crystal structure falls within the 3 m point group, characterized by a three-fold rotational symmetry around the [0001] *c*-axis (typically referred to as the *z*-axis) and mirror symmetry across three planes spaced 60° apart [86,87]. In simpler terms, LN has a trigonal crystal symmetry that lacks inversion symmetry, a crucial attribute for second-order bulk nonlinear effects. LN has a wide transparency range from 350 nm to

5 μm, spanning across the visible, near-infrared, and mid-infrared spectra. Its relatively high refractive index (roughly 2.2 at 1550 nm) enables the formation of high index contrast waveguides on various amorphous and crystalline substrates, such as silica and sapphire. With a high Curie temperature (1210 °C), LN maintains a stable ferroelectric phase, making it compatible with several fabrication processes and operating conditions. Unlike silicon, LN is a non-centrosymmetric crystal with large second-order nonlinear coefficients, making it ideal for optical wavelength conversion and photon-pair generation. Moreover, its exceptional Pockels coefficient (r_{33} = 31 pm/V) makes LN the preferred material for electro-optic modulators, a key component in telecommunication networks [88]. These combined features—broad optical transparency, strong nonlinearity, and robust fabrication tolerance—explain why LN is ubiquitous in nonlinear optical devices such as frequency converters and modulators. These properties make also LN an ideal material for studying THz emission through optical rectification (OR) mechanism in dielectrics. Indeed, LN strong second-order nonlinearity and polar lattice vibrations offer a promising platform for converting optical pulses into THz pulses via OR, which will be discussed next. OR, a second-order nonlinear optical process, is one of the most widely used methods for generating THz radiation [89,90]. OR is a nonlinear optical process in which an intense optical beam (labeled as the pump) is utilized to generate a low-frequency charge motion in a nonlinear medium, emitting in turn THz radiation. In practical terms, when a femtosecond laser pulse passes through a nonlinear crystal like LN, the pulse is an ultrafast intensity envelope that can induce a DC or very low-frequency polarization wave in the crystal; this time-varying polarization at THz frequencies radiates as a burst of THz electromagnetic waves. Thus, OR can be viewed as a special case of difference frequency generation where the difference between frequency components of a broadband pulse produces a continuum of THz output. Thin-film lithium niobate (TFLN) platforms have demonstrated significant advantages in this context, mainly related to the reduction of the generated THz signal self-absorption with respect to bulk crystals in which strong intrinsic absorption significantly reduces the detected THz [81,91,92]. In bulk LN, the emitted THz wave must traverse a long path in the material and is heavily attenuated by LN phonon absorption bands; a thin-film geometry minimizes this propagation loss, allowing more of the generated THz to escape. LN exhibits vibrational lattice modes (i.e., optical phonons) which can interact with incident electromagnetic radiation in the THz range. This interaction alters the dispersion properties of the electromagnetic wave propagating through the crystal, resulting in a hybrid mode known as a phonon–polariton (PhP). It has recently been demonstrated that optical phonons are responsible for a substantial improvement of THz generation between 2 and 8 THz in $LiNbO_3$ (LN) samples [93]. Interestingly, the optimal thickness to benefit from the PhP enhancement of the nonlinear response is estimated to be less than 2 μm. This finding indicates that there is a sweet spot in LN thickness: a sub-micron film is thick enough to efficiently generate THz radiation but thin enough to avoid excessive absorption, leveraging phonon–polariton resonances to boost the output in the 2–8 THz range.

The generation and detection of THz radiation require precise experimental setups. Figure 2 depicts the experimental configuration for THz generation using x-cut $LiNbO_3$, highlighting the use of pump lasers, parabolic mirrors, and electro-optic detection via gallium phosphide (GaP) crystals, which characterize the generated THz radiation via electro-optical sampling technique [94]. The time domain traces are shown in the bottom panel of Figure 2a. In particular, in [93] the polarization of the pump signal and of the probe pulse are tuned to achieve polarization-resolved measurements, thus addressing the various tensorial components of the nonlinear optical response of LN. Four combinations for the pump and THz polarization have been considered to isolate the contributions of the individual nonlinear coefficients. By measuring THz signals for various polarization alignments (denoted in compact notation as zzz, zyy, yyy, yzy where the first letter indicates the THz signal polarization, while the others identify the pump polarization relative to the LN crystal axes), the experiment could disentangle which tensor components of LN second-order susceptibility were involved in the THz generation. Interestingly, the emitted THz radiation polarized along the z-axis (in compact notation zzz and zyy) possesses a broad spectrum, while THz waves polarized along the y-axis (i.e., yyy and yzy) display a sharper peak around 4 THz and a dip at 6 THz, as shown in Figure 2b. These features are attributed to the ionic enhancement of the nonlinear response in the LN film. In other words, when the THz field is aligned with the crystal optical axis (z), the phonon–polariton effects broaden the emission spectrum. Conversely, for THz polarization along y (perpendicular to the polar axis), the interaction with specific optical phonon modes causes a resonance peak (near 4 THz) and an attenuation (near 6 THz), revealing how the lattice vibrations selectively amplify or suppress certain THz frequencies. To confirm the significant impact of ionic contributions on the nonlinear response, frequency-domain numerical simulations using the finite-element method in COMSOL have been conducted in [93], focusing on the optical-to-THz conversion mechanism. By slightly adjusting the magnitude scale of the nonlinear coefficients an excellent agreement between experiments and simulations has been obtained. This close match between simulation and experiment, achieved after accounting for the frequency-dependent behavior of LN's nonlinear coefficients, reinforces the interpretation that phonon dynamics strongly influence the THz generation process. The analysis in [93] also reports OR measurements at several different pump powers and on LN samples with different thicknesses to provide additional validation that the detected THz signal in the experiments originates from the optical rectification process. As a characteristic of second-order nonlinear processes, the generated signal power scales quadratically with the pump power. However, when keeping the pump power constant, increasing the LN sample thickness leads to a significant deviation from this quadratic trend due to THz absorption in LN. As the LN thickness grows, the spectral power density in the lower frequency range becomes dominant, ultimately suppressing THz generation above 4 THz. This is consistent with stronger re-absorption of higher-frequency THz components (near LN phonon resonance) in thicker samples—longer path lengths inside LN allow the material's intrinsic absorption to attenuate the high-frequency portion of the THz pulse. Notably, in [93] also the case of a LN rod (with 12 μm length, 2 μm width and 500 nm height) has been reported demonstrating that the THz emission is boosted by almost a factor of five compared to the unstructured film. Patterning the LN into a discrete rod introduces geometrical resonances or improves out-coupling that significantly enhances the THz output beyond the uniform thin-film case. These findings highlight the powerful capability of phonon-resonance engineering in precisely tuning the characteristics of generated THz radiation at sub-wavelength dimensions. Moreover, the observed enhancement in the rod-structured system sets the stage for the next section of this review, which explores in detail the THz generation from metasurfaces. If a single isolated LN rod can increase emission fivefold, an array of such many resonant structures—i.e., a metasurface—could potentially yield even larger enhancements and additional control over the THz waves, as we discuss in the next Session. We recall that hybrid gold THz antennas integrated with thin-film lithium niobate circuits have shown promising potential for tailored THz sources operating in the lower THz frequency range (<1 THz) [81]. With advancements in nano-structuring LN, recent theoretical proposals for THz radiation generation have emerged in hybrid LN/Si coupled-waveguide structure, [95] alongside experimental approaches leveraging topological confinement in laser-written LN slabs [96]. Apart from LN and AlGaAs that are deeply presented in this manuscript, conventional semiconducting crystals used for THz-wave generation and detection include Zinc Telluride (ZnTe) and Gallium Phosphide

(GaP). The latter have much lower electro-optic coefficients compared to LN, but are widely used because of the possibility for phase matching that is not directly possible with LN and other inorganic crystals with a high electro-optic coefficient [97,98]. Let us also mention that Perovskite, Barium Borate, and Quartz may find applications in THz generation such as those reported in [99–101]. Alongside widely recognized inorganic nonlinear optical crystals commonly utilized as THz emitters [102], organic nonlinear electro-optic crystals have gained growing attention due to their higher macroscopic nonlinearity and lower dielectric constant, making them promising candidates for efficient and broadband THz wave generation. In this context, in recent decades, extensive research has been conducted on various electro-optic crystals; however, only a limited number of crystals exhibiting an electro-optic response exceeding 50 pm/V have been identified [103–107].

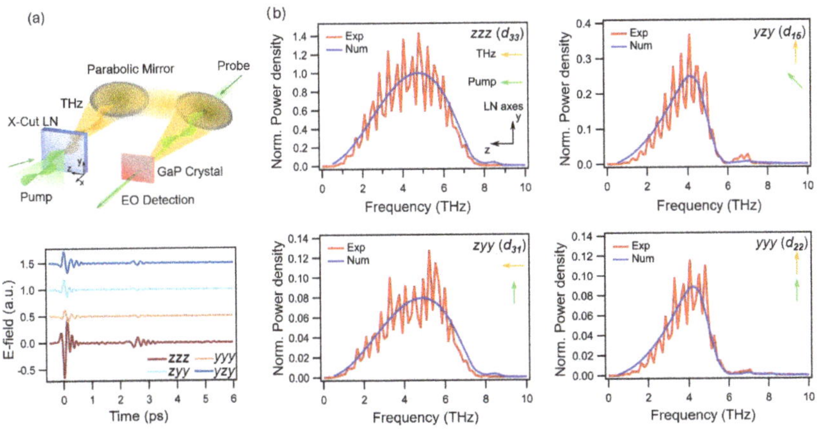

Figure 2. (**a**) (**Top**) experimental setup and (**bottom**) time-domain signals for THz generation using x-cut $LiNbO_3$ slab (500 nm thick). (**b**) THz emission spectra calculated numerically (blue) or derived from the experiments (red). Adapted from [93].

4. THz Generation from Resonant Dielectric Metasurfaces

In this section, we introduce the concept of dielectric metasurfaces for efficient THz generation via difference frequency generation (DFG) process. DFG is a nonlinear process generating beams with differing frequencies of the input beams. For example, if two infrared laser beams at frequencies ω_1 and ω_2 (with $\omega_2 > \omega_1$) interact in a nonlinear medium, DFG can produce radiation at the THz frequency $\omega_3 = \omega_2 - \omega_1$. Indeed, a metasurface can be defined as an engineered material composed of a repetition of sub-wavelength scale structures usually called meta-atoms, designed to manipulate and control electromagnetic waves [108–111]. In practice, a metasurface is an ultrathin (often planar) array of nanoantennas or resonators whose individual scattering responses can be tailored. By arranging these meta-atoms in a specific pattern, one can achieve desired effects on an incoming wavefront, such as focusing light, altering its polarization, or in our context, facilitating nonlinear frequency conversion. Among the others, AlGaAs-based metasurfaces have earned significant acclaim in photonics due to their optical properties, high versatility, and sizable nonlinear optical response [112–115]. With a broad transparency range spanning from the visible to the infrared spectrum, AlGaAs is adaptable to a multitude of photonic applications, while its high refractive index facilitates precise control over light, enhancing the manipulation of amplitude, phase, and polarization. Additionally, its low absorption losses, particularly in the near-infrared region, ensure efficient light interaction, improving the final device performance. Another advantage of AlGaAs lies in its compatibility with standard CMOS fabrication techniques, enabling scalable manufacturing and accessible integration into existing systems [116–120]. In other words, AlGaAs nanostructures can be fabricated using processes similar to those employed for silicon microelectronics, which is important for mass production and integration. These attributes position AlGaAs metasurfaces as key players in different research areas such as nonlinear optics, sensing, and communications [121–124]. Indeed, AlGaAs metasurfaces have already shown efficient second-harmonic generation (frequency doubling) and other nonlinear effects, highlighting their potential to serve as active elements in sensors or as compact frequency converters in optical communication links. Interestingly, a recently published study has meticulously explored how an AlGaAs meta-atom can be employed for THz applications, achieving a generation efficiency several orders of magnitude higher than that of previously proposed metallic structures [7]. In particular, a fully dielectric transceiver is proposed. Through the use of a near-IR pump beam, the optical-domain information signal is converted into the THz frequency band. Notably, adjusting the frequency of the pump beam allows for coverage of the entire THz spectral range. Crucially, this approach directly translates the information into THz radiation without requiring additional components, ensuring compatibility with various modulation schemes and signal formats. This is significant because it implies a very simple transmitter: an optical data signal is fed into the nonlinear meta-atom and comes out as a broadcast THz signal carrying the same data, with no need for electronic oscillators or separate mixers. The optimized AlGaAs nanoantenna can achieve a conversion efficiency of up to 10^{-7} W^{-1} at approximately 11 THz, see Figure 3, while an efficiency on the order of 10^{-7} is still quite low in absolute terms, it represents a substantial improvement relative to earlier nanoscale THz emitters. This result showcases the potential of a single sub-wavelength AlGaAs resonator to serve as a highly miniaturized THz source. In more detail, in [7] an AlGaAs nanodisk with diameter and height equal to 400 nm supporting a magnetic dipolar resonance around 1550 nm is considered as the meta-atom. Moreover, the AlGaAs permittivity and the $\chi^{(2)}_{ijk}$ in the THz range are modeled. These quantities are indeed necessary to perform the nonlinear DFG simulation at $\omega_3 = \omega_2 - \omega_1$ where the nonlinear currents in the AlGaAs nanocylinder volume can be written as follows:

$$J_i(\omega_3) = -i\omega_3\epsilon_0\chi^{(2)}_{ijk}(\omega_3)[E_j(\omega_2)E^*_k(\omega_1) + E_k(\omega_2)E^*_j(\omega_1)] \tag{1}$$

where the subscripts i, j, k are associated with the Cartesian coordinates. Instead, the DFG efficiency is defined as follows:

$$\eta_{DFG} = P^{\omega_3}_{rad}/P_iP_s \tag{2}$$

where the numerator is the power radiated at THz, and the denominator is the product of the incident powers of the information signal and pump beam, respectively. Figure 3b displays the AlGaAs nanodisk η_{DFG} from 4 to 14 THz. The results are obtained by fixing ω_2 and by varying ω_1 to cover the desired THz range at the generated ω_3. Interestingly, one can observe four peaks in the THz emission: two of them are located at the spectral position of the AlGaAs transverse-optical (TO) phonon frequencies, whereas another couple of peaks are in exact correspondence with the localized surface phonon-polaritons (LSPhP) resonances of the nanopillar. This result demonstrates the key role of surface phonon–polaritons in the THz generation from all-dielectric nonlinear nanoantennas. In other words, when the difference frequency matches either a bulk lattice vibration mode (TO phonon) or a surface-confined phonon–polariton mode in the nanodisk, the THz output is significantly enhanced, producing a peak in efficiency. The interested reader may find more information in [7]. It should be highlighted that the obtained THz efficiency can be useful in many communication applications. As an example, the inset of Figure 3b reports the capability of the proposed platform to convert an information signal in the IR region with a bandwidth of 160 GHz into the THz region, around 11 THz, without significant distortion. This approach is equally applicable to other materials that exhibit second-order nonlinearity, which allows access to different spectral regions [8]. Using a different material for the meta-atom, one could target lower THz frequencies—for instance, LN meta-atoms might be optimized for the 1–5 THz range, thus expanding the versatility of the DFG approach.

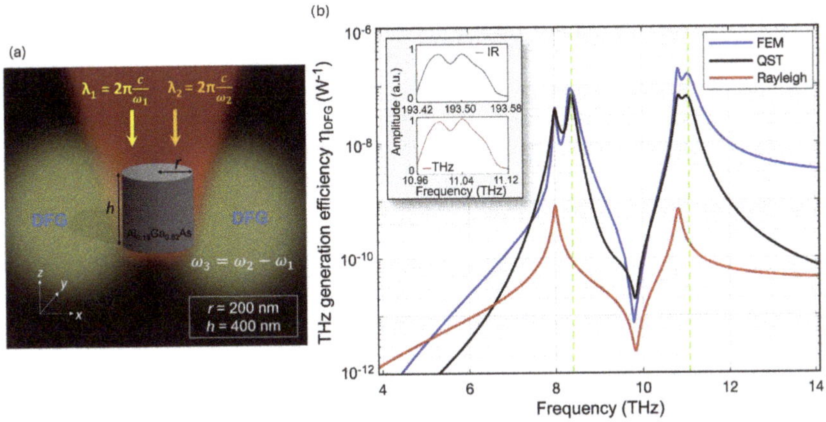

Figure 3. (**a**) The AlGAs nanodisk considered as the meta-atom is excited by two different infrared pumps at λ_1 and λ_2, respectively. (**b**) The computed THz efficiency achieved through DFG of the input beams as obtained from Finite Element Method (FEM) simulation (blue), quasi static (QST) reduced model with (black) and without contribution from transverse-optical phonon permittivity (red). Adapted from [7].

After having presented the THz generation mechanism via DFG in a single meta-atom, we continue our analysis considering the THz generation from a dielectric metasurface made of AlGaAs nanodisks. In this context, recent efforts have been reported in [17,75,125]. It should be noted that, so far in this paragraph discussion, we considered two independent sources with an infinitesimally narrow frequency band as input signals that nonlinearly combine to generate the THz radiation. From a theoretical perspective, this is certainly the easiest scenario to consider. However, real sources typically have pulses with non-negligible spectral bandwidth. Therefore, it is relevant to consider the different spectral components of the incident signal which are responsible for the so-called intra-pulse DFG, i.e., OR. An experimental demonstration of this concept is reported in [125], where different metasurfaces composed of AlGaAs nanodisks with a different elliptical basis are investigated. The pump consists of a femtosecond laser (80 MHz repetition rate with 140 fs pulse duration) that delivers *p*-polarized light on the samples with an angle of incidence of $\pi/4$ rad. The measured THz power spectral density (PSD) is in strong agreement with the theoretical simulations, as reported in Figure 4. For the modeling of the incident beam, the authors consider a Gaussian pulse centered around 234 THz (1280 nm) with a bandwidth of 2 THz. For estimating the total THz emission behavior coming from the metasurface, a discretized version of the incident Gaussian beam spectrum is considered. This spectrum is sampled with a finite number of frequency steps N (N = 18) thus allowing the DFG calculation due to the weighted spectral components of the incident beam. In other words, the broadband pulse was discretized into 18 narrowband slices, and all pairwise mixes of those slices that differ by a given THz frequency were computed. In this way, the simulation of the total THz generation is divided into several DFG problems, which can be modeled following the same procedure reported in Equation (1), where only the bulk nonlinearities are considered. The final THz efficiency at a specific ω_n is obtained as the sum of all the DFG processes that combine all the possible spectral components which are ω_n distant in frequency. Interestingly, the excellent agreement with the experiment is achieved by considering only the bulk nonlinear contributions in the simulations. Therefore, for these pillar dimensions, the contribution of surface nonlinearities is negligible. However, as observed in thinner GaAs membranes, surface nonlinear contributions have also been shown to play a role in THz generation. For instance, in a 160 nm thick GaAs metasurface excited above the material bandgap, a significant portion of the THz field is reported to be likely generated at the surfaces [17]. In that scenario, pumping above the bandgap likely creates ultrafast photocurrents or depletion field dynamics at the GaAs surface, which become a source of THz radiation. This serves as a reminder that, depending on material and excitation conditions (especially in centrosymmetric media or at high photon energies), surface effects might dominate.

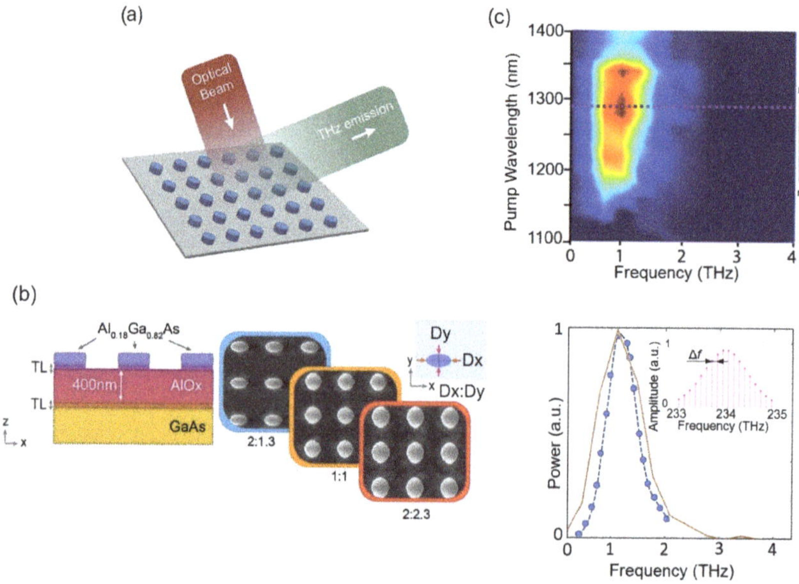

Figure 4. (**a**) Sketch of the experiment. A dielectric metasurface is illuminated by an infrared optical beam and generates a THz emission through an intra-pulse DFG process. (**b**) **Left panel**: the fabricated dielectric metasurfaces are constituted by AlGaAs nanocylinders over a 400 nm thick AlOx substrate. TL—transition layers (about 90 nm thick). **Right panel**: Scanning Electron Microscopy images of three different metasurfaces (top view). (**c**) **Top panel**: the experimental power spectral density coming from the metasurface. **Bottom panel**: the comparison between the measurement (orange curve) and the simulation (blue line). The inset represents the considered incident spectrum. Adapted from [125].

Thus, all-dielectric metasurfaces can act as an efficient and customizable platform for THz pulse generation. The design, based on cylindrical AlGaAs resonators, achieves a remarkable 40-fold enhancement in THz emission efficiency compared to a bare substrate. This enhancement factor underscores how nanostructuring a material into resonant elements can dramatically improve THz yield by concentrating optical energy and engineering emission properties, rather than using an unpatterned film. Beyond the aspect of efficiency, the metasurface allows precise control over THz radiation characteristics by adjusting the elliptical cross-section of the nanocylinders. Additionally, tuning the optical excitation wavelength enables direct manipulation of the THz phase [125]. This unique capability for spatio-temporal structuring of the THz wavefront underscores the potential of all-dielectric metasurfaces for nonlinear frequency conversion, overcoming the limitations of plasmonic and non-transparent systems. With further optimization, these engineered nanoresonator surfaces could replace traditional bulk crystals for THz generation, paving the way for advancements in THz photonics. One can envision ultrathin THz emitters integrated onto photonic chips, using metasurfaces to produce and direct THz beams without the need for centimeter-thick nonlinear crystals. This would represent a significant step toward compact, on-chip THz systems.

5. Perspectives

The experimental demonstration of THz generation from a metasurface shown so far is mainly limited in efficiency by constraints imposed by the spectral width of the used pulsed source. Indeed, the finite and relatively narrow bandwidth of the experimental pulses (around 2 THz) does not allow for exploiting the AlGaAs $\chi^{(2)}$ resonances (located around 8 and 11 THz). A possible future development that may improve the efficiency of the nonlinear process could be to explore these spectral regions using sources with sufficiently wide pulses. For instance, using an even shorter pump pulse or a specially shaped multi-color pump that covers 8–11 THz in difference-frequency content could drive the AlGaAs metasurface at its peak nonlinear response, thereby producing stronger THz output. State-of-the-art ultrafast laser technology (e.g., optical parametric amplifiers or chirped-pulse difference frequency generation) can already reach such broadband excitation. However, in [9] it was demonstrated that the zeros of nonlinear susceptibility can also have important implications in the field of analog computing. Curiously, as highlighted in Figure 5, the AlGaAs $\chi^{(2)}$ shows a 'zero-crossing' and almost linear frequency dependence of its real part around 5.6 THz with a negligible imaginary part. In [9], a fully analytic analysis is reported when an AlGaAs thin metasurface is excited by two different signals. It is shown that under certain conditions and, in particular, when the bandwidth of the input electric field envelope is smaller than the THz carrier frequency f_{THz}, the DFG output signal is proportional to the first derivative of the product of the incident fields, and it oscillates at frequency f_{THz}. This analog computation capability reveals that the metasurface is not just generating THz for communications or spectroscopy, but actually processing information (in this case, detecting the temporal edges or rate-of-change of the input signals' envelopes) in the THz domain. Such a functionality is reminiscent of an optical analog computer that can; for example, detect motion or transitions (since taking a derivative accentuates rapid changes in a signal), which is indeed noted as the main function required in motion detection applications.

To further validate the analytical framework, a Comsol Multiphysics simulation is also presented in [9] where one optical pump is a continuous-wave (CW) signal, while the second, representing the information signal, has a specific modulation bandwidth. If the CW signal is frequency separated from the information signal by approximately 5.6 THz—which is the AlGaAs zero-crossing frequency—the nonlinearly emitted THz electric field corresponds to the time derivative of the information signal. The proposed metasurface is composed of a square array of nanodisks with a radius of 160 nm, a height of 400 nm, and a period equal to 400 nm, positioned over a low refractive index substrate. In the simulation, a CW source at 1030 nm and an optical signal modulated by a rectangular pulse illuminate the device at an oblique incidence of $\pi/4$ rad. For these design parameters, the metasurface has a magnetic dipolar resonance around 1030 nm, leading to a two-fold enhancement of the event detection nonlinear THz field compared to the unpatterned AlGaAs slab. Remarkably, the output THz field exhibits two distinct peaks corresponding to the steps of the information signal. All the presented discussion demonstrates that dielectric metasurfaces serve as a viable flat-optic platform for efficiently generating and modulating THz radiation. This opens up a new paradigm where a THz nonlinear metasurface could be engineered to process information at the speed of light, potentially detecting motion or performing edge enhancement in imaging, entirely via optical means. Additionally, we have briefly reported their capability to perform analog computing operations such as the first derivative, which is the main function required in motion detection applications. This discussion demonstrates how the future of THz technology lies in the mixture between metasurface engineering and material properties together with the possibility of integration of multifunctional platforms. Materials such as hybrid perovskites, transition metal dichalcogenides, and engineered dielectric surfaces appear to be very promising for nonlinear generation, offering sizable efficiency, and suitable spectral coverage [126–130]. For example, halide perovskites can be tailored in composition to enhance their second-order or third-order nonlinearities and can be processed into microstructures, making them attractive for THz applications. Similarly, two-dimensional materials and other emerging semiconductors might be incorporated into metasurfaces to extend THz generation into frequency ranges or functionalities not accessible with LN or AlGaAs alone. Future metasurfaces are expected to integrate tunable functionalities for the dynamic control of the scattered light in terms of efficiency, polarization, and shape. By combining THz generation, manipulation, and detection on a single chip, these technologies might be able to make significant progress in many different fields such as medical diagnostics, security screening, and high-speed communications. Imagine a single chip that can emit a tailored THz pulse, detect the reflected signal, and analyze it, all within an integrated metasurface framework. Such a device could revolutionize THz imaging scanners, making them more portable and sensitive, or enable ultra-fast data links at THz frequencies by integrating transmitters and receivers in one photonic package. In summary, the trajectory of research suggests that versatile, efficient THz metasurface devices will likely bridge the gap between fundamental laboratory demonstrations and real-world applications in the near future.

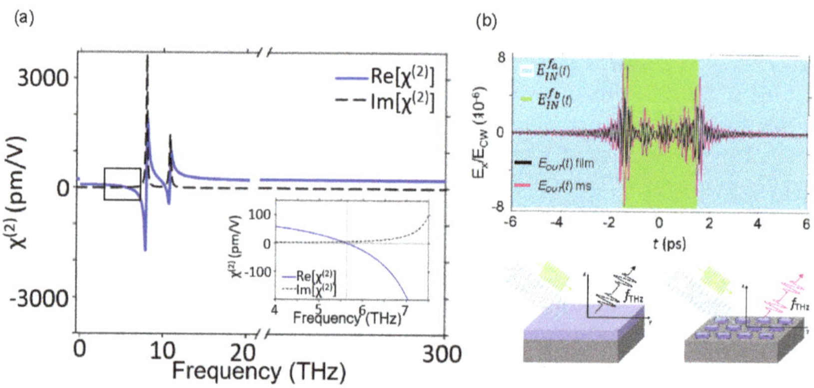

Figure 5. (**a**) Real (blue) and imaginary (black dashed) parts of AlGaAs $\chi^{(2)}$. The inset highlights the region where the real part crosses zero. (**b**) THz field coming from the AlGaAs metasurface (magenta) in comparison to an unpattered AlGaAs thin film (black) when the inputs are a CW pump (light-blue background) and an information signal consisting of a rectangular pulse (green area). Adapted from [9].

6. Conclusions

The convergence of nonlinear optics, advanced materials, and metasurface engineering has ushered in a new era of THz technology. Innovations in optical rectification, resonant metasurfaces, and LSPhP have addressed traditional challenges while unlocking new possibilities for compact, efficient THz devices. With applications ranging from communication to analog computing, the THz spectrum is set to become a cornerstone of next-generation photonic systems, while THz waves were once confined to niche laboratory setups, the developments surveyed in this review indicate that broadly tunable, powerful, and integrable THz sources may soon be feasible, enabling practical systems in various domains. Despite significant progress, challenges remain, including enhancing conversion efficiency, minimizing losses, and integrating metasurfaces with compact and scalable technologies. Future research will further refine these structures, paving the way for practical and high-performance THz devices. For example, achieving higher THz output power will require not only materials with larger nonlinear coefficients or broader bandwidth lasers, but also clever thermal management and out-coupling strategies to handle high optical intensities and extract THz radiation efficiently. Similarly, incorporating THz metasurfaces into portable systems will demand innovative packaging and possibly co-integration with electronics for drive and read-out. Through ongoing advancements, THz metasurfaces are destined to become key elements of next-generation THz technologies, bridging the gap between fundamental research and real-world applications. As researchers continue to improve device performance and address engineering hurdles, we anticipate that the concepts outlined here—from thin-film nonlinear crystals to flat nonlinear optics and analog computation—will possibly translate into tangible THz components in communication networks, imaging devices, and sensing systems, thus strengthening the role of THz photonics in the technological landscape.

Author Contributions: Conceptualization, F.H. and D.R.; methodology, F.H.; writing—original draft preparation, F.H. and D.R.; writing—review and editing, F.H., D.R., O.S., L.C., P.F., A.T., A.L., U.A.L., G.D.V. and C.D.A.; visualization, F.H.; supervision, O.S., L.C., C.D.A. and D.R.; funding acquisition D.R., L.C., G.D.V. and C.D.A. All authors have read and agreed to the published version of the manuscript.

Funding: The authors acknowledge financial support by the European Union–Next Generation EU, Mission 4 Component 1—PRIN 2022 project GRACE6G (2022H7RR4F) CUP D53D23001250001, PRIN 2022 PNRR project FLAIRS (P2022RFF9K) CUP D53D23016160001, PRIN 2022 project NO LIMITHz (2022BC5BW5) CUP D53D23001140001, PRIN 2020 project METEOR (2020EY2LJT), PRIN 2022 project HOTMETA (2022LENW33), and PNRR RESTART project SMART METASURFACES ADVANCING RADIO TECHNOLOGY-SMART-CUP E63C22002040007. This work was partially supported by the European Union under the Italian National Recovery and Resilience Plan (NRRP) of NextGenerationEU, of partnership on "Telecommunications of the Future" (PE00000001—program "RESTART"), S2 SUPER—Programmable Networks, Cascade project PRISM—CUP: C79J24000190004. This work was partially supported by the European Commission Horizon 2020 H2020-FETOPEN2018-2020 project METAFAST (899673).

Institutional Review Board Statement: Not applicable.

Informed Consent Statement: Not applicable.

Data Availability Statement: The data used to support the findings of this study are available from the corresponding author upon request.

Acknowledgments: The authors extend their sincere gratitude to Giuseppe Leo, Marco Peccianti, Michele Celebrano, Tal Ellenbogen and Andrea Toma for their invaluable insights and thoughtful discussions.

Conflicts of Interest: The authors declare no conflicts of interest.

References

1. Davies, G.; Linfield, E. Bridging the terahertz gap. *Phys. World* **2004**, *17*, 37. [CrossRef]
2. Pang, X.; Ozolins, O.; Jia, S.; Zhang, L.; Schatz, R.; Udalcovs, A.; Bobrovs, V.; Hu, H.; Morioka, T.; Sun, Y.T.; et al. Bridging the terahertz gap: Photonics-assisted free-space communications from the submillimeter-wave to the mid-infrared. *J. Light. Technol.* **2022**, *40*, 3149–3162. [CrossRef]
3. Han, R.; Hu, Z.; Wang, C.; Holloway, J.; Yi, X.; Kim, M.; Mawdsley, J. Filling the gap: Silicon terahertz integrated circuits offer our best bet. *IEEE Microw. Mag.* **2019**, *20*, 80–93. [CrossRef]
4. Tzydynzhapov, G.; Gusikhin, P.; Muravev, V.; Dremin, A.; Nefyodov, Y.; Kukushkin, I. New real-time sub-terahertz security body scanner. *J. Infrared Millim. Terahertz Waves* **2020**, *41*, 632–641. [CrossRef]
5. Markl, D.; Ruggiero, M.T.; Zeitler, J.A. Pharmaceutical applications of terahertz spectroscopy and imaging. *Eur. Pharm. Rev.* **2016**, *21*, 45–50.
6. Federici, J.F.; Schulkin, B.; Huang, F.; Gary, D.; Barat, R.; Oliveira, F.; Zimdars, D. THz imaging and sensing for security applications—Explosives, weapons and drugs. *Semicond. Sci. Technol.* **2005**, *20*, S266. [CrossRef]
7. Leon, U.A.; Rocco, D.; Carletti, L.; Peccianti, M.; Maci, S.; Della Valle, G.; De Angelis, C. THz-photonics transceivers by all-dielectric phonon-polariton nonlinear nanoantennas. *Sci. Rep.* **2022**, *12*, 4590. [CrossRef] [PubMed]
8. Arregui Leon, U.; Carletti, L.; Rocco, D.; De Angelis, C.; Della Valle, G. THz Generation via Optical Rectification in Nanomaterials: Universal Modeling Approach and Effective $\chi^{(2)}$ Description. *Laser Photonics Rev.* **2024**, *18*, 2300669. [CrossRef]
9. Arregui Leon, U.; Franceschini, P.; Sergaeva, O.; Tognazzi, A.; Rocco, D.; Carletti, L.; de Ceglia, D.; Della Valle, G.; De Angelis, C. Event detection via THz generation with flat nonlinear optics. *Opt. Mater. Express* **2025**, *15*, 307–318. [CrossRef]
10. Zheludev, N.I.; Kivshar, Y.S. From metamaterials to metadevices. *Nat. Mater.* **2012**, *11*, 917–924. [CrossRef]
11. Tao, H.; Bingham, C.; Strikwerda, A.; Pilon, D.; Shrekenhamer, D.; Landy, N.; Fan, K.; Zhang, X.; Padilla, W.; Averitt, R. Highly flexible wide angle of incidence terahertz metamaterial absorber: Design, fabrication, and characterization. *Phys. Rev. B Condens. Matter Mater. Phys.* **2008**, *78*, 241103. [CrossRef]
12. Grady, N.K.; Heyes, J.E.; Chowdhury, D.R.; Zeng, Y.; Reiten, M.T.; Azad, A.K.; Taylor, A.J.; Dalvit, D.A.; Chen, H.T. Terahertz metamaterials for linear polarization conversion and anomalous refraction. *Science* **2013**, *340*, 1304–1307. [CrossRef] [PubMed]
13. Liu, L.; Zhang, X.; Kenney, M.; Su, X.; Xu, N.; Ouyang, C.; Shi, Y.; Han, J.; Zhang, W.; Zhang, S. Broadband metasurfaces with simultaneous control of phase and amplitude. *Adv. Mater.* **2014**, *26*, 5031–5036. [CrossRef] [PubMed]
14. Keiser, G.; Karl, N.; Liu, P.; Tulloss, C.; Chen, H.T.; Taylor, A.J.; Brener, I.; Reno, J.; Mittleman, D. Nonlinear terahertz metamaterials with active electrical control. *Appl. Phys. Lett.* **2017**, *111*, 121101. [CrossRef]
15. Seren, H.R.; Zhang, J.; Keiser, G.R.; Maddox, S.J.; Zhao, X.; Fan, K.; Bank, S.R.; Zhang, X.; Averitt, R.D. Nonlinear terahertz devices utilizing semiconducting plasmonic metamaterials. *Light Sci. Appl.* **2016**, *5*, e16078. [CrossRef]
16. Lepeshov, S.; Gorodetsky, A.; Krasnok, A.; Rafailov, E.; Belov, P. Enhancement of terahertz photoconductive antenna operation by optical nanoantennas. *Laser Photonics Rev.* **2017**, *11*, 1600199. [CrossRef]
17. Hale, L.L.; Jung, H.; Gennaro, S.D.; Briscoe, J.; Harris, C.T.; Luk, T.S.; Addamane, S.J.; Reno, J.L.; Brener, I.; Mitrofanov, O. Terahertz pulse generation from GaAs metasurfaces. *ACS Photonics* **2022**, *9*, 1136–1142. [CrossRef]
18. Minerbi, E.; Keren-Zur, S.; Ellenbogen, T. Nonlinear metasurface Fresnel zone plates for terahertz generation and manipulation. *Nano Lett.* **2019**, *19*, 6072–6077. [CrossRef]
19. Dong, T.; Li, S.; Manjappa, M.; Yang, P.; Zhou, J.; Kong, D.; Quan, B.; Chen, X.; Ouyang, C.; Dai, F.; et al. Nonlinear THz-nano metasurfaces. *Adv. Funct. Mater.* **2021**, *31*, 2100463. [CrossRef]
20. Tymchenko, M.; Gomez-Diaz, J.S.; Lee, J.; Belkin, M.; Alù, A. Highly-efficient THz generation using nonlinear plasmonic metasurfaces. *J. Opt.* **2017**, *19*, 104001. [CrossRef]
21. McDonnell, C.; Deng, J.; Sideris, S.; Ellenbogen, T.; Li, G. Functional THz emitters based on Pancharatnam-Berry phase nonlinear metasurfaces. *Nat. Commun.* **2021**, *12*, 30. [CrossRef] [PubMed]
22. Dragoman, D.; Dragoman, M. Terahertz fields and applications. *Prog. Quantum Electron.* **2004**, *28*, 1–66. [CrossRef]
23. Pawar, A.Y.; Sonawane, D.D.; Erande, K.B.; Derle, D.V. Terahertz technology and its applications. *Drug Invent. Today* **2013**, *5*, 157–163. [CrossRef]
24. Zhong, S. Progress in terahertz nondestructive testing: A review. *Front. Mech. Eng.* **2019**, *14*, 273–281. [CrossRef]
25. Leitenstorfer, A.; Moskalenko, A.S.; Kampfrath, T.; Kono, J.; Castro-Camus, E.; Peng, K.; Qureshi, N.; Turchinovich, D.; Tanaka, K.; Markelz, A.G.; et al. The 2023 terahertz science and technology roadmap. *J. Phys. D Appl. Phys.* **2023**, *56*, 223001. [CrossRef]
26. Jiang, W.; Zhou, Q.; He, J.; Habibi, M.A.; Melnyk, S.; El-Absi, M.; Han, B.; Di Renzo, M.; Schotten, H.D.; Luo, F.L.; et al. Terahertz communications and sensing for 6G and beyond: A comprehensive review. *IEEE Commun. Surv. Tutorials* **2024**, *26*, 2326–2381. [CrossRef]
27. Chen, H.; Sarieddeen, H.; Ballal, T.; Wymeersch, H.; Alouini, M.S.; Al-Naffouri, T.Y. A tutorial on terahertz-band localization for 6G communication systems. *IEEE Commun. Surv. Tutorials* **2022**, *24*, 1780–1815. [CrossRef]

28. Shafie, A.; Yang, N.; Han, C.; Jornet, J.M.; Juntti, M.; Kürner, T. Terahertz communications for 6G and beyond wireless networks: Challenges, key advancements, and opportunities. *IEEE Netw.* **2022**, *37*, 162–169. [CrossRef]
29. Han, C.; Wu, Y.; Chen, Z.; Wang, X. Terahertz communications (TeraCom): Challenges and impact on 6G wireless systems. *arXiv* **2019**, arXiv:1912.06040.
30. Serghiou, D.; Khalily, M.; Brown, T.W.; Tafazolli, R. Terahertz channel propagation phenomena, measurement techniques and modeling for 6G wireless communication applications: A survey, open challenges and future research directions. *IEEE Commun. Surv. Tutor.* **2022**, *24*, 1957–1996. [CrossRef]
31. Wang, J.; Wang, C.X.; Huang, J.; Chen, Y. 6G THz propagation channel characteristics and modeling: Recent developments and future challenges. *IEEE Commun. Mag.* **2022**, *62*, 56–62. [CrossRef]
32. Alsharif, M.H.; Albreem, M.A.; Solyman, A.A.A.; Kim, S. Toward 6G communication networks: Terahertz frequency challenges and open research issues. *Comput. Mater. Contin.* **2021**, *66*, 2831–2842. [CrossRef]
33. Azari, M.M.; Solanki, S.; Chatzinotas, S.; Bennis, M. THz-empowered UAVs in 6G: Opportunities, challenges, and trade-offs. *IEEE Commun. Mag.* **2022**, *60*, 24–30. [CrossRef]
34. Jepsen, P.U.; Cooke, D.G.; Koch, M. Terahertz spectroscopy and imaging—Modern techniques and applications. *Laser Photonics Rev.* **2011**, *5*, 124–166. [CrossRef]
35. Beard, M.C.; Turner, G.M.; Schmuttenmaer, C.A. Terahertz spectroscopy. *J. Phys. Chem. B* **2002**, *106*, 7146–7159. [CrossRef]
36. Plusquellic, D.F.; Siegrist, K.; Heilweil, E.J.; Esenturk, O. Applications of terahertz spectroscopy in biosystems. *ChemPhysChem* **2007**, *8*, 2412–2431. [CrossRef]
37. Baxter, J.B.; Guglietta, G.W. Terahertz spectroscopy. *Anal. Chem.* **2011**, *83*, 4342–4368. [CrossRef] [PubMed]
38. Globus, T.; Woolard, D.; Khromova, T.; Crowe, T.; Bykhovskaia, M.; Gelmont, B.; Hesler, J.; Samuels, A. THz-spectroscopy of biological molecules. *J. Biol. Phys.* **2003**, *29*, 89–100. [CrossRef]
39. Dexheimer, S.L. *Terahertz Spectroscopy: Principles and Applications*; CRC Press: Boca Raton, FL, USA, 2017.
40. Zaytsev, K.; Dolganova, I.; Chernomyrdin, N.; Katyba, G.; Gavdush, A.; Cherkasova, O.; Komandin, G.; Shchedrina, M.; Khodan, A.; Ponomarev, D.; et al. The progress and perspectives of terahertz technology for diagnosis of neoplasms: A review. *J. Opt.* **2019**, *22*, 013001. [CrossRef]
41. Yan, Z.; Zhu, L.G.; Meng, K.; Huang, W.; Shi, Q. THz medical imaging: From in vitro to in vivo. *Trends Biotechnol.* **2022**, *40*, 816–830. [CrossRef]
42. Vafapour, Z.; Keshavarz, A.; Ghahraloud, H. The potential of terahertz sensing for cancer diagnosis. *Heliyon* **2020**, *6*, e05623. [CrossRef] [PubMed]
43. Li, R.; Li, C.; Li, H.; Wu, S.; Fang, G. Study of automatic detection of concealed targets in passive terahertz images for intelligent security screening. *IEEE Trans. Terahertz Sci. Technol.* **2018**, *9*, 165–176. [CrossRef]
44. Cheng, Y.; Qiao, L.; Zhu, D.; Wang, Y.; Zhao, Z. Passive polarimetric imaging of millimeter and terahertz waves for personnel security screening. *Opt. Lett.* **2021**, *46*, 1233–1236. [CrossRef]
45. Choi, M.K.; Bettermann, A.; Van Der Weide, D. Potential for detection of explosive and biological hazards with electronic terahertz systems. *Philos. Trans. R. Soc. Lond. Ser. A Math. Phys. Eng. Sci.* **2004**, *362*, 337–349. [CrossRef]
46. Trofimov, V.A.; Varentsova, S.A. A possible way for the detection and identification of dangerous substances in ternary mixtures using THz pulsed spectroscopy. *Sensors* **2019**, *19*, 2365. [CrossRef]
47. Gowen, A.A.; O'Sullivan, C.; O'Donnell, C.P. Terahertz time domain spectroscopy and imaging: Emerging techniques for food process monitoring and quality control. *Trends Food Sci. Technol.* **2012**, *25*, 40–46. [CrossRef]
48. Wietzke, S.; Jördens, C.; Krumbholz, N.; Baudrit, B.; Bastian, M.; Koch, M. Terahertz imaging: A new non-destructive technique for the quality control of plastic weld joints. *J. Eur. Opt.-Soc.-Rapid Publ.* **2007**, *2*, 07013. [CrossRef]
49. Küter, A.; Reible, S.; Geibig, T.; Nüßler, D.; Pohl, N. THz imaging for recycling of black plastics. *Tech. Mess.* **2018**, *85*, 191–201. [CrossRef]
50. Jelali, M.; Papadopoulos, K. Inline Inspection of Packaged Food Using Microwave/Terahertz Sensing—An Overview with Focus on Confectionery Products. *Processes* **2024**, *12*, 712. [CrossRef]
51. Hao, D.; Liu, J.; Zou, P.; Zhang, Y.; Moro, R.; Ma, L. All-dielectric Metasurfaces and Their Applications in the Terahertz Range. *Laser Photonics Rev.* **2024**, *18*, 2301210. [CrossRef]
52. Tan, L.; Wang, D.; Xu, K.D. Terahertz metamaterials for spectrum modulation: Structural design, materials and applications. *Mater. Des.* **2024**, *244*, 113217. [CrossRef]
53. He, J.; He, X.; Dong, T.; Wang, S.; Fu, M.; Zhang, Y. Recent progress and applications of Terahertz metamaterials. *J. Phys. D Appl. Phys.* **2022**, *55*, 123002. [CrossRef]
54. Lee, W.S.L.; Atakaramians, S.; Withayachumnankul, W. Terahertz Metasurfaces, Metawaveguides, and Applications. In *More-Than-Moore Devices and Integration for Semiconductors*; Iacopi, F., Balestra, F., Eds.; Springer International Publishing: Cham, Switzerland, 2023; pp. 127–156.

55. Leng, J.; Peng, J.; Jin, A.; Cao, D.; Liu, D.; He, X.; Lin, F.; Liu, F. Investigation of terahertz high Q-factor of all-dielectric metamaterials. *Opt. Laser Technol.* **2022**, *146*, 107570. [CrossRef]
56. Bai, J.; Shen, P.; Wang, S.; Xu, W.; Shen, W.; Chang, S. A High-Q Terahertz Metamaterials Absorber for Refractive Index Sensing. *Phys. Status Solidi B* **2023**, *260*, 2200444. [CrossRef]
57. Zhou, S.; Shen, Z.; Kang, R.; Ge, S.; Hu, W. Liquid Crystal Tunable Dielectric Metamaterial Absorber in the Terahertz Range. *Appl. Sci.* **2018**, *8*, 2211. [CrossRef]
58. Bi, K.; Yang, D.; Chen, J.; Wang, Q.; Wu, H.; Lan, C.; Yang, Y. Experimental demonstration of ultra-large-scale terahertz all-dielectric metamaterials. *Photon. Res.* **2019**, *7*, 457–463. [CrossRef]
59. Ma, Z.; Hanham, S.M.; Albella, P.; Ng, B.; Lu, H.T.; Gong, Y.; Maier, S.A.; Hong, M. Terahertz All-Dielectric Magnetic Mirror Metasurfaces. *ACS Photonics* **2016**, *3*, 1010–1018. [CrossRef]
60. Kadlec, F.; Kužel, P.; Coutaz, J.L. Optical rectification at metal surfaces. *Opt. Lett.* **2004**, *29*, 2674–2676. [CrossRef]
61. Kadlec, F.; Kužel, P.; Coutaz, J.L. Study of terahertz radiation generated by optical rectification on thin gold films. *Opt. Lett.* **2005**, *30*, 1402–1404. [CrossRef]
62. Ramakrishnan, G.; Planken, P.C. Percolation-enhanced generation of terahertz pulses by optical rectification on ultrathin gold films. *Opt. Lett.* **2011**, *36*, 2572–2574. [CrossRef]
63. Welsh, G.H.; Hunt, N.T.; Wynne, K. Terahertz-pulse emission through laser excitation of surface plasmons in a metal grating. *Phys. Rev. Lett.* **2007**, *98*, 026803. [CrossRef] [PubMed]
64. Welsh, G.H.; Wynne, K. Generation of ultrafast terahertz radiation pulses on metallic nanostructured surfaces. *Opt. Express* **2009**, *17*, 2470–2480. [CrossRef] [PubMed]
65. Berry, C.W.; Wang, N.; Hashemi, M.R.; Unlu, M.; Jarrahi, M. Significant performance enhancement in photoconductive terahertz optoelectronics by incorporating plasmonic contact electrodes. *Nat. Commun.* **2013**, *4*, 1622. [CrossRef]
66. Jin, Y.; Reno, J.L.; Kumar, S. Phase-locked terahertz plasmonic laser array with 2 W output power in a single spectral mode. *Optica* **2020**, *7*, 708–715. [CrossRef]
67. Neu, J.; Schmuttenmaer, C.A. Tutorial: An introduction to terahertz time domain spectroscopy (THz-TDS). *J. Appl. Phys.* **2018**, *124*, 231101. [CrossRef]
68. Reimann, K. Table-top sources of ultrashort THz pulses. *Rep. Prog. Phys.* **2007**, *70*, 1597. [CrossRef]
69. Averitt, R.; Taylor, A.J. Ultrafast optical and far-infrared quasiparticle dynamics in correlated electronmaterials. *J. Phys. Condens. Matter* **2002**, *14*, R1357. [CrossRef]
70. Fang, M.; Niu, K.; Huang, Z.; Sha, W.E.; Wu, X.; Koschny, T.; Soukoulis, C.M. Investigation of broadband terahertz generation from metasurface. *Opt. Express* **2018**, *26*, 14241–14250. [CrossRef]
71. Luo, L.; Chatzakis, I.; Wang, J.; Niesler, F.B.; Wegener, M.; Koschny, T.; Soukoulis, C.M. Broadband terahertz generation from metamaterials. *Nat. Commun.* **2014**, *5*, 3055. [CrossRef]
72. Kowerdziej, R.; Jaroszewicz, L.; Olifierczuk, M.; Parka, J. Experimental study on terahertz metamaterial embedded in nematic liquid crystal. *Appl. Phys. Lett.* **2015**, *106*, 092905. [CrossRef]
73. Khan, M.I.; Fraz, Q.; Tahir, F.A. Ultra-wideband cross polarization conversion metasurface insensitive to incidence angle. *J. Appl. Phys.* **2017**, *121*, 045103. [CrossRef]
74. Tal, M.; Keren-Zur, S.; Ellenbogen, T. Nonlinear plasmonic metasurface terahertz emitters for compact terahertz spectroscopy systems. *ACS Photonics* **2020**, *7*, 3286–3290. [CrossRef]
75. Keren-Zur, S.; Tal, M.; Fleischer, S.; Mittleman, D.M.; Ellenbogen, T. Generation of spatiotemporally tailored terahertz wavepackets by nonlinear metasurfaces. *Nat. Commun.* **2019**, *10*, 1778. [CrossRef]
76. Shcherbakov, M.R.; Neshev, D.N.; Hopkins, B.; Shorokhov, A.S.; Staude, I.; Melik-Gaykazyan, E.V.; Decker, M.; Ezhov, A.A.; Miroshnichenko, A.E.; Brener, I.; et al. Enhanced third-harmonic generation in silicon nanoparticles driven by magnetic response. *Nano Lett.* **2014**, *14*, 6488–6492. [CrossRef] [PubMed]
77. Di Francescantonio, A.; Zilli, A.; Rocco, D.; Vinel, V.; Coudrat, L.; Conti, F.; Biagioni, P.; Duò, L.; Lemaître, A.; De Angelis, C.; et al. All-optical free-space routing of upconverted light by metasurfaces via nonlinear interferometry. *Nat. Nanotechnol.* **2024**, *19*, 298–305. [CrossRef] [PubMed]
78. Liu, S.; Sinclair, M.B.; Saravi, S.; Keeler, G.A.; Yang, Y.; Reno, J.; Peake, G.M.; Setzpfandt, F.; Staude, I.; Pertsch, T.; et al. Resonantly enhanced second-harmonic generation using III–V semiconductor all-dielectric metasurfaces. *Nano Lett.* **2016**, *16*, 5426–5432. [CrossRef]
79. Meng, Y.; Chen, Y.; Lu, L.; Ding, Y.; Cusano, A.; Fan, J.A.; Hu, Q.; Wang, K.; Xie, Z.; Liu, Z.; et al. Optical meta-waveguides for integrated photonics and beyond. *Light Sci. Appl.* **2021**, *10*, 235. [CrossRef]
80. Wang, C.; Li, Z.; Kim, M.H.; Xiong, X.; Ren, X.F.; Guo, G.C.; Yu, N.; Lončar, M. Metasurface-assisted phase-matching-free second harmonic generation in lithium niobate waveguides. *Nat. Commun.* **2017**, *8*, 2098. [CrossRef]
81. Herter, A.; Shams-Ansari, A.; Settembrini, F.F.; Warner, H.K.; Faist, J.; Lončar, M.; Benea-Chelmus, I.C. Terahertz waveform synthesis in integrated thin-film lithium niobate platform. *Nat. Commun.* **2023**, *14*, 11. [CrossRef]

82. Hu, L.; Wang, B.; Guo, Y.; Du, S.; Chen, J.; Li, J.; Gu, C.; Wang, L. Quasi-BIC Enhanced Broadband Terahertz Generation in All-Dielectric Metasurface. *Adv. Opt. Mater.* **2022**, *10*, 2200193. [CrossRef]
83. Sun, G.; Wang, Y.; Cui, Z.; Xie, R.; Zhao, X. Enhanced terahertz high-harmonic generation from high-Q quasi-bound states in the continuum empowered by permittivity-broken metasurface. *Appl. Phys. Lett.* **2024**, *124*, 111704. [CrossRef]
84. Liu, Y.; Xu, Y.; Yu, B.; Liu, W.; Zhang, Z.; Cheng, H.; Chen, S. Terahertz Metasurfaces for Polarization Manipulation and Detection: Principles and Emerging Applications. *Adv. Phys. Res.* **2025**, *4*, 2400100. [CrossRef]
85. Hu, F.; Li, L.; Liu, Y.; Meng, Y.; Gong, M.; Yang, Y. Two-plasmon spontaneous emission from a nonlocal epsilon-near-zero material. *Commun. Phys.* **2021**, *4*, 84. [CrossRef]
86. Weis, R.; Gaylord, T. Lithium niobate: Summary of physical properties and crystal structure. *Appl. Phys. A* **1985**, *37*, 191–203. [CrossRef]
87. Volk, T.; Wöhlecke, M. Polarization Reversal and Ferroelectric Domains in LiNbO3 Crystals. In *Lithium Niobate: Defects, Photorefraction and Ferroelectric Switching*; Springer: Berlin/Heidelberg, Germany, 2009; pp. 153–212.
88. Zhu, D.; Shao, L.; Yu, M.; Cheng, R.; Desiatov, B.; Xin, C.; Hu, Y.; Holzgrafe, J.; Ghosh, S.; Shams-Ansari, A.; et al. Integrated photonics on thin-film lithium niobate. *Adv. Opt. Photonics* **2021**, *13*, 242–352. [CrossRef]
89. Hebling, J.; Almasi, G.; Kozma, I.Z.; Kuhl, J. Velocity matching by pulse front tilting for large-area THz-pulse generation. *Opt. Express* **2002**, *10*, 1161–1166. [CrossRef]
90. Hirori, H.; Blanchard, F.; Tanaka, K.J.A.P.L. Single-cycle terahertz pulses with amplitudes exceeding 1 MV/cm generated by optical rectification in $LiNbO_3$. *Appl. Phys. Lett.* **2011**, *98*, 091106. [CrossRef]
91. Jang, D.; Sung, J.H.; Lee, S.K.; Kang, C.; Kim, K.Y. Generation of 0.7 mJ multicycle 15 THz radiation by phase-matched optical rectification in lithium niobate. *Opt. Lett.* **2020**, *45*, 3617–3620. [CrossRef]
92. Boyd, G.; Pollack, M. Microwave nonlinearities in anisotropic dielectrics and their relation to optical and electro-optical nonlinearities. *Phys. Rev. B* **1973**, *7*, 5345. [CrossRef]
93. Carletti, L.; McDonnell, C.; Arregui Leon, U.; Rocco, D.; Finazzi, M.; Toma, A.; Ellenbogen, T.; Della Valle, G.; Celebrano, M.; De Angelis, C. Nonlinear THz generation through optical rectification enhanced by phonon–polaritons in lithium niobate thin films. *ACS Photonics* **2023**, *10*, 3419–3425. [CrossRef]
94. Wu, Q.; Zhang, X.C. Free-space electro-optic sampling of terahertz beams. *Appl. Phys. Lett.* **1995**, *67*, 3523–3525. [CrossRef]
95. Yang, J.; Wang, C. Efficient terahertz generation scheme in a thin-film lithium niobate-silicon hybrid platform. *Opt. Express* **2021**, *29*, 16477–16486. [CrossRef] [PubMed]
96. Wang, J.; Xia, S.; Wang, R.; Ma, R.; Lu, Y.; Zhang, X.; Song, D.; Wu, Q.; Morandotti, R.; Xu, J.; et al. Topologically tuned terahertz confinement in a nonlinear photonic chip. *Light Sci. Appl.* **2022**, *11*, 152. [CrossRef] [PubMed]
97. Liu, H.; Bai, W.; Feng, J.; Jie, W. The synthesis of large-diameter ZnTe crystal for THz emitting and detection. *J. Cryst. Growth* **2017**, *475*, 115–120. [CrossRef]
98. Wilson, D.J.; Schneider, K.; Hönl, S.; Anderson, M.; Baumgartner, Y.; Czornomaz, L.; Kippenberg, T.J.; Seidler, P. Integrated gallium phosphide nonlinear photonics. *Nat. Photonics* **2020**, *14*, 57–62. [CrossRef]
99. Ponseca, C.S., Jr.; Arlauskas, A.; Yu, H.; Wang, F.; Nevinskas, I.; Duda, E.; Vaicaitis, V.; Eriksson, J.; Bergqvist, J.; Liu, X.K.; et al. Pulsed terahertz emission from solution-processed lead iodide perovskite films. *ACS Photonics* **2019**, *6*, 1175–1181. [CrossRef]
100. Valverde-Chávez, D.A.; Cooke, D.G. Multi-cycle terahertz emission from β-barium borate. *J. Infrared Millim. Terahertz Waves* **2017**, *38*, 96–103. [CrossRef]
101. Balos, V.; Wolf, M.; Kovalev, S.; Sajadi, M. Optical rectification and electro-optic sampling in quartz. *Opt. Express* **2023**, *31*, 13317–13327. [CrossRef]
102. Xu, L.; Zhang, X.C.; Auston, D. Terahertz beam generation by femtosecond optical pulses in electro-optic materials. *Appl. Phys. Lett.* **1992**, *61*, 1784–1786. [CrossRef]
103. Marder, S.R.; Perry, J.W.; Yakymyshyn, C.P. Organic salts with large second-order optical nonlinearities. *Chem. Mater.* **1994**, *6*, 1137–1147. [CrossRef]
104. Pan, F.; Knöpfle, G.; Bosshard, C.; Follonier, S.; Spreiter, R.; Wong, M.; Günter, P. Electro-optic properties of the organic salt 4-N, N-dimethylamino-4′-N′-methyl-stilbazolium tosylate. *Appl. Phys. Lett.* **1996**, *69*, 13–15. [CrossRef]
105. Yang, Z.; Mutter, L.; Stillhart, M.; Ruiz, B.; Aravazhi, S.; Jazbinsek, M.; Schneider, A.; Gramlich, V.; Guenter, P. Large-size bulk and thin-film stilbazolium-salt single crystals for nonlinear optics and THz generation. *Adv. Funct. Mater.* **2007**, *17*, 2018–2023. [CrossRef]
106. Coe, B.J.; Harris, J.A.; Asselberghs, I.; Clays, K.; Olbrechts, G.; Persoons, A.; Hupp, J.T.; Johnson, R.C.; Coles, S.J.; Hursthouse, M.B.; et al. Quadratic Nonlinear Optical Properties of N-Aryl Stilbazolium Dyes. *Adv. Funct. Mater.* **2002**, *12*, 110–116. [CrossRef]
107. Kim, P.J.; Jeong, J.H.; Jazbinsek, M.; Choi, S.B.; Baek, I.H.; Kim, J.T.; Rotermund, F.; Yun, H.; Lee, Y.S.; Günter, P.; et al. Highly Efficient Organic THz Generator Pumped at Near-Infrared: Quinolinium Single Crystals. *Adv. Funct. Mater.* **2012**, *22*, 200–209. [CrossRef]

108. Hu, J.; Bandyopadhyay, S.; Liu, Y.h.; Shao, L.y. A review on metasurface: From principle to smart metadevices. *Front. Phys.* **2021**, *8*, 586087. [CrossRef]
109. Huang, L.; Zhang, S.; Zentgraf, T. Metasurface holography: From fundamentals to applications. *Nanophotonics* **2018**, *7*, 1169–1190. [CrossRef]
110. Deng, Z.L.; Li, G. Metasurface optical holography. *Mater. Today Phys.* **2017**, *3*, 16–32. [CrossRef]
111. Arbabi, A.; Arbabi, E.; Horie, Y.; Kamali, S.M.; Faraon, A. Planar metasurface retroreflector. *Nat. Photonics* **2017**, *11*, 415–420. [CrossRef]
112. Rocco, D.; Locatelli, A.; Carletti, L.; Vincenti, M.A.; De Angelis, C. Nonlinear asymmetric imaging with AlGaAs metasurface. *Opt. Express* **2024**, *32*, 11673–11680. [CrossRef]
113. Qiu, Y.; Yan, D.; Li, X.; Zhang, L.; Li, J. Highly efficient second harmonic generation assisted by the quasi-bound states in the continuum from AlGaAs meta-gratings. *Opt. Commun.* **2023**, *546*, 129772. [CrossRef]
114. Camacho-Morales, R.; Rahmani, M.; Kruk, S.; Wang, L.; Xu, L.; Smirnova, D.A.; Solntsev, A.S.; Miroshnichenko, A.; Tan, H.H.; Karouta, F.; et al. Nonlinear generation of vector beams from AlGaAs nanoantennas. *Nano Lett.* **2016**, *16*, 7191–7197. [CrossRef] [PubMed]
115. Gili, V.F.; Carletti, L.; Locatelli, A.; Rocco, D.; Finazzi, M.; Ghirardini, L.; Favero, I.; Gomez, C.; Lemaître, A.; Celebrano, M.; et al. Monolithic AlGaAs second-harmonic nanoantennas. *Opt. Express* **2016**, *24*, 15965–15971. [CrossRef] [PubMed]
116. Liu, T.; Fang, X.; Xiao, S. Tuning nonlinear second-harmonic generation in AlGaAs nanoantennas via chalcogenide phase-change material. *Phys. Rev. B* **2021**, *104*, 195428. [CrossRef]
117. Marino, G.; Rocco, D.; Gigli, C.; Beaudoin, G.; Pantzas, K.; Suffit, S.; Filloux, P.; Sagnes, I.; Leo, G.; De Angelis, C. Harmonic generation with multi-layer dielectric metasurfaces. *Nanophotonics* **2021**, *10*, 1837–1843. [CrossRef]
118. Liu, S.; Keeler, G.A.; Reno, J.L.; Sinclair, M.B.; Brener, I. *III–V Semiconductor Nanoresonators—A New Strategy for Passive, Active, and Nonlinear All-Dielectric Metamaterials*; Technical Report; Sandia National Lab. (SNL-NM): Albuquerque, NM, USA, 2016.
119. Romeira, B.; Borme, J.; Fonseca, H.; Gaspar, J.; Nieder, J.B. Efficient light extraction in subwavelength GaAs/AlGaAs nanopillars for nanoscale light-emitting devices. *Opt. Express* **2020**, *28*, 32302–32315. [CrossRef]
120. Gandolfi, M.; Carletti, L.; Tognazzi, A.; Cino, A.C.; De Angelis, C.; Guasoni, M. Near to short wave infrared light generation through AlGaAs-on-insulator nanoantennas. *Opt. Express* **2023**, *31*, 31051–31060. [CrossRef]
121. Cruciano, C.; Rocco, D.; Genco, A.; Tognazzi, A.; Locatelli, A.; Carletti, L.; Fedorov, A.; Trovatello, C.; Di Blasio, G.; Bargigia, I.; et al. Shaping the Emission Directivity of Single Quantum Dots in Dielectric Nanodisks Exploiting Mie Resonances. *ACS Nano* **2025**, *19*, 3500–3509. [CrossRef]
122. Baboux, F.; Moody, G.; Ducci, S. Nonlinear integrated quantum photonics with AlGaAs. *Optica* **2023**, *10*, 917–931. [CrossRef]
123. Coudrat, L.; Boulliard, G.; Gérard, J.M.; Lemaître, A.; Degiron, A.; Leo, G. Unravelling the nonlinear generation of designer vortices with dielectric metasurfaces. *Light Sci. Appl.* **2025**, *14*, 51. [CrossRef]
124. Gandhi, H.K.; Rocco, D.; Carletti, L.; De Angelis, C. Gain-loss engineering of bound states in the continuum for enhanced nonlinear response in dielectric nanocavities. *Opt. Express* **2020**, *28*, 3009–3016. [CrossRef]
125. Peters, L.; Rocco, D.; Olivieri, L.; Arregui Leon, U.; Cecconi, V.; Carletti, L.; Gigli, C.; Della Valle, G.; Cutrona, A.; Totero Gongora, J.S.; et al. Resonant Fully dielectric metasurfaces for ultrafast Terahertz pulse generation. *Adv. Opt. Mater.* **2024**, *12*, 2303148. [CrossRef]
126. Zhou, Y.; Huang, Y.; Xu, X.; Fan, Z.; Khurgin, J.B.; Xiong, Q. Nonlinear optical properties of halide perovskites and their applications. *Appl. Phys. Rev.* **2020**, *7*, 041313. [CrossRef]
127. He, Y.; Li, N.; Feng, Y.; Li, X.; Liu, D.; Huang, J.; Zhou, R.; Wu, M.; Miao, L.; Zhao, C. Broadband nonlinear optical modulator with 2D organic-inorganic hybrid perovskite nanocrystals. *IEEE J. Sel. Top. Quantum Electron.* **2023**, *29*, 1–8. [CrossRef]
128. Marjanowska, A.; El Karout, H.; Guichaoua, D.; Sahraoui, B.; Płóciennik, P.; Zawadzka, A. Topography and nonlinear optical properties of thin films containing iodide-based hybrid perovskites. *Nanomaterials* **2023**, *14*, 50. [CrossRef]
129. Li, H.; Diao, M.; Boukhvalov, D.W.; Ke, Y.; Humphrey, M.G.; Zhang, C.; Huang, Z. Prominent Nonlinear Optical Absorption in SnS2-Based Hybrid Inorganic—Organic Superlattice. *Adv. Funct. Mater.* **2024**, *34*, 2400077. [CrossRef]
130. Huang, Y.; Zhu, L.; Yao, Z.; Zhang, L.; He, C.; Zhao, Q.; Bai, J.; Xu, X. Terahertz surface emission from layered MoS2 crystal: Competition between surface optical rectification and surface photocurrent surge. *J. Phys. Chem. C* **2018**, *122*, 481–488. [CrossRef]

photonics

MDPI

Communication

Localized Effects in Graphene Oxide Systems: A Pathway to Hyperbolic Metamaterials

Grazia Giuseppina Politano

Department of Environmental Engineering, University of Calabria, 87036 Rende, CS, Italy; grazia.politano@unical.it

Abstract: Graphene oxide (GO) has emerged as a carbon-based nanomaterial providing a different pathway to graphene. One of its most notable features is the ability to partially reduce it, resulting in graphene-like sheets through the elimination of oxygen-including functional groups. In this paper, the effect of localized interactions in an Ag/GO/Au multilayer system was studied to explore its potential for photonic applications. GO was dip-coated onto magnetron-sputtered silver, followed by the deposition of a thin gold film to form an Ag/GO/Au structure. Micro-Raman Spectroscopy, SEM and Variable Angle Ellipsometry (VASE) measurements were performed on the Ag/GO/Au structure. An interesting behavior of the GO deposited on magnetron-sputtered silver with the formation of Ag nanostructures on top of the GO layer is reported. In addition to typical GO bands, Micro-Raman analysis reveals peaks such as the 1478 cm^{-1} band, indicating a transition from sp^3 to sp^2 hybridization, confirming the partial reduction of GO. Additionally, calculations based on effective medium theory (EMT) highlight the potential of Ag/GO structures in hyperbolic metamaterials for photonics. The medium exhibits dielectric behavior up to 323 nm, transitions to type I HMM between 323 and 400 nm and undergoes an Epsilon Near Zero and Pole (ENZP) transition at 400 nm, followed by type II HMM behavior.

Keywords: graphene oxide; optical properties; micro Raman

1. Introduction

Photonic metamaterials [1], which enable exact manipulation of light at subwavelength scales, have greatly revolutionized optical technologies over the last years. These materials' special qualities—like improved light–matter interactions, negative refraction and superlensing—have created new opportunities for energy harvesting, imaging and sensing [2]. Recent advancements in metasurface-based photonic devices have further expanded these possibilities by demonstrating improved optical field control, higher efficiency and enhanced damage thresholds, making them particularly suitable for high-power laser applications [3]. Moreover, recent research on terahertz metamaterial absorbers based on Dirac semimetals has shown exceptional tunability and sensitivity [4].

These advances have been made possible in large part by the creation of novel materials and enhanced nanofabrication methods [5]. In addition to optimal designs, the creation of new materials with adjustable optical characteristics and scalable manufacturing techniques are also necessary to achieve such breakthroughs. Because of their remarkable electrical and optical properties, graphene and its derivatives, including graphene oxide (GO) [6], have turned out to be attractive options for incorporation into photonic metamaterials.

Graphene and GO differ in terms of structure and properties. Graphene consists of carbon atoms arranged in a two-dimensional honeycomb lattice with sp^2 hybridization,

 https://doi.org/10.3390/photonics12020121

which provides it with exceptional electrical conductivity, mechanical strength and thermal properties [7]. Graphene can be synthesized using four main methods. The first method involves chemical vapor deposition (CVD) [8]. The second method is micromechanical exfoliation, commonly known as the 'Scotch tape' [9] or peel-off technique, which originated from earlier work on exfoliating patterned graphite. The third approach utilizes epitaxial growth on insulating substrates like silicon carbide (SiC) [9], while the fourth involves the production of colloidal suspensions [10]. Although graphene has earned a lot of interest due to its many uses, the difficulties associated with mass manufacture prevent it from being used in technology on a significant scale [11]. As an alternative to graphene, GO offers a more approachable and adaptable platform for real-world uses. GO contains both sp^2 and sp^3 hybridized carbon atoms due to the presence of oxygen-containing functional groups such as epoxy, hydroxyl and carboxyl. These functional groups enhance its dispersibility in water but significantly reduce its electrical conductivity compared to pristine graphene [12]. GO can be converted into reduced graphene oxide (RGO) by eliminating oxygen-including functional groups, resulting in graphene-like sheets with restored properties, regaining its electrical conductivity and optical characteristics despite its insulating nature brought on by the breakage of its conjugated electronic structure [13]. However, conventional chemical reduction techniques frequently include dangerous chemicals like hydrazine, which emphasizes the need for alternative scalable and sustainable strategies [14]. In the development of new materials for photonic applications, where environmentally friendly and adjustable procedures are becoming more and more crucial, these alternative reduction techniques may be crucial [15]. Concerning the catalytic reduction of GO, Wu et al. [16] demonstrated that Ag nanoparticles can facilitate the visible-light-driven photocatalytic reduction of GO through the mechanism of Surface Plasmon Resonance (SPR). Furthermore, other important advancements have been reported in this research field. For example, Zhuo et al. [17] described a method for the large-scale production of graphene through the room-temperature reduction of GO, utilizing metal nanoparticles as catalysts. In addition, the γ-ray-assisted synthesis of silver nanoparticle-decorated RGO/Ag represents a promising green approach for the development of advanced materials in energy storage applications, particularly symmetric supercapacitors [18]. The fabrication process has a particular influence on the design of metamaterials for practical applications [19,20]. Magnetron sputtering [21] is a scalable and repeatable method that ensures the consistency required for advanced photonic metamaterials by producing thin films with nanoscale accuracy.

This work focuses on the magnetron sputtering deposition of a multilayer Ag/GO/Au system. Micro-Raman spectroscopy reveals that localized plasmonic interactions between gold and silver enhance electromagnetic field localization at the interface, potentially modifying the optical behavior of GO. This approach highlights a possible alternative to traditional chemical methods for reducing GO, meeting the criteria of being inexpensive and environmentally friendly.

When the Ag/GO interface in the system is examined more closely, some features become apparent. The appearance of silver nanostructures on the GO layer is demonstrated by SEM measurements, generating "hot spots" that are essential for Surface-Enhanced Raman Scattering (SERS). Additionally, effective medium theory (EMT) [22] calculations demonstrate that the GO/Ag combination is a promising candidate for integration into hyperbolic metamaterials. Such metamaterials enable unique optical properties, including negative refraction and enhanced light–matter interactions, opening the door for innovative applications in photonic technologies.

2. Materials and Methods

Deposition of silver thin films (10 nm) was carried out on glass substrates using a DC magnetron sputtering system (Edwards Auto306, West Sussex, UK) under an argon partial pressure of 4.5×10^{-2}, with a base vacuum of 10^{-5} mbar and a cathode power of 10 W for 75 s. The glass substrates were pre-cleaned with a "piranha solution" (hydrogen peroxide and sulfuric acid).

GO films (8 nm) were deposited on the silver thin films using a dip-coating process at 0.33 mm/s. A 2 g/L GO dispersion in water, sourced from "Punto Quantico", was used. Gold thin films (15 nm) were then deposited onto the GO/Ag/glass substrates using the same sputtering system at 4.2×10^{-2} mbar, with a base vacuum of 10^{-5} mbar and a cathode power of 30 W for 55 s. A schematic of the Ag/GO/Au structure is shown in Figure 1.

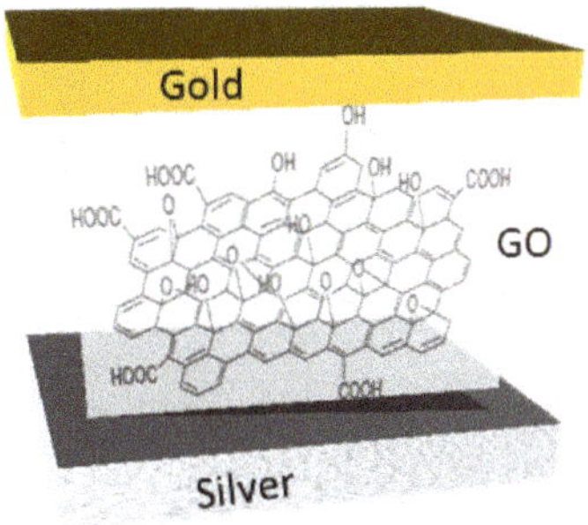

Figure 1. Schematic illustration of the samples.

Micro-Raman spectroscopy was carried out using a Horiba-Jobin Yvon LabRam HR system (Horiba-Jobin Yvon Srl, Piscataway, NJ, USA) with a 532 nm laser and a 100× Mplan Olympus objective. An OD2 filter (1% transmission) was used to minimize thermal effects. SEM analysis was performed with a FEI Quanta FEG 400 ESEM microscope (Eindhoven, The Netherlands).

Optical properties were estimated via Variable Angle Spectroscopic Ellipsometry (VASE) using a Woollam M2000 F (Woollam Co., Lincoln, NE, USA) rotating compensator ellipsometer, covering the photon energy range of [1.3–4.1] eV.

3. Results and Discussion

3.1. Micro-Raman Measurements

The visual investigation of the samples (Figure 2A) reveals that two types of GO deposition are possible. The "light" zones can be associated with a monolayer or few-layer growth. In contrast, the remaining areas appear in varying shades of dark colors.

The formation of some "hot spots" is observed, usually in the "dark" regions of this sample. Hot spots are highly concentrated areas of strong local field enhancement attributed to localized surface plasmon resonance (LSPR) [23]. These regions, found in the gaps of metallic nanostructures, are known to significantly amplify SERS signals [24]. An example of this phenomenon can be observed in Figure 2B. The D and G bands of GO are visible at about 1365 cm^{-1} and 1595 cm^{-1}, correspondingly. Moreover, it is possible to see the second order D + G band ($\approx$2912 cm^{-1}). In addition, several sharp peaks appear at about 1077 cm^{-1}, 1122 cm^{-1}, 1240 cm^{-1}, 1286 cm^{-1}, 1478 cm^{-1} and 1976 cm^{-1}. All these distinct peaks are associated with localized vibrations of molecular groups bound to the GO network and in proximity to metal particles or nanostructures. These "hot spots" arise from the intense plasmonic coupling between the metal and the molecule, leading to an important enhancement of the Raman signal through the SERS effect. In particular, the peak at about 1478 cm^{-1} can be due to a CH2 deformation vibration; indeed, Aunkor et al. [25] reported that the spectra of RGO display peaks that can be attributed to the CH2 stretching vibration. This observation might confirm that the atomic frame of sp^2 carbon has arisen in RGO, thus indicating the transition from sp^3 (oxidized regions) to sp^2 (graphitic regions) hybridization [25].

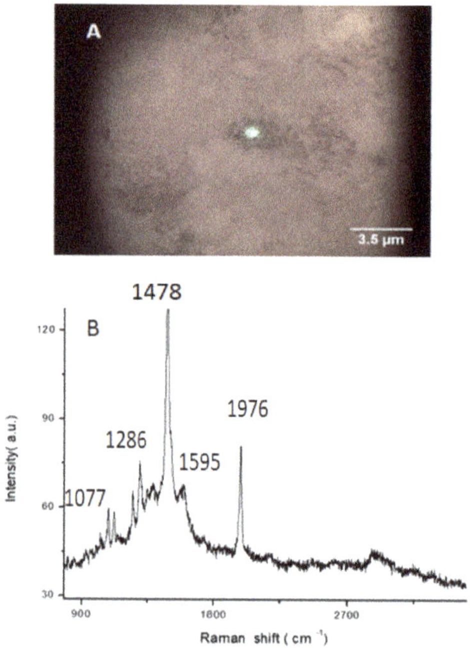

Figure 2. Optical image of a typical region of Ag/Graphene oxide/Au structure. Green laser spot on a dark region (**A**). Raman spectrum on a hot spot of the samples at low laser intensity (OD2 filter) (**B**).

It was observed that the sharp peaks associated with *hot spots* disappeared under high laser power, suggesting thermal effects. To minimize this influence, low-power laser settings were used during the measurements (OD2 filter). While plasmonic effects play a role in enhancing the Raman signal, the potential for laser-induced thermal degradation must also be considered when interpreting these results.

3.2. Morphological SEM Analysis

In Figure 3, sections of the Glass/Ag/GO/Au layer structure were obtained using a secondary electron sensor. In the images, it is possible to see some 40/50 nm Ag nanoparticles covered by a thin film of different chemical compositions that flexibly conform to the nanostructured substrate.

In Figure 4, the glass/Ag/GO sample was studied without the gold coating to explore the Ag/GO interface of the more complex glass/Ag/GO/Au structure.

Figure 4A shows a topography image of the glass/Ag/GO structure acquired by the secondary electron sensor. Some objects (supposedly Ag nanoparticles of 40/50 nm diameters) are visible on the GO surface. Figure 4C shows a magnification of Figure 4A. The same objects were observed with the Back Scattering sensor that highlights an atomic number difference between superficial (brighter) objects, presumably Ag nanoparticles, on a less luminous surface of GO (Figure 4B). Figure 4D shows a magnification of Figure 4B.

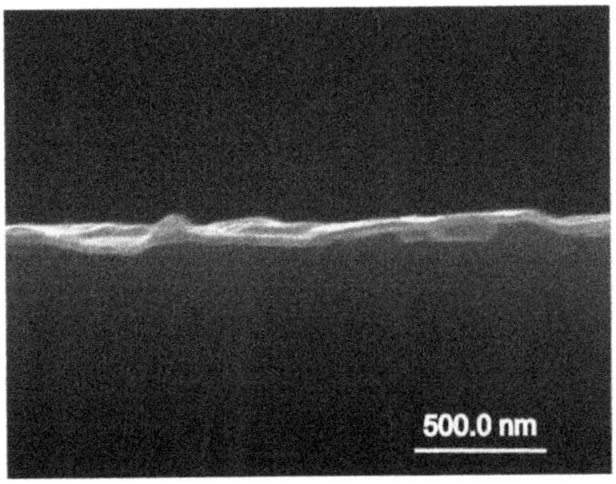

Figure 3. Higher magnification ESEM image of glass/Ag/GO/Au layer samples. It is possible to see some Ag nanoparticles below the graphene oxide and the Au thin films obtained using a secondary electron detector.

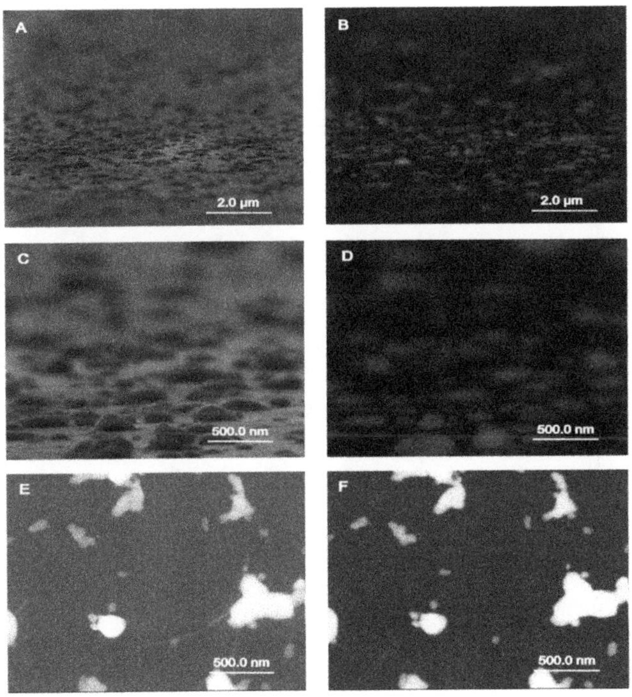

Figure 4. ESEM images of the surface topography of a glass/Ag/graphene oxide sample. Images (**A**,**C**,**E**) were obtained using a secondary electron detector; images (**B**,**D**,**F**) were acquired by a backscattered electron detector. Some objects, supposedly Ag nanoparticles of 40/50 nm diameters, are visible on the graphene oxide surface (**A**) and related magnification (**C**). The backscattered electron detector highlights the atomic number difference between superficial (brighter) objects, presumably Ag nanoparticles, and a less luminous graphene oxide background film. The Ag nanoparticles pierce the graphene oxide layer in some places (**E**,**F**); in (**E**), we can also see the cleavage lines, which start where the Ag nanoparticles break the graphene oxide layer.

An example of cleavage across the Ag/GO interface is visible in Figure 4E (see, in particular, the scratches due to the silver nanoparticles below the GO film). Figure 4F reports the corresponding backscattering image.

The glass/Ag/GO system is very complex, and there are zones of samples in which the GO layer adjusts elastically to the Ag nanoparticles.

Figure 5A, for instance, shows the topography acquired by the secondary electron sensor on a surface of GO dip-coated on 10 nm magnetron-sputtered Ag. Objects of about 40/50 nm diameter are visible on the surface, but they are thought to be a deformation of the GO film induced by Ag nanoparticles grown in the underlying surface, which did not perforate the GO layer, as confirmed by the corresponding image. Figure 5B was acquired by the backscattering sensor. The behavior of the complex interface Ag/GO, never reported before, requires further studies because a greater understanding could open the way for several applications. The dispersion of Ag nanoparticles within the thin Ag film, along with their potential detachment from the surface, leads to varying interactions with molecular groups. In this study, this phenomenon may explain the formation of the "hot spots" detected using micro-Raman spectroscopy.

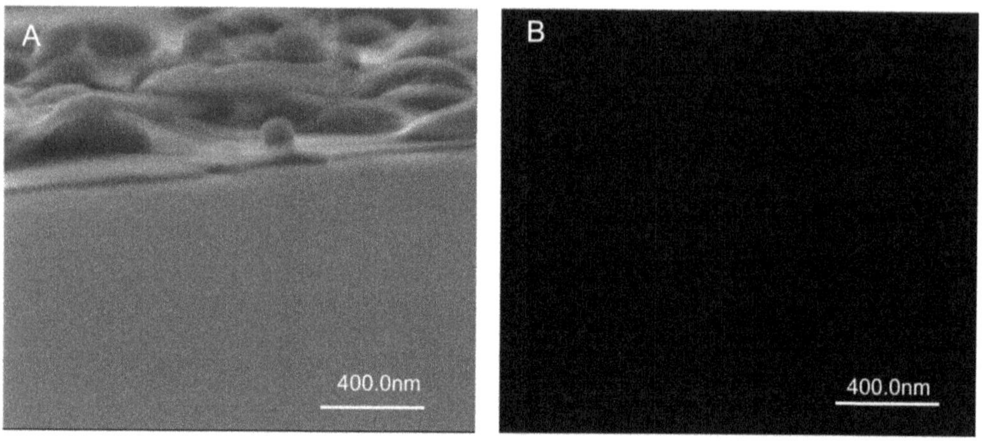

Figure 5. ESEM images of a glass/Ag/graphene oxide sample acquired by a secondary (**A**) and a backscattered (**B**) electron detector. The images show the same morphology as those reported in Figure 4, but the backscattering image reveals that the chemical composition of the surface is homogeneous. In this case, the Ag nanoparticles can curve the graphene oxide film, but they do not pierce the graphene oxide layer.

3.3. Variable Angle Spectroscopic Ellipsometry Measurements

VASE was used to check the average thickness of the thin films. The analysis of the ellipsometric data requires a multilayer optical model of the samples.

The ψ and Δ spectra were measured at different angles of incidence (50°–70°) to increase the accuracy of layer modeling.

The model was obtained using the superposition of a layer of glass (1 mm), Ag (10 nm) and EMA (Effective Medium Approximation) model between the optical constants of GO and that of silver (8 nm) and Au (15 nm).

In Figure 6A,B, the generated and experimental data of the ψ and Δ spectra are reported for different angles of incidents in the [1.3–4.1] eV photon energy range for the glass/Ag/GO/Au structure.

It is worth noting that the spot size of the light beam used in spectroscopic ellipsometry is typically several millimeters, which means a bigger spot size in comparison with Raman spectra and SEM analysis. However, important information can be deduced from Figure 6A,B.

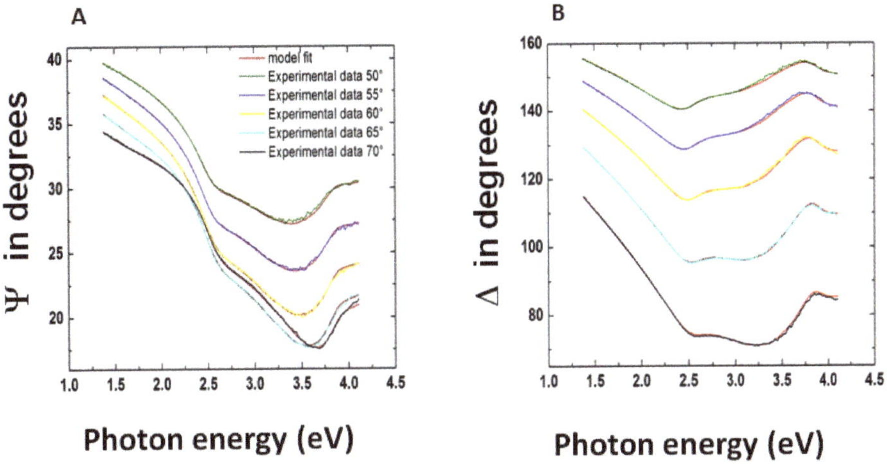

Figure 6. Experimental and model-generated ellipsometric ψ (**A**) and Δ (**B**) data fits at five angles of incidences for the glass/Ag/GO/Au structure.

In Figure 7A, the estimated dispersion laws of the GO sample deposited with the dip-coating technique on magnetron-sputtered Ag/glass substrate by ellipsometry characterization are reported in the [1.3–4.1] eV photon energy range.

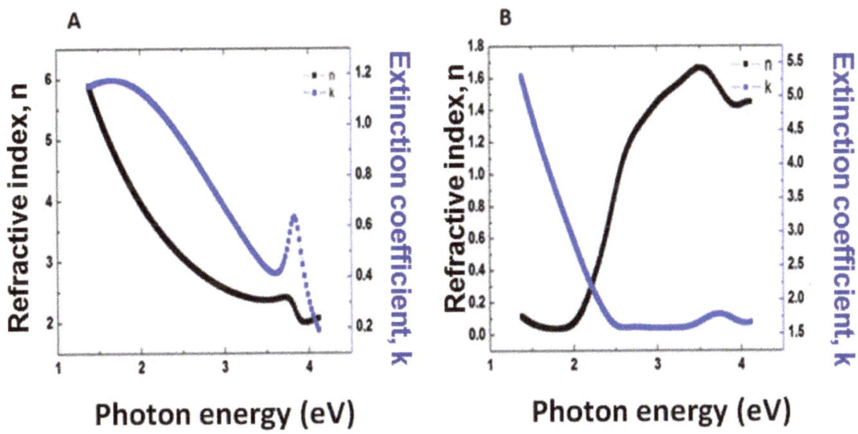

Figure 7. Estimated dispersion laws of the glass/Ag/GO structure (**A**) and of the glass/Ag/GO/Au structure (**B**). The curves show the index of refraction (black lines) and the extinction coefficient (blue lines).

The model was obtained using the superposition of a layer of glass (1 mm), Ag (10 nm) and an EMA model, which mixes the Cauchy optical constants of GO and those of silver (8 nm). The EMA model was selected due to the nanostructured nature of the interface, and its validity has been supported by the consistency between modeled and experimental ellipsometric data. The model indicates the formation of a layer of GO with mixed silver features; for example, we find an oscillator at 3.8 eV, which can be attributable to the volume plasmon of Ag [26]. The mixed properties between silver and GO are consistent with the observed behavior in SEM images. In Figure 7B, the dispersion laws are reported in the [1.3–4.1] eV photon energy range. The optical constants are very similar to those of gold [27]; hence, the thin gold film adjusts elastically on the glass/Ag/GO sample (as can be seen in Figure 3) without modifying its overall optical behavior.

3.4. Potential Application in Hyperbolic Metamaterials

A potential application has been identified for films composed of GO dip-coated on magnetron-sputtered silver as hyperbolic metamaterials.

A hyperbolic metamaterial [28] is an anisotropic medium characterized by uniaxial dielectric tensor components:

$\varepsilon_{xx} = \varepsilon_{yy} = \varepsilon_{\parallel}$ and $\varepsilon_{yy} = \varepsilon_{\perp}$, which can be approximated using effective medium theory (EMT):

$$\varepsilon_{\parallel} = \frac{t_m\varepsilon_m + t_d\varepsilon_d}{t_m + t_d} \quad (1)$$

$$\varepsilon_{\perp} = \frac{\varepsilon_m\varepsilon_d(t_m + t_d)}{t_m\varepsilon_d + t_d\varepsilon_m} \quad (2)$$

In Equations (1) and (2), (t_d, ε_d) and (t_m, ε_m) are the thickness and dielectric permittivity of dielectric and metal, respectively.

Figure 8 presents an EMT simulation based on the optical constants of 20 nm thick graphene oxide (GO) and 20 nm thick magnetron-sputtered silver. The results demonstrate the theoretical feasibility of achieving the rare coexistence of two opposite anisotropies, type I and type II, within the same metamaterial. This coexistence is characterized by a zero dielectric or metal gap separating the two anisotropic states.

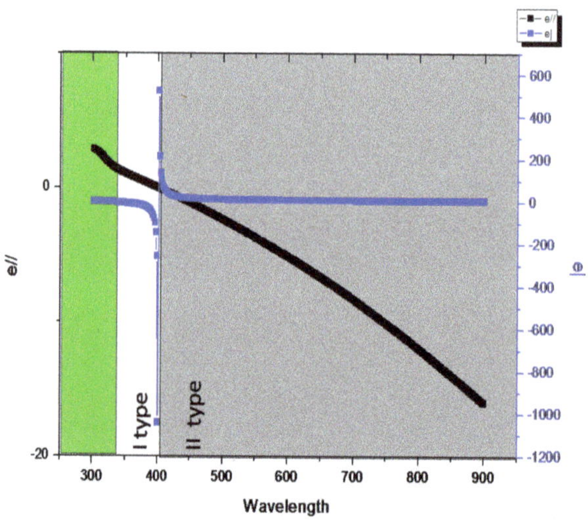

Figure 8. The effective medium theory (EMT) applied to alternating GO/Ag multilayers. The black curve represents the real part of the parallel component of epsilon, while the blue curve corresponds to the perpendicular component of the entire structure.

For type I hyperbolic metamaterials HMM ($\varepsilon_{\parallel} > 0$ and $\varepsilon_{\perp} < 0$), the structure exhibits dielectric properties in the xy plane and metallic properties along the $\varepsilon_{\perp}$ direction, while the isofrequency surfaces assume the shape of an open-bounded hyperboloid. In this configuration, the isofrequency surfaces take the form of an open, bounded hyperboloid. Conversely, in type II HMM, the behavior is reversed and the isofrequency surface transitions to a continuous hyperboloid.

As illustrated in Figure 8, the medium exhibits dielectric behavior up to a wavelength of 323 nm (green area), as both $\varepsilon_{\parallel}$ and $\varepsilon_{\perp}$ are positive. Between 323 nm and 400 nm (white area), $\varepsilon_{\parallel} > 0$ while $\varepsilon_{\perp} < 0$, resulting in the emergence of a type I HMM region. At λ = 400 nm, the ε_{NZP} (Epsilon Near Zero and Pole) behavior obviously manifests a strong discontinuity in $\varepsilon_{\perp}$, passing from a very high negative value (virtually $-\infty$) to a very high positive one (virtually $+\infty$) while simultaneously, $\varepsilon_{\parallel} = 0$. The HMM transitions to type II behavior at $\lambda > 400$ nm (grey area).

The results presented here emphasize the critical role of the reduction process in tuning the optical properties of the Ag/GO/Au structure. Raman spectroscopy confirms the transition from sp^3 to sp^2 hybridization, indicating partial reduction of GO, which is essential for enhancing conductivity and optimizing its performance in photonic applications. However, despite the environmental advantages of the plasmonic reduction approach compared to conventional chemical methods, achieving uniformity and reproducibility on a larger scale is still a limiting factor. In addition, the theoretical model confirms the potential of the application of Ag/GO structures in hyperbolic metamaterials; nevertheless, practical implementation of these metamaterials faces challenges related to fabrication precision and consistency, which must be addressed to fully exploit their potential in applications such as imaging, sensing and waveguiding at the nanoscale.

4. Conclusions

This study explores the potential of Ag/GO/Au structures for photonic applications, particularly their suitability as hyperbolic metamaterials. Gold is used to ensure chemical stability and enhance plasmonic coupling with silver. Micro-Raman analysis detects additional peaks beyond the typical GO bands, including a peak at 1478 cm^{-1} attributed to CH_2 deformation vibrations, indicating the transition from sp^3 (oxidized) to sp^2 (graphitic) carbon hybridization. SEM images reveal that the Au film conforms elastically to Ag nanostructures (~40–50 nm), potentially contributing to the observed "hot spots" in Raman measurements. The dispersion properties of the structure align with the SEM observations, with the Au layer maintaining the overall optical behavior, while the GO/Ag interface presents complexity. Ellipsometry suggests a GO layer incorporating silver features, with an oscillator at 3.8 eV linked to Ag volume plasmons. Theoretical analysis indicates that by optimizing Ag and GO thicknesses, these structures can act as multilayer hyperbolic metamaterials, exhibiting simultaneous type I and type II anisotropies, separated by a zero dielectric or metal gap. While still theoretical, these findings present significant experimental challenges and promising applications in plasmonics and metamaterials.

Funding: This research received no external funding.

Institutional Review Board Statement: Not applicable.

Informed Consent Statement: Not applicable.

Data Availability Statement: The data are contained within the article.

Conflicts of Interest: The author declares no conflict of interest.

Abbreviations

The following abbreviations are used in this manuscript:

GO	Graphene oxide
EMT	Effective medium theory
RGO	Reduced graphene oxide
VASE	Variable Angle Spectroscopic ellipsometry
EMA	Effective Medium Approximation
ENZP	Epsilon Near Zero and Pole

References

1. Wang, Z.; Cheng, F.; Winsor, T.; Liu, Y. Optical chiral metamaterials: A review of the fundamentals, fabrication methods anc applications. *Nanotechnology* **2016**, *27*, 412001. [CrossRef] [PubMed]
2. Ma, W.; Cheng, F.; Xu, Y.; Wen, Q.; Liu, Y. Probabilistic Representation and Inverse Design of Metamaterials Based on a Dee Generative Model with Semi-Supervised Learning Strategy. *Adv. Mater.* **2019**, *31*, 1901111. [CrossRef] [PubMed]
3. Wang, Q.; Fang, Y.; Meng, Y.; Hao, H.; Li, X.; Pu, M.; Ma, X.; Luo, X. Vortex-field enhancement through high-threshold geometri metasurface. *Opto-Electron. Adv.* **2024**, *7*, 240112. [CrossRef]
4. Zeng, Z.; Liu, H.; Zhang, H.; Cheng, S.; Yi, Y.; Yi, Z.; Wang, J.; Zhang, J. Tunable ultra-sensitive four-band terahertz sensors basec on Dirac semimetals. *Photonics Nanostruct. Fundam. Appl.* **2025**, *63*, 101347. [CrossRef]
5. Chi, T.; Somers, P.; Wilcox, D.A.; Schuman, A.J.; Iyer, V.; Le, R.; Gengler, J.; Ferdinandus, M.; Liebig, C.; Pan, L.; et al. Tailorec thioxanthone-based photoinitiators for two-photon-controllable polymerization and nanolithographic printing. *J. Polym. Sci. Par B Polym. Phys.* **2019**, *57*, 1462–1475. [CrossRef]
6. Politano, G.G. Optimizing Graphene Oxide Film Quality: The Role of Solvent and Deposition Technique. *C* **2024**, *10*, 90. [CrossRef
7. Urade, A.R.; Lahiri, I.; Suresh, K.S. Graphene Properties, Synthesis and Applications: A Review. *JOM* **2023**, *75*, 614–630. [CrossRef [PubMed]
8. Saeed, M.; Alshammari, Y.; Majeed, S.A.; Al-Nasrallah, E. Chemical Vapour Deposition of Graphene—Synthesis, Characterisation and Applications: A Review. *Molecules* **2020**, *25*, 3856. [CrossRef]
9. Han, Y.-C.; Yin, S.-H.; Zheng, J.-R.; Hu, Y.-F.; Sun, L.; Zhang, L.; Tian, Z.-Q.; Yi, J. Epitaxial Growth of Graphene on SiC by Therma Shock Annealing Within Seconds. *Adv. Funct. Mater.* **2024**, *34*, 2307298. [CrossRef]
10. Lee, S.J.; Huh, H.K.; Kwon, D.H. Energy dissipation of graphene colloidal suspension droplets impacting on solid substrates *RSC Adv.* **2014**, *4*, 7216–7224. [CrossRef]
11. Politano, G.G. Optical Properties of Graphene Nanoplatelets on Amorphous Germanium Substrates. *Molecules* **2024**, *29*, 4089 [CrossRef] [PubMed]
12. Mohammed, S. Graphene oxide: A mini-review on the versatility and challenges as a membrane material for solvent-based separation. *Chem. Eng. J. Adv.* **2022**, *12*, 100392. [CrossRef]
13. Kurian, M. Recent progress in the chemical reduction of graphene oxide by green reductants–A Mini review. *Carbon Trends* **2021** *5*, 100120. [CrossRef]
14. Cao, K.; Tian, Z.; Zhang, X.; Wang, Y.; Zhu, Q. Green preparation of graphene oxide nanosheets as adsorbent. *Sci. Rep.* **2023**, *13* 9314. [CrossRef] [PubMed]
15. Joshi, R.; De Adhikari, A.; Dey, A.; Lahiri, I. Green reduction of graphene oxide as a substitute of acidic reducing agents for supercapacitor applications. *Mater. Sci. Eng. B* **2023**, *287*, 116128. [CrossRef]
16. Wu, T.; Liu, S.; Luo, Y.; Lu, W.; Wang, L.; Sun, X. Surface plasmon resonance-induced visible light photocatalytic reduction of graphene oxide: Using Ag nanoparticles as a plasmonic photocatalyst. *Nanoscale* **2011**, *3*, 2142–2144. [CrossRef] [PubMed]
17. Zhuo, Q.; Gao, J.; Peng, M.; Bai, L.; Deng, J.; Xia, Y.; Ma, Y.; Zhong, J.; Sun, X. Large-scale synthesis of graphene by the reduction of graphene oxide at room temperature using metal nanoparticles as catalyst. *Carbon N. Y.* **2013**, *52*, 559–564. [CrossRef]
18. Kodous, A.S.; Taha, E.O.; El-Maghraby, D.F.; Hassana, A.A.; Atta, M.M. Gamma radiation assisted green synthesis of hesperidin-reduced graphene oxide nanocomposite targeted JNK/SMAD4/MMP2 signaling pathway. *Sci. Rep.* **2024**, *14*, 11535. [CrossRef]
19. Fan, Z.; Li, B.; Zhou, S.; Huang, G. Terahertz Meta-Mirror with Scalable Reflective Passband by Decoupling of Cascaded Metasurfaces. *Photonics* **2024**, *11*, 796. [CrossRef]
20. Zhang, Y.; Feng, R.; Shi, B.; Li, X.; Gao, Y.; Gao, W.; Jia, Q.; Sun, F.; Cao, Y.; Ding, W. Ultra-Compact Reflective Waveguide Mode Converter Based on Slanted-Surface and Subwavelength Metamaterials. *Photonics* **2024**, *11*, 838. [CrossRef]
21. Sasani Ghamsari, M. Development of Thin Film Fabrication Using Magnetron Sputtering. *Metals* **2023**, *13*, 963. [CrossRef]
22. Shekhar, P.; Atkinson, J.; Jacob, Z. Hyperbolic metamaterials: Fundamentals and applications. *Nano Converg.* **2014**, *1*, 14. [CrossRef]
23. Mayer, K.M.; Hafner, J.H. Localized Surface Plasmon Resonance Sensors. *Chem. Rev.* **2011**, *111*, 3828–3857. [CrossRef]
24. Thuy An, N.; Kieu Thi Ta, H.; Van Hoang, D.; Phung, V.-D.; Hoa Thi Tran, N.; Thang Phan, B. Multilayer Graphene Oxide-Silver Nanoparticles for Stable, Highly Sensitive, and Reusable SERS Platforms. *ChemNanoMat* **2023**, *9*, e202200516. [CrossRef]

25. Aunkor, M.T.H.; Mahbubul, I.M.; Saidur, R.; Metselaar, H.S.C. The green reduction of graphene oxide. *RSC Adv.* **2016**, *6*, 27807–27828. [CrossRef]
26. Zacharias, P.; Kliewer, K.L. Dispersion relation for the 3.8 eV volume plasmon of silver. *Solid State Commun.* **1976**, *18*, 23–26. [CrossRef]
27. Johnson, P.B.; Christy, R.W. Optical Constants of the Noble Metals. *Phys. Rev. B* **1972**, *6*, 4370–4379. [CrossRef]
28. Cai, W.; Shalaev, V.M. *Optical Metamaterials*; Springer: Berlin/Heidelberg, Germany, 2010; Volume 10.

photonics

MDPI

Article

Design of a Tunable Metamaterial Absorption Device with an Absorption Band Covering the Mid-Infrared Atmospheric Window

Zongliang He [1], Dong Fang [2,*] and Yougen Yi [3,*]

[1] Yunnan Radioactive Environment Supervision Station, Kunming 650093, China; hezongl@126.com
[2] Faculty of Materials Science and Engineering, Kunming University of Science and Technology, Kunming 650093, China
[3] College of Physics, Central South University, Changsha 410083, China
* Correspondence: fangdong@kmust.edu.cn (D.F.); yougenyi@csu.edu.cn (Y.Y.); Tel./Fax: +86-0731-8887611 (Y.Y.)

Abstract: We propose a highly efficient broadband tunable metamaterial infrared absorption device. The design is modeled using the three-dimensional finite element method for the absorption device. The results show that the absorption device captures over 90% of the light in the wavelength range from 6.10 μm to 17.42 μm. We utilize VO_2's phase change property to adjust the absorption device, allowing the average absorption level to vary between 20.61% and 94.88%. In this study, we analyze the electromagnetic field distribution of the absorption device at its peak absorption point and find that the high absorption is achieved through both surface plasmon resonance and Fabry–Perot cavity resonance. The structural parameters of the absorption device are fine-tuned through parameter scanning. By comparing our work with previous studies, we demonstrate the superior performance of our design. Additionally, we investigate the polarization angle and incident angle of the absorption device and show that it is highly insensitive to these factors. Importantly, the simple structure of our absorption device broadens its potential uses in photodetection, electromagnetic stealth, and sensing.

Keywords: wideband; high absorption; tunable; infrared band; atmospheric window; wide angle

1. Introduction

Infrared is one of the invisible rays of the sun and is also an electromagnetic wave [1,2]. It has a lower frequency than visible light and is also known as infrared radiation. Since its discovery by Herschel in 1800, infrared radiation has been studied continuously. In later studies, it has been found that infrared radiation can interact with many molecules and show strong heating effects at the large scale [3]. Moreover, infrared radiation has three atmospheric windows [4–6]. Water vapor, CO_2, and ozone in the Earth's atmosphere absorb electromagnetic waves, and the electromagnetic band with a higher transmission rate is called the atmospheric window [7,8]. This part of infrared radiation that passes through the atmosphere to reach the Earth has a big impact on human activities. The question of how to control this part of electromagnetic waves has gained much attention. Initially, researchers used the natural properties of materials to modulate this part of electromagnetic waves, but this method has many drawbacks such as narrow bandwidth, low efficiency, and large size [9,10]. This has become the main problem limiting the control of infrared electromagnetic waves. With the improvement of manufacturing processes, the materials

https://doi.org/10.3390/photonics12020148

studied are no longer limited to natural materials, but efforts are made to create materials that do not exist in nature to achieve electromagnetic control [11]. Such artificial materials are called metamaterials [12–14].

Metamaterials are materials with special structures and specific designs with sub-wavelength unit scales that are capable of modifying or controlling the behavior of electromagnetic radiation in a very significant way [15]. They allow energy such as light, sound, and electromagnetic waves to interact with the material in an extraordinary way through their carefully designed structures. Metamaterials have witnessed tremendous progress over the past few decades, capturing widespread interest owing to their extraordinary properties and vast potential for diverse applications [16–18]. Metamaterials often exhibit unconventional properties such as reverse refraction, negative refraction, negative refractive indexes, and superscattering. Among them, a metamaterial absorption device is a metamaterial-based device that is characterized by a high absorption capacity for electromagnetic waves in a specific frequency range. This device utilizes a combination of the special structure of the metamaterial and the wave-absorbing material to effectively capture and absorb electromagnetic radiation at specific wavelengths. The operating principle of metamaterial absorption devices involves microscopic elements in their structure, such as micro- and nanostructures or electromagnetic resonators [19–21]. They achieve the efficient absorption of electromagnetic waves by tuning the electromagnetic response of the material. By rationally designing the structural parameters and material composition of metamaterial absorption devices, the selective absorption and suppression of electromagnetic waves in specific frequency bands can be realized [22]. Metamaterial absorption devices have a wide range of application prospects. Researchers have now applied metamaterials to electromagnetic stealth [23], biosensors [24], energy harvesting [25], and so on. However, by reviewing the previous studies, we find that there are fewer studies on metamaterial absorbers with absorption bands covering the infrared atmospheric window. And most of the metamaterial infrared absorbers have a narrow absorption bandwidth and the absorption characteristics are not tunable [26,27]. Therefore, the design of a metamaterial absorber with a broadband absorption band covering the infrared atmospheric window and with an adjustable absorption rate is extremely promising. Regarding the tuning of the absorption rate, we use phase change materials for this purpose. Phase change materials are a special type of material that can exhibit dramatic changes in physical properties in response to changes in temperature, pressure, or other external conditions [28,29]. Such materials can change rapidly from one phase to another, for example, from solid to liquid, liquid to gas, or vice versa. The special feature of phase change materials is that the phase change process is accompanied by a large amount of heat absorption or exothermic phenomena and that the energy required for the phase change is relatively small. This means that phase change materials have an efficient ability to store and release energy when absorbing or releasing energy. This property gives phase change materials the potential for a wide range of applications in many fields. Based on this, we have investigated and proposed a mid-infrared metamaterial absorption device. At an elevated temperature of 342 K, it achieves seamless absorption of over 90% across the eloquent 6.10 μm to 17.42 μm (11.32 μm) spectrum, boasting an average absorption efficacy of 94.88%. It is obvious that the operating band of our designed absorption device contains the mid-infrared band atmospheric window (8–14 μm). Not only that, but our absorption device also has tunable absorption. Its average absorption rate in the working band can be tuned within 20.61–94.88%. It is well known that the sensitivity to the polarization angle and incidence angle of electromagnetic waves has a non-negligible effect on optoelectronic devices. Through the study, we found that our designed absorption device has good insensitivity to polarization angle and incidence angle. Not only that, but the absorption device we designed also only has a three-layer structure, which is easy to fabricate. Therefore, it has more application potential in infrared stealth, optoelectronic detection, and energy harvesting. At the same time, this tunable absorption device designed by us is also more capable of meeting the application requirements in different scenarios. This study used a general analytical approach and only explored the absorption effect of the absorber device. We can also use fractional heat transport models to enhance the analysis of transient heat dissipation in infrared absorbing devices. For example, the modified Caputo fractional derivative can capture the memory effect in thermal diffusion, especially under pulsed laser irradiation. Furthermore, the size-dependent thermal conductivity observed in 2D materials suggests that the fractional model may be useful for optimizing nanostructured absorber layers with controlled defects. Moreover, the fractional heat transport model has many applications, not only in the thermal management of infrared absorbers, but it also has advantages in material design and thermal transport modulation. In future studies, the use of fractional heat transport models will be focused on to increase the depth of the work.

2. Structural Design and Parameters

In this endeavor, we have crafted a metamaterial infrared absorption device apparatus with a flexibly adjustable absorption capacity that spans the mid-infrared atmospheric range from 8 μm to 14 μm. We used FDTD solutions simulation software (Lumerical FDTD Solutions 2020) for simulation. The absorption device's structure is shown in Figure 1. Our proposed absorption device is a three-layer structure. Its bottom layer is a titanium substrate. The middle layer is a silica dielectric layer. The top layer is a dart-shaped VO_2 resonator layer. VO_2 undergoes a reversible phase transition from the insulating to the metallic state at ~68 °C, which significantly alters its optical and electrical properties. The tunable absorption of electromagnetic waves can be achieved using its phase transition properties. Moreover, in the infrared band, VO_2 can effectively absorb infrared radiation at specific wavelengths, thus achieving efficient infrared absorption. Titanium, with its high reflectivity, is used as a substrate material to enhance the overall performance of the IR absorption device. SiO_2, with its low refractive index, is used as a dielectric layer to reduce the reflection loss of the incident light, thus improving the absorption efficiency. In Figure 1, H3 is the thickness of the titanium substrate, H2 is the thickness of the middle SiO_2 layer, H1 is the thickness of the top VO_2 layer, P is the period length of the absorber unit structure, L is the horizontal distance between two adjacent dart edges of the dart-shaped VO_2 structure, and W is the length of each dart edge. Specific values for these structural parameters are given in Table 1.

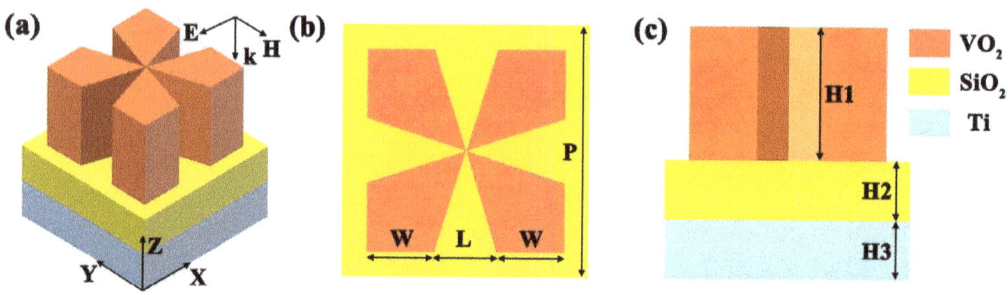

Figure 1. (**a**) Unit structure of the absorption unit. (**b**) A panoramic perspective of the unit structure of the absorption device. (**c**) Profile view unveiling the unit's structure in the absorption device.

Table 1. Structural parameters of the absorption device.

Parameter Name	P	L	W	H1	H2	H3
Parameter value (μm)	3.03	0.76	0.82	1.38	0.50	0.50

The effect of the variation in these structural parameters on the absorption device will be specifically analyzed subsequently. The modulation of our designed absorption device is mainly realized through the dart-like VO_2 layer. This is because the substance VO_2 has a phase change effect [30]. At temperatures below 340 K, VO_2 exhibits a monoclinic crystalline phase in the insulating state. At a temperature of 340 K, VO_2's phase transition temperature starts to transform from a monoclinic phase to a tetragonal phase in the metal state. At a temperature of 342 K, VO_2 transforms to the metal state completely. During the phase transition of VO_2, its conductivity changes. Therefore, we can realize the regulation of the absorption rate by changing the temperature and thus the conductivity of VO_2. The insulating component in the VO_2 used in the design can be understood as a dielectric with dielectric constant $\varepsilon_D = 9$. For the metal component in VO_2, the dielectric constant can be expressed using the Drude model as shown in Equation (1) [31,32] as follows:

$$\varepsilon_M(\omega,\sigma) = \varepsilon_\infty - \frac{\omega_P^2(\sigma)}{\omega(\omega + i/\tau)} \tag{1}$$

where ω is the frequency of the infrared radiation. ε_∞ is the high-frequency limiting dielectric constant with a value of 12. τ is the relaxation time, which can be described using Equation (2). $\omega_P(\sigma)$ is the plasma frequency, and the value at σ can be expressed using Equation (3). Equations (2) and (3) are expressed as follows:

$$\tau = \frac{m^* \mu}{e} \tag{2}$$

$$\omega_P(\sigma) = \sqrt{\frac{Ne^2}{\varepsilon_0 m^*}} \tag{3}$$

In Equation (2), $m^* = 2m_e$ is the effective mass. m_e is the mass of the free electron. μ is the carrier mobility and is 2 $\text{cm}^{-3}/V \cdot S$. e is the charge of the free electron. Thus, $\tau = 2.2fs$. In Equation (3) N is the carrier concentration inside the VO_2 medium, noted as $8.7 \times 10^{21}\text{cm}^3$. ε_0 is the dielectric constant in vacuum. In addition, the volume fraction f of the metal component in VO_2 versus temperature can be described by the Boltzmann function, as shown in Equation (4) as follows:

$$f = f_{\max}\left(1 - \frac{1}{1 + \exp[(T - T_0)/\Delta T]}\right) \tag{4}$$

Here, $f_{\max} = 0.95$ is the maximum value that can be reached by the volume fraction of the metal component in VO_2 [33]. $T_0 = 68\ ^\circ\text{C}$ is the critical temperature of the phase transition at elevated temperatures of VO_2, which corresponds to the critical temperature of the phase transition at reduced temperatures 62 °C. $\Delta T = 2\ ^\circ\text{C}$ is the transition temperature of the phase transition. For the VO_2 composite system, the dielectric function ε_c can be described using the Bruggeman effective medium theory, as shown in Equation (5) [34,35] as follows:

$$\varepsilon_c = \frac{1}{4}\left\{\varepsilon_D(2-3f) + \varepsilon_M(3f-1) + \sqrt{[\varepsilon_D(2-3f) + \varepsilon_M(3f-1)]^2 + 8\varepsilon_D\varepsilon_M}\right\} \tag{5}$$

The definition for each parameter in Equation (5) has been given previously and will not be repeated here. According to Equations (1)–(5), the relationship between VO_2 complex permittivity and conductivity can be given by Equation (6) [36] as follows:

$$\sigma_{VO_2} = i\varepsilon_0\omega(1-\varepsilon_c) \tag{6}$$

Based on the derivation of the above equation, we can obtain the conductivity at different temperatures during the phase transition of VO_2. We give the conductivity at some temperatures during the phase transition in Figure 2.

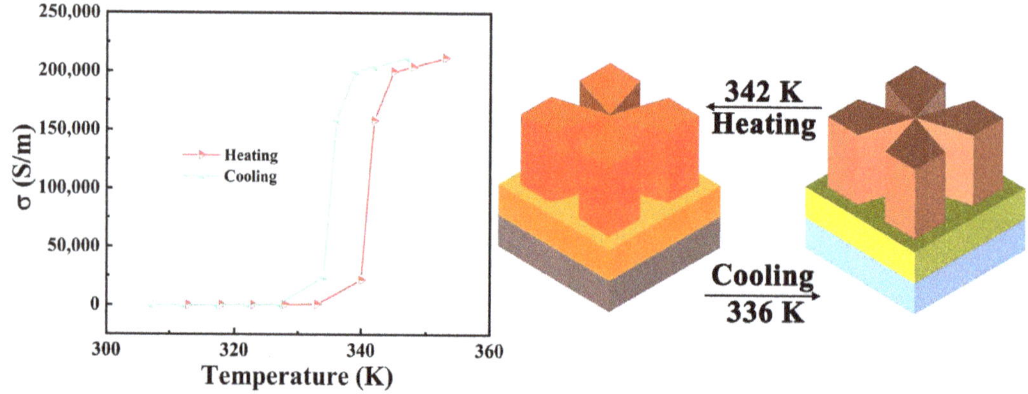

Figure 2. VO_2's conductivity at different temperatures.

When the electromagnetic wave is incident into the absorbing device, the wave number at this point can be expressed by Equation (7) because the material has loss, as follows:

$$k = \omega\sqrt{\mu\varepsilon} \tag{7}$$

Here, ω is the angular frequency. μ is the equivalent magnetic permeability of the absorbing device. ε is the equivalent dielectric constant of the absorbing device. At this point, the electromagnetic wave is absorbed during propagation, as shown in Equation (8) as follows:

$$\vec{E(r)} = \vec{E_0}e^{i(\vec{k}\cdot\vec{r}-\omega t)}e^{-\alpha\cdot\vec{r}} \tag{8}$$

Here, α is the attenuation constant, which is related to the loss characteristics of the material. We use Perfectly Matched Layer (PML) boundary conditions in the Z direction when designing the absorber device. Periodic boundary conditions are used in the X and Y directions.

In this paper, the proposed absorption device is simulated by the three-dimensional time-domain finite element difference method. We use a planar light source with a wavelength band of 6–19 μm (polarized in the X direction) incident vertically along the k direction into the absorption device. The permittivity of VO_2 at different wavelengths is obtained using the Drude model. The dielectric constants of Ti and SiO_2 at different wavelengths are obtained from Palik [37]. For the absorption device, the absorption rate can be described using Equation (9) [38–40] as follows:

$$A(\omega) = 1 - R(\omega) - T(\omega) \tag{9}$$

where ω is the frequency of the light emitted by the light source used. $A(\omega)$ represents the absorption rate of the absorbing device. $R(\omega)$ is the reflection rate of the absorbing device for the incident electromagnetic wave. $T(\omega)$ is the transmittance rate of the absorbing device for the incident electromagnetic wave. The transmittance of the absorption device for the electromagnetic wave is related to the skinning depth of the material. The thickness of the Ti substrate we use here far exceeds the skinning depth. Therefore, the transmittance $T(\omega)$ can be considered as 0. The absorptivity equation of the absorption device is then simplified to $A(\omega) = 1 - R(\omega)$ [41]. Not only that, but the absorptivity is also related to the equivalent impedance of the absorbing device, as shown in Equation (10) [42] as follows:

$$A(\omega) = 1 - \left|\frac{1-Z}{1+Z}\right|^2 \tag{10}$$

Z in Equation (8) is the equivalent impedance of the absorbing device. Z can be calculated using S-parameter extraction.

To enable practical implementation, we have outlined a fabrication methodology for the proposed absorption device [43]. Initially, a 0.5 μm titanium (Ti) film is deposited onto a silicon (Si) substrate via magnetron sputtering. Subsequently, a 0.5 μm silicon dioxide (SiO_2) layer is deposited onto the Ti film using ion beam sputtering. Following this, a 1.38 μm vanadium dioxide (VO_2) layer is deposited onto the SiO_2 film through magnetron sputtering. Finally, a dart-shaped VO_2 structure is fabricated employing electron beam evaporation in conjunction with photolithography techniques.

3. Calculation and Discussion

Figure 3a shows the absorption characteristic curve of the absorption device at 342 K temperature. After calculation, we found that its absorption bandwidth reaches 11.32 μm above 90%, and the average absorption in the working band (6.10–17.42 μm) is 94.88%. Figure 3b shows the absorption curves of the absorption device at different temperatures during warming. By calculating the absorption rate in the working band, we see that the absorption device has the lowest average absorption rate at 318 K, which is 20.61%. The average absorption at 339.5 K is 33.85%. It is 35.54% at 340 K, 62.62% at 340.5 K, 82.32% at 341 K, 92.08% at 341.5 K, and 89.99% at 345 K. The absorption device's best absorption effect occurs at 342 K. The average absorption rate at this time is 94.88%. Our designed absorption device's absorption rate can be therefore regulated by changing the temperature. The average absorption rate can be regulated between 20.61% and 94.88%. VO_2 has a phase transition temperature of 340 K, which is critical for its thermochromic properties. For the optimum performance of an infrared absorber device, the operating temperature should be maintained within a range that allows VO_2 to transition between insulating and metallic phases. Typically, this range is between 30 °C and 100 °C, depending on the specific design and application. Below 30 °C, VO_2 remains in the insulating phase; above 100 °C, VO_2 may degrade or lose the desired properties. Absorption devices also have requirements for the radiant power of the incident light. In the low power range (<10 mW/cm^2), VO_2 can operate normally, the phase change behavior is reversible, and the material does not degrade. In the medium power range (10–100 mW/cm^2), the temperature of VO_2 increases significantly with increasing incident light power. If the temperature exceeds the phase transition temperature but is below the degradation temperature, VO_2 may temporarily lose its properties (e.g., dynamic optical properties), but this can be restored upon cooling. In the high power range (>100 mW/cm^2), where the incident optical power is too high, the temperature of VO_2 may exceed its degradation temperature, resulting in irreversible oxidation or structural damage of the material, which can lead to the loss of its properties. Taken together, the incident light power was controlled in the range of 10–50 mW/cm^2 to avoid overheating and degradation [44,45].

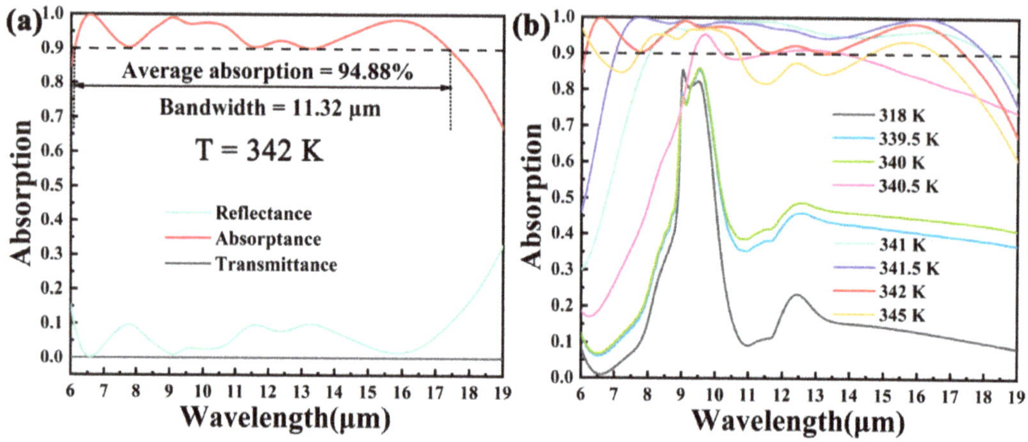

Figure 3. (**a**) Absorption, reflectance, and transmittance curves of the absorbing device at 342 K. (**b**) Absorption curves of the absorption device at different temperatures.

It is well known that the peaks of the curves generally play an important role in the analysis. We have therefore investigated the absorption curve's five absorption peaks in the working band in Figure 4a. By calculation, we obtained the corresponding wavelengths at these five absorption peaks. Among them, F_1 = 6.57 μm, F_2 = 9.09 μm, F_3 = 10.12 μm, F_4 = 12.36 μm, and F_5 = 15.85 μm. The corresponding absorption values at these absorption peaks are 99.71%, 98.96%, 97.46%, 92.25%, and 98.38%, respectively. We calculate the effective impedance of the absorbing device by extracting the S-parameters of the absorbing device in our simulation. In Figure 4b, we give the effective impedance curve of the absorbing device. When the value of effective impedance is 1 + 0i, perfect absorption (100% absorption) can be realized [46]. Here, we bring the calculated effective impedance value into Equation (8) to calculate the peak absorption point. By comparison, it is found that the simulated absorption peak point remains consistent with the peak point calculated from the effective impedance. For broadband absorption in the absorption device, these absorption peaks play a decisive role. It is due to the superposition of multiple absorption peaks that more than 90% broadband absorption can be achieved.

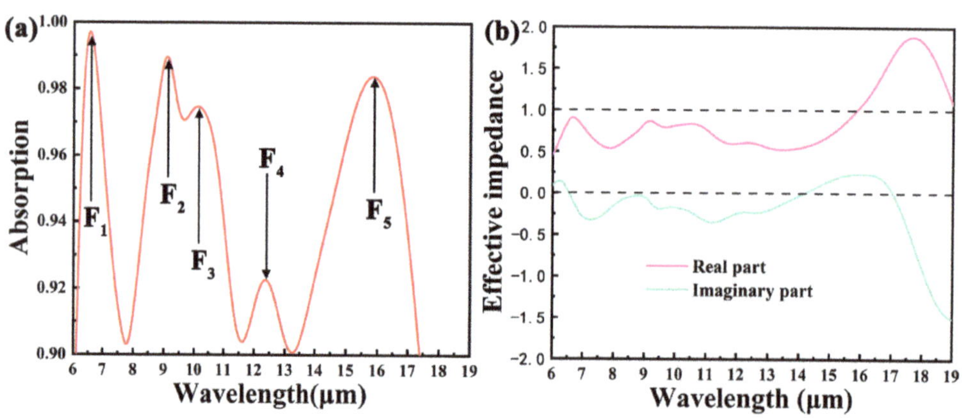

Figure 4. (**a**) Absorption peaks of 90% or more absorption curves. (**b**) Effective impedance of the absorption device.

In this section, we analyze the electromagnetic fields of the absorption device at the wavelengths corresponding to the absorption peaks to explain the broadband absorption mechanism of our designed absorption device. The electromagnetic field distributions at these five absorption peaks are shown in Figure 5a–o. In Figure 5a, it is observed that at F_1, the electric field is primarily concentrated at the edge of the dart-shaped structure's blade in VO_2. Moving to Figure 5f, the electric field is predominantly focused in the upper section of the dart structure. The distribution of the magnetic field at F_1 is illustrated in Figure 5k, where it is evident that the magnetic field is concentrated mainly at the upper end of the dart structure, the air on the surface, and the SiO_2 dielectric layer. The reason for the electromagnetic field distribution at F_1 is due to the fact that VO_2 is in the metallic phase at this point. The frequency of the incident electromagnetic wave is equal to the collective oscillation frequency of free electrons, which excites the localized surface plasmon resonance (LSPR) of the VO_2 layer [47]. When the incident electromagnetic wave reaches the interface between the metallic structure and the dielectric layer, the frequency of photons at this point matches the frequency of the propagating surface plasmon resonance [48,49]. Therefore, the propagating surface plasmon resonance (LSPR) is generated here. Moreover, an electric field exists in the dielectric layer at F_1. This is due to the fact that electromagnetic waves are absorbed as they are reflected back and forth in the dielectric layer [50]. This is usually interpreted as the Fabry-Polo cavity resonance (FP-R). Therefore, the absorption peaks at F_1 are mainly due to LSPR, PSPR, and FP-R. Figure 5b,g show the electric field distribution at F_2. In the figure, it is evident that the electric field is concentrated at the central intersection of the dart-shaped VO_2 layer, with the electric field also present along the edges of the dart. Furthermore, at this point, the electric field extends across the entire thickness of the VO_2 layer. Moving to Figure 5l, it is evident that the magnetic field at F_2 is primarily concentrated at the surface intersection of the dart-like VO_2 layer. Taken together, the main reason for the generation of the absorption peak at F_2 is the PSPR. Figure 5c,h show the electric field at F_3. At this time, the electric field is concentrated at the intersection with the dart structure as well as at the tip. Moreover, the electric field is distributed throughout the thickness of the dart-like VO_2 layer. As seen in Figure 5m, the magnetic field is concentrated at this wavelength at the intersection of the surface of the dart-like VO_2 layer as well as at the interface between the ends of the SiO_2 dielectric layer and the Ti substrate. It is analyzed that the absorption peaks at F_3 are due to PSPR and LSPR [51]. Figure 5d,i show the electric field distribution at F_4. As depicted in the figure, the electric field is concentrated at the intersection of the dart-like VO_2 layer, extending through its entire thickness. Additionally, there is an electric field distribution along the edges of the dart. In Figure 5n, the magnetic field is primarily focused at the surface intersection of the dart-like VO_2 layer, indicating that the absorption peak at F_4 is mainly due to the plasmon-enhanced SPPR (Surface Plasmon Polariton Resonance) [52]. Figure 5e,j demonstrate the electric field distribution at F_5, which resembles that at F_2, yet with a slightly different magnetic field distribution. In Figure 5o, the magnetic field at F_5 is concentrated at the lower intersection of the dart-shaped VO_2 layer. By combining the electric field distribution, it can be inferred that the absorption peak at F_5 is predominantly generated by the plasmon-enhanced SPPR. This section focuses on analyzing the mechanism underlying the absorption peaks of the absorption device. And these absorption peaks' existence is the basis of broadband absorption that is generated by the absorption device.

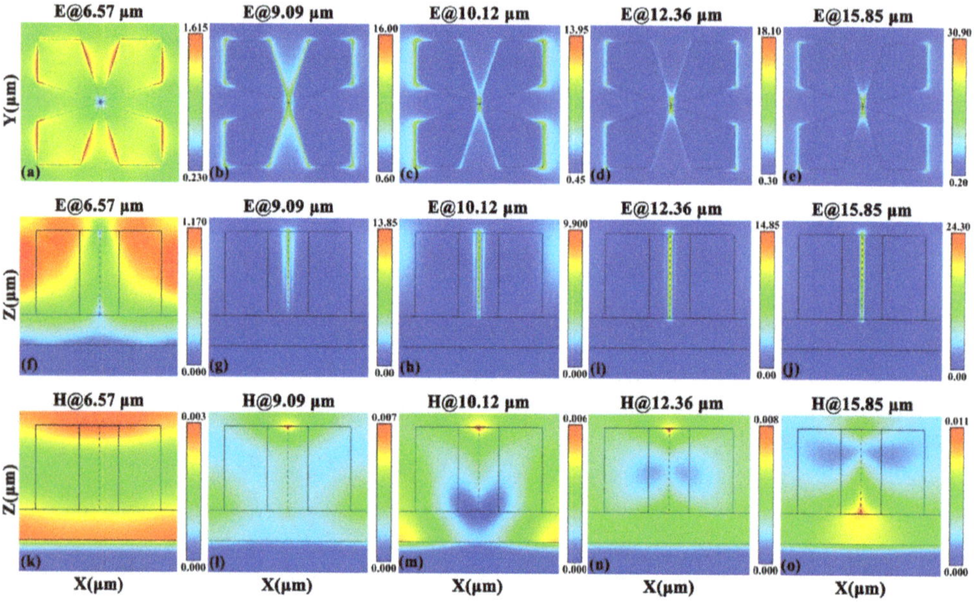

Figure 5. (**a–e**) Electric field distribution in the XOY plane at the absorption peak of the absorption device. (**f–j**) Electric field distribution in the XOZ plane at the absorption peak for the absorption device. (**k–o**) Magnetic field distribution of the absorption device in the XOZ plane at the absorption peak.

In the preceding section, we explored the electromagnetic field distribution of the absorption device. Here, we delve into the impact of the shape parameters of the absorption device. Figure 6a illustrates the influence of the period length (P) of the unit structure of the absorption device on its absorption rate. The absorption peaks in the absorption curves experience a blueshift with an increase in period length, as depicted in the figure. For period lengths of P = 2.83 μm and 2.93 μm, the absorption device fails to achieve continuous absorption exceeding 90%. Conversely, at P = 3.13 μm and 3.23 μm, the absorption bandwidth of over 90% absorption narrows compared to that at P = 3.03 μm. Hence, we have selected P = 3.03 μm as the optimal cycle length for the unit structure of the absorption device. Figure 6b demonstrates the impact of the blade length (L) of the top dart-shaped VO_2 layer of the absorption device on its absorption rate. Absorption rates above 90% continuous absorption were not attainable at L = 0.66 μm, L = 0.86 μm, and L = 0.96 μm. Additionally, the absorption bandwidth at L = 0.56 μm is narrower than that at L = 0.76 μm. Consequently, L = 0.76 μm has been identified as the optimal parameter. Figure 6c showcases the effect of the distance (W) between neighboring dart blades of the dart-shaped VO_2 layer of the absorption device on its absorption. Notably, at W = 0.62 μm and 0.72 μm, the absorption bandwidth narrows compared to that at W = 0.82 μm. Furthermore, at W = 0.92 μm and 1.02 μm, the absorption rate mostly remains below 90%. Thus, W = 0.82 μm has been selected as the optimal parameter for the absorption device.

In the preceding section, we examined the shape parameters of the absorption device. In this section, we will delve into the thickness parameters of the absorption device. Figure 7a,b display the absorption spectra of the absorption device for varying thicknesses of the VO_2 layer and SiO_2 layer, respectively. From Figure 7a, it is observed that the absorption bandwidth of the absorption device exceeds 90% as the VO_2 thickness (H1) increases from 1.18 μm to 1.38 μm. However, the absorption bandwidth narrows as H1 is further increased to 1.58 μm. Consequently, the optimal parameter for the VO_2 thickness H1 is determined to be 1.38 μm. Figure 7b illustrates the impact of SiO_2 thickness on the absorption rate. The absorption bandwidth of the absorption device exceeds 90% initially as the SiO_2 thickness increases from H2 = 0.20 μm to H2 = 0.80 μm, reaching optimal absorption characteristics at 0.50 μm. Therefore, the optimal parameter for SiO_2 thickness (H2) is chosen as 0.50 μm. Furthermore, upon comparing Figure 7a,b, it is noted that the absorption peaks of the absorption device undergo red-shifting as the thicknesses of both the VO_2 and SiO_2 layers increase. This red-shifting phenomenon is attributed to interference theory [53,54]. An increase in the thickness of the absorption device results in a greater phase difference in the electromagnetic wave, leading to a lengthening of the wavelength during resonance. In conclusion, the structural parameters of the absorption device have been optimized through an analysis of absorption rates at different structural parameters.

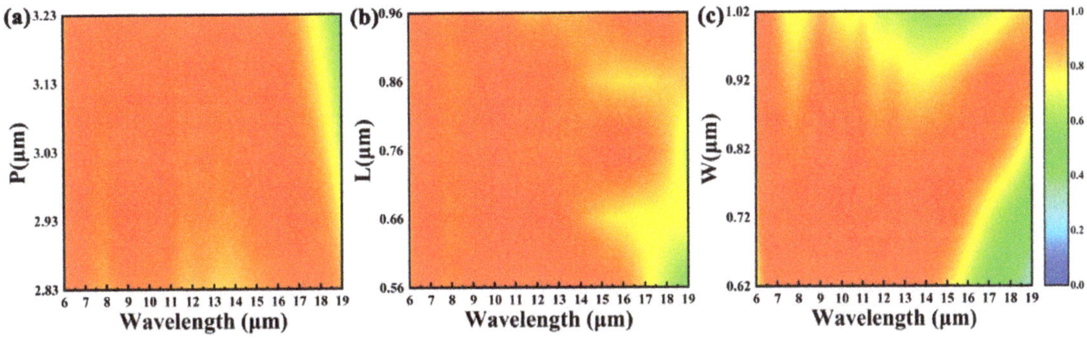

Figure 6. (**a**) Absorption spectra of the absorbing device at different cycle lengths. (**b**) Absorption spectra of the absorption device at different distances between neighboring darts. (**c**) Absorption spectra of the absorption device at different dart lengths.

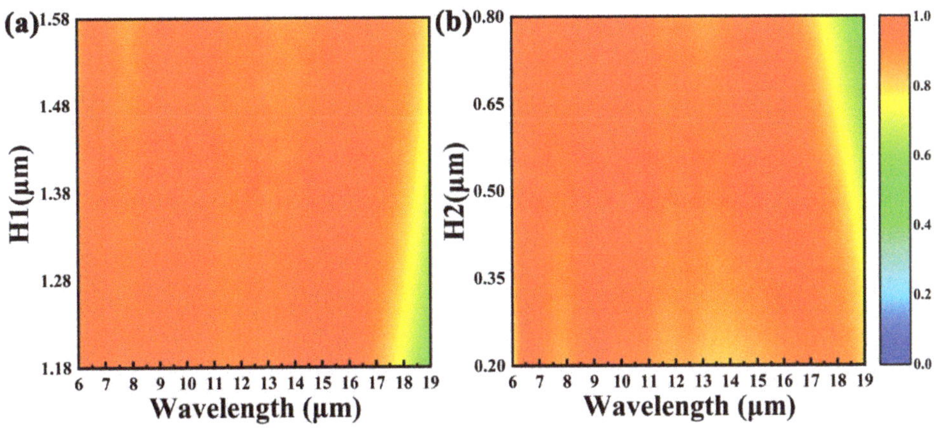

Figure 7. (**a**) Absorption spectra of the absorption device at different VO_2 thicknesses. (**b**) Absorption spectra of the absorption device at different SiO_2 thicknesses.

Table 2 demonstrates the comparison between previous studies and our design [55–58]. The comparison reveals that the absorption device of our design has a wider absorption bandwidth. Moreover, the absorption band of the absorption device covers the whole 8–14 μm. Our design also has a higher average absorptivity and the absorptivity is tunable. Not only that, but our design also has a simpler structure and is easier to fabricate compared to other designs. Therefore, our design has better absorption performance than previous designs.

Table 2. Comparison of our designed absorption device with previous designs.

Reference	Absorption Band	Average Absorption in the Operating Band	Shape of Structure Used	Adjustable Characteristics
[55]	8.5–11.5	88%	Stacked nanodisk structure	Not possessed
[56]	7.5–13.25	91.7%	Cross-Ti resonator	Not possessed
[57]	6.5–11.5	95.7%	Nanocylindrical arrays	Adjustable absorption mode
[58]	7.96–14.16	96.5%	Nanocube	Not possessed
Our Design	6.10–17.42	94.88%	Dart-shaped VO_2 structure	Average absorption rate regulated within 20.61–94.88%

The sensitivity to the angle of incidence and polarization plays a crucial role in the functionality of the device [59–63]. In contrast, the vector behavior of this infrared absorbing system of our design is characterized precisely by a polarization-insensitive response. This is attributed to the highly symmetric structure of the surface of the proposed infrared absorbing device. Due to the geometrical symmetry, light of different polarization directions excites similar electromagnetic resonance modes, and the electric field distribution and resonance modes are homogeneous, independent of the incident polarization direction. We show in Figure 8a,b the electric field distributions of the absorbing device in TE and TM modes for an electromagnetic wavelength of 6.57 µm. As shown, the surface electric field maps of the absorber device in TE and TM modes are rotationally symmetric and have the same intensity. This indicates that the electric field vectors exhibit the same pattern for TE polarization and TM polarization, which suggests that the electromagnetic response of the absorber device is consistent across polarization states. We scanned the polarization angle in 5° steps using FDTD simulation software. The scanning results are shown in Figure 8c. The results further confirm that the absorption spectra of randomly polarized light are almost identical, thus validating the polarization-independent nature of the system.

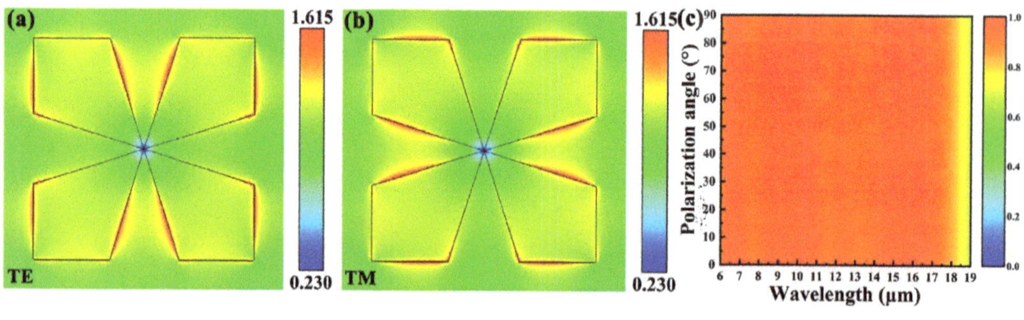

Figure 8. (**a**) Surface electric field distribution of the absorbing device in TE mode at an incident wavelength of 6.57 µm. (**b**) Surface electric field distribution of the absorbing device in TM mode at an incident wavelength of 6.57 µm. (**c**) Absorption spectra of the absorption device at different polarization angles (0–90°).

In this section, we explore the effect of the angle of incidence of electromagnetic waves on the absorbing device in different modes. Figure 9a shows the absorption spectra of the absorber device in TE mode for different incidence angles. It is observed that the absorption bandwidth remains relatively uniform in the range of 0–30° incidence angle. Subsequently, the absorption bandwidth shortens between 30° and 50° with very little change in absorption. As the angle of incidence exceeds 50°, especially between 50° and 60°, the absorption bandwidth and absorbance of the absorber decreases. Figure 9b,c show that the excitation position of the electric field on the surface of the absorber device does not change significantly as the angle of incidence increases, but the excitation intensity changes. Figure 9d illustrates the effect of different incidence angles on the absorber device for TM mode, which is similar to the trend observed in Figure 9a for TE mode. Figure 9e,f are also similar to the features in TE mode. Based on the above analyses, it can be concluded that the electromagnetic wave excites similar electromagnetic resonance (EMR) modes at varying angles of incidence, with only a slight change in intensity. Therefore, our designed absorber device is able to maintain excellent absorption characteristics even at larger incidence angles. This performance indicates the high potential applicability of our design [64–66].

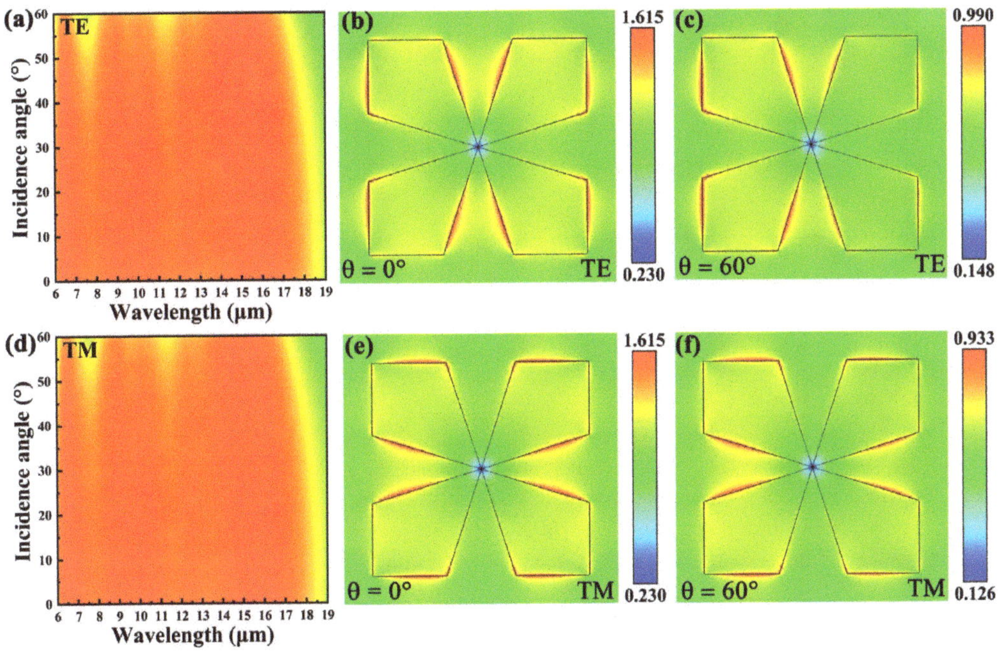

Figure 9. (**a**) Absorption spectra of the absorbing device in TE mode for different angles of incidence (0–60°). (**b**) Surface electric field distribution of the absorption device at $\theta = 0°$ incidence in TE mode. (**c**) Surface electric field distribution of the absorption device in TE mode at $\theta = 60°$ incidence. (**d**) Absorption spectra of the absorption device in TM mode at different angles of incidence (0–60°). (**e**) Surface electric field distribution of the absorption device in TM mode at $\theta = 0°$ incidence. (**f**) Surface electric field distribution of the absorption device in TM mode at $\theta = 60°$ incidence.

In this study, we initially considered other surface shapes as well, such as circular or square structures. We exemplify several shapes in Figure 10 and give their absorption curves. As shown in the figure, the dart surface shape we designed consistently exhibits excellent performance in terms of absorption bandwidth and absorption efficiency. The sharp edges and corners of the surface structure of this dart shape enhance the local electromagnetic field strength, leading to stronger photomatter interactions [67,68]. This property is essential for achieving high absorption rates, as it maximizes the dissipation of energy within the metasurface structure. Moreover, this dart-like structure has a high degree of symmetry, making it inherently independent of polarization. This is a key feature for infrared absorption applications as it ensures consistent performance regardless of the polarization state of the incident light [69–71]. This property is particularly advantageous in practical applications where the polarization of the incident radiation may change.

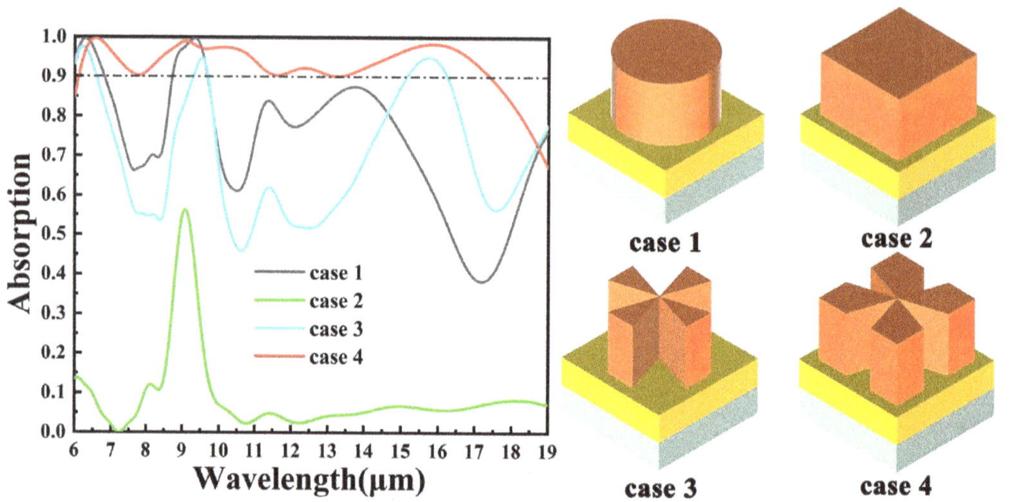

Figure 10. Different surface shapes and corresponding absorption curves.

4. Conclusions

In this study, an infrared metamaterial absorption device has been designed. Working at a temperature of 342 K, the absorption bands with over 90% absorption cover the range from 6.10 μm to 17.42 μm, with an absorption bandwidth of 11.32 μm that includes the mid-infrared atmospheric window (8–14 μm). The average absorption of the absorption device within this band reaches an impressive 94.88%. By changing the temperature to adjust the conductivity of VO_2 within the device, the absorption rate of our designed device can be fine-tuned, allowing for an adjustable average absorption rate in the operating band ranging from 20.61% to 94.88%. The broadband absorption mechanism was explained by analyzing the electromagnetic field of the absorption device in the design. Structural parameters were optimized through parameter scanning techniques. Furthermore, a study on the sensitivity of the absorption device to both incident angle and polarization angle was conducted, showing the device's strong insensitivity in these aspects. We strongly believe that our proposed design has great potential for future applications in energy harvesting, sensing, and electromagnetic stealth.

Author Contributions: Conceptualization, Z.H. and D.F.; data curation, Z.H., D.F. and Y.Y.; formal analysis, Z.H. and D.F.; methodology, Z.H., D.F. and Y.Y.; resources, Z.H., D.F. and Y.Y.; software, Y.Y.; data curation, Z.H., D.F. and Y.Y.; writing—original draft preparation, Z.H. and D.F.; writing—review and editing, Z.H., D.F. and Y.Y. All authors have read and agreed to the published version of the manuscript.

Funding: The authors are grateful for the support by the Major Special Projects in Yunnan Province (202302AF050001).

Institutional Review Board Statement: Not applicable.

Informed Consent Statement: Not applicable.

Data Availability Statement: Publicly available datasets were analyzed in this study. These data can be found at [https://www.lumerical.com/] (accessed on 1 January 2020).

Conflicts of Interest: The authors declare no conflicts of interest.

References

1. Fei, Z. Electrically detecting infrared light. *Nat. Mater.* **2018**, *17*, 950–951. [CrossRef] [PubMed]
2. Liu, J.J.; Yang, X.X.; Xu, Q.L.; Chang, R.G.; Wu, Z.H.; Shen, H.B. Unraveling the efficiency losses and improving methods in quantum dot-based infrared up-conversion photodetectors. *Opto-Electron. Sci.* **2024**, *3*, 230029. [CrossRef]
3. Yan, C.C.; Che, Z.L.; Yang, W.Y.; Wang, X.D.; Liao, L.S. Deep-red and near-infrared organic lasers based on centrosymmetric molecules with excited-state intramolecular double proton transfer activity. *Opto-Electron. Adv.* **2023**, *6*, 230007. [CrossRef]
4. Alecu, A.; Albu, C.; Badea, G.-I.; Alionte, A.; Enache, A.-A.; Radu, G.-L.; Litescu, S.-C. Infrared Laser-Assisted Extraction of Bioactive Compounds from *Rosa canina* L. *Int. J. Mol. Sci.* **2025**, *26*, 992. [CrossRef]
5. Li, W.; Yi, Y.; Yang, H.; Cheng, S.; Yang, W.X.; Zhang, H.; Yi, Z.; Yi, Y.; Li, H. Active Tunable Terahertz Band-width Absorber Based on single layer Graphene. *Commun. Theor. Phys.* **2023**, *75*, 045503. [CrossRef]
6. Argirusis, N.; Achilleos, A.; Alizadeh, N.; Argirusis, C.; Sourkouni, G. IR Sensors, Related Materials, and Applications. *Sensors* **2025**, *25*, 673. [CrossRef]
7. Gu, Z.F.; Gao, Y.X.; Zhou, K.S.; Ge, J.; Xu, C.; Xu, L.; Rahmani, M.; Jiang, R.; Chen, Y.; Liu, Z.; et al. Surface-patterned chalcogenide glasses with high-aspect-ratio microstructures for long-wave infrared metalenses. *Opto-Electron. Sci.* **2024**, *3*, 240017. [CrossRef]
8. Wang, H.Y.; Ma, R.; Liu, G.D.; Wang, L.L.; Lin, Q. Optical force conversion and conveyor belt effect with coupled graphene plasmon waveguide modes. *Opt. Express* **2023**, *31*, 32422. [CrossRef]
9. Rane, S.; Prabhu, S.; Chowdhury, D.R. Physics and applications of terahertz metagratings. *Opto-Electron. Sci.* **2024**, *3*, 230049. [CrossRef]
10. Luo, M.H.; Hu, J.Y.; Li, Y.M.; Bai, W.D.; Zhang, R.L.; Lin, Q.; Wang, L.L. Anapole-assisted ultra-narrow-band lattice resonance in slotted silicon nanodisk arrays. *J. Phys. D Appl. Phys.* **2023**, *56*, 375102. [CrossRef]
11. Nurzed, B.; Saha, N.; Millward, J.M.; Niendorf, T. 3D Metamaterials Facilitate Human Cardiac MRI at 21.0 Tesla: A Proof-of-Concept Study. *Sensors* **2025**, *25*, 620. [CrossRef]
12. Kang, D.H.; Heo, H.S.; Yang, Y.; Seong, J.; Kim, H.; Kim, J.; Rho, J. Liquid crystal-integrated metasurfaces for an active photonic platform. *Opto-Electron. Adv.* **2024**, *7*, 230216. [CrossRef]
13. Li, Z.T.; Cheng, S.B.; Zhang, H.F.; Yang, W.X.; Yi, Z.; Yi, Y.G.; Wang, J.Q.; Ahmad, S.; Raza, R. Ultrathin broadband terahertz metamaterial based on single-layer nested patterned graphene. *Phys. Lett. A* **2025**, *534*, 130262. [CrossRef]
14. Li, Y.M.; Tan, C.X.; Hu, J.Y.; Bai, W.D.; Zhang, R.L.; Lin, Q.; Zhang, Y.; Wang, L.L. Ultra-narrow band perfect absorbance induced by magnetic lattice resonances in dielectric dimer metamaterials. *Results Phys.* **2022**, *39*, 105730. [CrossRef]
15. Berhe, A.M.; As'ham, K.; Al-Ani, I.; Hattori, H.T.; Miroshnichenko, A.E. Strong coupling and catenary field enhancement in the hybrid plasmonic metamaterial cavity and TMDC monolayers. *Opto-Electron. Adv.* **2024**, *7*, 230181. [CrossRef]
16. Kowerdziej, R.; Ferraro, A.; Zografopoulos, D.C.; Caputo, R. Soft-Matter-Based Hybrid and Active Metamaterials. *Adv. Opt. Mater.* **2022**, *10*, 2200750. [CrossRef]
17. Esfandiari, M.; Lalbakhsh, A.; Shehni, P.N.; Jarchi, S.; Ghaffari-Miab, M.; Mahtaj, H.N.; Reisenfeld, S.; Alibakhshikenari, M.; Koziel, S.; Szczepanski, S. Recent and emerging applications of Graphene-based metamaterials in electromagnetics. *Mater. Des.* **2022**, *221*, 110920. [CrossRef]
18. Lee, Y. Metamaterials and Their Devices. *Crystals* **2025**, *15*, 119. [CrossRef]
19. Fan, J.X.; Li, Z.L.; Xue, Z.Q.; Xing, H.Y.; Lu, D.; Xu, G.; Gu, J.; Han, J.; Cong, L. Hybrid bound states in the continuum in terahertz metasurfaces. *Opto-Electron. Sci.* **2023**, *2*, 230006. [CrossRef]
20. Li, B.X.; Liu, M.L.; Wen, R.Q.; Wei, Y.; Zeng, L.L.; Deng, C.S. Dynamic control of Fano-like interference in the graphene periodic structure. *J. Phys. D Appl. Phys.* **2023**, *56*, 115104. [CrossRef]
21. He, M.Y.; Wang, Q.Q.; Zhang, H.; Xiong, J.; Liu, X.P.; Wang, J.Q. Analog electromagnetic induced transparency of T-type Si-based metamaterial and its applications. *Phys. Scr.* **2024**, *99*, 035506. [CrossRef]
22. Phan, T.N.K.; Kato, K.; Takano, K.; Fujioka, S.; Nakajima, M. Spectral Shape Control of Laser-Induced Terahertz Waves from Micro Split-Ring Resonators Made of Metallic Nanostructures. *Photonics* **2024**, *11*, 1209. [CrossRef]
23. Wang, J.Q.; Sun, J.Y.; Sun, S.; Zhang, H.; Wang, Q.Q.; Yang, J.Y.; Mei, Y.W. Numerical simulation of electromagnetically induced transparency in composite metamaterial. *Phys. Scr.* **2025**, *100*, 025512. [CrossRef]
24. Zeng, Z.L.; Liu, H.F.; Zhang, H.F.; Cheng, S.B.; Yi, Y.G.; Yi, Z.; Wang, J.Q.; Zhang, J.G. Tunable ultra-sensitive four-band terahertz sensors based on Dirac semimetals. *Photonics Nanostruct.-Fundam. Appl.* **2025**, *63*, 101347. [CrossRef]
25. Xiao, Y.F.; Ma, C.; Sun, T.Y.; Song, Q.J.; Bian, L.; Yi, Z.; Hao, Z.Q.; Tang, C.J.; Wu, P.H.; Zeng, Q.D. Investigation of a high-performance solar absorber and thermal emitter based on Ti and InAs. *J. Mater. Chem. A* **2024**, *12*, 29145. [CrossRef]
26. Ogawa, S.; Kimata, M. Metal-Insulator-Metal-Based Plasmonic Metamaterial Absorbers at Visible and Infrared Wavelengths: A Review. *Materials* **2018**, *11*, 458. [CrossRef]
27. Avitzour, Y.; Urzhumov, Y.A.; Shvets, G. Wide-angle infrared absorber based on a negative-index plasmonic metamaterial. *Phys. Rev. B* **2009**, *79*, 045131. [CrossRef]

28. Liang, S.; Cheng, S.; Zhang, H.; Yang, W.; Yi, Z.; Zeng, Q.; Tang, B.; Wu, P.; Ahmad, S.; Sun, T. Structural color tunable intelligent mid-infrared thermal control emitter. *Ceram. Int.* **2024**, *50*, 23611–23620. [CrossRef]
29. Twaróg, R.; Szatkowski, P.; Pielichowska, K. Phase Change Materials in Electrothermal Conversion Systems: A Review. *Energies* **2025**, *18*, 569. [CrossRef]
30. Li, W.; Cheng, S.; Zhang, H.; Yi, Z.; Tang, B.; Ma, C.; Wu, P.; Zeng, Q.; Raza, R. Multi-functional metasurface: Ul-tra wideband/multi-band absorption switching by adjusting guided mode resonance and local surface plasmon resonance effects. *Commun. Theor. Phys.* **2024**, *76*, 065701. [CrossRef]
31. Gao, H.; Fan, X.H.; Wang, Y.X.; Liu, Y.C.; Wang, X.G.; Xu, K.; Deng, L.; Zeng, C.; Li, T.; Xia, J.; et al. Multi-foci metalens for spectra and polarization ellipticity recognition and reconstruction. *Opto-Electron. Sci.* **2023**, *2*, 220026. [CrossRef]
32. Zeng, L.L.; Li, B.X.; Wen, R.Q.; Zhang, X.J. Plasmonic Sensor Based on Multi Fano Resonance in Inverse T Shape Structure for Detection of CO2 Concentration. *IEEE Photonics J.* **2023**, *15*, 2201805. [CrossRef]
33. Hu, J.Y.; Tan, C.X.; Bai, W.D.; Li, Y.M.; Lin, Q.; Wang, L.L. Dielectric nanocavity-coupled surface lattice resonances for high-efficiency plasmonic sensing. *J. Phys. D Appl. Phys.* **2022**, *55*, 075105. [CrossRef]
34. Liu, Y.C.; Ma, X.M.; Chao, K.; Sun, F.; Chen, Z.; Shan, J.; Chen, H.; Zhao, G.; Chen, S. Simultaneously realizing thermal and electromagnetic cloaking by multi-physical null medium. *Opto-Electron. Sci.* **2024**, *3*, 230027. [CrossRef]
35. Li, W.X.; Cheng, S.B.; Yi, Z.; Zhang, H.F.; Song, Q.J.; Hao, Z.Q.; Sun, T.Y.; Wu, P.H.; Zeng, Q.D.; Raza, R. Advanced optical reinforcement materials based on three-dimensional four-way weaving structure and metasurface technology. *Appl. Phys. Lett.* **2025**, *126*, 033503. [CrossRef]
36. He, C.; Zhao, D.; Fan, F.; Zhou, H.; Li, X.; Li, Y.; Li, J.; Dong, F.; Miao, Y.-X.; Wang, Y.; et al. Pluggable multitask diffractive neural networks based on cascaded metasurfaces. *Opto-Electron. Adv.* **2024**, *7*, 230005. [CrossRef]
37. Palik, E.D. *Handbook of Optical Constants of Solids*; Academic Press: Cambridge, MA, USA, 1998; Volume 3.
38. Cheng, S.B.; Li, W.X.; Zhang, H.F.; Akhtar, M.N.; Yi, Z.; Zeng, Q.D.; Ma, C.; Sun, T.Y.; Wu, P.H.; Ahmad, S. High sensitivity five band tunable metamaterial absorption device based on block like Dirac semimetals. *Opt. Commun.* **2024**, *569*, 130816. [CrossRef]
39. Liu, M.L.; Li, B.X.; Zeng, L.L.; Wei, Y.; Wen, R.Q.; Zhang, X.J.; Deng, C.S. Dynamic tunable narrow-band perfect absorber for fiber -optic communication band based on liquid crystal. *J. Phys. D Appl. Phys.* **2023**, *56*, 505102. [CrossRef]
40. Yang, C.; Luo, M.H.; Ju, X.W.; Hu, J.Y. Ultra-narrow dual-band perfect absorber based on double-slotted silicon nanodisk arrays. *J. Phys. D Appl. Phys.* **2024**, *57*, 345104. [CrossRef]
41. Mao, Y.; Zhang, H.; Xiong, J.; Liu, X.P.; Wang, Q.Q.; Wang, J.Q. Controlling of spontaneous emission of quantum dots based on hyperbolic metamaterials. *J. Phys. D Appl. Phys.* **2024**, *57*, 255111. [CrossRef]
42. Nan, T.; Zhao, H.; Guo, J.; Wang, X.; Tian, H.; Zhang, Y. Generation of structured light beams with polarization variation along arbitrary spatial trajectories using tri-layer metasurfaces. *Opto-Electron. Sci.* **2024**, *3*, 230052. [CrossRef]
43. Yu, Z.; Li, M.; Xing, Z.; Gao, H.; Liu, Z.; Pu, S.; Mao, H.; Cai, H.; Ma, Q.; Ren, W.; et al. Genetic algorithm assisted meta-atom design for high-performance metasurface optics. *Opto-Electron. Sci.* **2024**, *3*, 240016. [CrossRef]
44. Yang, C.; Lin, Q.; Du, W.J.; Wang, L.L.; Liu, G.D. Bi-tunable absorber based on borophene and VO2 in the optical telecommunication band. *J. Opt. Soc. Am. B* **2022**, *39*, 2969–2974. [CrossRef]
45. Xu, T.; Zhang, W.M.; Song, Q.J.; Yi, Z.; Ma, C.; Cheng, S.B.; Hao, Z.Q.; Sun, T.Y.; Wu, P.H.; Tang, C.J.; et al. Thermotunable mid-infrared metamaterial absorption material based on combined hollow cylindrical VO_2 structure. *Surf. Interfaces* **2024**, *52*, 104868. [CrossRef]
46. Li, B.X.; Zeng, L.L.; Wen, R.Q.; Zhang, X.J. Dynamic Regulation of Multiple Fano Resonances Based on Liquid Crystal. *IEEE Photonics J.* **2023**, *15*, 2200506. [CrossRef]
47. Chen, Z.Y.; Cheng, S.B.; Zhang, H.F.; Yi, Z.; Tang, B.; Chen, J.; Zhang, J.G.; Tang, C.J. Ultra wideband absorption absorber based on Dirac semimetallic and graphene metamaterials. *Phys. Lett. A* **2024**, *517*, 129675. [CrossRef]
48. Hu, J.Y.; Bai, W.D.; Tan, C.X.; Li, Y.M.; Lin, Q.; Wang, L.L. Highly electric field enhancement induced by anapole modes coupling in the hybrid dielectric-metal nanoantenna. *Opt. Commun.* **2022**, *511*, 127987. [CrossRef]
49. Tan, Z.Q.; Lin, Q.; Du, W.J.; Wang, L.L.; Liu, G.D. Simultaneously enhance electric and magnetic Purcell factor by strong coupling between toroidal dipole quasi-BIC and electric dipole. *J. Appl. Phys.* **2025**, *137*, 033103. [CrossRef]
50. Wan, M.; Li, C.; Bao, D.; Wang, J.; Lu, K.; Qu, Z.; Gao, H. Terahertz CMOS High-Sensitivity Sensor Based on Hybridized Spoof Surface Plasmon Resonator. *Photonics* **2025**, *12*, 102. [CrossRef]
51. Jo, Y.; Park, H.; Yoon, H.; Kim, I. Advanced biological imaging techniques based on metasurfaces. *Opto-Electron. Adv.* **2024**, *7*, 240122. [CrossRef]
52. Xiang, T.; Sun, Z.; Wang, L.L.; Lin, Q.; Liu, G.D. Polarization independent perfect absorption of borophene metamaterials operating in the communication band. *Phys. Scr.* **2024**, *99*, 085519. [CrossRef]
53. Wang, X.Y.; Lin, Q.; Wang, L.L.; Liu, G.D. Dynamic control of polarization conversion based on borophene nanostructures in optical communication bands. *Phys. Scr.* **2024**, *99*, 085531. [CrossRef]

54. Yang, J.P.; Tang, A.; Lian, B.W.; Xiong, Z.J.; Ma, Z.B.; Gu, N.T. Optimal multi-spectrum polarization detection with full-Stokes in a single channel by using standard retarders. *Measurement* **2025**, *242*, 116052. [CrossRef]
55. Hou, E.; Meng, D.; Liang, Z.; Xiong, Y.; Yang, F.; Tang, Y.; Fan, Y.; Qin, Z.; Shi, X.; Zhang, Y.; et al. Mid-wave and long-wave infrared dual-band stacked metamaterial absorber for broadband with high refractive index sensitivity. *Appl. Opt.* **2020**, *59*, 2695–2700. [CrossRef]
56. Qin, Z.; Meng, D.; Yang, F.; Shi, X.; Liang, Z.; Xu, H.; Smith, D.R.; Liu, Y. Broadband long-wave infrared metamaterial absorber based on single-sized cut-wire resonators. *Opt. Express* **2021**, *29*, 20275–20285. [CrossRef]
57. Lian, Y.; Li, Y.; Lou, Y.; Liu, Z.; Jiang, C.; Hu, Z.; Wang, J. Adjustable Trifunctional Mid-Infrared Metamaterial Absorber Based on Phase Transition Material VO_2. *Nanomaterials* **2023**, *13*, 1829. [CrossRef]
58. Yang, F.; Liang, Z.; Shi, X.; Zhang, X.; Meng, D.; Dai, R.; Zhang, S.; Jia, Y.; Yan, N.; Li, S.; et al. Broadband long-wave infrared metamaterial absorbers based on germanium resonators. *Results Phys.* **2023**, *51*, 106660. [CrossRef]
59. Li, Z.T.; Li, X.; Liu, G.D.; Wang, L.L.; Lin, Q. Analytical investigation of unidirectional reflectionless phenomenon near the exceptional points in graphene plasmonic system. *Opt. Express* **2023**, *31*, 30458. [CrossRef]
60. Tang, C.J.; Nie, Q.M.; Cai, P.G.; Liu, F.X.; Gu, P.; Yan, Z.D.; Huang, Z.; Zhu, M.W. Ultra-broadband near-infrared absorption enhancement of monolayer graphene by multiple-resonator approach. *Diam. Relat. Mater.* **2024**, *141*, 110607. [CrossRef]
61. Ma, R.; Zhang, L.G.; Zeng, Y.; Liu, G.D.; Wang, L.L.; Lin, Q. Extreme enhancement of optical force via the acoustic graphene plasmon mode. *Opt. Express* **2023**, *31*, 482723. [CrossRef]
62. Medhat, M.; Malek, C.; Tlija, M.; Abukhadra, M.R.; Bellucci, S.; Elsayed, H.A.; Mehaney, A. One-Dimensional Photonic Crystals Comprising Two Different Types of Metamaterials for the Simple Detection of Fat Concentrations in Milk Samples. *Nanomaterials* **2024**, *14*, 1734. [CrossRef] [PubMed]
63. Wang, Q.S.; Fang, Y.; Meng, Y.; Hao, H.; Li, X.; Pu, M.; Ma, X.; Luo, X. Vortex-field enhancement through high-threshold geometric metasurface. *Opto-Electron. Adv.* **2024**, *7*, 240112. [CrossRef]
64. Chen, Y.; Zhang, S.M.; Tian, Y.; Li, C.; Huang, W.; Liu, Y.; Jin, Y.; Fang, B.; Hong, Z.; Jing, X. Focus control of wide-angle metalens based on digitally encoded metasurface. *Opto-Electron. Adv.* **2024**, *7*, 240095. [CrossRef]
65. Yan, X.F.; Lin, Q.; Wang, L.L.; Liu, G.D. Tunable strong plasmon–exciton coupling based modulator employing borophene and deep subwavelength perovskite grating. *J. Phys. D Appl. Phys.* **2023**, *56*, 435106. [CrossRef]
66. Huang, Z.; Liu, G.D.; Wang, L. Active modulation of quasi-bound state in the continuum based on bulk Dirac semimetals metamaterial. *Appl. Phys. Express* **2022**, *15*, 032006. [CrossRef]
67. Hatsuoka, R.; Yamasaki, K.; Wada, K.; Matsuyama, T.; Okamoto, K. Tunable Plasmon Resonance in Silver Nanodisk-on-Mirror Structures and Scattering Enhancement by Annealing. *Nanomaterials* **2024**, *14*, 1559. [CrossRef]
68. Zhang, Y.X.; Lin, Q.; Yan, X.Q.; Wang, L.L.; Liu, G.D. Flat-band Friedrich-Wintgen bound states in the continuum based on borophene metamaterials. *Opt. Express* **2024**, *32*, 10669–10678. [CrossRef]
69. Xiao, Y.T.; Chen, L.W.; Pu, M.B.; Xu, M.F.; Zhang, Q.; Guo, Y.; Chen, T.; Luo, X. Improved spatiotemporal resolution of anti-scattering super-resolution label-free microscopy via synthetic wave 3D metalens imaging. *Opto-Electron. Sci.* **2023**, *2*, 230037. [CrossRef]
70. Ling, Z.X.; Zeng, Y.; Liu, G.D.; Wang, L.L.; Lin, Q. Unified model for plasmon-induced transparency with direct and indirect coupling in borophene-integrated metamaterials. *Opt. Express* **2022**, *30*, 21966. [CrossRef]
71. Zheludev, N.I.; Kivshar, Y.S. From metamaterials to metadevices. *Nat. Mater.* **2012**, *11*, 917–924. [CrossRef]

photonics

MDPI

Article

A High-Sensitivity Graphene Metasurface and Four-Frequency Switch Application Based on Plasmon-Induced Transparency Effects

Aijun Zhu [1,2,3,4], Mengyi Zhang [1], Weigang Hou [5], Lei Cheng [3,4,*], Cong Hu [1,*] and Chuanpei Xu [1]

1 Guangxi Key Laboratory of Automatic Detecting Technology and Instruments, Guilin University of Electronic Technology, Guilin 541004, China; zbluebird@guet.edu.cn (A.Z.); ztli@mails.guet.edu.cn (M.Z.); xcp@guet.edu.cn (C.X.)
2 Guangxi Key Laboratory of Brain-Inspired Computing and Intelligent Chips, Guangxi Normal University, Guilin 541004, China
3 Jincheng Research Institute of Opto-Mechatronics Industry, Jincheng 048000, China
4 Shanxi Key Laboratory of Advanced Semiconductor Optoelectronic Devices and Integrated Systems, Jincheng 048000, China
5 Institute of Intelligent Communication and Network Security, Chongqing University of Posts and Telecommunications, Chongqing 400065, China; houwg@cqupt.edu.cn
* Correspondence: chenglei@jcgjd.org.cn (L.C.); hucong@guet.edu.cn (C.H.)

Abstract: In this paper, we propose the use of a monolayer graphene metasurface to achieve various excellent functions, such as sensing, slow light, and optical switching through the phenomenon of plasmon-induced transparency (PIT). The designed structure of the metasurface consists of a diamond-shaped cross and a pentagon graphene resonator. We conducted an analysis of the electric field distribution and utilized Lorentz resonance theory to study the PIT window that is generated by the coupling of bright-bright modes. Additionally, by adjusting the Fermi level of graphene, we were able to achieve tunable dual frequency switching modulators. Furthermore, the metasurface also demonstrates exceptional sensing performance, with sensitivity and figure of merit (FOM) reaching values of 3.70 THz/RIU (refractive index unit) and 22.40 RIU-1, respectively. As a result, our numerical findings hold significant guiding significance for the design of outstanding terahertz sensors and photonic devices.

Keywords: plasmon induced transparency (PIT); high-sensitivity; graphene; metasurface

1. Introduction

Electromagnetically induced transparency (EIT) is a phenomenon that occurs when different energy levels in an atomic system interfere destructively. This results in the generation of narrow transmission peaks in the spectrum. EIT has potential applications in optical sensing [1–3], optical modulation [4–6], and all-optical logic devices [7]. However, implementing EIT requires extremely low temperatures and strict experimental conditions with rare gases. To overcome these limitations, various metal-based metamaterial structures, such as waveguides [8] and split ring resonators [9], have been proposed and proven to achieve the same effect known as "plasmon-induced transparency" (PIT). Like EIT, PIT has great potential in the development of new optical sensors [10–12], modulators [13,14], and encoders [15,16]. However, these metal-based devices have fixed spectral responses and operating frequencies, making it difficult to generate tunable PIT windows. On the other hand, traditional PIT structures suffer from limited plasma lifetime and high Ohmic losses. These drawbacks significantly hinder the practical application of tunable PIT devices.

Metasurface is a two-dimensional (2D) metamaterial with few planar artificial layered micro/nanostructures, exhibiting novel manipulation of EM and light waves [17,18]. By adjusting the geometric shape of micro/nanostructures with sub-wavelength dimensions, the singular properties of metasurfaces can be reasonably designed [19,20]. Based on metasurfaces, some anomalous physical phenomena have been achieved, such as polarization detection and chiral light sources [21], spectral imaging [22], holography [23], and anomalous reflection [24]. However, the manufacturing process of the metasurface will have an impact on its resonant frequency and transmittance. Factors such as etching deviation, material thickness, and defects at the edge can all affect the performance of the metasurface [25]. Moreover, breakthroughs are also needed in the mass production of metasurfaces. Deep UV lithography, nanoimprint lithography, and self-assembly-based manufacturing processes have the potential to mass-produce cost-effective and environmentally friendly metasurfaces [26].

Graphene has been intensively studied for applications in nanoelectronics and nanophotonics. There have been numerous reports regarding the advantages of graphene, which is fascinating for applications in nanoelectronics due to its high mechanical strength, electronic mobility, and thermal conductivity [27]. Graphene metasurfaces are two-dimensional planar structures that consist of specially arranged artificial units. These structures have the ability to flexibly control the amplitude, phase, and polarization of incident electromagnetic waves. In comparison to three-dimensional metamaterials, the fabrication process for graphene metasurfaces is simpler and more flexible. This provides a significant advantage for the integration and miniaturization of plasma devices. In addition, the bias voltage on graphene can be adjusted between 0 and 3 V to regulate the extrinsic electromagnetic transmission (EET) of graphene hybrid metasurfaces. Moreover, the transmission amplitude and operating frequency band of EET can be controlled by changing the relative position, thereby controlling the coupling of two graphene hybrid resonators [28]. Numerous studies on graphene metasurfaces have been conducted in the field of PIT research. The application of graphene metasurfaces to PIT effect has been extended to various devices, such as sensors, photoelectric switches, and slow light devices. For instance, in 2023, Liu et al. proposed a metasurface composed of graphene split rings to achieve multi-frequency asynchronous optoelectronic switching [29]. In 2022, Xie et al. proposed a single-layer terahertz graphene metasurface that achieves tunable PIT effect in two vertical polarization directions and exhibits a good slow light effect [30]. Furthermore, Wang et al. proposed a tunable multifunctional graphene metasurface to achieve high refractive index sensitivity and slow light devices [31]. These works demonstrate the ability to simultaneously achieve high-sensitivity sensors and optical switches with excellent performance.

In this paper, we present the design of a graphene metasurface based on the terahertz PIT effect. The metasurface is composed of a diamond-shaped cross and pentagon graphene located at the edge. Through the coupling of bright-bright modes, we observe that as the Fermi level increases, the PIT transparent window shifts towards the high-frequency region. We conducted a systematic analysis by modulating the Fermi level, carrier mobility, and incident and polarization angles, exploring its multifunctional applications such as sensors and optical switches. Finally, we evaluate the performance of the switch by calculating parameters such as modulation depth, extinction ratio, and insertion loss. We believe that this designed structure holds tremendous potential for applications in tunable optical switches and sensors.

2. Model and Design

The graphene metasurface that achieves the PIT effect is shown in Figure 1a. The incident wave propagates along the z-axis, with the polarization electric field oriented along the x-axis. Figure 1b provides a top view and detailed geometric parameters of the graphene unit structure. The length of the diamond-shaped graphene cross is represented as r1, and it is set to 2.6 μm. The width of the cross, denoted as r2, is set to 0.6 μm. The pentagon structure surrounding it has a side length of L, which is equal to 1.5 μm. The right leg of the isosceles triangle, cut off by the square, has a length of d, which is set to 0.9 μm. Additionally, in Figure 1b, the silica layer has a thickness of 100 nm, and the graphene thickness is set to 1 nm. The dielectric constant of silica is assigned a value of 3.9. To simplify the analysis, periodic boundary conditions are applied in both the x and y directions, with a periodicity of Px = Py = 7 μm.

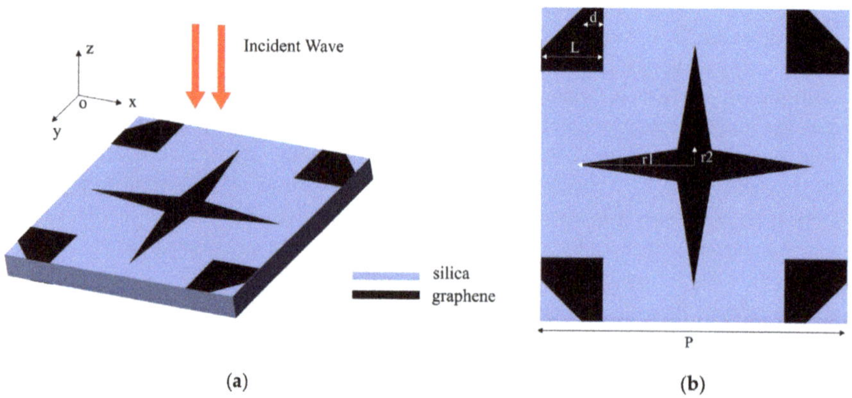

Figure 1. (**a**) schematic diagram of unit structure; (**b**) vertical view of cell structure.

With the rapid advancements in photonics and metamaterials technologies, it is now possible to fabricate graphene metamaterials. There are several methods available for producing uniform single-layer graphene, such as micro-mechanical exfoliation, SiC thermal decomposition, and chemical vapor deposition (CVD) [32]. Among these methods, CVD is particularly popular for its ability to produce graphene with high conductivity and field-effect mobility. Additionally, electron beam lithography offers high resolution, making it suitable for patterning graphene shapes [33]. Therefore, the fabrication of our proposed graphene metamaterials involves two steps. First, a uniform monolayer of graphene is grown on a silica surface using CVD technology. Then, a standard exfoliation process is used to transform the large-area single-layer graphene into a series of isolated single-layer graphene patches.

Graphene is a unique material made up of carbon atoms arranged in a two-dimensional honeycomb structure. On the surface of single-layer graphene, plasmon polaritons can travel, and their behavior can be modified by adjusting the Fermi level. This adjustment can be made by either connecting an electrode to the graphene or changing its carrier concentration through doping. The surface conductivity of graphene is a complex property that includes both interband and intraband conductivities. It is derived from the Kubo equation [34–36]:

$$\sigma_g(\omega) = \sigma_{\text{intra}}(\omega) + \sigma_{\text{inter}}(\omega) \tag{1}$$

$$\sigma_{\text{intra}}(\omega) = \frac{2e^2 k_B T}{\pi\hbar^2}\frac{i}{\omega + \frac{i}{\pi}}\ln\left[2\cosh\left(\frac{E_F}{2k_B T}\right)\right] \tag{2}$$

$$\sigma_{\text{inter}}(\omega) = \frac{e^2}{4\hbar^2}\left[\frac{1}{2} + \frac{1}{\pi}\arctan\left(\frac{\hbar\omega - 2E_F}{2k_BT}\right) - \frac{i}{2\pi}\ln\frac{(\hbar\omega + 2E_F)^2}{(\hbar\omega - 2E_F)^2 + 4(k_BT)^2}\right] \tag{3}$$

In the formula, e represents the electron charge, $\hbar$ is the reduced Planck constant, k_B is the Boltzmann constant, ω refers to the angular frequency, T is the room temperature set at 300 K, and E_F denotes the Fermi level. In the terahertz region, the Kubo equation disregards the interband conductivity by applying the Pauli exclusion principle. Simplifying the conductivity of single-layer graphene, we assume that E_F greatly exceeds both $\hbar\omega$ and k_BT [37,38]:

$$\sigma(\omega) = \frac{ie^2E_F}{\pi\hbar^2(\omega + i\tau^{-1})} \tag{4}$$

Among these parameters, τ represents the relaxation time and is given by the equation $\tau = \mu E_F/(eV_F^2)$, where μ is the carrier mobility and V_F is the Fermi velocity. In our study, linearly polarized plane waves are incident on the surface of graphene along the z-direction. To further analyze the properties of graphene, we calculated the propagation constant (β) by solving Maxwell's equations and applying electromagnetic field boundary conditions. The expression for the propagation constant of single-layer graphene is given as follows [39]:

$$\beta = k_0\sqrt{\varepsilon_d - \left(\frac{2\varepsilon_d}{\eta_0\sigma}\right)^2} \tag{5}$$

In the formula, k_0 represents the wavenumber in free space, η_0 represents the intrinsic impedance in free space, and ε_d represents the relative dielectric constant of silicon dioxide. Figure 2 illustrates the use of the Drude model to plot the dielectric constant of graphene within the 0–7 terahertz range. Figure 2a,b depicts the real and imaginary parts of the dielectric constant of graphene at different chemical potentials, respectively. The thickness of graphene is 1 nm. It is clear that within the 0–7 terahertz range, both the real and imaginary parts of the dielectric constant of graphene are relatively large. Additionally, as the chemical potential increases, the dielectric constant of graphene also increases.

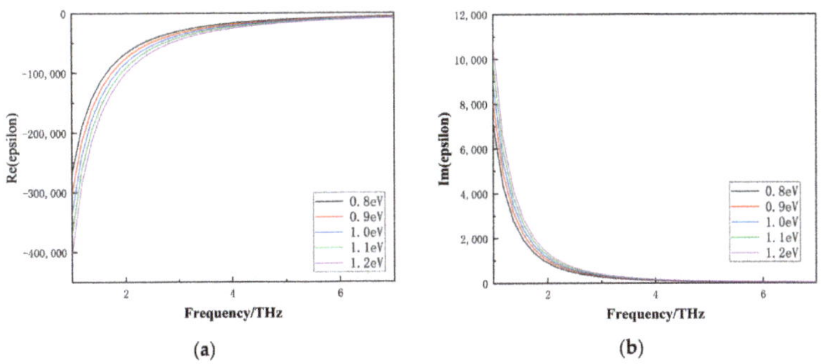

Figure 2. (**a**) the real parts of permittivity for graphene in different chemical potentials.; (**b**) the imaginary parts of permittivity for graphene in different chemical potentials.

To validate the proposed structure, the transmission spectrum was simulated using CST STUDIO SUITE 2018, which uses the finite-integration time-domain (FITD) method [40]. At the same time, in order to shorten the simulation time while ensuring the accuracy of the simulation, we use an adaptive tetrahedral mesh of appropriate size. The frequency range is set to 1–9 THz. Periodic boundary conditions are applied in the x and y directions, while open boundary conditions are applied in the z direction. Using a tetrahedral grid, a total of about 10,000 grids were divided.

3. Results and Discussions

The transmission spectrum of the structure we designed is shown in Figure 3a. The figure also includes the spectra of a single diamond-shaped cross graphene resonator and pentagon graphene located at the edge, which were studied and compared. In our proposed graphene metasurface, we can observe a significant PIT phenomenon at 6 terahertz, indicated by the blue line in the figure. The black line represents a single transmission peak generated solely by the diamond-shaped cross graphene monomer structure, while the red line represents a single transport valley generated solely from pentagon graphene unit structure. The transmission valleys of the diamond-shaped cross and pentagon graphene resonators are located at 5.3 and 6.7 terahertz, respectively, as two bright modes. The electric field distribution at each resonance point is shown in Figure 3b–d. For structures consisting only of pentagon and diamond-shaped graphene, each resonant cavity directly interacts with the incident wave, appearing as two dipoles in Figure 3b,d. However, in composite structures, the electric field redistributes, and strong coupling occurs between the two resonant cavities, as shown in Figure 3c. As a result, the PIT transparent window is caused by the destructive interference of the two bright modes.

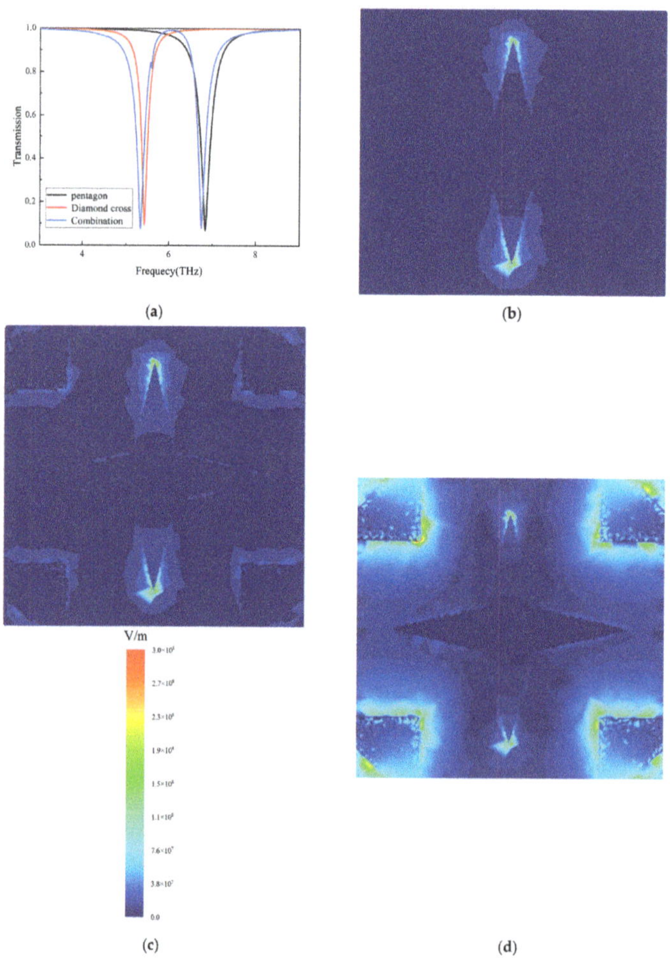

Figure 3. (**a**) transmission spectrum of graphene metasurface; (**b–d**) the normalized electric field distribution at the frequency of 5.3 THz, 6.0 THz, and 6.7 THz.

To provide a clearer understanding of the formation process of the PIT transparent window, we utilized the Lorentz oscillation coupling model to fit the parameters into the simulation results. In this coupling model, we represent the incident plane wave as E, the resonant cavity of mode 1 as $\tilde{M}_1$, and the resonant cavity of mode 2 as $\tilde{M}_2$. As per the definition, the Lorentz oscillation coupling model, when coupled in bright mode, can be expressed as follows [41]:

$$\begin{bmatrix} \omega - \omega_1 + i\gamma_1 & \tilde{k} \\ \tilde{k} & \omega - \omega_2 + i\gamma_2 \end{bmatrix} \begin{bmatrix} \tilde{M}_1 \\ \tilde{M}_2 \end{bmatrix} = \begin{bmatrix} g_1\tilde{E} \\ g_2\tilde{E} \end{bmatrix} \tag{6}$$

Due to the energy dissipation of metamaterial structures being primarily determined by the imaginary part of magnetic permeability, the transmittance of graphene metamaterial structures can be expressed simply as follows:

$$T(\omega) = 1 - \left|\frac{\tilde{M}_1}{\tilde{E}}\right|^2 = 1 - \left|\frac{\left(g_1(\omega - \omega_2 + i\gamma_2) - g_2\tilde{k}\right)\tilde{E}}{(\omega - \omega_1 + i\gamma_1)(\omega - \omega_2 + i\gamma_2) - \tilde{k}^2}\right|^2 \tag{7}$$

The numerical fitting results are illustrated by the blue line in Figure 4. The fitted parameters are as follows: $\tilde{E}$ = 0.5988, ω_1 = 5.329, ω_2 = 6.714, g_1= 0.0061, g_2 = 1.2823, k = 0.186, γ_1 = 0.1073, and γ_2 = 0.1036. This figure demonstrates a strong correspondence between the Lorentz oscillation coupling model and the numerical simulation results. This finding further corroborates the reliability and accuracy of the simulated PIT curve.

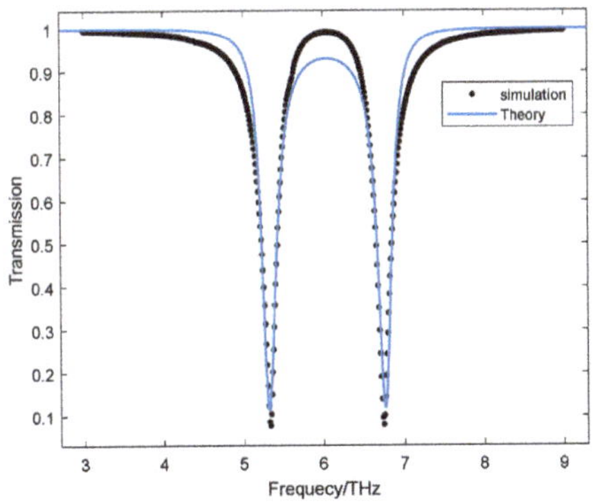

Figure 4. The transmission of our proposed graphene metasurface in THz region. The black dotted line indicates numerical simulation result and the blue solid line represents the Lorentz oscillation coupling model calculation result.

Figure 5a,b analyze the tunable transmission spectra of our designed graphene metasurface at different Fermi levels. As the Fermi level increases from 0.8 to 1.2 eV, the entire transmission spectrum shifts towards the high-frequency region. This phenomenon is consistent with the relationship between frequency (f) and Fermi level (E_f) $f \propto \sqrt{\frac{E_f\alpha_0}{2\pi^2\hbar cL}}$ [30]. According to the formula $\alpha_0 = e^2/\hbar$, where α_0 is the structural constant of graphene, and L is the length of the graphene ribbon, the resonant frequency will increase with the increase of the Fermi level. This allows for the tunable function of the device using different Fermi levels. To adjust the Fermi level and carrier concentration of the entire graphene in the laboratory, a metal electrode will be installed in the graphene layer in order to excite different Fermi levels.

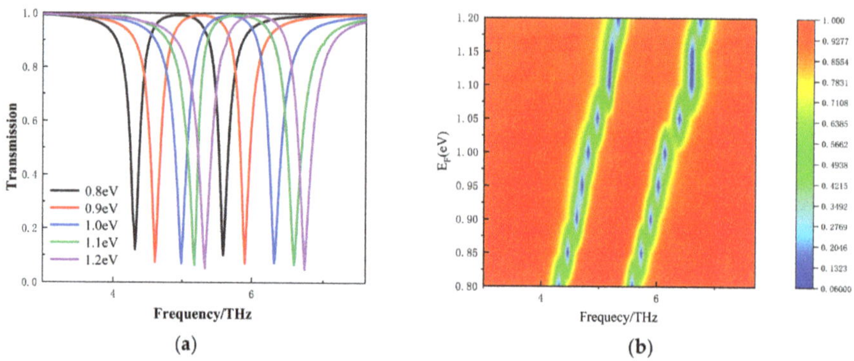

Figure 5. (**a**,**b**) transmittance spectra with different Fermi energies in the THz region.

The relationship between the gate voltage Ev and its controlled graphene Fermi level μc could be expressed as shown in Equation (8) [42,43].

$$\mu_c = \hbar V_F \sqrt{\frac{\pi \varepsilon_0 \varepsilon_d E_v}{e d_0}} \quad (8)$$

In the equation, $\hbar$, V_F, ε_0, ε_d, e, d_0 are the reduced Planck constant, Fermi velocity of graphene, dielectric constant of vacuum, dielectric constant of substrate, electron charge, and substrate thickness, respectively. The Fermi velocity of graphene is 1.0×10^6 m/s.

Our proposed graphene metasurface can achieve a four-frequency switching modulator. In this example, the lower limit of the Fermi level is set to E_F = 0.9 eV, and the upper limit is set to E_F = 1.2 eV, which serves as the reference. We assume a carrier mobility of graphene of 1.8 m^2/vs. As shown in Figure 6, the frequencies of the four-frequency optical switch are set at 4.6 terahertz, 5.3 terahertz, 5.9 terahertz, and 6.7 terahertz, respectively. The state of the optical switch changes simultaneously with the Fermi level. Therefore, these four optical switches are synchronized. Modulation depth (MD = (Ton-Toff)/(Ton) × 100%), insertion loss (IL = $-10\log_{10}$(Ton)), and extinction ratio (ER = $10\log_{10}$(Ton/Toff)) are three important parameters for evaluating the performance of switch modulators. The modulation depth (MD) values calculated for the four switches are 92.15%, 94.79%, 92.82%, and 95.04%, respectively. The corresponding insertion losses (IL) are 0.19 dB, 0.04 dB, 0.06 dB, and 0.08 dB, respectively. The corresponding extinction ratios (ER) are 11.05 dB, 12.83 dB, 11.44 dB, and 13.00 dB. Table 1 presents a performance comparison between our design and designs proposed in other recent literature. From Table 1, it is evident that our designed optical switch outperforms others in terms of modulation depth, insertion loss, and extinction ratio.

Table 1. Comparison of performance with other terahertz optical switch.

MD (%)	IL (dB)	ER (dB)	Ref
95.6	0.315	/	[44]
94.3	/	12.43	[45]
94.5	1.360	7.77	[46]
93.3	0.250	11.75	[47]
89.0	0.55	/	[48]
95.04	0.08	13.00	This work

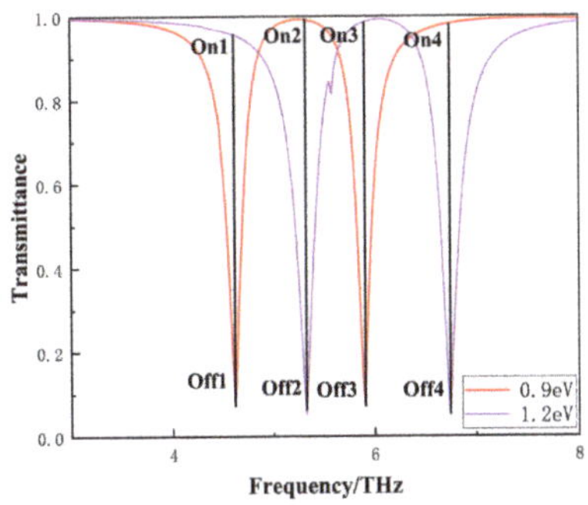

Figure 6. Four-frequency switching modulator modulation of the transmittance spectra when E_F is set as 0.9 eV and 1.2 eV.

In addition, Figure 7 shows the sensing characteristics of the graphene metasurface proposed by us under different surrounding media. Figure 7a shows that as the refractive index (n) increases from 1.0 to 1.5, the PIT spectrum shifts towards the low-frequency region. This shift exhibits a good linear relationship with the refractive index. The performance of the metamaterial structure as a sensor is evaluated using sensitivity, defined as S = Δf/Δn, where Δ f represents the frequency shift of the transmission peak/dip, and Δn represents the shift of the refractive index. By calculating the slope in Figure 7b, we obtained sensitivity values of 2.63 THz/RIU and 3.70 THz/RIU for the two transmission dips, and 3.16 THz/RIU for the transmission peak. Another performance metric used to describe metamaterial sensors is the figure of merit (FOM), which can be determined using FOM = S/FWHM. The FWHM represents the full width at half maximum. After calculation, the designed sensor achieved an FOM of 22.397 RIU-1. Table 2 compares the performance of our designed sensor with other literature in recent years. From Table 2, it is evident that our designed sensor exhibits more sensitive sensing performance and a higher FOM.

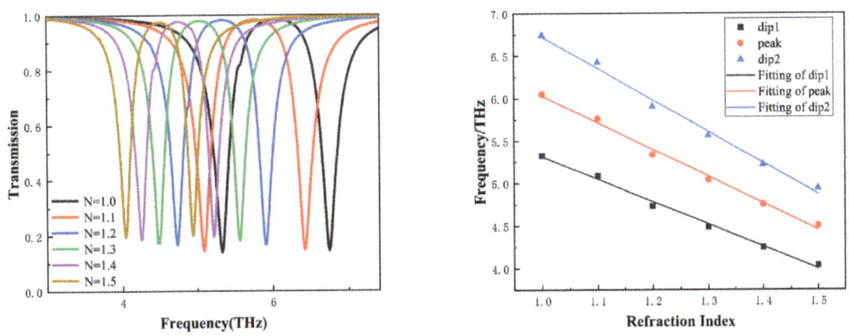

Figure 7. (**a**) transmission spectra with different refraction indices of the surrounding medium; (**b**) linear fit of resonant peak with different refractive indices.

We discussed the modulation of the PIT transparent window in Figure 8a,b with different structural parameters r1 and r2. For graphene, we chose the Fermi level E_F = 1.2 eV and carrier mobility μ = 1.5 m^2/vs. In Figure 8a, as r1 increases, dip1 redshifts and the PIT transparent window widens. In Figure 8b, as r2 increases, dip1 blue shifts and the PIT transparent window narrows. These phenomena indicate that the size of r1 and r2 will affect the resonant frequency of dip1. The LC resonance theory can explain these phenomena. The periodic graphene units in metasurfaces can be viewed as a combination of capacitive, inductive, and resistive elements. The resonant frequency of these graphene units with incident light is inversely proportional to the length of graphene and directly proportional to its width.

Table 2. Comparison of sensing performance in our design with other graphene-based sensors.

Structure	Sensitivity (THz/RIU)	*FOM* (max)	Ref
Graphene	1.1	/	[49]
Graphene	3.4269	21.92	[50]
Graphene	0.7928	8.12	[31]
Graphene	1.7134	6.998	[51]
Graphene	1.40	17.30	[10]
Graphene	1.8424	44.27	[52]
Graphene	1.21	2.75	[53]
Graphene	3.70	22.40	This work

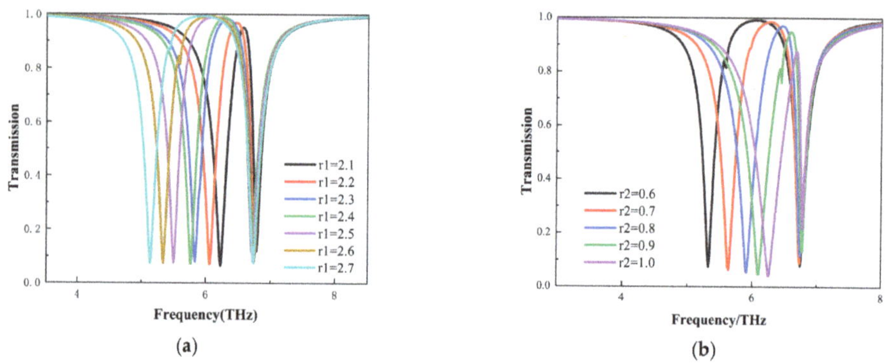

Figure 8. (**a**) transmission spectra with r1 from 2.1 μm to 2.7 μm; (**b**) transmission spectra with r2 from 0.6 μm to 1.0 μm.

Moreover, we also discuss the influence of the incident angle on metasurfaces. In experiments, it is challenging to achieve a situation where the incident wave is completely perpendicular to the metasurface. The incident wave has a certain angle with the z-axis, which may affect the observed data to some extent. To investigate the effect of the incident angle on the properties of metasurfaces, we varied the incident angle to study the changing characteristics of metasurfaces. We analyzed the impact of oblique incidence on plasmon-induced transparency in Figure 9a,b. Let us denote the incident angle of the electromagnetic wave as α, representing the angle between the incident direction and the z-axis. When the incident angle is adjusted from 0° to 60°, the transmittance is minimally affected. However, when the incident angle exceeds 60°, the transmittance decreases at all frequencies within the range of 2–9 THz, while the position of the PIT window remains unchanged. This effect is attributed to the reduced coupling between these two bright modes as the incident angle increases. Therefore, our proposed graphene metasurface design demonstrates insensitivity to changes in the incident angle within the terahertz range. This characteristic allows for the application of our proposed metasurface in experimental environments with incident angle errors and provides greater adaptability.

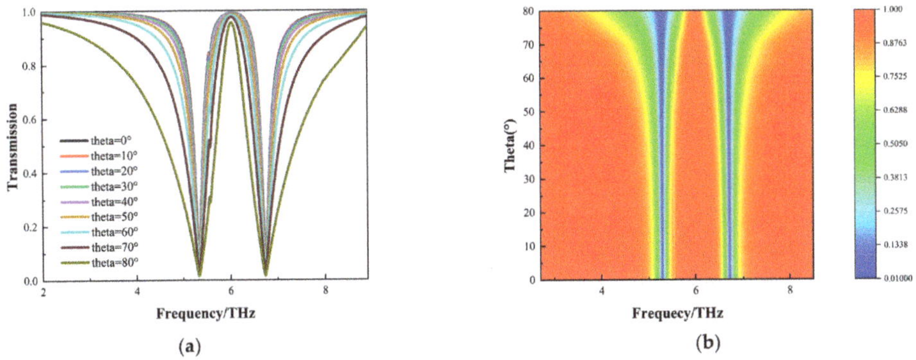

Figure 9. (**a**,**b**) transmission spectra with various incident angles.

4. Conclusions

In summary, we have designed a graphene metasurface that achieves a tunable PIT effect. This effect is caused by the coupling of bright modes in each graphene resonant cavity. Transparent windows can be accurately fitted using the Lorentz model. By varying the Fermi levels, carrier mobility, and oblique incidence angles, we can effectively generate tunable PIT effects. Additionally, we have discussed the impact of different Fermi levels on the transmission spectra of our proposed metasurface. Furthermore, our graphene metasurface can function as a four-frequency switch modulator and a high-sensitivity sensor in the terahertz band. In terms of optical switches, we have achieved a maximum MD of 95.04%, a minimum IL of 0.08 dB, and a maximum ER of 13 dB. As for sensing capabilities, we have achieved a maximum sensitivity of 3.70 THz/RIU and a maximum FOM of 22.40 RIU^{-1}. These physical properties demonstrate that our metasurface not only exhibits excellent sensing performance, but also enables high-performance multi-frequency switching. Therefore, this study holds significant potential for the application of tunable metasurfaces. Furthermore, the research results may pave the way for the design and manufacture of ultra-sensitive sensors, tunable filters, and other related optoelectronic devices.

Author Contributions: Conceptualization, A.Z. and M.Z.; methodology, A.Z.; investigation, L.C.; writing—original draft, A.Z. and M.Z.; writing—review & editing, A.Z., M.Z. and C.H.; project administration, C.X.; Funding acquisition, W.H. All authors have read and agreed to the published version of the manuscript.

Funding: This work was supported in part by the National Natural Science Foundation of China (62461011), Guangxi Natural Science Foundation (2025GXNSFAA069299), Guangxi Key Laboratory of Automatic Detecting Technology and Instruments (YQ22110), Shanxi Province Science and Technology Major Program (202201030201009), the Open Project Program of Shanxi Key Laboratory of Advanced Semiconductor Optoelectronic Devices and Integrated Systems (2023SZKF04), and a grant from Guangxi Key Laboratory of Brain-inspired Computing and Intelligent Chips (BCIC-24-K1).

Institutional Review Board Statement: Not applicable.

Informed Consent Statement: Not applicable.

Data Availability Statement: The data supporting the findings of this study are available within the article.

Conflicts of Interest: The authors declare no conflicts of interest.

References

1. Huang, P.; Yao, Y.; Zhong, W.; Gu, P.; Yan, Z.; Liu, F.; Yan, B.; Tang, C.; Chen, J.; Zhu, M. Optical sensing based on classical analogy of double Electromagnetically induced transparencies. *Results Phys.* **2022**, *39*, 105732. [CrossRef]
2. Jamshidi, M.; Park, C.B.; Azhari, F. An EIT-based piezoresistive sensing skin with a lattice structure. *Mater. Des.* **2023**, *233*, 112227. [CrossRef]
3. Li, J.-S.; Xue, Y.-Y.; Guo, F.-L. Triple frequency bands terahertz metasurface sensor based on EIT and BIC effects. *Opt. Commun* **2023**, *554*, 130225. [CrossRef]
4. Foong, L.K.; Shabani, M.; Sharghi, A.; Reihanisaransari, R.; Al-Bahrani, M.; Le, B.N.; Khalilian, A. Electromagnetically induced transparency for efficient optical modulation in a graphene-dielectric metasurface with surface roughness. *Surfaces Interfaces* **2022**, *35*, 102423. [CrossRef]
5. Lu, Z.; Yuan, J.; Xu, G.; Luo, B.; Tan, J. Optically implemented deep terahertz switch based on perovskite film and electromagnetically induced transparency metasurface. *Opt. Laser Technol.* **2024**, *183*, 112303. [CrossRef]
6. Chen, Y.; Duan, G.; Xu, C.; Qin, X.; Zhao, Q.; Zhou, H.; Wang, B.-X. Triple-band graphene-based tunable electromagnetically induced transparency terahertz metamaterial with multi-frequency optical switching. *Diam. Relat. Mater.* **2024**, *143*, 110939. [CrossRef]
7. Wang, Z.; Yang, F.; Zhou, P.; Zhang, X.; Yang, Y.; Yan, X.; Gao, S.; Zong, M.; Wang, M.; Yao, H.; et al. Frequency-tunable hybrid metamaterial terahertz logic gate with liquid crystal based on electromagnetically induced transparency. *Results Phys.* **2023**, *56*, 107295. [CrossRef]
8. Wang, Y.; He, Z.; Cui, W.; Ren, X.; Li, C.; Xue, W.; Cao, D.; Li, G.; Lei, W. Tunable plasmon induced transparency in the ellipse-shaped resonators coupled waveguide. *Results Phys.* **2020**, *16*, 102981. [CrossRef]
9. Tassin, P.; Zhang, L.; Koschny, T.; Economou, E.N.; Soukoulis, C.M. Low-Loss Metamaterials Based on Classical Electromagnetically Induced Transparency. *Phys. Rev. Lett.* **2009**, *102*, 053901. [CrossRef]
10. Liang, D.; Chen, T. Optical modulated graphene metamaterial based on plasmon-induced transparency in the terahertz band: Application for sensing. *Diam. Relat. Mater.* **2022**, *131*, 109613. [CrossRef]
11. Guo, X.; Cong, J.; Li, C. Dynamic tunable multiple plasmon induced transparency sensor and optical switch in dual-polarization excitation in a terahertz graphene metamaterial. *Opt. Commun.* **2023**, *551*, 130058. [CrossRef]
12. Xiang, J.; Chen, T. A triple plasmon-induced transparency terahertz sensor based on graphene metamaterials. *Diam. Relat. Mater.* **2024**, *149*, 111608. [CrossRef]
13. Meng, Q.; Chen, F.; Xu, Y.; Cheng, S.; Yang, W.; Yao, D.; Yi, Z. Multi-frequency polarization and electro-optical modulator based on triple plasmon- induced transparency in monolayer graphene metamaterials. *Diam. Relat. Mater.* **2023**, *138*, 110216. [CrossRef]
14. Dai, C.; Li, B.; Zeng, L.; Wang, Q.; Chen, Z.; Zeng, Y.; Zhang, X.; Deng, C. Graphene modulator and 2-bit encoder based on plasma induced transparency effect. *Diam. Relat. Mater.* **2024**, *150*, 111715. [CrossRef]
15. Zhu, A.; Li, Z.; Hou, W.; Cheng, L.; Hu, C.; Zhao, T.; Xu, C.; Mahapatra, R. Double plasmon-induced transparency 3 bit graphene encoder. *Diam. Relat. Mater.* **2024**, *142*, 110800. [CrossRef]
16. Zhu, A.; Li, Z.; Hou, W.; Yang, X.; Cheng, L.; Hu, C.; Qiao, F.; Mahapatra, R. A 2-bit graphene encoder based on the plasmon-induced transparency effect and its sensing characteristics. *Results Phys.* **2024**, *59*, 107608. [CrossRef]
17. Yu, N.; Capasso, F. Flat optics with designer metasurfaces. *Nat. Mater.* **2014**, *13*, 139–150. [CrossRef]
18. Chen, S.; Li, Z.; Liu, W.; Cheng, H.; Tian, J. From Single-Dimensional to Multidimensional Manipulation of Optical Waves with Metasurfaces. *Adv. Mater.* **2019**, *31*, 1802458. [CrossRef]
19. Yu, N.; Genevet, P.; Kats, M.A.; Aieta, F.; Tetienne, J.-P.; Capasso, F.; Gaburro, Z. Light Propagation with Phase Discontinuities: Generalized Laws of Reflection and Refraction. *Science* **2011**, *334*, 333–337. [CrossRef]
20. Zhong, F.; Li, J.; Liu, H.; Zhu, S. 12-Photon Entanglement and Scalable Scattershot Boson Sampling with Optimal Entangled-Photon Pairs from Parametric Down-Conversion. *Phys. Rev. Lett.* **2018**, *120*, 243901. [CrossRef]
21. Su, Z.; Yang, Y.; Xiong, B.; Zhao, R.; Wang, Y.; Huang, L. Planar Chiral Metasurface Based on Coupling Quasi-Bound States in the Continuum. *Adv. Opt. Mater.* **2024**, *12*, 2303195. [CrossRef]
22. Shao, Y.; Su, Z.; He, H.; Jing, X.; Liu, Y.; Geng, G.; Li, J.; Wang, Y.; Huang, L. Multispectral imaging through metasurface with quasi-bound states in the continuum. *Opt. Express* **2024**, *32*, 23268–23279. [CrossRef] [PubMed]
23. Huang, L.; Chen, X.; Mühlenbernd, H.; Zhang, H.; Chen, S.; Bai, B.; Tan, Q.; Jin, G.; Cheah, K.-W.; Qiu, C.-W.; et al. Three-dimensional optical holography using a plasmonic metasurface. *Nat. Commun.* **2013**, *4*, 2808. [CrossRef]
24. Pu, M.; Chen, P.; Wang, C.; Wang, Y.; Zhao, Z.; Hu, C.; Huang, C.; Luo, X. Broadband anomalous reflection based on gradient low-Q meta-surface. *AIP Adv.* **2013**, *3*, 052136. [CrossRef]
25. Lopato, P.; Herbko, M.; Gora, P.; Mescheder, U.; Kovacs, A.; Filbert, A. Numerical Analysis of the Influence of Fabrication Process Uncertainty on Terahertz Metasurface Quality. *Electronics* **2023**, *12*, 2198. [CrossRef]

26. Seong, J.; Jeon, Y.; Yang, Y.; Badloe, T.; Rho, J. Cost-Effective and Environmentally Friendly Mass Manufacturing of Optical Metasurfaces Towards Practical Applications and Commercialization. *Int. J. Precis. Eng. Manuf. Technol.* **2023**, *11*, 685–706. [CrossRef]
27. Fan, Y.; Shen, N.; Zhang, F.; Zhao, Q.; Wu, H.; Fu, Q.; Wei, Z.; Li, H.; Soukoulis, C.M. Graphene Plasmonics: A Platform for 2D Optics. *Adv. Opt. Mater.* **2018**, *7*, 1800537. [CrossRef]
28. Nan, J.; Zhang, Y.; Xie, Y.; Li, Z.; Yang, R.; Xu, J.; Fu, Q.; Zhang, F.; Fan, Y. Coupling Controlled Dual-Band Tunable Electromagnetic Extraordinary Transmission in Graphene Hybrid Metasurfaces. *Adv. Electron. Mater.* **2023**, *9*, 2300065. [CrossRef]
29. Liu, Z.; Yang, G.; Luo, X.; Zhou, F.; Cheng, Z.; Yi, Z. Tunable quintuple plasmon-induced transparency and the effect of symmetry breaking based on monolayer graphene split rings metasurface. *Diam. Relat. Mater.* **2024**, *142*, 110786. [CrossRef]
30. Xie, Q.; Guo, L.; Zhang, Z.; Gao, P.; Wang, M.; Xia, F.; Zhang, K.; Sun, P.; Dong, L.; Yun, M. Versatile terahertz graphene metasurface based on plasmon-induced transparency. *Appl. Surf. Sci.* **2022**, *604*, 154575. [CrossRef]
31. Wang, Y.; Chang, B.; Xue, J.; Cao, X.; Xu, H.; He, H.; Cui, W.; He, Z. Sensing and slow light applications based on graphene metasurface in terahertz. *Diam. Relat. Mater.* **2022**, *123*, 108881. [CrossRef]
32. Xu, Y.; Xu, H.; Jiang, X.; Yin, J. Versatile Functionalization of the Micropatterned Hydrogel of Hyperbranched Poly(ether amine) Based on "Thiol-yne" Chemistry. *Adv. Funct. Mater.* **2014**, *24*, 1679–1686. [CrossRef]
33. Yang, R.; Zhang, L.; Wang, Y.; Shi, Z.; Shi, D.; Gao, H.; Wang, E.; Zhang, G. An Anisotropic Etching Effect in the Graphene Basal Plane. *Adv. Mater.* **2010**, *22*, 4014–4019. [CrossRef] [PubMed]
34. Jia, W.; Ren, P.-W.; Tian, Y.-C.; Fan, C.-Z. Dynamically tunable optical properties in graphene-based plasmon-induced transparency metamaterials. *Chin. Phys. B* **2019**, *28*, 026102. [CrossRef]
35. Chen, D.-C.; Li, H.-J.; Xia, S.-X.; Qin, M.; Zhai, X.; Wang, L.-L. Dynamically tunable electromagnetically-induced-transparency-like resonances in graphene nanoring and nanodisk hybrid metamaterials. *EPL Europhys. Lett.* **2017**, *119*, 47002. [CrossRef]
36. Tian, Y.-C.; Jia, W.; Ren, P.-W.; Fan, C.-Z. Tunable plasmon-induced transparency based on asymmetric H-shaped graphene metamaterials. *Chin. Phys. B* **2018**, *27*, 124205. [CrossRef]
37. Liu, T.; Wang, H.; Liu, Y.; Xiao, L.; Zhou, C.; Xu, C.; Xiao, S. Dynamically tunable electromagnetically induced transparency in a terahertz hybrid metamaterial. *Phys. E Low-Dimens. Syst. Nanostruct.* **2018**, *104*, 229–232. [CrossRef]
38. Zhou, F.; Wang, Y.; Zhang, X.; Wang, J.; Liu, Z.; Luo, X.; Zhang, Z.; Gao, E. Dynamically adjustable plasmon-induced transparency and switching application based on bilayer graphene metamaterials. *J. Phys. D Appl. Phys.* **2020**, *54*, 054002. [CrossRef]
39. Li, Y.; Xu, Y.; Jiang, J.; Ren, L.; Cheng, S.; Wang, B.; Zhou, X.; Wang, Z. Dual dynamically tunable plasmon-induced transparency and absorption in I-type-graphene-based slow-light metamaterial with rectangular defect. *Optik* **2021**, *246*, 167837. [CrossRef]
40. Hu, F.; Fan, Y.; Zhang, X.; Jiang, W.; Chen, Y.; Li, P.; Yin, X.; Zhang, W. Intensity modulation of a terahertz bandpass filter: Utilizing image currents induced on MEMS reconfigurable metamaterials. *Opt. Lett.* **2018**, *43*, 17–20. [CrossRef]
41. Tang, B.; Jia, Z.; Huang, L.; Su, J.; Jiang, C. Polarization-Controlled Dynamically Tunable Electromagnetically Induced Transparency-Like Effect Based on Graphene Metasurfaces. *IEEE J. Sel. Top. Quantum Electron.* **2020**, *27*, 4700406. [CrossRef]
42. Chen, H.; Zhang, Z.; Yang, J.; Zhou, Z. Controlled and tunable plasmon-induced transparency based on graphene metasurfaces in atmospheric windows. *Diam. Relat. Mater.* **2022**, *127*, 109210. [CrossRef]
43. Agravat, D.; Patel, S.K.; Almawgani, A.H.M.; Alsuwian, T.; Armghan, A.; Daher, M.G. Investigation of a Novel Graphene-Based Surface Plasmon Resonance Solar Absorber to Achieve High Absorption Efficiency Over a Wide Spectrum of Wavelengths, from Ultraviolet to Infrared. *Plasmonics* **2023**, *19*, 1071–1083. [CrossRef]
44. Zhu, J.; Xiong, J. Tunable terahertz graphene metamaterial optical switches and sensors based on plasma-induced transparency. *Measurement* **2023**, *220*, 113302. [CrossRef]
45. Qin, Y.; Zhou, F.; Liu, Z.; Zhang, X.; Zhuo, S.; Luo, X.; Ji, C.; Yang, G.; Zhou, Z.; Sun, L.; et al. Triple plasmon-induced transparency and dynamically tunable electro-optics switch based on a multilayer patterned graphene metamaterial. *J. Opt. Soc. Am. A* **2022**, *39*, 377–382. [CrossRef]
46. Meng, Q.; Chen, F.; Xu, Y.; Cheng, S.; Yang, W.; Yao, D.; Yi, Z. Tunable terahertz double plasmon induced-transparency based on monolayer patterned graphene structure. *Photonics Nanostruct. Fundam. Appl.* **2023**, *54*, 101132. [CrossRef]
47. Wu, Z.; An, P.; Ding, M.; Qi, Y.; Zhang, L.; Han, S.; Lian, D.; Chen, C.; Yang, X. Tunable Electromagnetically Induced Transparent Window of Terahertz Metamaterials and Its Sensing Performance. *Appl. Sci.* **2022**, *12*, 7057. [CrossRef]
48. Li, J.; Weng, J.; Li, J.; Chen, S.; Guo, Z.; Xu, P.; Liu, W.; Wen, K.; Qin, Y. Dynamic manipulation of plasmon induced transparency with parallel-orthometric graphene strips structure. *Results Phys.* **2022**, *40*, 105816. [CrossRef]
49. Li, G. Tunable Fano resonance for advanced sensing in graphene-based metasurface. *Diam. Relat. Mater.* **2024**, *142*, 110823. [CrossRef]
50. Li, Z.; Yang, N.; Liu, Y.; Zhong, Z.; Song, C.; He, Z.; Cui, W.; Xue, W.; Li, L.; Li, C.; et al. Tunable plasmonic optical responses and the sensing application in graphene-based metasurface. *Diam. Relat. Mater.* **2022**, *126*, 109071. [CrossRef]
51. He, Z.; Cui, W.; Ren, X.; Li, C.; Li, Z.; Xue, W.; Zhang, B.; Zhao, R. Ultra-high sensitivity sensing based on tunable plasmon-induced transparency in graphene metamaterials in terahertz. *Opt. Mater.* **2020**, *108*, 110221. [CrossRef]

52. Wang, M.; Cui, W.; Wang, Y.; Li, M.; Qie, X. Tunable fano-like resonances in graphene metasurface for quad-frequency optoelectronic sensing. *Phys. E Low-Dimens. Syst. Nanostruct.* **2025**, *168*, 116191. [CrossRef]
53. Tan, C.; Wang, S.; Li, S.; Liu, X.; Wei, J.; Zhang, G.; Ye, H. Cancer Diagnosis Using Terahertz-Graphene-Metasurface-Based Biosensor with Dual-Resonance Response. *Nanomaterials* **2022**, *12*, 3889. [CrossRef] [PubMed]

hv photonics

MDPI

Article

Optoplasmonics of Single-Walled Carbon Nanotube Thin Films

Chandra Mani Adhikari

Department of Chemistry, Physics & Materials Science, Fayetteville State University, Fayetteville, NC 28301, USA; cadhikari@uncfsu.edu

Abstract: An ultrathin film capable of exhibiting material properties across and around two different dimensions by bridging two-dimensionality frameworks, called a trans-dimensional (TD) material, can be an exceptional tool to tune various electronic and optoplasmonic properties of a system that are unattainable from either dimension. Taking an example of the planar periodic arrangement of single-walled carbon nanotube (SWCNT) TD films, we semi-analytically calculated their dynamical conductivities and dielectric responses as a function of the incident photon frequency and the SWCNT's radius using the many-particles Green's function formalism within the Matsubara frequency technique. The periodic array of SWCNTs has an anisotropic dielectric response, which is almost a constant and the same as that of the host dielectric medium in the perpendicular direction of the alignment of the SWCNT array due to the depolarization effect that SWCNTs have. However, the dielectric response functions depend on the incident photon energy in addition to the film's thickness, the SWCNT's sparseness, inhomogeneity, and the SWCNT's diameter. The energy difference between the resonant absorption peak and the plasmonic peak varies with the thickness of the film. Varying the length of the CNTs, we also observed that the exciton–plasmon coupling strength increases with the increase in length of the SWCNTs. The metallic SWCNT-containing films have comparatively pronounced plasmon resonance peaks at low photon energy than semiconducting SWCNT-containing films. Both metallic and semiconducting SWCNT-consisting films have negative refraction for a wide range of energy, making them good candidates for metamaterials.

Keywords: single-walled carbon nanotube; dielectric response; metamaterials; exciton–plasmon coupling; plasmonics; trans-dimensional material

1. Introduction

Since the first observation of a carbon nanotube (CNT) with a diameter of 50 nanometers in 1952 [1], many scientific studies have been conducted on the synthesis, characterization, and detailed investigation of the mechanical, thermal, electronic, optical, and plasmonic properties of carbon nanotubes. In 1975, Oberlin et al. observed a multi-walled CNT (MWCNT) as a hollow tube of rolled graphite in carbon fiber synthesis utilizing the chemical vapor deposition method decomposing benzene and hydrogen at high temperatures in the presence of a transition metal as a catalyst [2]. Dresselhaus et al. contributed to the theoretical investigation of carbon-based materials, namely carbon whiskers and graphite intercalation compounds, in the late 1970s [3,4]. The first evidence of CNT production as fiber in a carbon anode via the arc discharge technique was presented at the Biennial Conference of Carbon in 1978 [5]. MWCNTs were perceived as rolled graphene layers in a cylindrical geometry after Nesterenko et al.'s work on the thermoanalytical disproportionation of carbon monoxide to characterize carbon nanoparticles [6]. Howard Tennent received a US patent for producing cylindrical discrete carbon fibrils of diameters in the range of 3.5 to 70 nm [7]. The experimental discovery of MWCNTs by S. Iijima in arc-burned graphite rods and their imaging through a high-resolution electron beam technique of transmission electron microscopy has been considered a turning point of CNT research [8]. Within two years after the discovery, it was observed that some single-walled CNTs (SWCNTs) show exceptional conducting behavior [9,10]. Since then, CNTs have received significant attention among researchers in this field because of their high tensile strength (~order of 10^{11} Pa for SWCNTs [11]), higher electrical and thermal conductivities, optoplasmonic properties, and chirality-dependent metallicity.

The plasmon spectrum of SWCNTs is rich in plasmon resonance peaks. Semiconducting SWCNTs have plasmon modes in the infrared and visible ranges of a few hundred to thousands of terahertz (THz), while metallic SWCNTs have plasmon modes and comparatively intense plasmon peaks in the low-frequency range of only a few THz besides the other plasmon modes in the infrared and visible ranges, like semiconducting SWCNTs [12,13]. For a thin film of CNTs, the film's plasmonic behavior on its interaction with light results due to the collective excitons, which can be tuned by changing the radius, sparseness of CNTs, thickness of dielectric medium in which CNTs are immersed, and chirality of the SWCNTs [14–17].

Plasmonic interconnected circuits built from CNTs offer a new material platform for nanoscale electronic devices integrated with optoelectronic components as they offer high electrical conductivity, ballistic transportation, resistance to electromigration, and broadband response [18]. When target molecules are bound to a CNT's surface, the CNT can be used for gas sensing, chemical sensing, and bio-sensing as the plasmonic properties of the CNT have become sensitive in the presence of target molecules [19,20]. SWCNT films' optical transparency, excellent flexibility, and remarkable potency under mechanical stress make them good candidates for transparent conductive films for liquid crystal display and organic light-emitting diode displays, transparent electrodes in photovoltaic cells, and quantum dot light-emitting diodes [21,22]. The films can also be used in optical modulators and optoelectronic sensors. SWCNT films can be integrated into metamaterials for sub-wavelength imaging in super lenses, cloaking devices, beam-steering technologies, and electromagnetic interference shielding for aerospace and telecommunications [23]. Hybrid plasmonic–CNT structures obtained by incorporating one-dimensional (1D) CNT tubules with other 2D plasmonic materials such as MXenes and graphene are useful for fine-tuning various electrical, optical, thermal, and plasmonic properties as MXenes and graphene provide additional pathways for electron transport and enhance the coupling between MXenes' or graphene's localized surface plasmons and the electron-rich CNT network, amplifying the plasmonic response [24–26].

Various semi-analytical approaches such as ab initio calculations, density functional theory, molecular dynamics, and continuum mechanics have been reported as reliable methods for studying the structural property relation between SWCNTs and their electro-optical nature. One can also employ mathematically sophisticated analytical approaches such as fractional calculus calculations as presented in Ref. [27], where the authors show that the temporal dynamics of the heat conduction induced by light in SWCNT thin films present a nonlinear dependence on optical irradiance.

Any SWCNT of chiral indices (n, m) can be thought of as being a hollow cylinder prepared by rolling a graphene sheet along the specific chiral vector $\vec{C}$ given by $\vec{C} = n\,\vec{a}_1 + m\,\vec{a}_2$, where $\vec{a}_1$ and $\vec{a}_2$ represent the graphene lattice vectors. $\vec{a}_1$ and $\vec{a}_2$ serve as translation vectors. A scheme of rolling a graphene sheet to obtain different types of SWCNTs is presented in Figure 1. The radius and chiral angle of any SWCNT can be calculated as [28–30]

$$R = \frac{a_{cc}}{\pi}\sqrt{n^2 + nm + m^2}, \tag{1}$$

and

$$\theta = \cos^{-1}\left(\frac{n + m/2}{\sqrt{n^2 + nm + m^2}}\right), \tag{2}$$

where $a_{cc} \approx 1.42\text{Å}$ is the carbon-to-carbon distance in a graphene sheet. The chiral angle θ satisfies $0° \leq \theta \leq 30°$. θ equals 0° for a zigzag SWCNT with chiral indices $(n, 0)$ and 30° for an armchair SWCNT with chiral indices (n, n). The chiral angle θ satisfies $0° < \theta < 30°$ for chiral SWCNTs. All the armchair SWCNTs are metallic. However, the zigzag SWCNTs are metallic or small bandgap semiconductors based on whether the chiral index n is divisible by 3 or not, respectively. The chiral SWCNTs, whose chiral indices n and m are neither identical nor either one is zero, are quasi-metallic if the difference in chiral indices, $n - m$, is divisible by 3. Otherwise, they are semiconductive. Statistically, 1/3rd of SWCNTs are metallic, while the remaining 2/3rd are semiconducting with a small bandgap. Depending on the technological requirement, one can synthesize SWCNTs of a specific chirality.

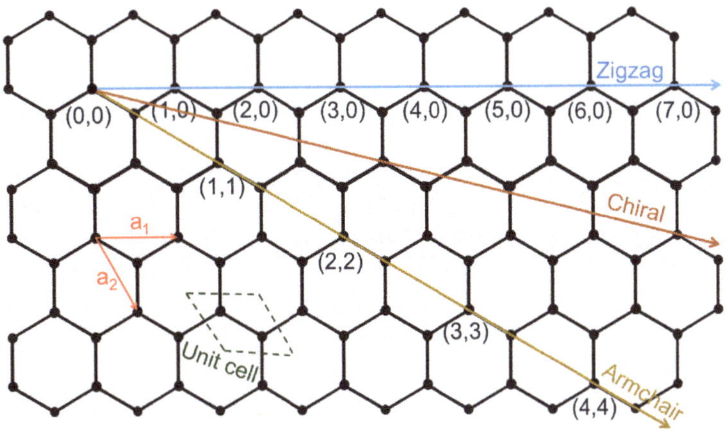

Figure 1. A schematic diagram showing the rolling direction for the most general chiral with chirality vector (n, m) and two special cases of achiral, namely zigzag for $(n, 0)$ and armchair (n, n) SWCNTs.

The literature is rich in the study of mechanical [31], electronic [32], optical [33–35], and plasmonic [36,37] behaviors of a single SWCNT. This paper extends our work to the thin and ultrathin films of homogeneous SWCNTs, which are oriented along the y-axis in periodic alignment. A schematic diagram of the theoretical model considered in this study is presented in Figure 2. Each SWCNT presented in the figure was plotted using Nanotube Modeler Software, Version 1.8 [38,39], which is a program to generate xyz-coordinates and plot interactive graphics for capped and uncapped nanotubes and nanocones for a chosen chirality, tube length, and bond length. The SWCNT array is immersed in a dielectric medium of thickness d and constant effective relative permittivity ε. The substrate and superstate have relative permittivities of ε_1 and ε_2, which are relatively much smaller in value than ε. Each SWCNT has a radius R and length L. The center-to-center inter-tube distance Δ satisfies $\Delta \geq 2R$ with the minimum ideal distance being the sum of the radii of the two closest tubes. Δ determines the dispersiveness of SWCNTs in the film, with a larger Δ signifying a highly dispersed SWCNT film. $L_\perp$ denotes the SWCNT film width. The SWCNTs have uniform electronic charge distribution throughout the surface of the tube.

The carrier concentration, mobility, and conductivity can be enhanced, thereby decreasing sheet resistance by treating SWCNT films with inorganic acids such as nitric, sulfuric, and hydrochloric acid (for example, see Ref. [40]). Acid-treated SWCNT films maintain a p-type nature with enhanced carrier mobility and low resistance, making them good candidates for hole-transporting layers in solar cells. Atomic impurities can induce localized surface plasmon resonance, while a vacancy acts as a localized charge trap, increases plasmonic decoherence, and broadens plasmon resonance by increasing scattering. This study considers the pristine SWCNTs with no impurities or defects.

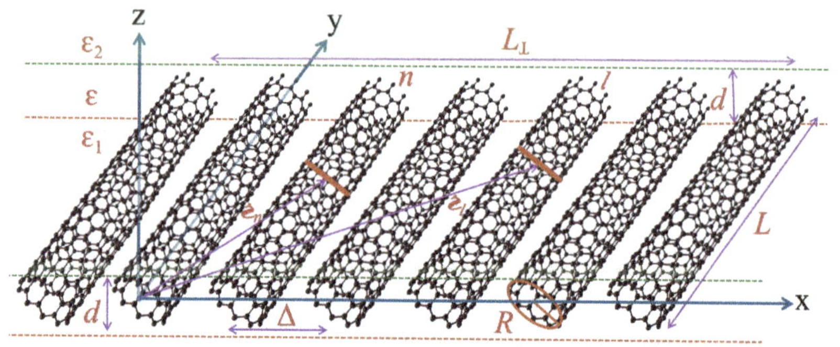

Figure 2. A schematic diagram showing a planar periodic array of identical non-chiral SWCNTs taken in our theoretical model.

Taking rings in nth SWCNT at position vector $\vec{r}_n$ and lth SWCNT at $\vec{r}_l$, the distance between the rings can be expressed as $r = |\vec{r}_n - \vec{r}_l|$. For a thin film with a thickness smaller than the distance between rings denoting unit cells of different SWCNTs as shown in Figure 2, i.e., $d \ll r$, the electrostatic (Coulomb) interaction between the rings of unit cells becomes so strong that it loses the dependency on its vertical component. As a result, the effective dimensionality of the SWCNT film decreases by one, making it a 2D problem retaining the vertical size, i.e., thickness, as a parameter [16,17]. Such a system with a relatively small thickness, better known as trans-dimensional material (TDM), offers an exceptional tool to tune various optoplasmonic properties that are unattainable from both 3D bulk and/or single-layered 2D counterparts.

To understand and utilize light–matter interactions in an SWCNT and its film at the nanoscale, evaluating and analyzing their dielectric response tensors are crucial, as dielectric response functions delineate collective oscillations of free electrons, excitons, and plasmons. The dielectric property-dependent surface plasmons and localized surface plasmon resonances occur on the film's surface, creating strong electromagnetic fields when free electrons resonate with incident photon [41,42]. We derived closed-form mathematical expressions for the dynamical (as a function of photon frequency) dielectric response functions of the SWCNT array as presented in Figure 2 using the many-particle Green's function technique by exploiting the Matsubara frequency approach, as discussed in Ref. [43].

We organize this paper as follows. We orient ourselves with a mathematical overview for dielectric responses for SWCNTs in Section 2. We first evaluate an SWCNT's dynamic conductivity, which is then used to evaluate its polarizability and dielectric response functions. A comparative study of the dynamical conductivity of two different types of SWCNTs is presented, followed by the dynamical permittivities and dynamical conductivity of an SWCNT used in the SWCNT array. The dielectric response functions of an array of identical SWCNTs are presented in Section 3. Therein, we set up the Hamiltonian of the system, determine the interaction Hamiltonian for the SWCNT film, and present the closed-form expressions for dielectric responses. Taking two different films, one containing metallic (9,0) SWCNTs and the other containing semiconducting (10,0) SWCNTs, as references, we discuss our research findings in Section 4. We present conclusions in Section 5. We have used Gaussian units throughout this paper unless otherwise stated.

2. Dielectric Responses of an Isolated SWCNT

When light interacts with a valence electron of carbon in an SWCNT, the electron becomes excited to the conduction band, leaving a hole behind, thereby creating an electron–hole pair called exciton and inducing an anisotropic polarization. The induced polarization is dominant along the SWCNT axis and negligible in the directions perpendicular to the SWCNT alignment due to the weakening of the response to incident photon along the width of the tube called the transverse depolarization effect [44,45]. The photon frequency and SWCNT radius-dependent axial polarizability $\alpha_{yy}(\omega, R)$ of an SWCNT can be expressed in terms of its axial surface conductivity $\sigma_{yy}(\omega + \mathrm{i}\delta)$ as [46]

$$\alpha_{yy}(\omega, R) = \mathrm{i}\frac{2\pi R}{\omega}\sigma_{yy}(\omega + \mathrm{i}\delta), \tag{3}$$

where δ is an infinitesimal frequency parameter. The dynamical conductivity of an SWCNT is estimated using the well-known $\vec{k} \cdot \vec{p}$ method of band structure calculations [45,47]. The conducting behavior of SWCNTs significantly varies with the chirality and divisibility of the difference between chiral indices by 3. For example, (9,0) SWCNT is metallic, while (10,0) SWCNT is a semiconducting one. The exciton–plasmon spectrum for the conductivity of metallic SWCNT significantly differs from the semiconducting one, as shown in Figure 3.

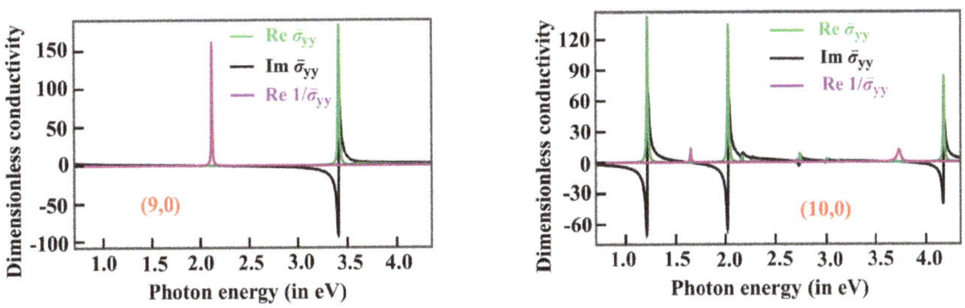

Figure 3. Dimensionless conductivities of (9,0) (**left panel**) and (10,0) (**right panel**) SWCNTs along the corresponding CNT axis as a function of photon energy (expressed in eV).

One may ask how the plasmon spectrum for the conductivity of an isolated SWCNT, measured by $\mathrm{Re}\,(1/\sigma_{yy})$, can have such intense resonance peaks. The quantity $\mathrm{Re}\,(1/\sigma_{yy})$ can be expressed as

$$\mathrm{Re}\,(1/\sigma_{yy}) = \frac{\mathrm{Re}\,\sigma_{yy}}{(\mathrm{Re}\,\sigma_{yy})^2 + (\mathrm{Im}\,\sigma_{yy})^2}, \tag{4}$$

where Re and Im denote the real and imaginary parts, respectively. At the energy value, in which the real part approaches zero, and the imaginary part switches the sign simultaneously, the $\mathrm{Re}\,(1/\sigma_{yy})$ quantity will have a significantly large value. As $\mathrm{Re}\,\sigma_{yy}$ can be smaller than unity when the imaginary part becomes zero, the $\mathrm{Re}\,(1/\sigma_{yy})$ can be larger than unity. The plasmon resonance appears at a lower photon energy than that of the exciton for metallic (9,0) SWCNT. The same is not true for semiconducting (10,0) SWCNT. In the chosen window of the photon energy, i.e., 0.75–4.25 eV, (10,0) SWCNT has three intense exciton resonance peaks, while the (9,0) SWCNT has only one resonance peak in this energy range for each real, imaginary, and energy-loss parts of the spectrum.

Knowing the conductivity of a single SWCNT, one can easily evaluate its dielectric response functions using [13]

$$\frac{\varepsilon_{yy}(\omega)}{\varepsilon} = 1 + \mathrm{i}\frac{4\pi\varrho S}{\omega}\sigma_{yy}(\omega + \mathrm{i}\delta), \tag{5}$$

where $S = 2\pi RL$ denotes the surface area of an SWCNT in which the charge is uniformly distributed, and ϱ is the number of SWCNTs per unit volume. For an isolated SWCNT, $\varrho = 1/(\pi R^2 L)$. Consequently, $\varrho S = 2/R$. One can normalize the axial surface conductivity $\sigma_{yy}(\omega + \mathrm{i}\delta)$ of an SWCNT as

$$\sigma_{yy}(\omega + \mathrm{i}\delta) = \frac{e^2}{2\pi\hbar}\,\overline{\sigma}_{yy}(x + \mathrm{i}\overline{\delta}) \tag{6}$$

such that $\overline{\sigma}_{yy}(x + \mathrm{i}\overline{\delta})$ is a dimensionless axial surface conductivity as a function of dimensionless energy $x = \hbar\omega/(2\gamma_0)$ and an infinitesimal dimensionless energy $\overline{\delta} = \hbar\delta/(2\gamma_0)$, where $\gamma_0 = 2.7\,\mathrm{eV}$ is the carbon atom's nearest neighbor overlap integral in an SWCNT. Substituting the value of $\sigma_{yy}(\omega + \mathrm{i}\delta)$ from Equation (6) to Equation (5), one obtains

$$\overline{\varepsilon}_{yy}(x) = \frac{\varepsilon_{yy}(x)}{\varepsilon} = 1 + \mathrm{i}\,\frac{2\,e^2}{\gamma_0 R}\,\frac{1}{x}\,\overline{\sigma}_{yy}(x + \mathrm{i}\overline{\delta})\,. \tag{7}$$

One can resolve the complex quantity $\overline{\varepsilon}_{yy}(x)$ presented in Equation (7) as

$$\mathrm{Re}\,\overline{\varepsilon}_{yy}(x) = 1 - \frac{1}{x}\,\frac{2\,e^2}{\gamma_0 R}\,\mathrm{Im}\,\overline{\sigma}_{yy}(x + \mathrm{i}\overline{\delta})\,, \tag{8a}$$

$$\mathrm{Im}\,\overline{\varepsilon}_{yy}(x) = \frac{2\,e^2}{\gamma_0 R}\,\frac{1}{x}\,\mathrm{Re}\,\overline{\sigma}_{yy}(x + \mathrm{i}\overline{\delta})\,, \tag{8b}$$

for the real $\mathrm{Re}\,\overline{\varepsilon}_{yy}(x)$ and imaginary $\mathrm{Im}\,\overline{\varepsilon}_{yy}(x)$ parts of dielectric response. The loss function $-\mathrm{Im}\,\frac{1}{\overline{\varepsilon}_{yy}(x)}$, determined from the plasmon spectra, is given by

$$-\mathrm{Im}\,\frac{1}{\overline{\varepsilon}_{yy}(x)} = \frac{\mathrm{Im}\,\overline{\varepsilon}_{yy}(x)}{\left[\mathrm{Re}\,\overline{\varepsilon}_{yy}(x)\right]^2 + \left[\mathrm{Im}\,\overline{\varepsilon}_{yy}(x)\right]^2}. \tag{9}$$

For illustrative purposes, the dimensionless response functions, namely longitudinal conductivities and longitudinal dielectric functions for an (11,0) SWCNT, are presented in Figure 4. $\mathrm{Re}\,1/\overline{\sigma}_{yy}(x + \mathrm{i}\overline{\delta})$ gives us the surface plasmon's density of states, which is nonzero when the imaginary part of $\overline{\sigma}_{yy}(x + \mathrm{i}\overline{\delta})$ vanishes and the real part of $\overline{\sigma}_{yy}(x + \mathrm{i}\overline{\delta})$ also approaches zero. The quantity $-\mathrm{Im}\,1/\overline{\varepsilon}_{yy}(x)$ is nonzero only if the $\mathrm{Re}\,\overline{\varepsilon}_{yy}(x)$ is zero and $\mathrm{Im}\,\overline{\varepsilon}_{yy}(x)$ approaches zero at the same time.

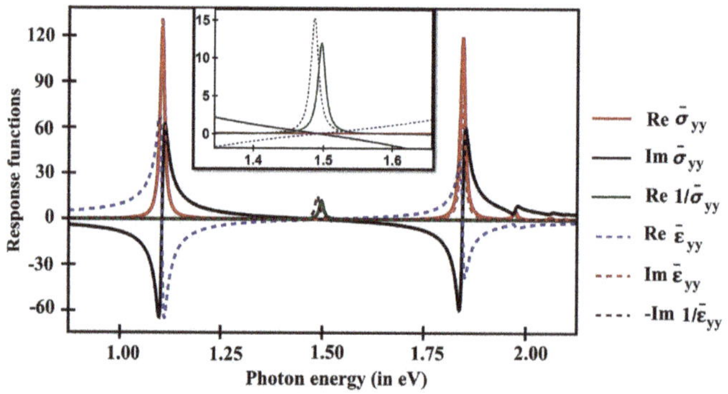

Figure 4. Dimensionless conductivities and dimensionless dielectric functions of an (11,0) SWCNT along the SWCNT axis plotted as a function of photon energy (expressed in eV). Only the first two excitonic peaks and a plasmonic peak are shown. For better visibility, curves $\mathrm{Re}\,\overline{\sigma}_{yy}(x + \mathrm{i}\overline{\delta})$ and $\mathrm{Im}\,\overline{\sigma}_{yy}(x + \mathrm{i}\overline{\delta})$ are scaled by 10^{-1}. In the inset, the magnified version of the curves near the resonances of $\mathrm{Re}\,1/\overline{\sigma}_{yy}(x + \mathrm{i}\overline{\delta})$ and $-\mathrm{Im}\,1/\overline{\varepsilon}_{yy}(x)$ are shown.

3. Dielectric Responses of SWCNT Films

From the principle of the second quantization, one can write the Hamiltonian of free exciton on the surface of the SWCNT as

$$\hat{H}_0 = \sum_{s,\vec{k}} E_s(\vec{k})\, b^{\dagger}_{s,\vec{k}} b_{s,\vec{k}}, \tag{10}$$

where $E_s(\vec{k})$ is an exciton energy in its s-subband. $\vec{k} = \vec{q} + \vec{k}_{\perp}$ is the electron's quasimomentum in the plane, where $\vec{q}$ is the quasimomentum component along the SWCNT alignment and $\vec{k}_{\perp}$ is along the perpendicular direction to the plane. The exciton annihilation $b_{s,\vec{k}}$ and creation $b^{\dagger}_{s,\vec{k}}$ operators follow boson statistics and satisfy the following identities:

$$\begin{aligned} \left[b_{s,\vec{k}}, b^{\dagger}_{s',\vec{k}'}\right] &= \delta_{ss'}\, \delta_{\vec{k}\vec{k}'}, \\ \left[b_{s,\vec{k}}, b_{s',\vec{k}'}\right] &= \left[b^{\dagger}_{s,\vec{k}}, b^{\dagger}_{s',\vec{k}'}\right] = 0. \end{aligned} \tag{11}$$

A complication arises when we have not just a single SWCNT but an array of SWCNTs as they interact via dipole–dipole interaction. The Hamiltonian interaction can be written as [17]

$$\hat{H}_{\text{int}} = \frac{1}{2} \sum_{\vec{k}ss'}{}' V_{ss'}(\vec{k}) \left(b_{s,\vec{k}} + b^{\dagger}_{s,-\vec{k}}\right) \left(b_{s',\vec{k}} + b^{\dagger}_{s',-\vec{k}}\right), \tag{12}$$

where

$$V_{ss'}(\vec{k}) = \frac{2m^*\, \omega_p^2(k)}{N_{2D} N a \Delta}\, X(s)\, T_{yy}(\vec{k})\, X(s'), \tag{13}$$

is the interaction potential with $X(s)$ being the transition dipole associated with s-subband excitation. $\omega_p(k)$, N_{2D}, a, and $T_{yy}(\vec{k})$ are the intraband plasma oscillation frequency, the surface electron density, the lattice translation period, and the Fourier transform of the SWCNTs' dipole–dipole interaction, respectively. The total Hamiltonian of the system is the sum of Equations (10) and (12):

$$\hat{H} = \hat{H}_0 + \hat{H}_{\text{int}}, \tag{14}$$

which can be analytically diagonalized using the Bogoliubov–Valatin transformation as presented in Ref. [48].

For the SWCNT array, the dielectric response is anisotropic, which, along the direction perpendicular to the SWCNT alignment, remains constant with the value of the dielectric constant of the host medium, while along the SWCNT alignment, it is given as [17]

$$\frac{\varepsilon_{yy}(q,\omega)}{\varepsilon} = 1 - \frac{2\, f_{\text{CN}}\, \sigma_{yy}(q,\omega)}{f_{\text{CN}}\, \sigma_{yy}(q,\omega) + \mathrm{i}\, \frac{\omega\varepsilon}{8\pi q\, I_0(qR)\, K_0(qR)} \left(1 + \frac{\varepsilon_1 + \varepsilon_2}{q\varepsilon d}\right)}, \tag{15}$$

where $f_{\text{CN}} = N_{\perp} V_{\text{CN}} / V$ is the ratio of the SWCNTs' volume to the volume of the film. f_{CN} measures the sparseness of SWCNTs and/or thickness of the SWCNT film. f_{CN} has a lower value for larger inter-tube distance and/or larger value of thickness. f_{CN} satisfies $0 < f_{\text{CN}} \le \pi/4 < 1$, implying that the screening of the host dielectric medium is always present in an SWCNT array even at the very tightly packed SWCNT array. The screening leads to broadening excitonic and plasmonic peaks and enhances exciton–plasmon coupling. $I_0(qR)$ and $K_0(qR)$ in Equation (15) are the zeroth-order modified cylindrical Bessel functions. The dielectric response of the SWCNT array is not only the function of photon energy and SWCNT radius similar to a single SWCNT, but it is also a function of the volumetric fraction f_{CN}, permittivities ε_1, ε_2, and ε of the substrate, superstrate, and the dielectric medium holding the SWCNT array and quasi-momentum along the SWCNT alignment. As the ratio of the physical quantities is the same as the dimensionality, the dielectric response function expressed in Equation (15) is a dimensionless quantity.

4. Discussion

For $\vec{k} = 0$, the right side of Equation (15) equals unity, implying that the dielectric response of the SWCNT array equals the host dielectric medium's permittivity, which makes the dielectric response function isotropic. For the nonzero value of $\vec{k}$, or $\vec{q}$, the yy component of the dielectric response is a complex function. The $\varepsilon_{yy}(q,\omega)$ for two different composite films, one consisting of a periodic array of (9,0) SWCNTs and the other consisting of (10,0) SWCNTs, are presented in Figure 5. The SWCNT layers are embedded in a dielectric medium of thickness $d = 2.5R$ and relative permittivity $\varepsilon = 20$, where R is the radius of the respective film's SWCNT. The fractional volume of the CNT in both these composites is $f_{CN} = \pi/5$. The real part $\mathrm{Re}\,\varepsilon_{yy}$ represents the exciton refraction spectrum, the imaginary $\mathrm{Im}\,\varepsilon_{yy}$ represents the exciton absorption spectrum, and $-1/\mathrm{Im}\,\varepsilon_{yy}$ represents the plasmon response. The plasmon mode has a resonance at the point in which $\mathrm{Re}\,\varepsilon_{yy}$ changes sign from negative to positive, and $\mathrm{Im}\,\varepsilon_{yy}$ approaches zero. The plasmon response is positive for the entire energy range, indicating that energy loss is unavoidable. Even in the wide range of energy window (up to ∼4.0 eV), only one resonance is present for the (9,0) SWCNT film. However, one witnesses two resonances for the (10,0) SWCNT film for photon energy up to ∼2.6 eV. The plasmon resonance appears at a lower energy than the excitonic resonance for the (9,0) SWCNT film. The same is not true for the (10,0) SWCNT film. The plasmon spectrum is comparatively more pronounced in the (9,0) than in the (10,0) SWCNT film.

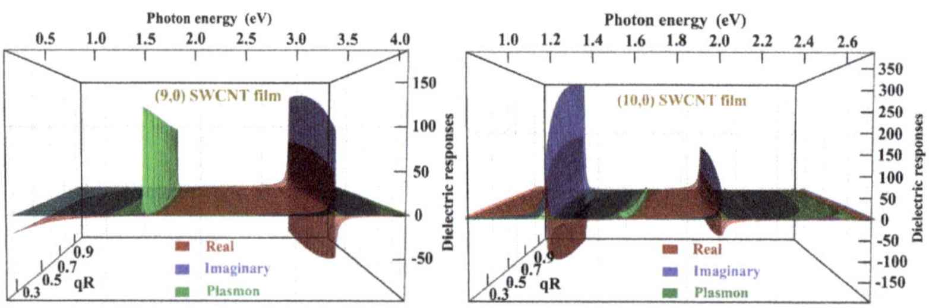

Figure 5. Dimensionless dielectric responses of a periodic array of (9,0) SWCNTs (**left panel**) and (10,0) SWCNTs (**right panel**) along their CNT axis as a function of momentum times the radius of the respective CNT qR and the photon energy (expressed in eV). The relative permittivity of the substrate, superstrate, and dielectric medium are taken as 1, 1, and 20, respectively. Only the first resonances of the response functions are shown for the (9,0) SWCNT film, while the first two resonances of the response functions are shown for the (10,0) SWCNT film. The fractional density parameter f_{CN} is $\pi/5$. The red-colored surface represents the $\mathrm{Re}\,\varepsilon_{yy}$, the blue-colored surface depicts the $\mathrm{Im}\,\varepsilon_{yy}$, and the negative of $\mathrm{Im}\,1/\varepsilon_{yy}$ is shown by the green-colored surface.

The real part $\mathrm{Re}\,\varepsilon_{yy}$ of the dielectric response function is negative for a large range of photon energy for both (9,0) and (10,0) SWCNT films, making both SWCNT films good candidates for hyperbolic metamaterials. The energy range, however, depends on the chirality of constituent SWCNTs. The semiconducting SWCNT films show hyperbolic metamaterial behavior in the near-infrared and visible spectrum. The metallic SWCNT films behave as hyperbolic metamaterials in low energy, including the microwave, in addition to the infrared and visible photon energy range as in semiconducting ones. The exciton–plasmon spectra of homogenous arrays of any metallic SWCNTs such as (12,0), (15,0), (18,0), etc. SWCNTs are expected to behave similarly to (9,0) SWCNT arrays. At the same time, resonance intensity and photon energy for resonances depend on the particular metallic SWCNT that the film consists of. The SWCNT films are made up of semiconducting SWCNT arrays such as (13,0), or (16,0), or (19,0), etc. SWCNTs have similar behavior to an array of (10,0) SWCNTs. The hyperbolic nature of the SWCNT array has been experimentally observed in various studies [49–52]. Pristine SWCNTs can be prepared in a controlled manner with desired chiral purity, density, and alignment as per their need in electronics [53,54]. This study opens new pathways for experimentalists to design and explore more on the SWCNT films discussed in this manuscript.

The exciton–plasmon coupling strength is stronger in the SWCNT array than in a single SWCNT. The inhomogeneity effect abruptly increases the exciton–plasmon coupling [15] for a mixture of SWCNTs of slightly different diameters. Comparing the dielectric responses for a single homogeneous array presented in Ref. [14], one can conclude that the SWCNT film containing more than a single array of identical SWCNTs has a more intense response function.

Substituting the expression of the volume of an isolated SWCNT and the composite volume, f_{CN} can be expressed as $f_{CN} = \pi R^2/d\Delta$, which depends on the thickness of the film and SWCNT sparseness. Choosing a fixed value of Δ, one can observe the effect of film thickness on the exciton–plasmon spectra, with varying values of f_{CN}. Note that f_{CN} is smaller for a larger value of d. Figure 6 shows a comparative study of dielectric responses for three different values of f_{CN} with fixed Δ.

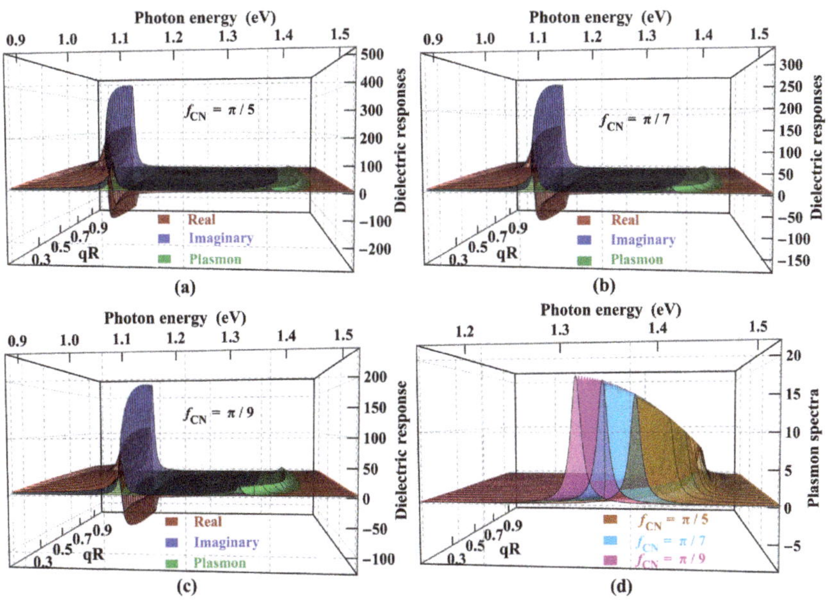

Figure 6. Dimensionless dielectric functions of a periodic array of (11,0) CNTs along the CNT axis as a function of momentum times radius of (11,0) CNT qR and the photon energy (expressed in eV). The relative permittivities of the substrate, superstrate, and dielectric medium are taken as 1, 1, and 15, respectively. Only the first resonances of the response functions are shown. Three different scenarios if the f_{CN} values, namely $\pi/5$ (**a**), $\pi/7$ (**b**), and $\pi/9$ (**c**), are taken. Figure (**d**) shows the comparison of plasmon spectra only for the cases chosen in (**a**–**c**).

As shown in Equation (15), the dielectric response function of the SWCNT array is electron quasi-momentum q-dependent. The minimum nonzero value that q can take is $2\pi/L$, where L is the length of an SWCNT. One can further explore how the length of an SWCNT plays a role in the minimum value of q. For $q = 2\pi/L$, the longitudinal component of the response functions reads

$$\frac{\varepsilon_{yy}(L',\omega)}{\varepsilon} = 1 - \frac{2\, f_{\mathrm{CN}}\, \sigma_{yy}(L',\omega)}{f_{\mathrm{CN}}\, \sigma_{yy}(L',\omega) + \mathrm{i}\, \frac{\omega\varepsilon L' R}{8\pi\, I_0(1/L')\, K_0(1/L')} \left(1 + \frac{\varepsilon_1+\varepsilon_2}{\varepsilon} \frac{L'R}{d}\right)}, \tag{16}$$

where $L' = L/(2\pi R)$ is the length of an SWCNT in terms of its circumference. See Figure 7 for the dielectric response function at the minimum nonzero value of momentum $q_{\mathrm{min}} = 2\pi/L$. The major contribution to the dielectric response comes from the q_{min} momentum. At the same time, one should not forget that $q = 2\pi n_x/L$ with $n_x > 1$ also has a nonzero contribution to the response function, which decays exponentially. The exciton and the plasmon resonances approach each other more closely for q_{min} than for the larger q. However, the intensity of each of the resonance peaks for q_{min} is smaller than the large q counterpart. Thus, the resonance intensity and exciton–plasmon coupling strength depend on the SWCNT length. The momentum q_{min} has a lower value for a longer SWCNT than the shorter one. As the CNT length increases, the plasmon resonance comes closer to the exciton resonance, strengthening the exciton–plasmon coupling.

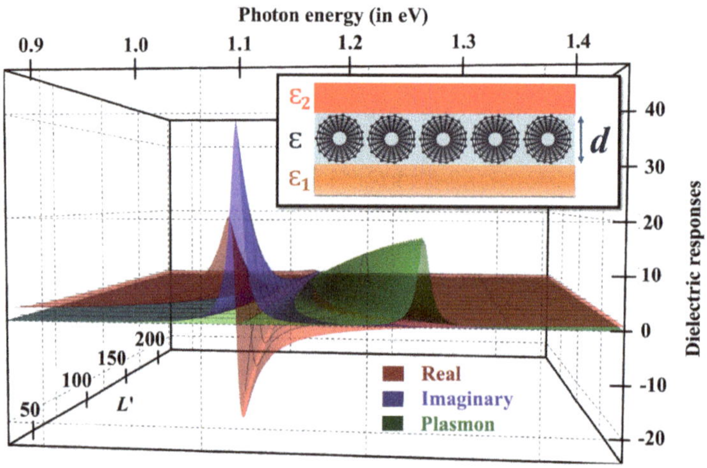

Figure 7. Dimensionless dielectric responses of an ultrathin array of (11,0) CNTs along the CNT axis as a function of photon energy (expressed in eV) and length of a CNT L' (expressed in terms of the circumference of an (11,0) CNT) for the mimimum nonzero momentum $q_{\mathrm{min}} = 2\pi/L$. The CNTs are tightly packed, and the thickness of the film equals the diameter of an (11,0) CNT, resulting in $f_{\mathrm{CN}} = \pi/4$. The inset shows the geometry related to the SWCNT film.

The resonance peaks become slightly less intense and appear at low photon energy with the increase in the radii of the SWCNTs, provided that the SWCNTs have the same chirality. For example, the film of the (10,0) SWCNT array has a more substantial resonance peak at higher photon energy than the film of the (13,0) SWCNT array. The theoretical model is based on an approximation that the host dielectric medium has larger relative permittivity than the substrate and superstrate. Thus, the substrate and superstrate should be chosen adequately based on the permittivity of the host dielectric material and vice versa. For example, if silicon or GaAs are taken as a host dielectric medium whose permittivities are, respectively, 11.7 and 13.1 [55], then the current study requires the substrate and superstrate to have relative permittivities of about 1 to 3, such as hexagonal boron nitride, fluorinated silica, etc. For high-dielectric host materials such as TiO_2, Ti_3C_2, and Ti_2C [56], one has a little more freedom of choice for the substrate and superstrate.

The SWCNT film is only a few nanometers thick, which warrants a brief discussion of its nanoscale effects and the advantages and disadvantages of such effects. At the nanoscale, conductivity can be tuned selectively with the chosen SWCNT of the desired chirality. One can use plasmonic resonance as a tool to enable surface plasmon-based sensing and optoelectronic applications. At the nanoscale, SWCNTs can sustain current densities much higher than copper. However, SWCNTs at the nanoscale have some limitations; as a result, using SWCNTs and their thin films in the current technology is not free from challenges. One of the significant challenges using the currently available synthesis methods is producing SWCNTs of a specific chirality and their scalability [57]. Although some claims on producing pure SWCNTs have been reported, some issues, such as removing amorphous carbon, washing out metallic catalysts after catalytic SWCNT synthesis, and yielding defect-free SWCNTs, remain [58]. As chirality control is usually an issue in SWCNT production, having a consistent film thickness, as this article proposed, needs extra work in the experimental setup, which may not only have an inhomogeneous mixing of SWCNTs of different chiralities, but also have tubular misalignment and inter-tube junction resistance due to phonon scattering. The phonon scattering at the inter-tube junction reduces the conductivity of the SWCNT film. In future studies, we will focus on the phonon scattering at inter-tube junctions and its effect on dielectric responses in the homogeneous SWCNT arrays and inhomogeneous arrays of SWCNTs of approximately the same diameters. Future work will concentrate on creating hybrid materials combining SWCNTs with other materials, such as graphene and polymers, in search of optoelectronic performance enhancements.

5. Conclusions

Using Green's function theory for many-body systems within the Matsubara frequency approach, we derived a closed form of mathematical expression for the dielectric response function of a periodic array of identical SWCNTs embedded in a dielectric medium with comparatively higher permittivities than substrates and superstates. The dielectric response functions of SWCNTs and its array are both anisotropic. The conducting nature of SWCNTs depends on the chirality resulting from turning graphene sheets into tubules. The SWCNT's photon energy and cylindrical tube's radius-dependent dynamical conductivity are calculated using the $\vec{k} \cdot \vec{p}$ method of the band structure theory. The dielectric response of the SWCNT film along the perpendicular direction of the SWCNT alignment is constant and matches the value of the host dielectric medium, while the same along the direction of SWCNT alignment takes a complex function, which depends on various parameters, namely photon energy, the radius of SWCNT, volumetric fraction of SWCNT f_{CN}, relative permittivity of the host medium of high dielectric constant, and the substrate and superstrate of low permittivities and dynamic conductivity of each SWCNT in the film. The expression we derived for the SWCNT film's dielectric response functions links the SWCNT film dielectric responses to the individual SWCNT's conductivity and the individual SWCNT's plasma frequency. The screening of the host dielectric medium is inevitable in an SWCNT array primarily because of its geometry.

The excitonic and plasmonic peaks come closer in the periodic array of SWCNTs than in an SWCNT, increasing the exciton–plasmon coupling. The real part $\mathrm{Re}\,\varepsilon_{yy}$ of the dielectric response function is negative for a vast range of photon energy in both cases of having an array of metallic and semiconducting SWCNTs, making them good candidates for hyperbolic metamaterials. This finding is consistent with a recent experimental observation of the hyperbolic metamaterial behavior for horizontally aligned films of SWCNTs. We revealed that the exciton–plasmon resonance peaks are closest to the minimum nonzero value of q, and the resonance intensity and exciton–plasmon coupling strength depend on the SWCNTs' length.

Funding: This research was funded by the 2024 Ralph E. Powe Junior Faculty Enhancement Awards, provided by Oak Ridge Associated Universities (ORAU).

Institutional Review Board Statement: Not applicable.

Informed Consent Statement: Not applicable.

Data Availability Statement: Data obtained in this research are included in this paper. Any raw data will be available by the author upon request.

Acknowledgments: The author gratefully acknowledges the support of the 2024 Ralph E. Powe Junior Faculty Enhancement Awards, provided by Oak Ridge Associated Universities (ORAU).

Conflicts of Interest: The author declares no conflicts of interest.

Abbreviations

The following abbreviations are used in this manuscript:

CNT	Carbon nanotube
SWCNT	Single-walled carbon nanotube
MWCNT	Multi-walled carbon nanotube
THz	Terahertz
1D	One-dimensional
2D	Two-dimensional
3D	Three-dimensional
TD	Trans-dimensional
TDM	Trans-dimensional material
ORAU	Oak Ridge Associated Universities

References

1. Radushkevich, L.V.; Lukyanovich, V.M. On the structure of carbon produced at thermal decomposition of carbon monoxide on an ironcontact. *J. Phys. Chem.* **1952**, *26*, 88–95.
2. Oberlin, A.; Endo, M.; Koyama, T. Filamentous growth of carbon through benzene decomposition. *J. Cryst. Growth* **1976**, *32*, 335–349. [CrossRef]
3. Dresselhaus, M.S.; Dresselhaus, G.; Eklund, P.C.; Chung, D.D.L. Lattice vibrations in graphite and intercalation compounds of graphite. *Mater. Sci. Eng.* **1977**, *31*, 141–152. [CrossRef]
4. Dresselhaus, M.S.; Dresselhaus, G.; Fischer, J.E. Graphite intercalation compounds: Electronic properties in the dilute limit. *Phys. Rev. B* **1977**, *15*, 3180.
5. Abrahamson, J.; Wiles, P.G.; Rhoades, B.L. Structure of carbon fibres found on carbon arc anodes. *Carbon* **1999**, *37*, 1873–1874. [CrossRef]
6. Nesterenko, A.M.; Kolesnik, N.F.; Akhmatov, Y.S.; Suhomlin, V.I.; Prilutskii, O.V. Osobennosti fazovogo sostava i struktury produktov vzaimodeistviya NiO i Fe_2O_3 s okis'yu ugleroda. *Izv. Akad. Nauk SSSR Seriya Met.* **1982**, 3, 12–17. Chemical Abstracts, 1982, v. 97, 201884t.
7. Tennent, H.G. Carbon Fibrils, Method for Producing Same and Compositions Containing Same. US Patent # 4663230, 5 May 1987.
8. Iijima, S. Helical microtubules of graphitic carbon. *Nature* **1991**, *354*, 56–58. [CrossRef]
9. Mintmire, J.W.; Dunlap, B.I.; White, C.T. Are fullerene tubules metallic? *Phys. Rev. Lett.* **1992**, *68*, 631–634. [CrossRef]
10. Iijima, S.; Ichihashi, T. Single-shell carbon nanotubes of 1-nm diameter. *Nature* **1993**, *363*, 603–605. [CrossRef]
11. Takakura, A.; Beppu, K.; Nishihara, T.; Fukui, A.; Kozeki, T.; Namazu, T.; Miyauchi, Y.; Itami, K. Strength of carbon nanotubes depends on their chemical structures. *Nat. Commun.* **2019**, *10*, 3040. [CrossRef]

12. Kempa, K.; Chura, R. Plasmons in Carbon Nanotubes. In *Low-Dimensional Systems: Theory, Preparation, and Some Applications*; NATO Science Series; Liz-Marzán, L.M., Giersig, M., Eds.; Springer: Dordrecht, The Netherlands, 2003; Volume 91.
13. Bondarev, I.V.; Meliksetyan, A.V. Possibility for exciton Bose-Einstein condensation in carbon nanotubes. *Phys. Rev. B* **2014**, *89*, 045414. [CrossRef]
14. Adhikari, C.M.; Bondarev, I.V. Optical response of ultrathin periodically aligned single-wall carbon nanotube films. *MRS Adv.* **2020**, *5*, 2685. [CrossRef]
15. Adhikari, C.M.; Bondarev, I.V. Controlled exciton–plasmon coupling in a mixture of ultrathin periodically aligned single-wall carbon nanotube arrays. *J. Appl. Phys.* **2021**, *129*, 015301. [CrossRef]
16. Keldysh, L.V. Coulomb interaction in thin semiconductor and semimetal films. *Engl. Transl. JETP Lett.* **1980**, *29*, 658.
17. Bondarev, I.V.; Adhikari, C.M. Collective Excitations and Optical Response of Ultrathin Carbon-Nanotube Films. *Phys. Rev. Appl.* **2021**, *15*, 034001. [CrossRef]
18. Liu, Y.; Zhang, J.; Liu, H.; Wang, S.; Peng, L.M. Electrically driven monolithic subwavelength plasmonic interconnect circuits. *Sci. Adv.* **2017**, *3*, e1701456. [CrossRef]
19. Gupta, B.D.; Pathak, A.; Semwal, V. Carbon-Based Nanomaterials for Plasmonic Sensors: A Review. *Sensors* **2019**, *19*, 3536. [CrossRef]
20. Duan, Q.; Liu, Y.; Chang, S.; Chen, H.; Chen, J.-H. Surface Plasmonic Sensors: Sensing Mechanism and Recent Applications. *Sensors* **2021**, *21*, 5262. [CrossRef]
21. Jiang, K. Chapter 4—Carbon Nanotubes for Displaying. In *Micro and Nano Technologies, Industrial Applications of Carbon Nanotubes*; Peng, H., Li, Q., Chen, T., Eds.; Elsevier: Amsterdam, The Netherlands, 2017; pp. 101–127. [CrossRef]
22. Singh, B.P.; Sikarwar, S.; Pandey, K.K.; Manohar, R.; Depriester, M.; Singh, D.P. Carbon Nanotubes Blended Nematic Liquid Crystal for Display and Electro-Optical Applications. *Electron. Mater.* **2021**, *2*, 466–481. [CrossRef]
23. Zecchi, S.; Cristoforo, G.; Piatti, E.; Torsello, D.; Ghigo, G.; Tagliaferro, A.; Rosso, C.; Bartoli, M. A Concise Review of Recent Advancements in Carbon Nanotubes for Aerospace Applications. *Micromachines* **2025**, *16*, 53. [CrossRef]
24. Serafinelli, C.; Fantoni, A.; Alegria, E.C.B.A.; Vieira, M. Hybrid Nanocomposites of Plasmonic Metal Nanostructures and 2D Nanomaterials for Improved Colorimetric Detection. *Chemosensors* **2022**, *10*, 237. [CrossRef]
25. Pyo, S.; Eun, Y.; Sim, J.; Kim, K.; Choi, J. Carbon nanotube-graphene hybrids for soft electronics, sensors, and actuators. *Micro Nano Syst. Lett.* **2022**, *10*, 9. [CrossRef]
26. Chaudhuri, K.; Alhabeb, M.; Wang, Z.; Shalaev, V.; Gogotsi, Y.; Boltasseva, A. Highly Broadband Absorber Using Plasmonic Titanium Carbide (MXene). *ACS Photonics* **2018**, *5*, 10. [CrossRef]
27. Hernández-Acosta, M.A.; Martines-Arano, H.; Soto-Ruvalcaba, L.; Martínez-González, C.L.; Martínez-Gutiérrez, H.; Torres-Torres, C. Fractional Thermal Transport and Twisted Light Induced by an Optical Two-Wave Mixing in Single-Wall Carbon Nanotubes. *Int. J. Therm. Sci.* **2020**, *147*, 106136. [CrossRef]
28. Obitayo, W.; Liu, T. A Review: Carbon Nanotube-Based Piezoresistive Strain Sensors. *J. Sens.* **2012**, *652438*. [CrossRef]
29. Kolahdouz, M.; Xu, B.; Nasiri, A.F.; Fathollahzadeh, M.; Manian, M.; Aghababa, H.; Wu, Y.; Radamson, H.H. Carbon-Related Materials: Graphene and Carbon Nanotubes in Semiconductor Applications and Design. *Micromachines* **2022**, *13*, 1257. [CrossRef]
30. Ando, T. The electronic properties of graphene and carbon nanotubes. *NPG Asia Mater.* **2009**, *1*, 17–21. [CrossRef]
31. Arash, B.; Wang, Q.; Varadan, V. Mechanical properties of carbon nanotube/polymer composites. *Sci. Rep.* **2014**, *4*, 6479. [CrossRef]
32. Liu, J. The Electrical Properties of Single-Walled Carbon Nanotubes. *J. Phys. Conf. Ser.* **2021**, *1748*, 052005. [CrossRef]
33. Nanot, S.; Hároz, E.H.; Kim, J.-H.; Hauge, R.H.; Kono, J. Optoelectronic Properties of Single-Wall Carbon Nanotubes. *Adv. Mater.* **2012**, *24*, 4977–4994. [CrossRef]
34. Rai, D.P.; Singh, Y.T.; Chettri, B.; Houmad, M.; Patra, P.K. A theoretical investigation of electronic and optical properties of (6,1) single-wall carbon nanotube (SWCNT). *Carbon Lett.* **2021**, *31*, 441–448. [CrossRef]
35. Preciado-Rivas, M.R.; Torres-Sánchez, V.A.; Mowbray, D.J. Optical absorption and energy loss spectroscopy of single-walled carbon nanotubes. *Phys. Rev. B* **2019**, *100*, 235429. [CrossRef]
36. Itas, Y.S.; Suleiman, A.B.; Ndikilar, C.E.; Lawal, A.; Razali, R.; Idowu, I.I.; Khandaker, M.U.; Ahmad, P.; Tamam, N.; Sulieman, A.; et al. Computational Studies of the Excitonic and Optical Properties of Armchair SWCNT and SWBNNT for Optoelectronics Applications. *Crystals* **2022**, *12*, 870. [CrossRef]
37. Nanot, S.; Thompson, N.A.; Kim, J.-H.; Wang, X.; Rice, W.D.; Hároz, E.H.; Ganesan, Y.; Pint C.L.; Kono, J. Single-Walled Carbon Nanotubes. In *Springer Handbook of Nanomaterials*; Springer Handbooks; Vajtai, R., Ed.; Springer: Berlin/Heidelberg, Germany, 2013.
38. Melchor, S.; Dobado, J.A. An algorithm for connecting two arbitrary carbon nanotubes. *J. Chem. Inf. Comput. Sci.* **2004**, *44*, 1639–1646. [CrossRef] [PubMed]
39. Johnson, J.E.; Speir, J.A. Quasi-equivalent Viruses: A Paradigm for Protein Assemblies. *J. Mol. Biol.* **1997**, *269*, 665–675. [CrossRef]

40. Dolafi Rezaee, M.; Dahal, B.; Watt, J.; Abrar, M.; Hodges, D.R.; Li, W. Structural, Electrical, and Optical Properties of Single-Walled Carbon Nanotubes Synthesized through Floating Catalyst Chemical Vapor Deposition. *Nanomaterials* **2024**, *14*, 965. [CrossRef]
41. Kosuda, K.; Bingham, J.; Wustholz, K.; Van Duyne, R.; Groarke, R. Nanostructures and surface-enhanced Raman Spectroscopy. In *Comprehensive Nanoscience and Nanotechnology*; Elsevier: Amsterdam, The Netherlands, 2016; pp. 117–152.
42. Vaghasiya, H.; Miclea, P.-T. Investigating Laser-Induced Periodic Surface Structures (LIPSS) Formation in Silicon and Their Impact on Surface-Enhanced Raman Spectroscopy (SERS). *Optics* **2023**, *4*, 538–550. [CrossRef]
43. Mahan, G.D. *Many-Particle Physics*, 3rd ed.; Kluwer Academic: New York, NY, USA, 2000.
44. Tasaki, S.I.; Maekawa, K.; Yamabe, T. π-band contribution to the optical properties of carbon nanotubes: Effects of chirality. *Phys. Rev. B* **1998**, *57*, 9301. [CrossRef]
45. Ando, T. Theory of Electronic States and Transport in Carbon Nanotubes. *J. Phys. Soc. Jpn.* **2005**, *74*, 777. [CrossRef]
46. Bondarev, I.V.; Woods, L.M.; Tatur, K. Strong exciton-plasmon coupling in semiconducting carbon nanotubes. *Phys. Rev. B* **2009**, *80*, 085407. [CrossRef]
47. Ando, T. Excitons in Carbon Nanotubes. *J. Phys. Soc. Jpn.* **1997**, *66*, 1066.
48. Adhikari, C.M.; Morris, D.M.; Noonan, T.W.; Neupane, T.; Lamichhane, B.R.; Gautam, B.R. Dispersion in Single-Wall Carbon Nanotube Film: An Application of Bogoliubov–Valatin Transformation for Hamiltonian Diagonalization. *Condens. Matter* **2023**, *8*, 53. [CrossRef]
49. Gao, W.; Doiron, C.F.; Li, X.; Kono, J.; Naik, G.V. Macroscopically Aligned Carbon Nanotubes as a Refractory Platform for Hyperbolic Thermal Emitters. *ACS Photonics* **2019**, *6*, 1602–1609.
50. Roberts, J.A.; Yu, S.-J.; Ho, P.-H.; Schoeche, S.; Falk, A.L.; Fan, J.A. Tunable Hyperbolic Metamaterials Based on Self-Assembled Carbon Nanotubes. *Nano Lett.* **2019**, *19*, 3131–3137. [CrossRef] [PubMed]
51. Schöche, S.; Ho, P.H.; Roberts, J.A.; Yu, S.J.; Fan, J.A.; Falk, A.L. Mid-IR and UV-Vis-NIR Mueller matrix ellipsometry characterization of tunable hyperbolic metamaterials based on self-assembled carbon nanotubes. *J. Vac. Sci. Technol. B* **2020**, *38*, 014015. [CrossRef]
52. Jerome, B.; Prasad, C.S.; Doumani, J.; Dewey, O.S.; Baydin, A.; Pasquali, M.; Kono, J.; Gao, W.; Alabastri, A.; Naik, G.V. Outcoupling hyperbolic modes from aligned carbon nanotube films. In *CLEO: QELS_Fundamental Science*; Optica Publishing Group: Washington, DC, USA, 2022.
53. Schlittler, R.R.; Seo, J.W.; Gimzewski, J.K.; Durkan, C.; Saifullah, M.S.M.; Welland, M.E. Single Crystals of Single-Walled Carbon Nanotubes Formed by Self-Assembly. *Science* **2001**, *292*, 1136–1139. [CrossRef]
54. Chen, Y.; Lyu, M.; Zhang, Z.; Yang, F.; Li, Y. Controlled Preparation of Single-Walled Carbon Nanotubes as Materials for Electronics. *ACS Cent. Sci.* **2022**, *8*, 1490–1505. [CrossRef] [PubMed]
55. Gao, J. *Heterojunction Bipolar Transistors for Circuit Design: Microwave Modeling and Parameter Extraction*; Wiley: Berlin, Germany, 2015.
56. Adhikari, C.M.; Dahal, D.; Kunwar, S.; Gautam, B.R. Ti_3C_2 and Ti_2C MXenes-Based Distributed Bragg Reflectors in Fabry Pérot Cavity's Resonance Tuning. *J. Electron. Mater.* **2025**, 1–10. [CrossRef]
57. Sakurai, S.; Inaguma, M.; Futaba, D.N.; Yumura, M.; Hata, K. A Fundamental Limitation of Small Diameter Single-Walled Carbon Nanotube Synthesis-A Scaling Rule of the Carbon Nanotube Yield with Catalyst Volume. *Materials* **2013**, *6*, 2633–2641. [CrossRef]
58. Daneshvar, F.; Chen, H.; Noh, K.; Sue, H.J. Critical challenges and advances in the carbon nanotube–metal interface for next-generation electronics. *Nanoscale Adv.* **2021**, *3*, 942–962. [CrossRef]

photonics

MDPI

Communication

Compact Reflective Metasurface: Production of Broadband Vortex Beams in Millimeter Waves

Asad Khan [1], Jinling Zhang [1,*], Muhammad Ishfaq [2], Ibrar Ahmad [3], Shahbaz Khan [1] and Kamlesh Kumar Soothar [1]

1 School of Electronic Engineering, Beijing University of Posts and Telecommunications, Beijing 100876, China; asadkhan@bupt.edu.cn (A.K.)
2 College of Electrical and Information Engineering, Lanzhou University of Technology, Lanzhou 730050, China
3 Department of Computer Science, University of Peshawar, Peshawar 25120, Pakistan
* Correspondence: zhangjl@bupt.edu.cn

Abstract: A low-profile reflectarray has been designed in the Ka-band to efficiently generate wideband orbital angular momentum (OAM) vortex beams. The proposed design employs a reflective phase-shifting patch etched onto a dielectric substrate, featuring a three-square loop structure intersected by two transverse dipoles. This unit cell achieves a 440° phase shift at 30 GHz with a minimal magnitude loss of (−0.25 dB), enabling high-efficiency reflectarray performance. The OAM vortex beam supports high-order phase distributions ($l = +1, +2, +3, +4$) modes, though fabrication and experimental validation focused on the +1 mode. Measurements confirm that the reflectarray produces a high-purity OAM vortex beam for +1 mode, covering the operational frequency range from 27 to 39 GHz, and achieving a 40% bandwidth with a peak gain of 23.39 dBi at 33 GHz and an aperture efficiency of 17.38%. These results demonstrate the ability of the reflectarray to produce broadband directive OAM beams with robust performance, making it ideal for Ka-band communication systems.

Keywords: vortex beam; metasurface; orbital angular momentum (OAM); aperture efficiency; wideband bandwidth

1. Introduction

Spectrum resources are becoming increasingly crowded as wireless communication technology develops [1]. Vortex beams carrying OAM have garnered significant attention in communication research due to their potential to increase transmission capacity [2]. In the microwave range, vortex waves are produced using a variety of techniques, including holographic substrates [3], spiral-phase plates [4], and array antennas [5]. Another approach in super-oscillation-based lenses offers the potential for high-resolution imaging with reduced sidebands, enabling more effective electromagnetic manipulation and improved wavefront control [6]. Vortex waves are identified by their spiral-shaped phase wavefronts. However, these techniques frequently encounter difficulties with prominent patterns or intricate feed networks. Reflective metasurfaces provide an alternative to these drawbacks in producing OAM vortex waves. Electromagnetic waves can have their phase wavefront, amplitude, frequency, and polarization state precisely controlled using metasurfaces [7]. Based on distinct concepts, various techniques can regulate the phase of reflected electromagnetic waves. The first technique, the propagation phase (or resonant phase), works with circular and linear polarized waves. It modifies the element's geometry to vary the phase. However, because of the resonant part's inherent dispersion, OAM generators that use this method frequently have narrow bandwidths [8]. Another method, the Pancharatnam–Berry (PB) phase, rotates the element to alter the phase. This method's generators typically offer broadband features, although they are limited to CP waves [9,10]. The detour phase method generates vortex waves but is limited by narrow bandwidth and compatibility only with linearly polarized (LP) waves, restricting their versatility in broader applications [11]. An independent LHCP- and RHCP-controlled translational transmission metasurface (TM) for dual-circularly polarized vortex beam generation is presented. Measurements confirm their efficacy; however, the challenges are polarization sensitivity and fabrication complexity [12]. To improve the performance of modern communication systems, a 1-bit reflecting RIS is proposed for the Ka-band, utilizing triangular patches and active switching elements. However, the 1-bit resolution limits its precision, and using active components could increase losses and power consumption [13]. Thermally switchable reflecting metasurfaces were created using Z-shaped resonators. Because of these resonators, the metasurface can produce numerous OAM (orbital angular momentum) modes by adjusting its properties in response to temperature variations. This means that the metasurface is adaptable for a range of applications because it can switch between different OAM beams by simply adjusting its temperature, which can limit stability and control in dynamic environments [14]. A wideband circularly polarized OAM vortex beam generator with $l = 1$ mode is presented using a reflectarray. With steady phase responses and sound polarization isolation, it is restricted to specific polarization modes (CP) [15]. A CCTA for OAM vortex wave generation with a double C-shaped grating metasurface is presented, which achieves sound transmission and 360° phase coverage. Its performance is validated through measurements, although issues like surface distortion and fabrication complexity still exist [16]. A compact system is produced by employing high-order modes to transform a single antenna element into an array. At 2.4 GHz, a patch antenna produces an OAM beam that is circularly polarized in the right direction and has a topological charge of OAM $l = -1$. Still, high-order modes often lead to more significant phase errors and reduced mode purity [17]. A spider-shaped unit cell is designed to generate OAM modes using a single-layer reflector that maintains high mode purity. The unit cell's multi-resonance characteristics enable a linear reflection phase response, allowing for broad bandwidth OAM generation, but complexity arises with this in unit cell design and fabrication [18]. A dual-polarized reflectarray is proposed for generating dual OAM beams with minimal inter-band interaction. Simulations and measurements confirm their feasibility, but there are issues with element coupling, bandwidth limitations, and the absence of dynamic beam control [19]. The compact reflectarray metasurface has a planar design, stability, and polarization insensitivity. Though it has potential for development, its limited beam coverage and low gain make it unsuitable for applications requiring high-performance beam shaping or wide-area coverage [20]. A compact cascaded metasurface system is introduced for efficient OAM vortex wave synthesis with customized SAM and polarization conversion. Its design enhances integration and enables multi-mode OAM generation, but challenges remain in alignment sensitivity and fabrication complexity [21]. The design integrates the reflection polarization conversion metasurface with the propagation phase and employs a low Q-factor C-shaped element. In addition to supporting linear and circularized polarized waves, the recommended OAM generator exhibits extensive bandwidth coverage in the microwave range, spanning from 12 to 18 GHz (40%) operations. Still, a broader bandwidth can increase the overall susceptibility to phase errors and losses [22].

Although these metasurfaces are innovative and effective in their particular domains, they face significant challenges in high gain, operational bandwidth, aperture efficiency, flexibility, and the cost and complexity of their design and fabrication processes. Overcoming these limitations is critical, and future research should focus on improving aperture efficiency and gain, achieving wider bandwidths, enhancing polarization versatility, and simplifying manufacturing processes to achieve practical and cost-effective deployment in real-world applications. This paper introduces a phase-shifting element design to enhance the mode purity, gain, aperture efficiency, and operational bandwidth of an OAM reflectarray antenna. The aperture phase distribution of the reflectarray vortex beam is precisely controlled using a variable-size adjustment technique. At the center frequency of 30 GHz, the proposed unit cell shows a low magnitude loss of −0.25 dB and achieves a phase range of 440°. Its symmetric structure enables dual-polarization functionality. Simulation results confirm that the reflectarray achieves improved mode purity, high aperture efficiency, and significant gain across a wide frequency range.

2. Element Design and Analysis

The dimensions of the reflectarray element are shown in Figure 1a shows the front view of the element, highlighting its geometric configuration and planar structure, and Figure 1b shows the side view of the element, highlighting the layer stack-up, including metallic patch, the dielectric substrate, an air gap, and a ground plane. The element has a periodicity (P) of 5 mm ($\lambda/2$ at 30 GHz) to ensure effective interaction with electromagnetic waves. The air gap (H_1) and the substrate thickness (H) are both 0.5 mm, contributing to the overall phase response and ensuring structural stability. The parameters are defined as follows: the patch width (W_p) is 0.28 mm, $L = M - W_p$, L_p is an adjustable parameter used to achieve the desired phase shifts, and $M = L_p - W_p$. The elements are fabricated on an FR4 substrate with a relative permittivity of $\varepsilon_r = 2.65$ and a low loss tangent ($\tan\delta = 0.001$), supporting compact element design while maintaining wide bandwidth. The moderate dielectric constant ensures efficient impedance matching and phase accuracy. At the same time, the low-loss tangent minimizes dielectric losses, preserving signal integrity and efficiency at high frequencies, making FR4 a suitable choice for Ka-band OAM vortex beam generation.

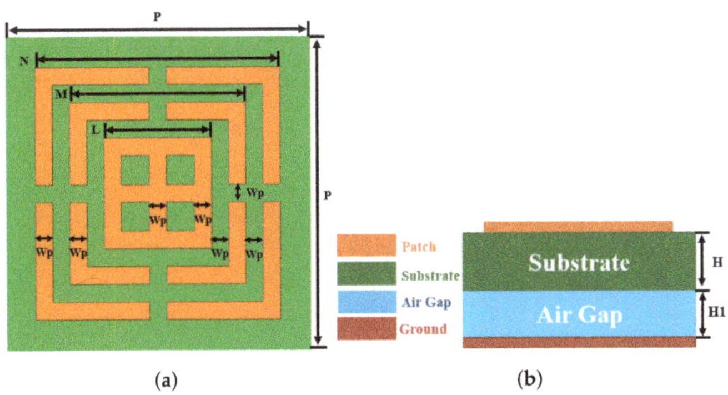

Figure 1. Unit cell geometry: (**a**) front view. (**b**) side view.

The electromagnetic properties of the reflectarray element are analyzed using ANSYS HFSS 2023 R2 software. A comprehensive parametric analysis is performed to optimize the element dimensions by changing the value of A from 3.5 mm to 5 mm. Figure 2 shows the reflective phase and the magnitude loss of the unit cell dependent on the parameter L_p at the center frequency. The analysis reveals that varying L_p allows the unit cell's phase response to be extended up to 440° with a −0.25 dB reflection loss. Therefore, the proposed single-layer design achieves a reflective phase response greater than a full

2π cycle, significantly enhancing the element's linear phase bandwidth. The unit cell bandwidth is determined by [23].

$$\text{BW} = \frac{2(f_u - f_l)}{f_u + f_l}, \quad (1)$$

where f_u indicates the highest frequency where the system operates effectively and f_l indicates the lowest frequency where the system operates effectively. Figure 3 shows the phase and magnitude loss of the element at various frequencies. The phase response curves at various frequencies are nearly symmetrical and show good linearity, a key feature for wideband reflectarray design. So, it is essential to consider the reflecting phases and magnitudes of the element under normal and oblique wave incidences. Figure 4 shows the reflective phases and magnitudes of the element at various incidence angles. The differences in reflective phases across these angles are minimal. As a result, the reflective phase under normal incidence serves as the foundation for the reflectarray design.

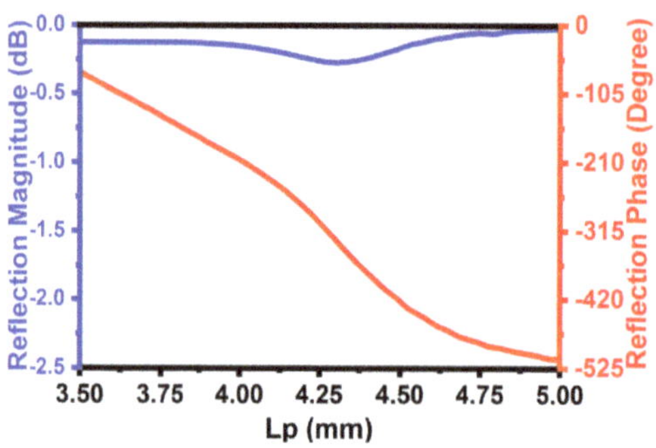

Figure 2. Unit cell reflective phase and magnitude.

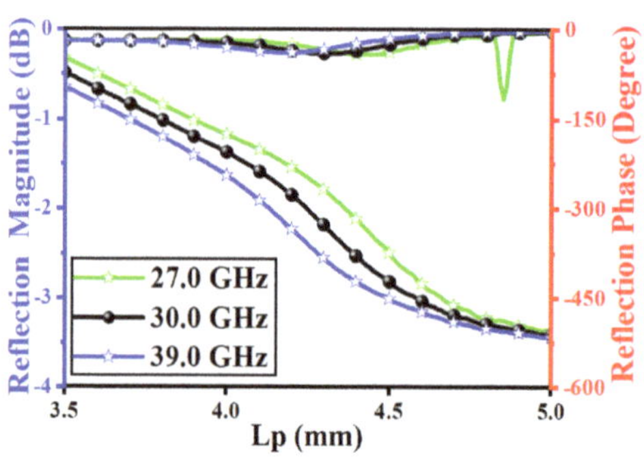

Figure 3. Unit cell reflective phase and magnitude at different frequencies.

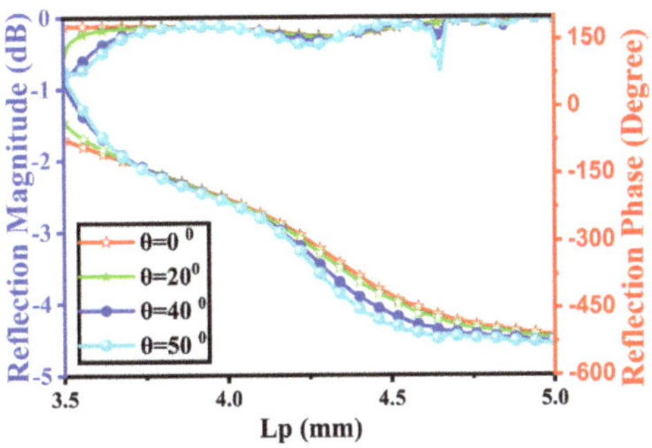

Figure 4. Unit cell reflective phase and magnitude at different incident angles.

3. Advanced Reflectarray Design and Evaluation Techniques

The geometry of the antenna goes through a structured process. First, element simulations are performed to obtain the phase-shifting curve. Then, we select the focal diameter ratio (F/D) and the feed position based on beam width and aperture size. After that, we derive the compensated phase distribution using the aperture phase optimization. Finally, we calculate the geometric parameters of the reflectarray by converting the compensated phase into unit geometry with the help of the phase-shifting curve. We can fine-tune the reference phase to optimize the array beam performance. This method designs a square-shaped $10\lambda \times 10\lambda$ dielectric reflectarray consisting of 20×20 elements at the Ka-band. The overall design of the reflectarray antenna is illustrated in Figure 5.

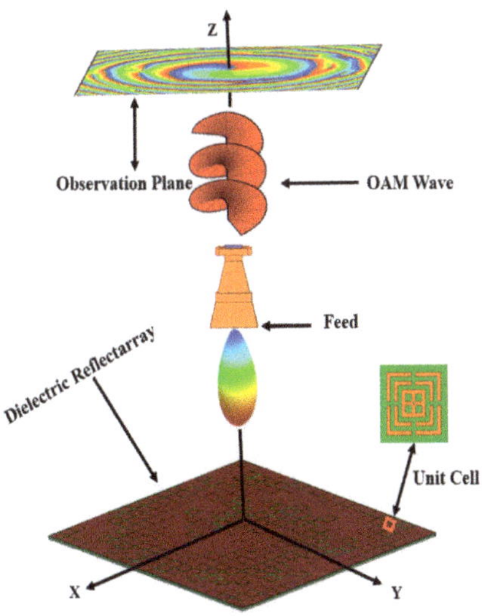

Figure 5. Design of OAM generating reflective surface.

The structure comprises $A \times B$ elements, which are explained by a horn antenna for operation. Here, A and B represent the number of elements along the X and Y axes. The reflective phase for each ABth reflectarray element, required to produce an OAM vortex wave, is determined as follows [24]:

$$\delta(A,B) = \frac{2\pi}{\lambda}\left(\sqrt{x^2+y^2+F^2} - F\right) + l \cdot \arctan\left(\frac{y}{x}\right) \quad (2)$$

Here, F is the distance between the geometric center of the reflecting metasurface and the phase center of the horn, λ represents the free-space wavelength, and (x, y) is the unit elements' location coordinates. The variable l represents the preferred OAM mode number. The aperture efficiency depends on the feed distance, and the reflectarray's feed source is a standard waveguide horn. Reducing F can reduce spillover loss but also reduce the reflectarray's illumination efficiency. To maximize gain, an optimal value of F is established at 3λ. The designed unit structure's phase shift is sufficient to generate a reflectarray that produces OAM beams with the required mode and beam direction. To achieve OAM modes ($l = +1, +2, +3, +4$) at 30 GHz, a software-based approach was used to calculate the overall phase distributions for the reflectarray for all four modes.

4. Simulated and Measured Results

The simulated phase and magnitude of the electric field produced by the proposed reflectarray are shown in Figure 6. In the reflection spectra, the phase patterns show a complete 2π phase shift around the reflectarray's center. In the meantime, the magnitude distribution forms a characteristic doughnut shape with a central singularity, a key feature of orbital angular momentum (OAM) vortex beams, confirming the predicted behavior of the OAM modes ($l = +1, +2, +3, +4$). These results verify that the reflectarray successfully produces the desired OAM vortex beam and demonstrates its ability to radiate an OAM vortex wave efficiently in the specified mode.

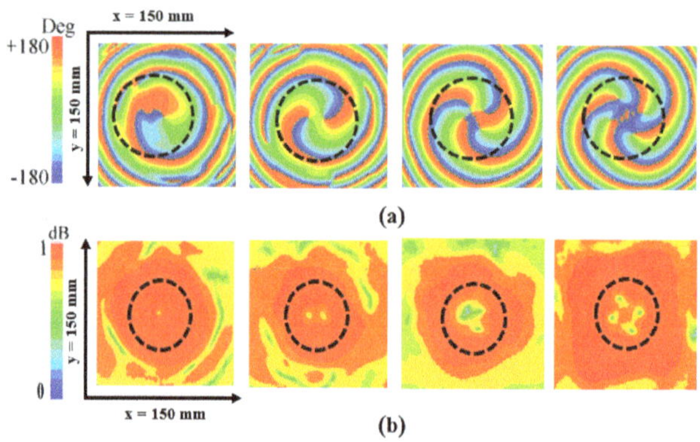

Figure 6. Simulated E-field distribution of modes ($l = +1, +2, +3, +4$) at 30 GHz. (**a**) Phase distribution. (**b**) Magnitude distribution.

The mode purity of each mode is shown in Figure 7. The mode purity was evaluated using a numerical Fourier transform of the aperture phase function, as defined in [25] below, to validate the accuracy of the proposed OAM waves:

$$P_{l_q} = \frac{1}{2\pi} \int_0^{2\pi} \Psi(\psi) e^{-kl_q\psi}\, d\psi \quad (3)$$

$$\text{Purity} = \frac{|P_{l_q}|}{\sum_{w=-\infty}^{\infty} |P_{l_w}|} \quad (4)$$

At an operating frequency of 30 GHz, the results of the simulations indicate impressive mode purities: 97% for mode +1, 94% for mode +2, 88% for mode +3, and 84% for mode +4. These confirm the generation of OAM waves for the modes ($l = +1, +2, +3, +4$). Simulated phase and magnitude responses from 27 to 39 GHz are presented in Figure 8. Furthermore, the purity of the mode consistently remains above 78% for the +1 mode across a frequency range, highlighting the wideband performance of the proposed design.

This demonstrates the design's ability to generate high-purity OAM waves over a wide range, as seen in Figure 9. Orbital angular momentum (OAM) waves are inherently resilient to feed obstructions due to their helical phase fronts, which distribute phase and radiation uniformly across the wave. This feature reduces the impact of localized disturbances on mode purity, as the spiral structure effectively preserves the wave's overall phase coherence and energy integrity. As a result, the generated OAM modes maintain high purity and performance, ensuring the precision and reliability of the system in advanced communication applications.

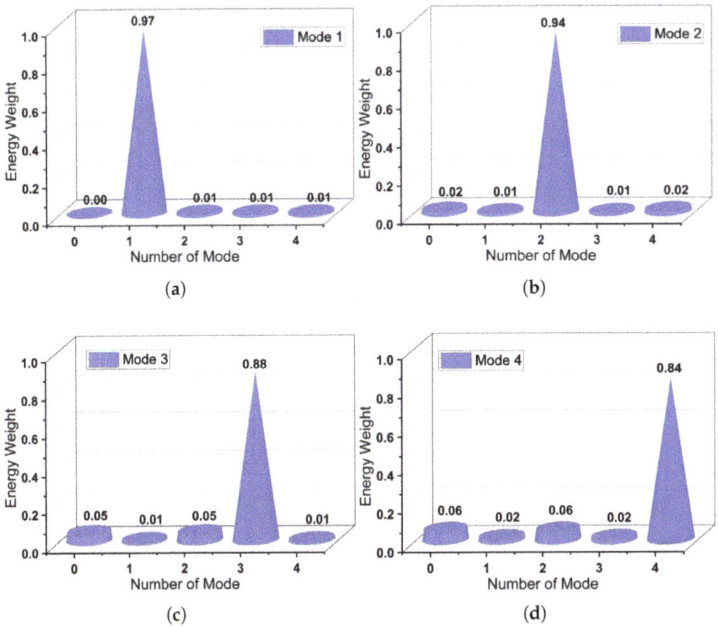

Figure 7. Simulated OAM modes purity for the proposed reflectarray metasurfaces at 30 GHz. (**a**) $l = +1$. (**b**) $l = +2$. (**c**) $l = +3$. (**d**) $l = +4$.

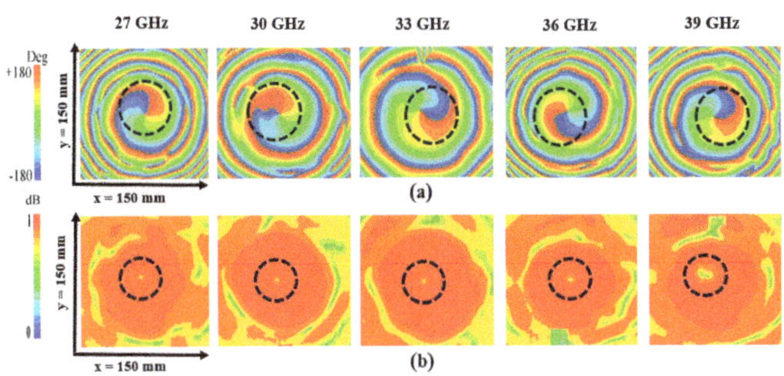

Figure 8. Simulated E-field response of the reflectarray for mode $l = +1$ at different frequencies: (**a**) phase; (**b**) magnitude.

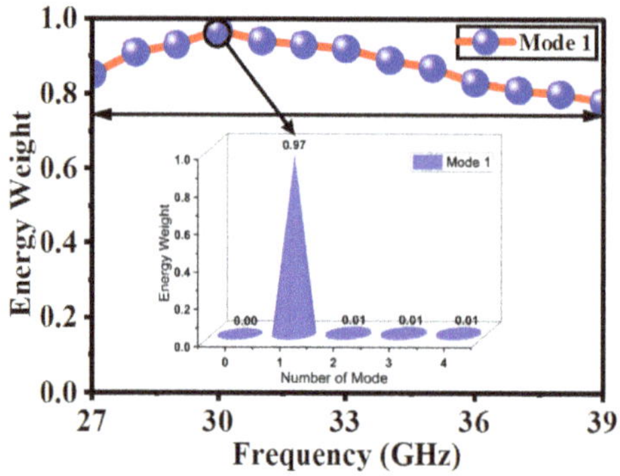

Figure 9. Reflectarray-simulated OAM mode purity for $l = +1$ over 27–39 GHz spectrum.

Figure 10 shows the fabricated reflective metasurface designed with OAM mode $l = +1$. (a) A perspective view highlights the overall structure, while (b) a close-in view shows the unit cell patterns and fabrication intricacies. The prototype and measurement setup is illustrated in Figure 11. Measured phase and magnitude responses from 27 to 39 GHz are presented in Figure 12, showing how well the proposed design performs over a wide range of frequencies. The measured results exhibit a single-arm spiral phase distribution, typical of orbital angular momentum (OAM) beams, and an intensity profile displaying the characteristic doughnut-shaped pattern. This indicates that the reflectarray continuously generates an excellent OAM beam over the entire frequency spectrum in the +1 mode. The results confirm that the design is suitable for wideband OAM beam-generating applications, as it maintains steady performance with decisive phase and magnitude characteristics.

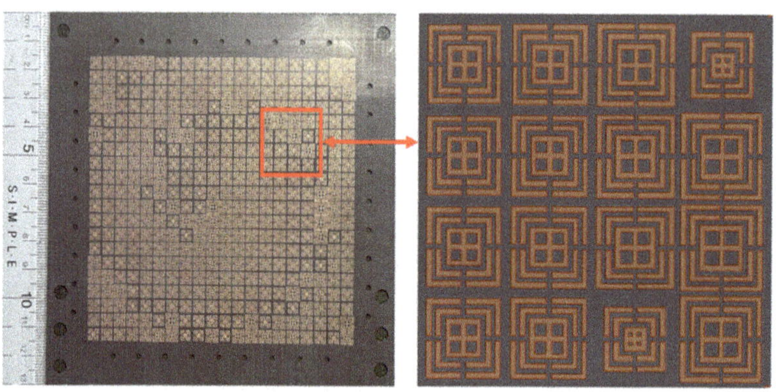

(a) (b)

Figure 10. Fabricated model of reflective metasurface with OAM mode $l = +1$: (**a**) overall perspective view and (**b**) close-up view.

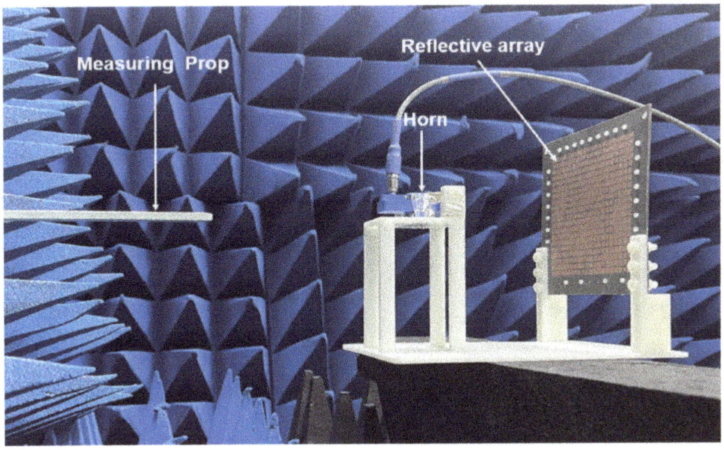

Figure 11. The measurement setup for OAM-generated reflective surface.

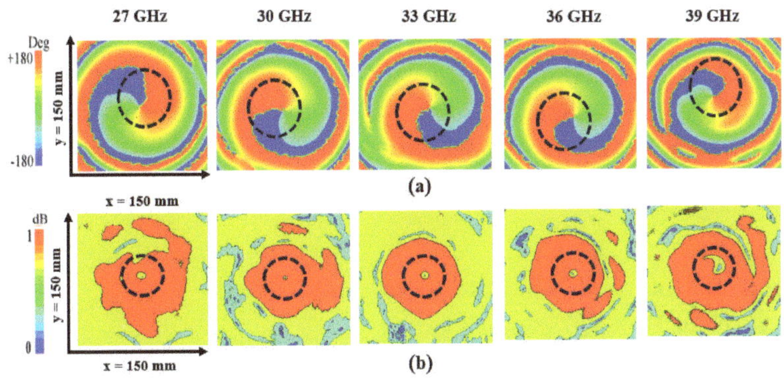

Figure 12. Measured E-field response of the reflectarray for mode $l = +1$ at different frequencies: (**a**) phase; (**b**) magnitude.

Normalized radiation patterns at 30 GHz, derived from measurements and simulations, are presented in Figure 13a,b. These patterns exhibit distinct conical beams in both H-plane and E-plane, characterized by a pronounced null at the center. The 2D radiation patterns show the significant orthogonal planes; the center null remains consistent across the E-plane and the H-plane at the operating frequency.These results confirm the efficient and reliable generation of the OAM +1 mode.

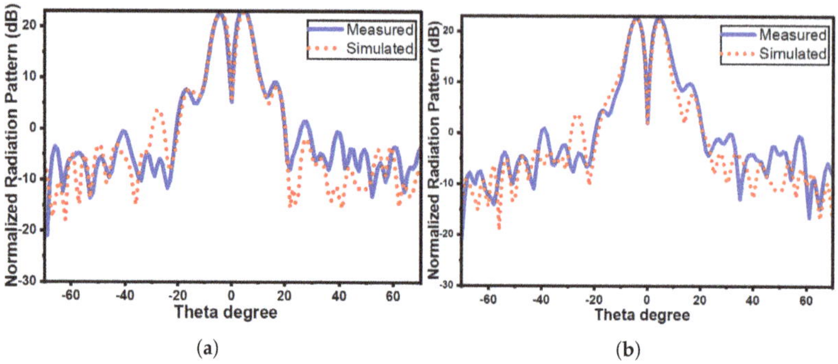

Figure 13. Analysis of simulated and measured 2-D radiation pattern at the center frequency: (**a**) H-plane; (**b**) E-plane.

The measured gain and aperture efficiency of the proposed reflectarray at various frequencies are presented in Figure 14. The reflectarray shows an exceptional performance at 33 GHz, achieving a peak gain of 23.39 dBi. At this gain, the equivalent aperture efficiency is 17.38%, indicating effective conversion of the available aperture area into radiated power. Furthermore, the reflectarray achieves a wide OAM bandwidth of 40%, covering a wide range of frequencies from 27 to 39 GHz. This extensive bandwidth highlights the reflectarray's ability to sustain consistent OAM mode generation and excellent performance across a significant portion of the spectrum, making it suitable for various applications within this range.

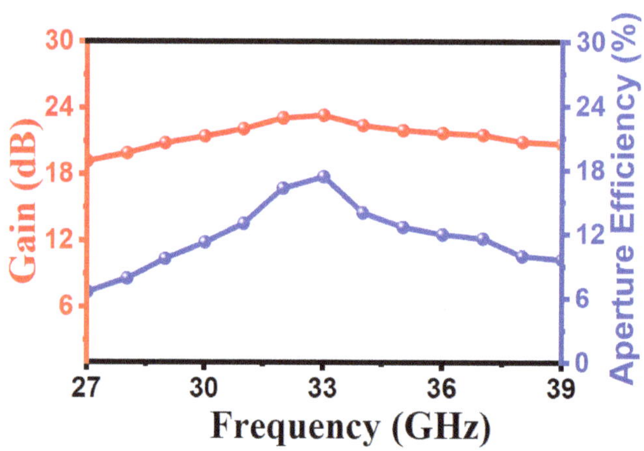

Figure 14. Measured gain and aperture efficiency at different frequencies.

Table 1 shows the performance of the proposed OAM reflective metasurface compared to state-of-the-art OAM antennas. Previous studies, such as [15] (10 GHz), [18] (5.75 GHz), [20] (5.8 GHz), and [26] (10 GHz), have primarily focused on generating OAM beams in the microwave range. In contrast, this work advances OAM technology into the millimeter-wave spectrum (27–39 GHz) to meet the growing demand for higher bandwidth and resolution in applications such as 5G/6G networks and radar systems. While the authors of [22] reported a 40% bandwidth at 15 GHz, the efficiency and gain remain lower than this design's. Similarly, ref. [27], operating at 30 GHz, targeted millimeter-wave frequencies and achieved a bandwidth of 21.7% and a peak gain of 20.5 dBi, but the profile was very high. The proposed reflectarray offers a 40% bandwidth and a peak gain of 23.39 dBi while maintaining an aperture efficiency of 17.78%, ensuring stable and efficient OAM beam generation across the Ka-band. Additionally, its low-profile structure provides a compact and practical solution for high-frequency communication and sensing applications by enhancing phase stability and mode purity. The proposed metasurface effectively bridges the gap between microwave and millimeter-wave OAM technologies, making it well-suited for next-generation wireless and radar systems.

Table 1. Performance comparison: This work vs. existing OAM antennas.

Ref.	Frequency	Antenna Type	Element Phase	Profile	Mode	Mode Purity	OAM BW	Gain dBi	Aperture Efficiency
[15]	10 GHz	Reflectarray	400°	$0.13\lambda_0$	1	N/A	20%	19.9	7.2%
[18]	5.75 GHz	Reflectarray	750°	$0.12\lambda_0$	1	50%	N/A	N/A	N/A
[20]	5.8 GHz	Reflectarray	377°	$0.11\lambda_0$	1	N/A	N/A	15.4	22.6%
[22]	15 GHz	Reflectarray	360°	$0.18\lambda_0$	1	N/A	40%	18.42	N/A
[26]	10 GHz	Reflectarray	360°	$0.10\lambda_0$	1	N/A	40%	20	14.35%
[27]	30 GHz	Reflectarray	360°	$0.5\lambda_0$	1	62.6%	21.7%	20.5	N/A
This Work	30 GHz	Reflectarray	440°	$0.10\lambda_0$	1	78%	40%	23.39	17.78%

N/A: not available; BW: bandwidth.

5. Conclusions

This work has developed a compact, low-profile reflectarray metasurface for wideband OAM wave generation in the Ka-band. The innovative unit cell design, comprising square loops with dipole cuts, achieves a 440° phase shift with negligible magnitude loss, enabling efficient reflection and OAM beam synthesis. Experimental validation for the +1 mode demonstrates exceptional performance, including a 40% bandwidth (27–39 GHz), a peak gain of 23.39 dBi, and an aperture efficiency of 17.38%. The proposed reflectarray's high gain, wide bandwidth, and excellent mode purity position make it a promising solution for advanced Ka-band applications, such as 5G/6G millimeter-wave communications, satellite links, and radar systems requiring multiplexed OAM channels. Future work will explore reconfigurable unit cells with tunable materials like liquid crystals, varactor diodes, and graphene-based metasurfaces to enhance the reflectarray's adaptability. These advancements could enable real-time OAM beam tuning, frequency agility, and adaptive mode selection, enhancing its versatility for wireless communication applications.

Author Contributions: Investigation, A.K. and M.I.; confirmation, A.K., J.Z., I.A., and K.K.S.; software, A.K., K.K.S., and M.I.; methodology, A.K. and S.K.; formal analysis, A.K., M.I., J.Z., and I.A.; data curation, A.K. and M.I.; writing—original draft preparation, A.K.; writing—review and editing, A.K., J.Z., and M.I.; supervision, J.Z.; funding acquisition, S.K. and A.K. All authors have read and agreed to the published version of the manuscript.

Funding: This work was supported by the National Natural Science Foundation of China (NSFC) under Project 62271063.

Institutional Review Board Statement: Not applicable.

Informed Consent Statement: Not applicable.

Data Availability Statement: Data are contained within the article.

Acknowledgments: The authors are thankful to the Beijing Key Laboratory of Space Ground Interconnection and Convergence, Beijing University of Posts and Telecommunications (BUPT), for providing the fabrication and measurement setup.

Conflicts of Interest: The authors declare no conflicts of interest.

References

1. Essiambre, R.-J.; Kramer, G.; Winzer, P.J.; Foschini, G.J.; Goebel, B. Capacity limits of optical fiber networks. *J. Light. Technol.* **2010**, *28*, 662–701. [CrossRef]
2. Kawaguchi, H.; Kubo, S.; Nakamura, H. Orbital angular momentum of vortex fields in corrugated cylindrical waveguide hybrid mode. *IEEE Microw. Wirel. Technol. Lett.* **2022**, *33*, 118–121. [CrossRef]
3. Huang, K.; Liu, H.; Restuccia, S.; Mehmood, M.Q.; Mei, S.-T.; Giovannini, D.; Danner, A.; Padgett, M.J.; Teng, J.-H.; Qiu, C.-W. Spiniform phase-encoded metagratings entangling arbitrary rational-order orbital angular momentum. *Light. Sci. Appl.* **2018**, *7*, 17156. [CrossRef]
4. Hui, X.; Zheng, S.; Hu, Y.; Xu, C.; Jin, X.; Chi, H.; Zhang, X. Ultralow reflectivity spiral phase plate for generation of millimeter-wave OAM beam. *IEEE Antennas Wirel. Propag. Lett.* **2015**, *14*, 966–969. [CrossRef]
5. Liu, K.; Cheng, Y.; Li, X.; Jiang, Y. Passive OAM-based radar imaging with single-in-multiple-out mode. *IEEE Microw. Wirel. Compon. Lett.* **2018**, *28*, 840–842. [CrossRef]
6. Zheng, P.; Zhu, Z.; Pei, X.; Wu, Q.; Liang, H.; Chen, Y.; Li, J.; Xie, X. Compound super-oscillation lens for reflective confocal imaging. *Opt. Lasers Eng.* **2023**, *166*, 107572. [CrossRef]
7. Wang, Y.; Yuan, Y.; Yang, G.; Ding, X.; Wu, Q.; Jiang, Y.; Burokur, S.N.; Zhang, K. Perfect control of diffraction patterns with phase-gradient metasurfaces. *ACS Appl. Mater. Interfaces* **2022**, *14*, 16856–16865. [CrossRef]
8. Yu, S.; Li, L.; Shi, G.; Zhu, C.; Shi, Y. Generating multiple orbital angular momentum vortex beams using a metasurface in radio frequency domain. *Appl. Phys. Lett.* **2016**, *108*, 242401. [CrossRef]
9. Yuan, Y.; Sun, S.; Chen, Y.; Zhang, K.; Ding, X.; Ratni, B.; Wu, Q.; Burokur, S.N.; Qiu, C.-W. A fully phase-modulated metasurface as an energy-controllable circular polarization router. *Adv. Sci.* **2020**, *7*, 2001437. [CrossRef]
10. Akram, Z.; Li, X.; Qi, Z.; Aziz, A.; Yu, L.; Zhu, H.; Jiang, X.; Li, X. Wideband vortex beam reflectarray design using quarter-wavelength element. *IEEE Antennas Wirel. Propag. Lett.* **2019**, *18*, 1458–1462. [CrossRef]
11. Zhang, K.; Wang, Y.; Burokur, S.N.; Wu, Q. Generating dual-polarized vortex beam by detour phase: From phase gradient metasurfaces to metagratings. *IEEE Trans. Microw. Theory Tech.* **2021**, *70*, 200–209. [CrossRef]
12. Wang, Z.; Zhou, D.; Liu, Q.; Yan, M.; Wang, X. Dual-mode vortex beam transmission metasurface antenna based on linear-to-circular polarization converter. *Opt. Express* **2023**, *31*, 35632–35643. [CrossRef]
13. Khan, A.; Zhang, J.; Ishfaq, M.; Bilal, H.M.; Khan, S.; Soothar, K.K. Design and analysis of a 1-bit reflective metasurface for reconfigurable intelligent surface applications in the Ka-band. In Proceedings of the 2024 Asia Communications and Photonics Conference (ACP) and International Conference on Information Photonics and Optical Communications (IPOC), Beijing, China, 2–5 November 2024; IEEE: Piscataway, NJ, USA, 2024; pp. 1–5.
14. He, B.; Liu, J.; Cheng, Y.; Chen, F.; Luo, H.; Li, X. Broadband and thermally switchable reflective metasurface based on Z-shape InSb for terahertz vortex beam generation. *Phys. E* **2022**, *144*, 115373. [CrossRef]
15. Chen, G.-T.; Jiao, Y.-C.; Zhao, G. A reflectarray for generating wideband circularly polarized orbital angular momentum vortex wave. *IEEE Antennas Wirel. Propag. Lett.* **2018**, *18*, 182–186. [CrossRef]
16. Rao, J.; Wang, C.; Yu, H.; Xu, G.; Ren, X.; Zhao, L.; Li, Y.; Huang, Z. Conical conformal OAM-generating transmitarray with high transmission double C-shaped grating metasurface. *Opt. Express* **2024**, *32*, 34128–34140. [CrossRef] [PubMed]
17. Guo, C.; Zhao, X.; Zhu, C.; Xu, P.; Zhang, Y. An OAM patch antenna design and its array for higher-order OAM mode generation. *IEEE Antennas Wirel. Propag. Lett.* **2019**, *18*, 816–820. [CrossRef]
18. Rao, M.V.; Malik, J.; Yuvaraj, S.; Kartikeyan, M.V. Polarization-insensitive reflectarray for OAM beam generation over octave bandwidth for 5G applications. *AEÜ-Int. J. Electron. Commun.* **2023**, *170*, 154775. [CrossRef]
19. Meng, X.; Wu, J.; Wu, Z.; Qu, T.; Yang, L. Dual-polarized reflectarray for generating dual beams with two different orbital angular momentum modes based on independent feeds in C- and X-bands. *Opt. Express* **2018**, *26*, 23185–23195. [CrossRef]
20. Huang, H.-F.; Li, S.-N. High-efficiency planar reflectarray with small-size for OAM generation at microwave range. *IEEE Antennas Wirel. Propag. Lett.* **2019**, *18*, 432–436. [CrossRef]
21. Yang, P.; Yang, R.; Li, Y. Compact cascaded metasurface system for controlling the spin and orbital angular momentum of electromagnetic fields simultaneously. *Opt. Express* **2021**, *29*, 20229–20239. [CrossRef]
22. Han, D.-H.; Wei, X.-C.; Zhao, X.-W. Wideband OAM generator design using a low Q-factor and C-shaped element. *IEEE Microw. Wirel. Technol. Lett.* **2023**, *33*, 615–618. [CrossRef]
23. Nayeri, P.; Yang, F.; Elsherbeni, A.Z. *Reflectarray Antennas: Theory, Designs, and Applications*; John Wiley & Sons: Hoboken, NJ, USA, 2018.
24. Aieta, F.; Genevet, P.; Kats, M.A.; Yu, N.; Blanchard, R.; Gaburro, Z.; Capasso, F. Aberration-free ultrathin flat lenses and axicons at telecom wavelengths based on plasmonic metasurfaces. *Nano Lett.* **2012**, *12*, 4932–4936. [CrossRef] [PubMed]

25. Yao, E.; Franke-Arnold, S.; Courtial, J.; Barnett, S.; Padgett, M. Fourier relationship between angular position and optical orbital angular momentum. *Opt. Express* **2006**, *14*, 9071–9076. [CrossRef] [PubMed]
26. Noamadeh, Z.; Giden, I.H.; Aksoy, E. Polarization-insensitive, wideband reflective metasurface generating orbital angular momentum vortex beam. *Optik* **2024**, *311*, 171921. [CrossRef]
27. Li, B.; Jing, P.F.; Sun, L.Q.; Leung, K.W.; Lv, X. 3D printed OAM reflectarray using half-wavelength rectangular dielectric element. *IEEE Access* **2020**, *8*, 142892–142899. [CrossRef]

photonics

MDPI

Article

Wide-Angle, Polarization-Independent Broadband Metamaterial Absorber by Using Plasmonic Metasurface-Based Split-Circular Structure

Thanh Son Pham [1], Bui Xuan Khuyen [2,3], Vu Dinh Lam [2], Liangyao Chen [4] and Youngpak Lee [4,5,*]

1 School of Electrical and Electronic Engineering, Hanoi University of Industry, Hanoi 100000, Vietnam; sonpt@haui.edu.vn
2 Graduate University of Science and Technology, Vietnam Academy of Science and Technology, 18 Hoang Quoc Viet, Cau Giay, Hanoi 100000, Vietnam; khuyenbx@ims.vast.ac.vn (B.X.K.); lamvd@gust-edu.vast.ac.vn (V.D.L.)
3 Institute of Materials Science, Vietnam Academy of Science and Technology, 18 Hoang Quoc Viet, Cau Giay, Hanoi 100000, Vietnam
4 Department of Optical Science and Engineering, Fudan University, Shanghai 200433, China; lychen@fudan.ac.cn
5 Department of Physics and Quantum Photonic Science Research Center, Hanyang University, Seoul 133-791, Republic of Korea
* Correspondence: yplee@hanyang.ac.kr

Abstract: Absorption of electromagnetic waves in a broadband frequency range with polarization insensitivity and wide incidence angles is greatly needed in modern technological applications. Many methods using metamaterials have been suggested to address this requirement; they can be complex multilayer structures or use external electronic components. In this paper, we present a plasmonic metasurface structure that was simply fabricated using the standard printed circuit board technique but provided a high absorption above 90%, also covering a broadband frequency range from 12.30 to 14.80 GHz. This plasmonic metasurface consisted of structural unit cells composed of multiple split rings connected by a copper bar. Analysis, simulation, and measurement results showed that the metasurface also showed polarization-insensitive properties and maintained an absorption above 90% at incident angles up to 45 degrees. The suggested plasmonic metasurface is a fundamental design that can also be used to design the absorber in different frequency ranges and is able to adapt well to being fabricated at various scales.

Keywords: metamaterial absorber; plasmonic metasurface; broadband

1. Introduction

Materials with unique electromagnetic (EM) properties that can be engineered are being extensively researched and applied in various fields of science and technology today [1]. Metamaterials are among these, produced by arranging meta-atoms according to specific principles. We can develop a material structure with extraordinary EM characteristics that cannot be found in natural materials [2]. Metamaterials with properties such as negative permittivity (ε) or permeability (μ) have been studied in recent years and introduced in numerous potential applications, including invisibility cloaking, super lenses, energy harvesting, wireless power transmission, and especially perfect EM wave absorption [3–6]. The metamaterial perfect absorber (MPA) is one of the remarkable applications of metamaterials, made by designing material structures in such a way that, when interacting with EM waves, all or most of the incident waves are trapped and dissipated within the structure [7,8].

Thanks to its unique properties, the MPA can be applied in various fields. For instance, in sensing applications, MPAs can detect changes in the refractive index of the surrounding environment [9]. In radar applications, they can absorb EM waves in the K_u-band wavelength range, rendering objects invisible to military radar systems [10]. In energy harvesting, they enhance the absorption and conversion of incident EM waves into usable electrical energy [11]. Additionally, MPAs find applications in many other areas. Research on different MPA structures is highly diverse, depending on specific applications. The most fundamental and earliest studies focused on MPAs capable of fully absorbing EM waves at a single designed frequency [12]. Subsequently, those capable of absorbing waves at multiple frequency peaks were introduced as an extension of the single-peak absorption type [13]. As the demand for practical applications grows, research on MPAs has increasingly focused on designing absorbers with a wide absorption bandwidth across a frequency range [14]. To realize broadband absorption, several MPA design approaches have been proposed. Ullah et al. suggested external resistors to absorb energy, which reduced the Q-factor of the metamaterial unit cell and broadened the absorption bandwidth [15]. Zhang et al. proposed a multilayered metamaterial structure to extend the absorption bandwidth, based on the principle of superposition [16]. Liang et al. utilized a special material, a resistive film, to obtain the absorption over a frequency range from 4 to 24 GHz [17]. Several other studies by Feng et al. and You et al. utilized materials such as Ti_2C_3Tx MXene or doped silicon to develop broadband MPA structures operating in the THz frequency range [18,19].

One of the methods used to design the broadband MPAs is to use the spoof surface plasmon polariton (SSPP) effect in the GHz and THz frequency range. Materials with plasmonic effects have also been studied and found to have numerous applications in the optical wavelength region [20]. The SSPP effect in metamaterial structures was first proposed by Pendry et al. in 2004 [21]. Metamaterial structures that responded to EM waves in the GHz and THz frequency range exhibited behavior similar to SSPPs in the optical range [22,23]. The properties of SSPP, such as local field enhancement, strong dispersion, and propagation characteristics in the deep subwavelength regime for application in waveguiding, plasmonic circuits, conformal circuits, and diffractive neural networks, have been extensively studied in recent years [24–28]. These properties were highly effective in designing broadband metamaterial absorbers. Many broadband metamaterial absorber structures have been proposed, based on the SSPP effects in metamaterials [29,30]. For example, periodic arrangements of metallic strips with overlapping lengths were designed to achieve the superposition of neighboring resonance peaks [31]. This resulted in a wide absorption band for the metamaterial structure. However, these designs typically utilized longitudinal arrays to obtain the superposition along the propagation direction of incident EM waves. While this approach was highly effective in developing metamaterial structures with wide absorption bands, it required the structures to be in a 3D form [32]. This made them bulky and challenging to fabricate.

In this study, we propose a plasmonic metasurface design, based on the SSPP principle, which operates entirely on a flat metamaterial surface. This structure is capable of absorbing EM waves over a wide frequency range, from 12.30 to 14.80 GHz. The unit cell of the plasmonic metasurface consists of multiple split rings that have a circular shape, providing excellent responses to various polarizations and incident angles of incoming EM waves. Notably, our plasmonic metasurface was fabricated using standard PCB technology without specialized materials or external electrical components. The proposed structure was planar and fabricated from circular copper strips on an FR4 dielectric substrate, making it easy to manufacture with high stability. With an absorption exceeding 90% across a wide wavelength range in the K_u-band, this plasmonic metasurface has potential applications in radar stealth, satellite communication, and spectral imaging.

2. Design and Analysis

The proposed plasmonic metasurface in this paper was composed of ten concentric split rings with varying diameters (Ring-1 to Ring-10), as illustrated in Figure 1a. The entire structure was fabricated as a three-layer configuration, where the top layer consisted of a structure comprising ten concentric split rings to generate the SSPP effect. However, this particular concentric ring structure was specially divided into four separate segments, and the rings were interconnected by four metallic bars. Thanks to these connections, a portion of the rings could simultaneously contribute to the generation of multiple resonance frequencies when linked with other split rings of different diameters. The multiple resonance frequencies, generated by these pairs of split rings, induced superposition, forming a broad resonance band for the structure. Similar to the conventional MPA structures, this design adopted a sandwich configuration, with a dielectric layer of FR-4 in the middle, having a thickness of 3.2 mm, a dielectric constant of $\varepsilon = 4.3$, and a loss tangent of tan $\delta = 0.025$. The bottom layer was a continuous copper sheet that prevented EM waves from transmitting through the structure [33].

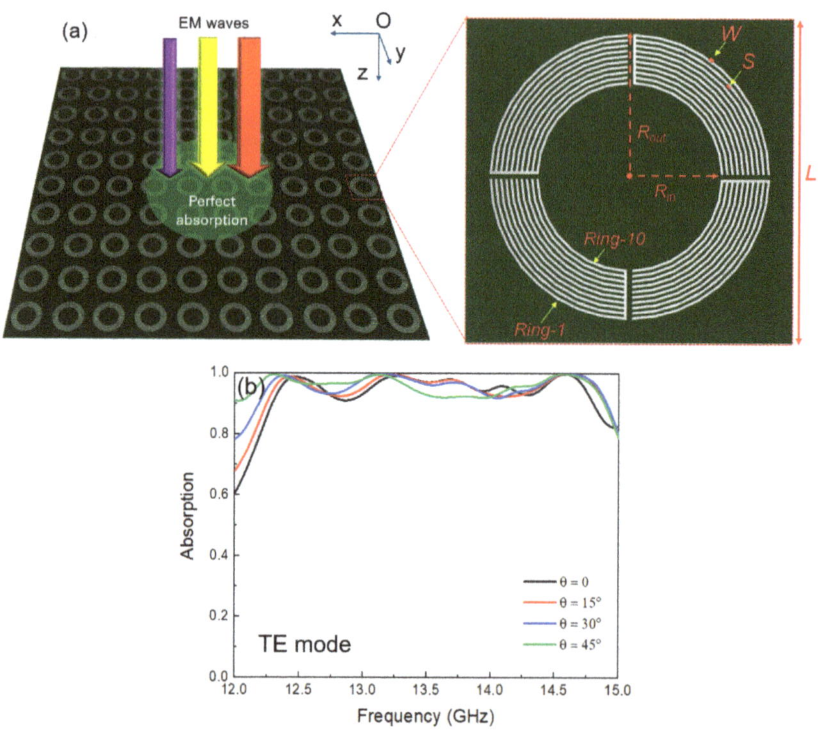

Figure 1. (**a**) Schematic illustration of the plasmonic metasurface and its unit cell, and (**b**) high absorption at various incident angles.

Figure 1b presents the simulated absorption spectrum of the proposed plasmonic metasurface at some incident angles of EM waves for the TE mode, ranging from 0 to 45°. The simulations were conducted using CST Studio Suite 2023 software with periodic boundary conditions applied along the *Ox* and *Oy* directions to model the 2D metamaterial structure [34]. In the CST Studio Suite 2023 software, this boundary condition can be selected by using the "unit cell" mode for the *Ox* and *Oy* dimensions, while "Et = 0" was chosen for the *Oz* axis (Z_{min}) at the bottom of the unit cell. The opposite side (Z_{max}) was configured in the "open" mode, serving as the wave source for the incoming wave. The results indicated that the plasmonic metasurface exhibited an absorption exceeding 90% within a broadband absorption range from 12.30 to 14.80 GHz.

The plasmonic metasurface, as illustrated in Figure 1, can be described in terms of an equivalent circuit model based on the transmission line theory [35]. The plasmonic metasurface is represented by a combination of circuit elements connected together, as shown in the circuit diagram in Figure 2. The impedance of free space is represented by Z_s. Subsequently, the circular structure is modeled by the *RLC* circuits. The dielectric substrate is represented by impedance Z_d. Finally, the continuous copper layer on the backside of the dielectric layer is considered a short-circuit component in the circuit model [36].

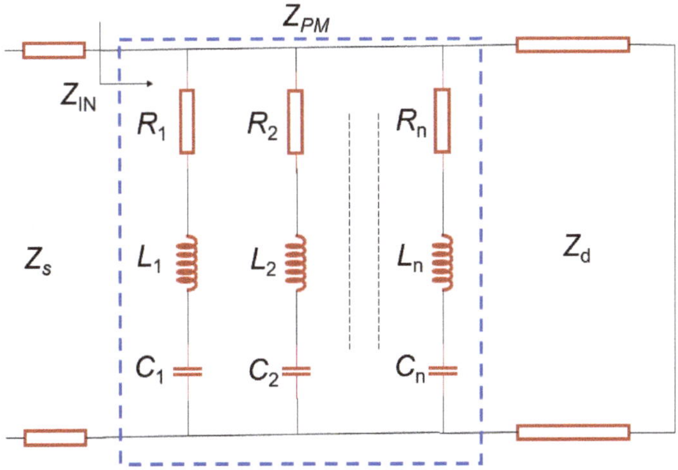

Figure 2. Equivalent circuit model for the plasmonic metasurface.

From the equivalent circuit, the input impedance of plasmonic metasurface can be calculated by using the following formula:

$$Z_{IN} = \frac{Z_{PM}Z_d}{Z_{PM} + Z_d}. \tag{1}$$

Z_{PM} is the impedance of the circular structure of the plasmonic metasurface and is calculated from the parallel RLC circuit components, as shown below.

$$\frac{1}{Z_{PM}} = \frac{1}{Z_1} + \frac{1}{Z_2} + \ldots \frac{1}{Z_n} \tag{2}$$

where

$$Z_i = R_i + j\omega L_i + \frac{1}{j\omega C_i} \tag{3}$$

with R_i, L_i, and C_i as the resistance, inductance, and capacitance of the *i*th branch, respectively. In addition, *n* is the total number of branches in the parallel circuit.

Consequently, the reflection coefficient of the plasmonic metasurface under the incidence of an EM wave can be determined with the following expression:

$$\Gamma = \frac{Z_{IN} - Z_s}{Z_{IN} + Z_s}. \tag{4}$$

According to the transmission line theory, to obtain the minimum reflection coefficient, the input impedance should match that of the free space, as mathematically expressed in Equation (4). The input impedance of the plasmonic metasurface is composed of those of the dielectric layer and the split circular structure. These split rings were positioned very close to one another, resulting in coupling effects between them. This caused the plasmonic metasurface to function similarly to a multilayer frequency-selective surface [30]. The final outcome was the existence of a broad frequency band where the plasmonic metasurface exhibited a low reflection coefficient and high absorption.

To further clarify the absorption characteristics of the plasmonic metasurface, we examined the response of the split-ring components within the structure. Figure 3a illustrates the absorption characteristics of the structure formed by pairs of adjacent split rings in the following order: Rings 1–2, 3–4, 5–6, 7–8, and 9–10, as depicted in Figure 1a. These paired split rings present the absorption spectra with two resonance peaks in a range of 12.30 to 14.80 GHz. At both end frequencies, a high absorption of over 95% is shown. The results indicate that, for the rings with smaller radii, the absorption peaks shift slightly toward higher frequencies, which can be explained by the fact that the total length of the split ring is shorter when the ring radius is smaller. It is noteworthy that the split-ring pairs can be formed not only by adjacent rings but also by any arbitrary pairing of rings. Consequently, a large number of coupled ring pairs can be generated. These coupled rings are capable of strongly absorbing EM waves across different frequency ranges. As a result, the plasmonic metasurface exhibits an overlap of multiple absorption peaks from various coupled ring pairs, leading to a broadband absorption spectrum.

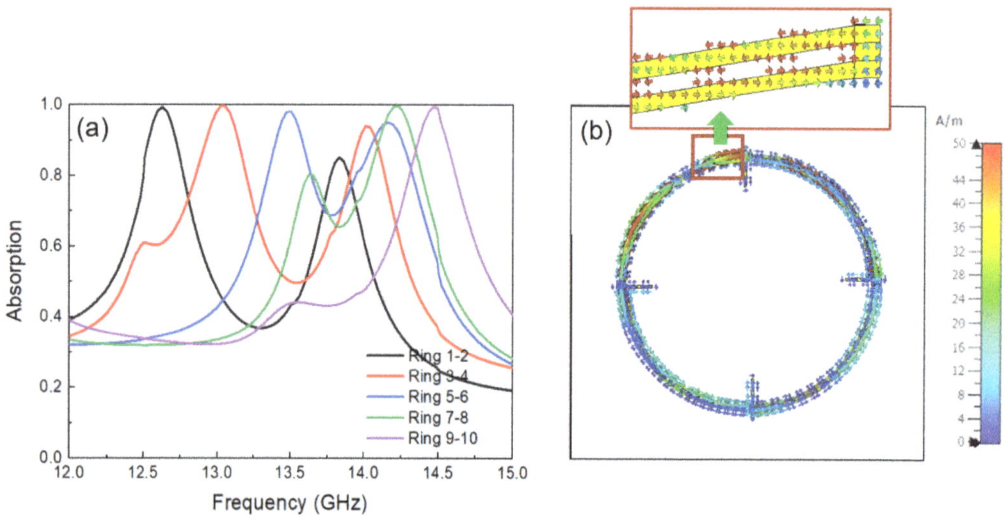

Figure 3. (**a**) Comparison of the absorption of coupled rings, and (**b**) surface current distribution on the adjacent coupled ring.

Figure 3b presents the surface current distribution on a coupled ring at the resonance frequency. It can be observed that strong currents flow on its surface, and notably, the current directions on the two branches of the adjacent rings are opposite. This phenomenon can be explained by the presence of a connected bar linking the two sections. To ensure a continuous current flow within the structure, the current directions on the two branches should be opposite. These results highlight the critical role of the connecting bars in this structure. Without them, the split-ring branches in the two rings would act independently, not allowing the existence of multiple interconnected resonance peaks. This interconnection is essential in forming a broadband absorption.

Two parameters, the strip width of ring (W) and the spacing between adjacent rings (S), are critical factors in the design of the plasmonic metasurface structure. These parameters influenced the coupling between neighboring rings as well as the overall length of split rings, thereby determining both the resonance frequency and the absorption efficiency of the structure. Therefore, we conducted an investigation on various values of W and S to identify the optimal dimensions. Figure 4a illustrates the simulated absorption performance of the plasmonic metasurface by varying the W value from 0.05 mm to 0.3 mm while keeping all the other parameters constant. The results indicate that, at $W = 0.1$ mm, the plasmonic metasurface shows an absorption exceeding 90% with the broadest frequency bandwidth. Similarly, Figure 4b presents the corresponding results by varying the S value from 0.05 to 0.3 mm while keeping the other parameters fixed. The results reveal that the optimal absorption performance, exceeding 90% over a wide frequency range from 12.30 GHz to 14.80 GHz, is obtained when $W = 0.1$ and $S = 0.1$ mm. Therefore, we selected $W = 0.1$ and $S = 0.1$ mm as the optimal dimensions for further studies. When observing the results in Figure 4a,b, it can be seen that, as the values of S and W increase, the broadband absorption region of the plasmonic metasurface becomes narrower. This phenomenon can be explained by the fact that the broadband absorption region of the plasmonic metasurface is formed, based on the overlap of the resonance regions of the split rings. When the spacing between the split rings increases, this overlap is reduced, thereby narrowing the broadband absorption frequency range as well as the absorption itself.

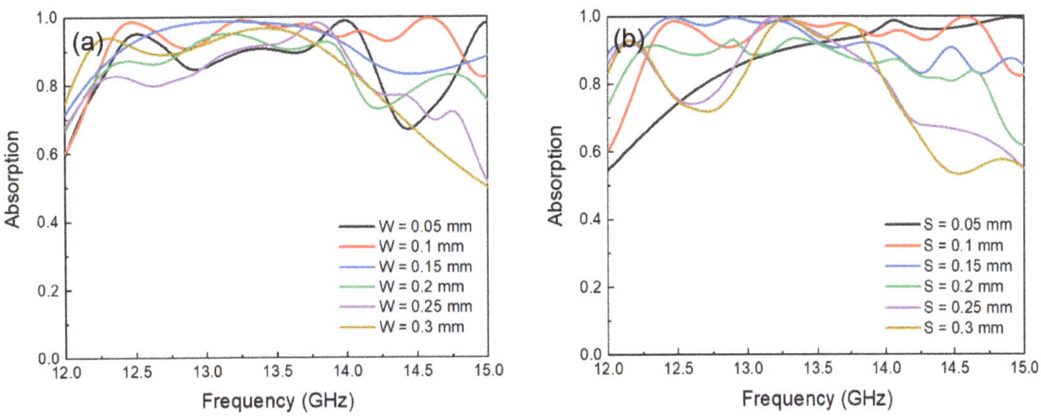

Figure 4. Comparison of the absorption of plasmonic metasurfaces with structural parameters $L = 18.7$, $R_{out} = 7$, and FR-4 thickness $t_{sub} = 3.2$ mm. (**a**) $S = 0.1$ and $W = 0.05$–0.3 mm. (**b**) $W = 0.1$ and $S = 0.05$–0.3 mm.

To better understand the underlying mechanisms responsible for the broadband absorption of the plasmonic metasurface, it is critical to examine the distributions of the E-field and surface current within the structure. Figure 5a–c provide a detailed visualization of the E-field distributions across the plasmonic metasurface at three representative frequencies: 12.5, 13.5, and 14.5 GHz. The components of this field represent the absolute value of the E-field on plane $z = 0$. The directions of the field are not considered here. Instead, we focus on observing the direction of the surface current flow within the structure later. It should be noted that, at all three frequencies, the metasurface showed an absorption exceeding 90%, as in Figure 4, highlighting its efficiency in capturing EM energy. However, the spatial distribution of the E-field reveals significant variation depending on the frequency, indicating that different regions of the metasurface contribute dominantly to the absorption at different frequencies.

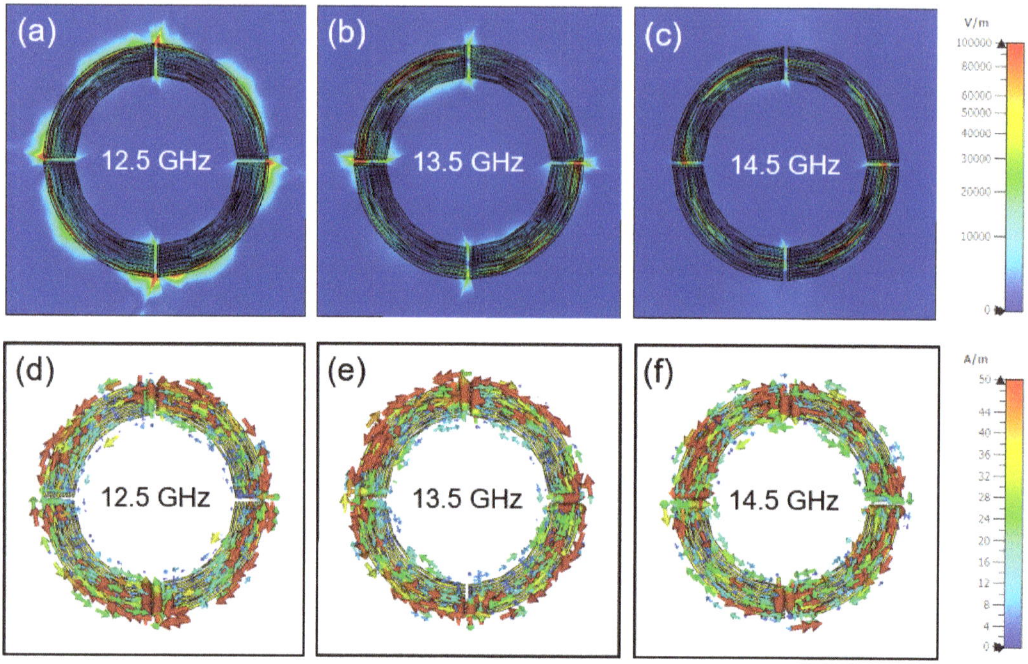

Figure 5. (**a–c**) E-field intensity distributions in the plasmonic metasurface at various frequencies. (**d–f**) Analysis on the surface current distribution of the plasmonic metasurface at the specific frequencies corresponding to the investigated E-field distributions.

At a frequency of 12.5 GHz, which corresponds to a wavelength of approximately 24 mm, the strongest E-field concentration is observed at the outermost cut circular segments of the plasmonic metasurface structure. These regions, located at the periphery of design, act as hotspots for field enhancement and play a critical role in the absorption process at lower frequencies. As the frequency increases to 13.5 GHz (wavelength ~ 22.2 mm) and further to 14.5 GHz (wavelength ~ 20.6 mm), the regions of maximum E-field intensity progressively shift inward, focusing on the smaller-diameter split rings closer to the center of the structure. This inward movement of the E-field intensity with an increasing frequency suggests a frequency-selective interaction between the EM wave and structural elements of the plasmonic metasurface, highlighting the role of geometric scaling in determining the resonant behavior of different regions. The results of the field distribution, which depend on the positions of the split rings within the unit cell, can be explained by the fact that each split ring with a different length resonates at a different frequency of the incident EM wave. The relationship between the length of the split ring and its resonance frequency can be expressed by the following formula [37]:

$$f_{ij} = \frac{c\sqrt{i^2 + j^2}}{2L\sqrt{\varepsilon_{eff}}} \tag{5}$$

where i and j are integers (0, 1, 2...), $i^2 + j^2 \neq 0$, and ε_{eff} is the effective dielectric constant.

Based on Equation (5), the operating frequency of the structure is inversely proportional to the length of the split ring. Therefore, the split rings located on the inner side respond to higher frequencies of the EM waves, and vice versa.

In addition to the E-field distributions, Figure 5e,f present the surface current distributions within the plasmonic metasurface at the same frequencies. These current distributions provide further insights into the physical mechanism that governs the broadband absorption. The regions exhibiting strong E-field intensities in Figure 5a–c are found to correspond to the areas with significant surface currents. Notably, the current distribution reveals that adjacent split rings form continuous current paths, resulting in a unique current flow pattern. Specifically, the currents in neighboring loops flow in opposite directions, developing localized magnetic dipoles that contribute to the absorption properties of the plasmonic metasurface. These opposite current directions are a direct consequence of the structural design with connecting bars.

A key feature of the metasurface design is the presence of connecting bars at the ends of split rings. These connecting bars play a crucial role in enabling current flow between adjacent rings. This connectivity facilitates the coupling and overlapping of resonances between different structural elements, thereby enhancing the capability of the metasurface to absorb EM waves across a wide frequency range. The coupling effect, combined with the structural geometry of the split rings, leads to the formation of multiple absorption peaks that overlap, resulting in the observed broadband absorption.

3. Experimental Verification for the Proposed Plasmonic Metasurface

The proposed plasmonic metasurface, which was analyzed above, was fabricated as shown in Figure 6. The metasurface consisted of a 10 × 10 array of unit cells, as described here and in Figure 1, with the following parameters: D = 187 mm, L = 18.7 mm, R_{out} = 7 mm, W = 0.1 mm, and S = 0.1 mm. The plasmonic metasurface was fabricated by using the standard PCB technology on an FR-4 substrate with a thickness of 3.2 mm. The copper layers on both the top and bottom surfaces of the structure had a thickness of 0.035 mm and conductivity of 5.96×10^7 S/m. FR-4 is a highly durable material resistant to environmental factors such as mechanical, temperature, and chemical factors, while the copper layer, manufactured by using the standard PCB technology, exhibits excellent adhesion to the FR-4 substrate. This method is widely employed in the fabrication of electronic circuits and antennas owing to its high durability [38]. The structure comprised only three layers, and with the dimensions of the unit cell as specified above, it was relatively straightforward to fabricate. This simplicity is one of the advantages of the proposed plasmonic metasurface by utilizing the SSPP effect, compared to multilayered or 3D structures commonly employed in MPA designs. The overall size of the fabricated plasmonic metasurface was 187 × 187 mm. The detailed geometrical parameters of the proposed plasmonic metasurface are presented in Table 1.

The plasmonic metasurface was characterized by using a Rohde & Schwarz ZNB20 Vector Network Analyzer (VNA) and a pair of wideband antennas operating in a frequency range of 12 to 18 GHz, as shown in Figure 7. The two antennas were connected to the two ports of the VNA, enabling the transmission of EM waves toward the structure and the reception of reflected waves. After the calibration process of the VNA, the reflection coefficient (S_{11}) of the plasmonic metasurface was determined through the transmission coefficient (S_{21}) from the transmitting antenna to the receiving one. Since the back side of the structure consisted of a continuous copper layer, EM waves could not transmit through the structure, and the absorption of the plasmonic metasurface was calculated by employing formula $A = 1 - |S_{11}|^2$. To measure the incident angle dependence of the plasmonic metasurface, the two antennas were rotated and moved along an arc-shaped support, ensuring that the incident angle θ_i equals the reflection angle θ_r.

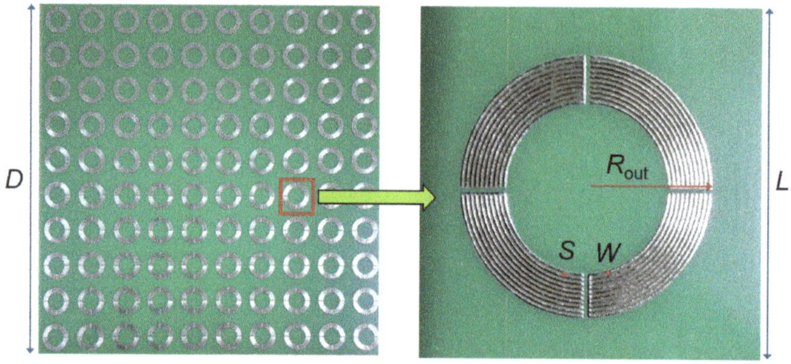

Figure 6. Fabricated plasmonic metasurface and its unit cell.

Table 1. Geometrical parameters of the plasmonic metasurface.

Structure Parameters	Value
Unit-cell periodicity (L)	18.7 mm
Dielectric thickness (t_d)	3.2 mm
Dielectric constant (ε)	4.3
Loss tangent (tan δ)	0.025
Metal thickness (t_m)	0.035 mm
Width of metal trace (W)	0.1 mm
Spacing between turns (S)	0.1 mm
Outer radius (R_{out})	7 mm
Number of rings (n)	10
Total size of plasmonic metasurface	187 × 187 mm

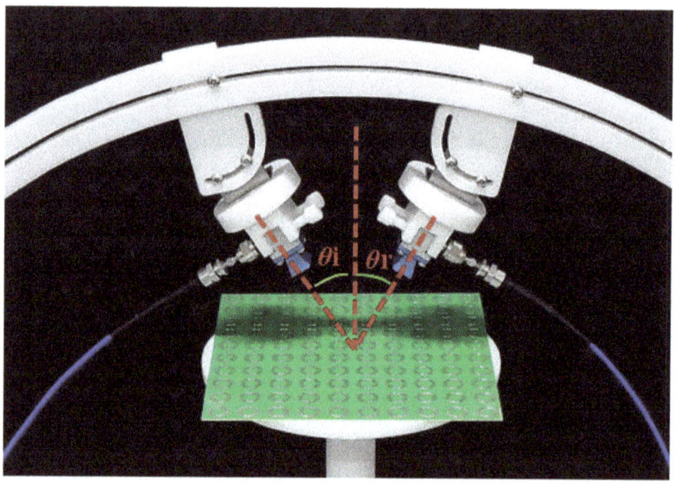

Figure 7. Experiment setup for measuring the absorption of the proposed plasmonic metasurface.

Figure 8a presents the measured absorption performance of the plasmonic metasurface for the TE-polarized waves at incidence angles ranging from 0 to 45°. The results demonstrate that the plasmonic metasurface realized an absorption over 90% of the incident wave within a frequency range of 12.40 to 14.75 GHz. At larger incidence angles, the absorption bandwidth becomes broader (e.g., 12.10 to 14.90 GHz at 45°) compared to smaller incidence angles. This could be attributed to the parallel component of the EM waves interacting with the plasmonic metasurface in a manner similar to 3D structures, where metal bars aligned with the direction of the incident EM wave contributed to the absorption. When compared to the absorption spectrum obtained from simulations, the measured values exhibit some discrepancies. However, the broadband absorption characteristic and high absorption exceeding 90% of the plasmonic metasurface are still preserved. The observed differences might stem from the fabrication process of the plasmonic metasurface, since small fabrication errors can influence the interactions between the split rings in the unit cell structure. Nevertheless, these interactions might compensate for each other, ensuring that the quality of the absorption spectrum of the plasmonic metasurface is not significantly affected. Moreover, the discrepancies might also derive from the measurement process, since we used a VNA to measure the reflection coefficient (S_{11}) of the plasmonic metasurface, from which the absorption performance was calculated. The antennas might have been placed at slightly misaligned positions, and certain unavoidable unwanted interactions could also contribute to these differences.

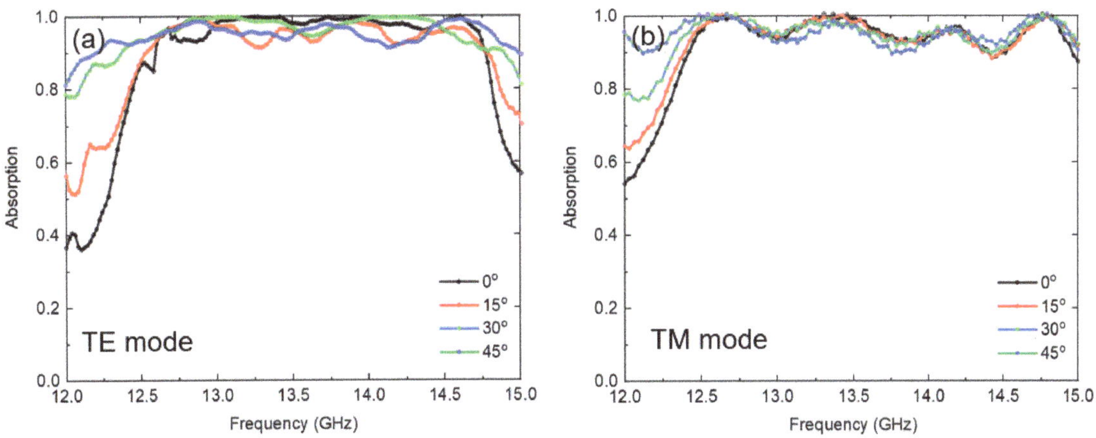

Figure 8. Measured absorption of the plasmonic metasurface at various incident angles for the (**a**) TE and (**b**) TM modes.

Figure 8b illustrates the absorption performance of the plasmonic metasurface for TM-polarized waves. Due to the design of the circular unit cell of the structure, the symmetry is relatively high, resulting in an absorption spectrum for the TM waves, which is quite similar to that of the TE mode. The metasurface provided an absorption over 90% in a frequency range of 12.40 to 14.90 GHz for the TM-polarized waves. Naturally, slight differences between the TE and TM modes remained, attributed to the inherent properties of the metasurface and the measurement inaccuracies discussed above.

For both TE and TM modes, the absorption spectrum is not completely flat at a level of 90% across the frequency range. Instead, it reaches 99% at certain frequencies for all incident angles. This behavior can be explained by the unequal interaction of the split-ring pairs forming the metasurface. Nevertheless, the lowest absorption level remains above 90%, making the metasurface suitable for applications requiring a high absorption efficiency.

4. Discussion

Metamaterials with ring and split-ring structures have been extensively studied for broadband absorption applications, yielding significant results [39–44]. However, the achievement of broadband absorption in previous studies often required the use of external electrical components, multilayer structures, or specialized materials. For instance, in Ref. [39], Li et al. obtained an absorption exceeding 90% at incident angles from 0 to 30° within a frequency range of 7.93 to 17.18 GHz using a double octagonal ring structure. However, this design necessitated the integration of four resistors into each unit cell, where the resistors functioned as energy absorption points, contributing to the broadband absorption performance of the structure. Similarly, in Ref. [40], the proposed absorber prototype demonstrated an absorption exceeding 80% across a frequency range of 5.46 to 14.75 GHz, with an incident angle tolerance up to 30°. To achieve this result, the metamaterial absorber was fabricated as a multilayer structure, incorporating two layers of double split-serration rings (DSSRs) sandwiched between three layers of modified epoxy with varying thicknesses. Additionally, several other structures have been proposed for broadband absorption. However, these designs typically required the use of four resistors per unit cell [41,42] or even eight ones per unit cell [43]. In cases where resistors are not employed, the proposed structures tended to rely on multilayer configurations and specialized materials, such as indium tin oxide combined with air gaps [44].

In other studies [45,46] that did not employ external electrical components or specialized materials, achieving broadband absorption still remained challenging. Instead, these studies often resulted in multi-peak absorption [45]. Occasionally, broadband absorption was achieved; however, the absorption bandwidth was typically less than 1 GHz [46]. In contrast, our proposed plasmonic design features a simple configuration based on a metal—dielectric–metal structure. The top copper layer was fabricated by using the standard PCB technology, which was highly straightforward and ensured high reproducibility. No resistors or external components are required in this plasmonic metasurface. Consequently, our proposed design achieved a broadband absorption within a range of 12.30 to 14.80 GHz. When compared to the previous results that used lumped elements/multilayer, our reported relative absorption bandwidth of 18.45% is indeed narrow. However, when compared to the publications without the use of lumped elements/multilayer, the 18.45% bandwidth is significantly larger, since the previous publications have reported values of 1.18%, 2.46%, and 6.46%, respectively (see Table 2). This approach offers significant advantages in terms of compactness, ease of fabrication, and cost-effectiveness. A detailed comparison of the performance of this work with previous studies is presented in Table 2.

Table 2. Comparison of the absorption performance with other works.

Ref.	Bandwidth (GHz)	Relative Absorption Bandwidth	Absorption	Lumped Elements/ Multilayer	Polarization (Both TE and TM?)	Incident Angle
[39]	7.93–17.18	73.67%	>90%	Yes/No	Both	0–20°
[40]	5.46–14.75	91.93%	>90%	No/Yes	Both	0–30°
[41]	6.80–11.80	53.76%	>70%	Yes/No	Both	0–30°
[43]	6.70–20.58	101.75%	>90%	Yes/Yes	Both	0–30°
[44]	6.21–19.31	102.66%	>90%	No/Yes	Both	0°
[45]	Peaks at 14.50, 16.50	n/a	>90%	No/No	Both	0, 45°, 90°
[46]	11.21–11.49 13.92–14.85 17.66–17.87	2.46% 6.46% 1.18%	>90%	No/No	Both	0–45°
This paper	12.30–14.80	18.45%	>90%	No/No	Both	0–45°

5. Conclusions

This study proposes a plasmonic metasurface structure capable of absorbing over 90% of EM waves in a broad frequency range from 12.30 to 14.80 GHz. Simulation and experimental studies demonstrated that the plasmonic metasurface maintained a high absorption and wide bandwidth for both TE and TM waves, for incident angles up to 45°. Notably, the broadband absorption was obtained without the use of external elements such as resistors or specialized materials. The structure was 2D and was easily fabricated using the standard PCB technology. Due to its simple and compact design, the plasmonic metasurface also exhibited high durability and was more practical for implementation compared to previously proposed structures that relied on complex fabrication techniques. Given its advantages, the plasmonic metasurface has potential for various applications, including radar stealth technology, satellite communications, surveillance radars, and many other promising fields.

Author Contributions: Conceptualization, T.S.P. and B.X.K.; methodology, T.S.P.; software, B.X.K.; validation, V.D.L. and Y.L.; formal analysis, T.S.P.; investigation, B.X.K.; resources, V.D.L.; data curation, B.X.K.; writing—original draft preparation, T.S.P.; writing—review and editing, Y.L., V.D.L., and L.C.; supervision, Y.L.; project administration, Y.L. and V.D.L. All authors have read and agreed to the published version of the manuscript.

Funding: This research was funded by the Vietnam Academy of Science and Technology, under grant number NCPTVL.03/24/25, supported by the Shanghai Science and Technology Commission, under grant number 24110714600, and funded by the Korea Evaluation Institute of Industrial Technology (Project No. 20016179).

Institutional Review Board Statement: Not applicable.

Informed Consent Statement: Not applicable.

Data Availability Statement: The data used and/or analyzed during the current study are available from the corresponding author upon reasonable request.

Conflicts of Interest: The authors declare no conflicts of interest.

References

1. Xiao, S.; Wang, T.; Liu, T.; Zhou, C.; Jiang, X.; Zhang, J. Active Metamaterials and Metadevices: A Review. *J. Phys. D Appl. Phys.* **2020**, *53*, 503002. [CrossRef]
2. Kadic, M.; Milton, G.W.; van Hecke, M.; Wegener, M. 3D Metamaterials. *Nat. Rev. Phys.* **2019**, *1*, 198–210. [CrossRef]
3. Nguyen, H.A.; Pham, T.S.; Tung, B.S.; Khuyen, B.X.; Le, D.T.; Vu, H.Y.; Vu, D.L.; Hien, N.T. Metamaterials Based on Hyperbolic-Graphene Composite: A Pathway from Positive to Negative Refractive Index at Terahertz. *Comput. Mater. Sci.* **2025**, *248*, 113574. [CrossRef]
4. Fowler, C.; Silva, S.; Thapa, G.; Zhou, J. High Efficiency Ambient RF Energy Harvesting by a Metamaterial Perfect Absorber. *Opt. Mater. Express OME* **2022**, *12*, 1242–1250. [CrossRef]
5. Hiep, L.T.H.; Bui, H.N.; Tung, B.S.; Lam, V.D.; Khuyen, B.X.; Pham, T.S. Enhanced Efficiency of Magnetic Resonant Wireless Power Transfer System Using Rollable and Foldable Metasurface Based on Polyimide Substrate. *Appl. Phys. A* **2024**, *130*, 521. [CrossRef]
6. Rakhshani, M.R.; Rashki, M. Metamaterial Perfect Absorber Using Elliptical Nanoparticles in a Multilayer Metasurface Structure with Polarization Independence. *Opt. Express OE* **2022**, *30*, 10387–10399. [CrossRef]
7. Bhati, R.; Malik, A.K. Multiband Terahertz Metamaterial Perfect Absorber for Microorganisms Detection. *Sci. Rep.* **2023**, *13*, 19685. [CrossRef]
8. Tung, B.S.; Viet, N.N.; Khuyen, B.X.; Pham, T.S.; Do, P.X.; Hoa, N.T.; Lam, V.D.; Tung, D.K. Multi-Band and Polarization-Insensitive Electromagnetically-Induced Transparency Based on Coupled-Resonators in a Metamaterial Operating at GHz Frequencies. *Phys. Scr.* **2024**, *99*, 115502. [CrossRef]
9. Shruti; Pahadsingh, S.; Appasani, B. Metamaterial-Based Terahertz Absorbers for Refractive Index Sensing: Types, Mechanism, and Applications. *Plasmonics* **2024**. [CrossRef]

10. Saadeldin, A.S.; Sayed, A.M.; Amr, A.M.; Sayed, M.O.; Hameed, M.F.O.; Obayya, S.S.A. Wideband Ultrathin and Polarization Insensitive Metamaterial Absorber for Ku-Band Applications. *J. Mater. Sci. Mater. Electron.* **2023**, *34*, 1797. [CrossRef]
11. Tan, T.; Yan, Z.; Zou, H.; Ma, K.; Liu, F.; Zhao, L.; Peng, Z.; Zhang, W. Renewable Energy Harvesting and Absorbing via Multi-Scale Metamaterial Systems for Internet of Things. *Appl. Energy* **2019**, *254*, 113717. [CrossRef]
12. Wang, C. Design, Calculation, and Measurement of a Single Resonant Mode Metamaterial Absorber and Its Tunable Performance. *Opt. Mater.* **2024**, *148*, 114810. [CrossRef]
13. Liu, Y.; Ma, W.-Z.; Wu, Y.-C.; Meng, D.; Cheng, Y.-Y.; Chen, Y.-S.; Liu, J.; Gu, Y. Multi-Peak Narrow-Band Metamaterial Absorber for Visible to near-Infrared Wavelengths. *Results Phys.* **2023**, *47*, 106374. [CrossRef]
14. Zheng, H.; Pham, T.S.; Chen, L.; Lee, Y. Metamaterial Perfect Absorbers for Controlling Bandwidth: Single-Peak/Multiple-Peaks/Tailored-Band/Broadband. *Crystals* **2024**, *14*, 19. [CrossRef]
15. Ullah, N.; Islam, M.S.; Hoque, A.; Yong, W.H.; Alrashdi, A.M.; Soliman, M.S.; Islam, M.T. An Efficient, Compact, Wide-Angle, Wide-Band, and Polarization-Insensitive Metamaterial Electromagnetic Energy Harvester. *Alex. Eng. J.* **2023**, *82*, 377–388. [CrossRef]
16. Zhang, C.; Yin, S.; Long, C.; Dong, B.W.; He, D.; Cheng, Q. Hybrid Metamaterial Absorber for Ultra-Low and Dual-Broadband Absorption. *Opt. Express OE* **2021**, *29*, 14078–14086. [CrossRef]
17. Liang, C.; Kong, X.; Wang, F.; Xu, R.; Fu, Y.; Pang, X.; Zhang, S.; Shen, X.; Zhao, L. A Broadband Perfect Metamaterial Absorber with Angle-Insensitive Characteristics. *J. Electromagn. Waves Appl.* **2023**, *37*, 401–410. [CrossRef]
18. Feng, S.; Yang, L.; Cai, B.; Yang, W.; Wu, L.; Cheng, Y.; Chen, F.; Luo, H.; Li, X. Tri-Band Terahertz Metamaterial Absorber Based on Structural $Ti_3C_2T_x$ MXene for Enhanced Sensing Application. *IEEE Sens. J.* **2024**, *24*, 28889–28896. [CrossRef]
19. You, X.; Upadhyay, A.; Cheng, Y.; Bhaskaran, M.; Sriram, S.; Fumeaux, C.; Withayachumnankul, W. Ultra-Wideband Far-Infrared Absorber Based on Anisotropically Etched Doped Silicon. *Opt. Lett. OL* **2020**, *45*, 1196–1199. [CrossRef]
20. Shen, Z.; Huang, D.; Lin, X. Dual-Band Chirality-Selective Absorbing by Plasmonic Metasurfaces with Breaking Mirror and Rotational Symmetry. *Opt. Express OE* **2023**, *31*, 35730–35741. [CrossRef]
21. Pendry, J.B.; Martín-Moreno, L.; Garcia-Vidal, F.J. Mimicking Surface Plasmons with Structured Surfaces. *Science* **2004**, *305*, 847–848. [CrossRef] [PubMed]
22. Pham, T.S.; Zheng, H.; Chen, L.; Khuyen, B.X.; Lee, Y. Wide-Incident-Angle, Polarization-Independent Broadband-Absorption Metastructure without External Resistive Elements by Using a Trapezoidal Structure. *Sci. Rep.* **2024**, *14*, 10198. [CrossRef] [PubMed]
23. Zhu, R.; Wang, J.; Sui, S.; Meng, Y.; Qiu, T.; Jia, Y.; Wang, X.; Han, Y.; Feng, M.; Zheng, L.; et al. Wideband Absorbing Plasmonic Structures via Profile Optimization Based on Genetic Algorithm. *Front. Phys.* **2020**, *8*, 231. [CrossRef]
24. Garcia-Vidal, F.J.; Fernández-Domínguez, A.I.; Martin-Moreno, L.; Zhang, H.C.; Tang, W.; Peng, R.; Cui, T.J. Spoof Surface Plasmon Photonics. *Rev. Mod. Phys.* **2022**, *94*, 025004. [CrossRef]
25. Tang, W.X.; Zhang, H.C.; Ma, H.F.; Jiang, W.X.; Cui, T.J. Concept, Theory, Design, and Applications of Spoof Surface Plasmon Polaritons at Microwave Frequencies. *Adv. Opt. Mater.* **2019**, *7*, 1800421. [CrossRef]
26. Cheng, Z.W.; Wang, M.; You, Z.H.; Ma, H.F.; Cui, T.J. Spoof Surface Plasmonics: Principle, Design, and Applications. *J. Phys. Condens. Matter* **2022**, *34*, 263002. [CrossRef]
27. Gu, Z.; Ma, Q.; Gao, X.; You, J.W.; Cui, T.J. Direct Electromagnetic Information Processing with Planar Diffractive Neural Network. *Sci. Adv.* **2024**, *10*, eado3937. [CrossRef]
28. Gao, X.; Gu, Z.; Ma, Q.; Chen, B.J.; Shum, K.-M.; Cui, W.Y.; You, J.W.; Cui, T.J.; Chan, C.H. Terahertz Spoof Plasmonic Neural Network for Diffractive Information Recognition and Processing. *Nat. Commun.* **2024**, *15*, 6686. [CrossRef]
29. Jidi, L.; Cao, X.; Gao, J.; Li, T.; Yang, H.; Li, S. Ultra-Wideband Absorber for Electromagnetic Waves under Large Incident Angle Based on Spoof Surface Plasmon Polaritons. *Opt. Mater. Express OME* **2021**, *11*, 3917–3929. [CrossRef]
30. Bai, J.; Yang, Q.; Liang, Y.; Gao, X. Broadband Frequency Selective Rasorber Based on Spoof Surface Plasmon Polaritons. *Micromachines* **2022**, *13*, 1969. [CrossRef]
31. Jiang, W.; Ma, H.; Wang, J.; Yang, J.; Yan, L.; Fan, Y.; Qu, S. Spoof Surface Plasmon Polaritons Realized by Unidirectional Carbon Fibers Arrays and Applications in Structure/Function Integrated Sandwich Structure. *Results Phys.* **2020**, *17*, 103081. [CrossRef]
32. Zhou, S.; Liang, X.; Xing, J.; Fan, Y.; Zhang, L.; Li, D.; Li, E.-P. Ultra-Broadband Metamaterial Absorbers Based on Spoof Surface Plasmon Polaritons Structure. *Microw. Opt. Technol. Lett.* **2022**, *64*, 489–495. [CrossRef]
33. Xu, J.; Zhao, Z.; Yu, H.; Yang, L.; Gou, P.; Cao, J.; Zou, Y.; Qian, J.; Shi, T.; Ren, Q.; et al. Design of Triple-Band Metamaterial Absorbers with Refractive Index Sensitivity at Infrared Frequencies. *Opt. Express OE* **2016**, *24*, 25742–25751. [CrossRef]
34. CST Studio Suite 3D EM Simulation and Analysis Software. Available online: https://www.3ds.com/products-services/simulia/products/cst-studio-suite/ (accessed on 10 October 2023).
35. Chen, H.; Yang, X.; Wu, S.; Zhang, D.; Xiao, H.; Huang, K.; Zhu, Z.; Yuan, J. Flexible and Conformable Broadband Metamaterial Absorber with Wide-Angle and Polarization Stability for Radar Application. *Mater. Res. Express* **2018**, *5*, 015804. [CrossRef]

36. Jeong, H.; Lim, S. Broadband Frequency-Reconfigurable Metamaterial Absorber Using Switchable Ground Plane. *Sci. Rep.* **2018**, *8*, 9226. [CrossRef]
37. Yoo, Y.J.; Kim, Y.J.; Hwang, J.S.; Rhee, J.Y.; Kim, K.W.; Kim, Y.H.; Cheong, H.; Chen, L.Y.; Lee, Y.P. Triple-Band Perfect Metamaterial Absorption, Based on Single Cut-Wire Bar. *Appl. Phys. Lett.* **2015**, *106*, 071105. [CrossRef]
38. Sandeep, D.R.; Sivanvitha, G.V. Design of Low-Cost Compact FR-4 Microstrip Patch Antenna for 5G, ISM, and X-Band Applications. In Proceedings of the 2024 IEEE Wireless Antenna and Microwave Symposium (WAMS), Visakhapatnam, India, 29 February–3 March 2024; pp. 1–5.
39. Li, S.; Gao, J.; Cao, X.; Li, W.; Zhang, Z.; Zhang, D. Wideband, Thin, and Polarization-Insensitive Perfect Absorber Based the Double Octagonal Rings Metamaterials and Lumped Resistances. *J. Appl. Phys.* **2014**, *116*, 043710. [CrossRef]
40. Li, S.-J.; Cao, X.-Y.; Gao, J.; Liu, T.; Zheng, Y.-J.; Zhang, Z. Analysis and Design of Three-Layer Perfect Metamaterial-Inspired Absorber Based on Double Split-Serration-Rings Structure. *IEEE Trans. Antennas Propag.* **2015**, *63*, 5155–5160. [CrossRef]
41. Nguyen, T.Q.H.; Nguyen, T.K.T.; Cao, T.N.; Nguyen, H.; Bach, L.G. Numerical Study of a Broadband Metamaterial Absorber Using a Single Split Circle Ring and Lumped Resistors for X-Band Applications. *AIP Adv.* **2020**, *10*, 035326. [CrossRef]
42. Shukoor, M.A.; Dey, S. A Novel Modified Circular Ring-Based Broadband Polarization-Insensitive Angular Stable Circuit Analog Absorber (CAA) for RCS Applications. *Int. J. Microw. Wirel. Technol.* **2023**, *15*, 440–453. [CrossRef]
43. Sambhav, S.; Ghosh, J.; Singh, A.K. Ultra-Wideband Polarization Insensitive Thin Absorber Based on Resistive Concentric Circular Rings. *IEEE Trans. Electromagn. Compat.* **2021**, *63*, 1333–1340. [CrossRef]
44. Zheng, Y.; Chen, K.; Jiang, T.; Zhao, J.; Feng, Y. Multi-Octave Microwave Absorption via Conformal Metamaterial Absorber with Optical Transparency. *J. Phys. D Appl. Phys.* **2019**, *52*, 335101. [CrossRef]
45. Hakim, M.L.; Alam, T.; Soliman, M.S.; Sahar, N.M.; Baharuddin, M.H.; Almalki, S.H.A.; Islam, M.T. Polarization Insensitive Symmetrical Structured Double Negative (DNG) Metamaterial Absorber for Ku-Band Sensing Applications. *Sci. Rep.* **2022**, *12*, 479. [CrossRef]
46. Hannan, S.; Islam, M.T.; Almutairi, A.F.; Faruque, M.R.I. Wide Bandwidth Angle- and Polarization-Insensitive Symmetric Metamaterial Absorber for X and Ku Band Applications. *Sci. Rep.* **2020**, *10*, 10338. [CrossRef]

photonics

MDPI

Article

Polarization-Insensitive Silicon Grating Couplers via Subwavelength Metamaterials and Metaheuristic Optimization

Jorge Parra

Nanophotonics Technology Center, Universitat Politècnica de València, Camino de Vera s/n, 46022 Valencia, Spain; jorpargo@ntc.upv.es

Abstract: Silicon photonics is the leading platform in photonic integrated circuits (PICs), enabling dense integration and low-cost manufacturing for applications such as data communications, artificial intelligence, and quantum processing, to name a few. However, efficient and polarization-insensitive fiber-to-PIC coupling for multipoint wafer characterization remains a challenge due to the birefringence of silicon waveguides. Here, we address this issue by proposing polarization-insensitive grating couplers based on subwavelength dielectric metamaterials and metaheuristic optimization. Subwavelength periodic structures were engineered to act as uniaxial homogeneous linear (UHL) materials, enabling tailored anisotropy. On the other hand, particle swarm optimization (PSO) was employed to optimize the coupling efficiency, bandwidth, and polarization-dependent loss (PDL). Numerical simulations demonstrated that a pitch of 100 nm ensures UHL behavior while minimizing leaky waves. Optimized grating couplers achieved coupling efficiencies higher than −3 dB and a PDL of below 1 dB across the telecom C-band (1530–1565 nm). Three optimization strategies were explored, balancing efficiency, the bandwidth, and the PDL while considering the Pareto front. This work establishes a robust framework combining metamaterial engineering with computational optimization, paving the way for high-performance polarization-insensitive grating couplers with potential uses in advanced photonic applications.

Keywords: polarization; grating coupler; subwavelength; particle swarm optimization; silicon photonics

1. Introduction

Along with electronics, which revolutionized the last century, photonics is becoming one of the technological pillars that will face current and forthcoming challenges for humanity. Among these challenges is the exponential growth of data every year, accelerated in recent years by the explosion of artificial intelligence [1–3], which in turn significantly increases energy consumption and raises concerns about the environmental sustainability of current infrastructures [4,5]. In this regard, light can address this challenge by encoding information in its "unlimited" bandwidth, thereby providing much more efficient energy consumption per bit [6–9]. Hence, to leverage light's benefits, photonic integrated circuits (PICs) pose a suitable technology to reduce costs, allow for dense integration, and implement different functionalities in the same system [10–15]. Among the different PIC platforms, silicon photonics is currently the most versatile due to its excellent electronic/optical integration and scalability, achieved by taking advantage of the complementary metal–oxide–semiconductor (CMOS) fabrication processes of the micro- and nano-electronic industry [16–18]. Moreover, the high refractive index contrast ($\Delta n \approx 2$)

between silicon (Si) and silicon dioxide (SiO_2) allows us to confine light in silicon wires with submicron cross-sections. Hence, a large number of emerging and disruptive applications have been demonstrated using silicon photonics, such as quantum processing [19–21], neuromorphic computing [22–25], LiDAR [26,27], and data communications [28], to name a few.

However, silicon photonics still faces some challenges that need to be addressed. The feasibility of efficient, versatile, and polarization-insensitive fiber-to-PIC coupling remains an open question due to the strong birefringence exhibited by silicon waveguides. Several coupling strategies have been proposed based either on edge or vertical coupling [29]. In the first case, the light is coupled by aligning the fiber and the waveguide on the lateral sides of the PIC using spot size converters [30–33]. In the latter, the fiber is positioned vertically to the top surface of the PIC, and the light beam is coupled to the silicon waveguide by varying the refractive index on the PIC surface using a device coined as a grating coupler [34–41]. Although edge coupling approaches provide low optical losses and negligible wavelength and polarization losses, alignment is critical. They are not as versatile as grating couplers. Therefore, grating couplers are the most popular solution since the structure integrates seamlessly into the PIC and can be placed at any chip position, facilitating multipoint wafer testing. However, the simplest forms suffer from polarization dependence loss (PDL) caused by their phase-matching operation, thereby limiting the system performance and stability against polarization fluctuations.

Therefore, efficient polarization-insensitive grating couplers for a silicon photonics platform are highly appealing [42]. Some approaches based on polarization diversity have been proposed in which a 2D grating coupler can be seen as a superposition of two orthogonally oriented 1D gratings [43–46]. However, on-chip light is polarized only along a particular axis. Usually, this approach increases the footprint of the PIC. More recent work has harnessed subwavelength dielectric structures [47–49]. A wide variety of subwavelength structures have been harnessed for free-space optics, such as metasurfaces [50–52]. In this case, such periodic structures behave as lossless metamaterials exhibiting a controlled and tailorable degree of anisotropy [53–55]. However, higher design flexibility comes at the cost of a larger design space, making finding the optimal geometrical parameters challenging. In this context, many optimization methods have been employed in nanophotonic devices, offering routes to leverage their full design potential [56].

In this work, we show through numerical simulation the potential of combining such dielectric metamaterials with a metaheuristic optimization method such as particle swarm optimization (PSO) [57] to achieve efficient coupling and polarization-insensitive grating couplers on the silicon-on-insulator (SOI) platform. We provide a comprehensive analysis to determine under what conditions Si/SiO_2 subwavelength periodic structures can be approximated to uniaxial homogeneous linear (UHL) materials. Moreover, considering the Pareto front, different configurations and strategies using the PSO algorithm are reported to benefit either the PDL or the coupling efficiency.

2. Principle of Polarization-Insensitive Grating Couplers Based on Subwavelength Si/SiO_2 Metamaterials

In standard silicon grating couplers, the relation between the grating's pitch and the working wavelength is given by the phase-matching condition [29]:

$$\Lambda = \frac{\lambda}{n_g - n_c \sin\theta} \tag{1}$$

where Λ is the pitch of the grating, λ the working wavelength, n_g the effective refractive index of the light in the grating, n_c the refractive index of the cladding material, and θ the angle of incidence of the light. The value of n_g can be expressed as

$$n_g = (1 - f)n_{wg} + fn_{etch} \tag{2}$$

where f is the fill factor of the etched region with respect to Λ, n_{wg} is the effective refractive index of the light in the unetched region of the slab waveguide, and n_{etch} is the effective refractive index of the light in the etched part of the slab waveguide. The value of n_{etch} will be higher for the fundamental mode of the transverse electric (TE) polarization than the fundamental mode of the transverse magnetic (TM) polarization in silicon slabs. Hence, PDL arises because the grating's phase-matching condition (Equation (1)) cannot be fulfilled for both polarizations and the same grating's geometry.

To enable the fulfillment of the polarization independence condition in the grating phase-matching equation [Equation (1)], the value of n_g must be the same for both polarizations. Thus, the fill factor must fulfill the following conditions:

$$\frac{1}{f} = 1 + \frac{n_{etch}^{TM} - n_{etch}^{TE}}{n_{wg}^{TE} - n_{wg}^{TM}} = 1 + \frac{\Delta n_{etch}}{\Delta n_{wg}}, \tag{3}$$

restricted by $0 < f < 1$ and $\Delta n_{wg} > 0$. As is shown later, in metamaterial-engineered grating couplers, it is possible to achieve $(n_{etch}^{TM} - n_{etch}^{TE}) > 0$ and thus a value for f between 0 and 1.

The proposed polarization-insensitive grating coupler is illustrated in Figure 1a. It consists of a one-dimensional grating along the light's propagation direction (z-axis) made up of unetched and etched silicon regions. The etched regions comprise a periodic stack with a Si/SiO_2 subwavelength structure (x-axis) that behaves as a dielectric metamaterial with UHL properties at a 1550 nm wavelength. Due to the UHL behavior of the periodic structure, the equivalent refractive index for the TE, n_{xx}, and TM, n_{yy}, polarizations will be different, as shown in Figures 1b and 1c, respectively. Moreover, the equivalent refractive index can be tailored to specific values by playing with the pitch, Λ_{SWG}, and the fraction of SiO_2, f_{SWG}, within a period. Additionally, for this work, we considered a silicon-on-insulator (SOI) platform with a 2 µm thick buried oxide layer (BOX), 1 µm thick cladding, and a fiber angle of 10°.

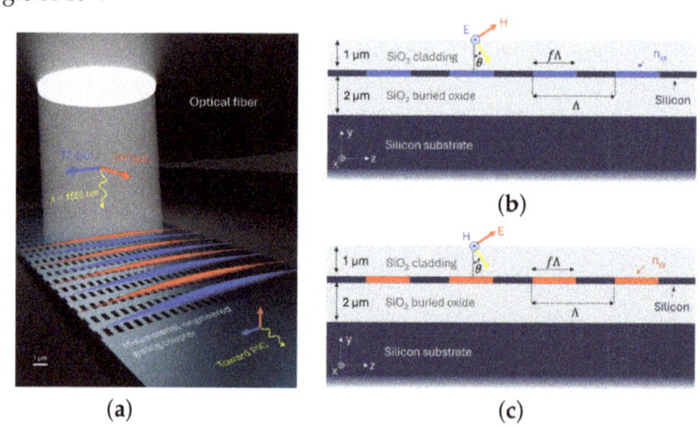

Figure 1. (**a**) 3D illustration of the proposed metamaterial-engineered silicon grating coupler with a polarization-insensitive response. The metamaterial is based on a subwavelength and periodic Si/SiO_2 stack. (**b**,**c**) Longitudinal cross-section of the grating coupler and the equivalent refractive index of the metamaterial under (**b**) TE and (**c**) TM polarization.

3. Applicability of Effective Medium Theory and Optimal Silicon Thickness

To facilitate the design of the grating coupler through numerical simulations, it was helpful to consider the subwavelength structure as a UHL medium. However, the effective medium theory (EMT) employed to carry out this approximation may not capture certain light phenomena, such as the formation of leaky waves at the interface of two media [58]. Therefore, it is convenient to analyze the impact of the subwavelength pitch in such terms. To this end, we hereinafter focus on the isolated subwavelength metamaterial to analyze its effective medium properties.

Let us consider a periodic structure formed by a stack of two materials with permittivities of ε_A and ε_B, where light impinges perpendicular to the periodicity. Although the structure in the grating coupler has a finite width, an infinite periodic approximation along the x-axis is valid since the width of grating couplers ($\sim$12 µm) is much larger than the working wavelength (1.55 µm) [29]. The periodic structure is depicted in Figure 2a, where a plane wave impinges perpendicular to a stack of material of thicknesses a and b with permittivities of ε_A and ε_B, respectively. The stack is periodic along the x-axis with a pitch of $\Lambda_{SWG} = a + b$ and extends infinitely along both the y- and z-axis. If the relation between the pitch and the incident wave's wavelength is small enough, then the periodic structures act as a UHL material with the following permittivity tensor:

$$\varepsilon_{SWG} = \begin{pmatrix} \varepsilon_{xx} & 0 & 0 \\ 0 & \varepsilon_{yy} & 0 \\ 0 & 0 & \varepsilon_{yy} \end{pmatrix}. \tag{4}$$

Thus, the equivalent metamaterial behaves as a uniaxial crystal with different permittivities depending on whether the light is TM-polarized, ε_{yy} (Figure 2b), or TE-polarized, ε_{xx} (Figure 2c).

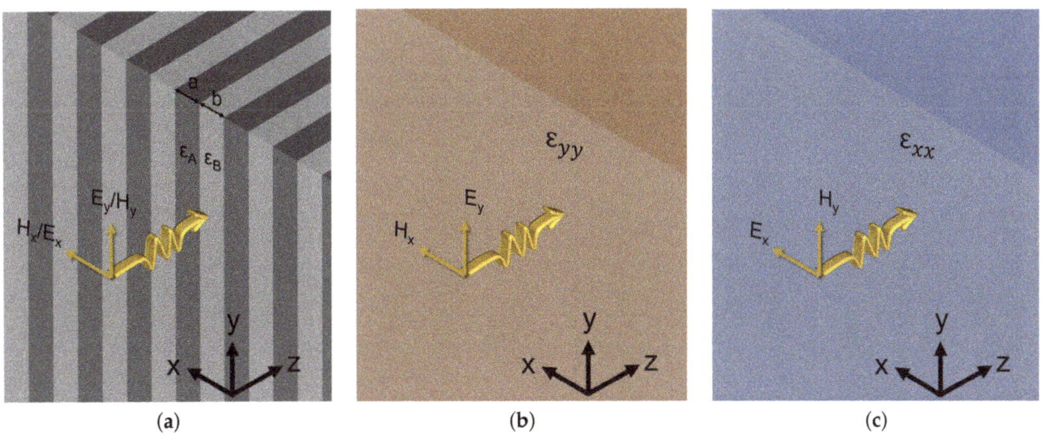

Figure 2. (**a**) Illustration of a plane wave with TE (E_x, H_y) or TM (E_y, H_y) polarization impinging along the z-axis from an isotropic homogeneous medium to a subwavelength one formed of a periodic stack of homogeneous materials with ε_A and ε_B permittivities. (**b**,**c**) Equivalent permittivity of the subwavelength medium when the incident wave is (**b**) TM- and (**c**) TE-polarized.

The values of ε_{xx} and ε_{yy} can be calculated by solving the dispersion relation equation of an infinite two-layer periodic structure [59–61]:

$$\cos(k_x \Lambda_{SWG}) = \cos(k_{z,a} a)\cos(k_{z,b} b) - \Delta \sin(k_{z,a} a)\sin(k_{z,b} b) \tag{5}$$

where

$$k_{z,a/b} = \frac{2\pi}{\lambda}\sqrt{\varepsilon_{a/b} - \varepsilon_{SWG}} \tag{6}$$

and Δ depends on the polarization as

$$\Delta^{TE} = \frac{1}{2}\left(\frac{\varepsilon_b k_{z,a}}{\varepsilon_a k_{z,b}} + \frac{\varepsilon_a k_{z,b}}{\varepsilon_b k_{z,a}}\right) \tag{7}$$

and

$$\Delta^{TM} = \frac{1}{2}\left(\frac{k_{z,a}}{k_{z,b}} + \frac{k_{z,b}}{k_{z,a}}\right). \tag{8}$$

Since the light propagates only along the z-axis, the term k_x equals zero.

Equation (5) has no analytical solution, and numeric root-finding methods are necessary. However, suppose the pitch–wavelength relation is small enough. In that case, it can be accurately approximated to the zero order of a Taylor expansion by using an EMT model such as the Maxwell Garnett model [62]. For this kind of structure, the approximations for both polarizations are the well-known Rytov's formulas [63], given by Equations (9) and (10) for TM and TE polarization, respectively. A second-order approximation can be utilized if the relation is not small enough. Nevertheless, obtaining the exact solution by solving Equation (5) is preferred to obtain much higher accuracy.

$$\varepsilon_{yy} = \frac{a}{\Lambda_{SWG}}\varepsilon_A + \frac{b}{\Lambda_{SWG}}\varepsilon_B \tag{9}$$

$$\frac{1}{\varepsilon_{xx}} = \frac{a}{\Lambda_{SWG}}\frac{1}{\varepsilon_A} + \frac{b}{\Lambda_{SWG}}\frac{1}{\varepsilon_B} \tag{10}$$

However, it is difficult to rigorously define the term "small enough" and thereby approximate the metamaterial as an equivalent UHL medium. On the other hand, by using Bragg's condition, it can be proved that if the pitch fulfills

$$\Lambda_{SWG} < \frac{\lambda}{\max(\sqrt{\varepsilon_A}, \sqrt{\varepsilon_B})} \tag{11}$$

then diffraction is not allowed, and the light travels through the subwavelength structure with a phase given by the dispersion relation expression [Equation (5)]. However, this does not mean the subwavelength structure behaves as a homogeneous material. If $\Lambda/\lambda \to 0$ is not accomplished, leaky waves could arise at the interface, which could tunnel through the grating structure if it is not long enough and affect its performance. Hence, the material cannot be treated as homogeneous.

Setting a value of Λ_{SWG} for which the subwavelength structure can be treated as a UHL medium is not trivial. Here, we have analyzed the following aspects: (i) the dispersion relation equation, (ii) the field profile inside the periodic structure, and (iii) the transmittance/reflectance values when it is interfaced with a homogeneous material. In the following analysis, we consider a stack comprised of Si ($\varepsilon_r = 12.08$) and SiO_2 ($\varepsilon_r = 2.085$) at $\lambda = 1550$ nm. Moreover, for the ease of calculations, the thickness of both layers is set to $a = b = \Lambda_{SWG}/2$.

3.1. Equivalent Refractive Index

The dispersion relation equation [Equation (5)] was first investigated by obtaining the real solutions for both polarizations as a function of the pitch, thus yielding an equivalent refractive index, n_{eq}. As shown in Figure 3a, for pitch values lower than $\lambda/n_{Si} = 445$ nm, only the zero order was allowed, in agreement with Equation (11), and the structure is in the subwavelength regime. It has to be noticed that the lower the pitch, the higher the birefringence between the effective refractive index of both polarizations. As discussed later, large birefringence values are beneficial for designing polarization-insensitive grating couplers. When the subwavelength condition is not fulfilled, high-order modes begin to arise. In contrast, when $\Lambda_{SWG} \to +\infty$, all the solutions converge to plane waves.

A comparison between the rigorous solution given by Equation (5) and Rytov's approximation [Equations (10) and (9)] is shown in Figure 3b. As can be noticed, the approximation was only accurate for pitch values lower than ∼100 nm. For higher values, it could only be used to estimate the values of the equivalent refractive index.

Regarding the imaginary solutions corresponding to leaky waves, we observed differences between those obtained from approximating the periodic stack as an equivalent homogeneous material and solving Equation (5). In a homogeneous medium, the imaginary solutions of n_{eq} are antisymmetric, i.e., $n_{eq,m} = -n_{eq,-m}$, where m is the mth imaginary solution. However, in a periodic subwavelength medium, the solutions given by Equation (5) are not antisymmetric and are only approximate when $\Lambda_{SWG} \leq 100$ nm.

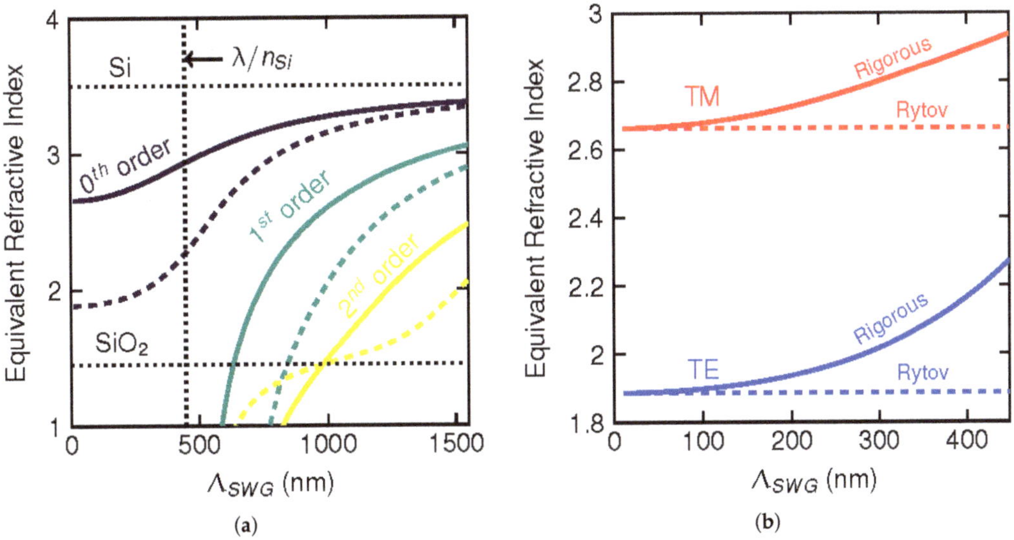

Figure 3. (**a**) Equivalent refractive index for TM- (solid line) and TE-polarized (dashed line) light as a function of the subwavelength pitch, calculated using Equation (5). (**b**) Comparison with Rytov's approximation [Equations (9) and (10)]. Equivalent refractive index for both polarizations as a function of the subwavelength pitch. Results were obtained for the different supported solutions at 1550 nm and for the Si/SiO_2 stack.

3.2. Field Profile

Recalling Maxwell's equations, a plane wave traveling through a homogeneous medium has only one component in both the E- and H-fields, with a constant value in the axis transverse (x-axis) to the propagation direction (z-axis). However, in a crosswise subwavelength periodic medium, two components (transverse and longitudinal) exist in the E- or H-field (depending on the polarization). The different profiles of the field components were obtained by implementing the transfer matrix method (TMM) in MATLAB 2020a [61].

The transverse components for TE and TM polarizations are depicted in Figures 4a and 4b, respectively, for different pitch values. The amplitude diminishes as the pitch is reduced, making it negligible for pitch values lower than 100 nm, thereby approximating a plane wave in a homogeneous medium.

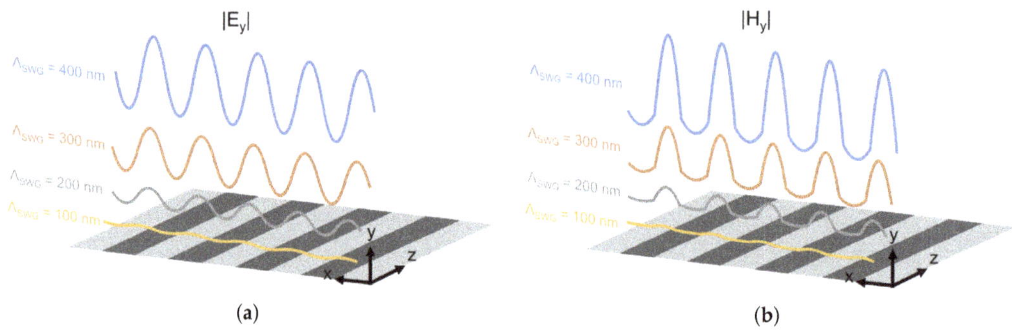

Figure 4. Field profiles obtained by TMM for different values of pitch. (**a**) TM and (**b**) TE polarization. Results are given for Si/SiO_2 stack at $\lambda = 1550$ nm.

On the other hand, the impact of the longitudinal component can be visualized by representing the ellipse of polarization. These ellipses are plotted for different pitch values in Figures 5a and 5b for TE (E_x and E_z) and TM (H_x and H_z) polarization, respectively. The ellipse was calculated at the interface of the Si/SiO_2 stack. The interface point was chosen since the longitudinal component attained its maximum value. As noted, for both polarizations, the polarization ellipse tended to shrink as the pitch was reduced, i.e., only the transverse component, as in a homogeneous medium, existed. Moreover, for TE polarization, this was more pronounced than for TM polarization because in the latter, $E_z \propto \varepsilon(x)$. On the other hand, the axial ratio was obtained as $|\phi_x/\phi_z|$ and is depicted in Figure 5c. For a plane wave, the axial ratio is infinity; however, in a subwavelength medium, this is not accomplished, as noted.

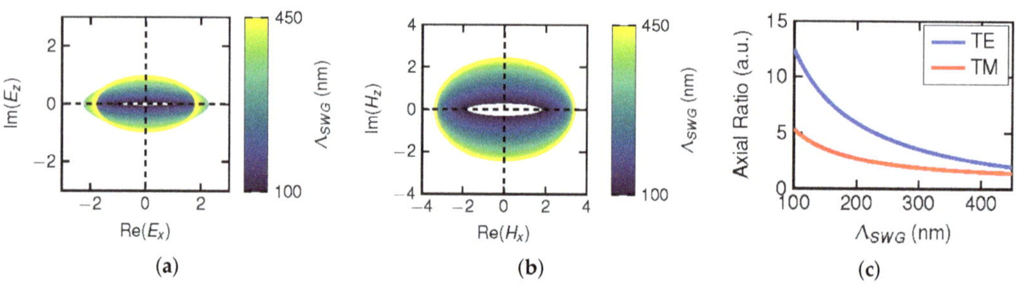

Figure 5. Ellipse of polarization for (**a**) TE and (**b**) TM polarization as a function of the pitch. (**c**) Axial ratio for both polarizations as a function of the pitch. Results were obtained for a Si/SiO_2 stack at $\lambda = 1550$ nm.

3.3. Interface with Homogeneous Material

A non-constant field along the x-axis must be considered when the subwavelength structure is interfaced with a homogeneous material. To fulfill boundary conditions, significant leaky waves may arise at the interface, thus making it inaccurate to treat the periodic stack as a homogeneous material.

To analyze this behavior, we considered the previous SiO_2/Si stack (medium II) interfaced with silicon (medium I). A plane wave traveled along the z-axis from medium I and impinged to medium II. The field profile at the interface of Si and the Si/SiO_2 stack and each medium was obtained using the eigenmode expansion method [64]. In contrast with other simulation methods such as the finite-difference time domain (FDTD), the utilization of eigenmode expansion allowed for discrimination between the propagating modes

($m = 0$) and the leaky modes ($m \neq 0$) and thus for gaining insight into the phenomena that occurred within the structure. Calculations were carried out for both polarizations and pitch values of 400 and 100 nm.

In Figure 6, the field distribution (x,z) is depicted for $\Lambda_{SWG} = 400$ nm of the $|E_y|$ (Figure 6a) and $|H_y|$ (Figure 6b) components for the TM- and TE-polarized waves, respectively. As can be noticed, leaky modes at the interface perturb the adjoint wave between the two media for both polarizations. However, for TE polarization, they extend further along the z-axis. Such leaky modes arise to satisfy the continuous condition of the tangential electric field at the interface between the two media. Therefore, if several homogeneous/subwavelength media are stacked along the z-axis, as in grating couplers, leaky waves may interact with the propagated wave, giving rise to optical perturbations. Consequently, the subwavelength medium cannot be approximated as being homogeneous. In contrast, if the value of Λ_{SWG} is reduced to 100 nm, leaky waves are negligible, as shown in Figure 7, and thus, a homogeneous approximation can be accurate.

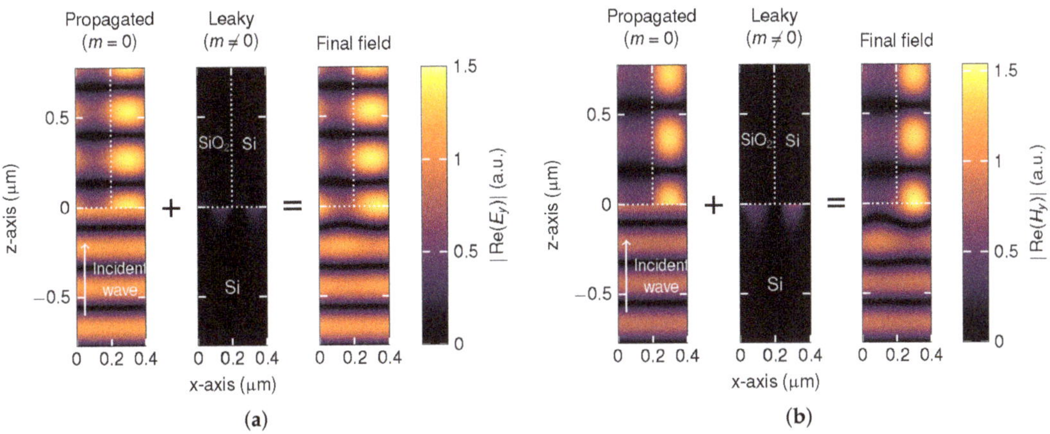

Figure 6. Field profiles (propagated + leaky) obtained through eigenmode expansion and mode-matching when a plane wave in a homogeneous medium made of Si (medium I) impinges into a crosswise subwavelength structure made of a Si/SiO_2 stack with a 400 nm pitch (medium II). The final field is the superposition of the propagated mode with the leaky waves that arise at the interface between mediums I and II. Results are given for (**a**) TM and (**b**) TE polarization at 1550 nm.

Therefore, based on these results, we chose $\Lambda_{SWG} = 100$ nm for the design of our polarization-insensitive grating coupler. Figure 8a shows the equivalent refractive index as a function of the filling factor using the zero-order approximation given by Rytov. As can be noticed, TM polarization exhibits a larger refractive index than TE polarization, which could enable the fulfillment of the polarization independence condition of the grating coupler. Moreover, the birefringence, Δn, can be as large as 0.8 for filling factors of around 40% and greater than 0.2 for almost the whole range (see Figure 8b).

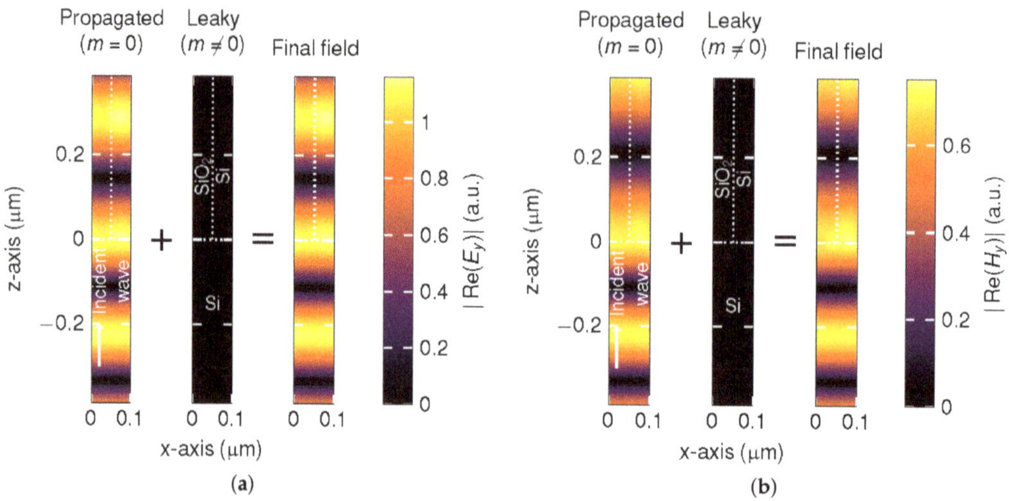

Figure 7. Field profiles (propagated + leaky) obtained through eigenmode expansion and mode-matching when a plane wave in a homogeneous medium made of Si (medium I) impinges into a crosswise subwavelength structure made of a Si/SiO_2 stack with a 100 nm pitch (medium II). The final field is the superposition of the propagated mode with the leaky waves that arise at the interface between mediums I and II. Results are given for (**a**) TM and (**b**) TE polarization at 1550 nm.

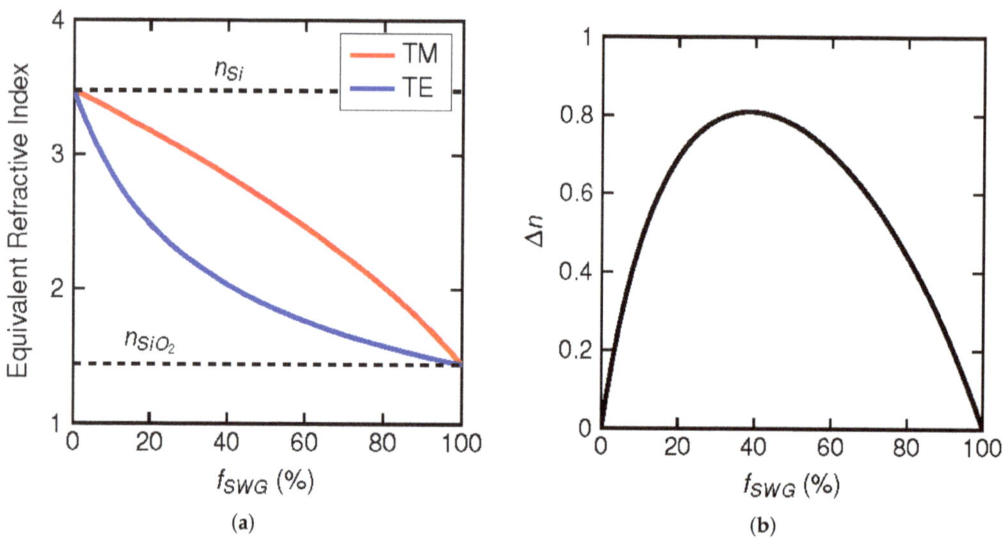

Figure 8. (**a**) Equivalent refractive index and (**b**) associated birefringence, Δn, as a function of the fill factor of the periodic structure. Results are given for a stack of SiO_2/Si and were obtained using Rytov's expressions [Equations (10) and (9)].

3.4. Optimal Thickness of Silicon

The thickness of the silicon layer plays a vital role in a grating coupler's performance. A thick silicon layer provides a broader range of tunability in the equivalent refractive index of the metamaterial. The best case or lower limit of the fill factor is given when Δn_{etch} attains its maximum value, i.e., when the subwavelength structure achieves the maximum birefringence ($f_{SWG} = 0.4$). For the latter, the values of Δn_{wg} and Δn_{etch} are depicted as a function of the waveguide thickness in Figure 9a. As the thickness increases from the standard 220 nm, the difference between both increments decreases until it reaches around 320 nm. The associated minimum value of f is shown in Figure 9b. Thick waveguide thicknesses may not be appropriate as they are highly multimodal and more sensitive to environmental variations such as the temperature. Thus, considering this trade-off, we chose 320 nm as the thickness for the silicon layer of the gratings. On the other hand, regarding the upper limit of the fill factor, this is achieved when $n_{etch}^{TM} - n_{etch}^{TE} = 0$. For a 320 nm thick waveguide, this yields a f_{SWG} value of between 7% and 99.6% (see Figure 8). Finally, minimum feature sizes below 10 nm can be obtained by electron beam lithography (EBL) [65,66] and are expected to be improved using photolithography in silicon photonic foundries. Hence, the upper and lower bounds of f_{SWG} were set to 0.1 and 0.9, respectively.

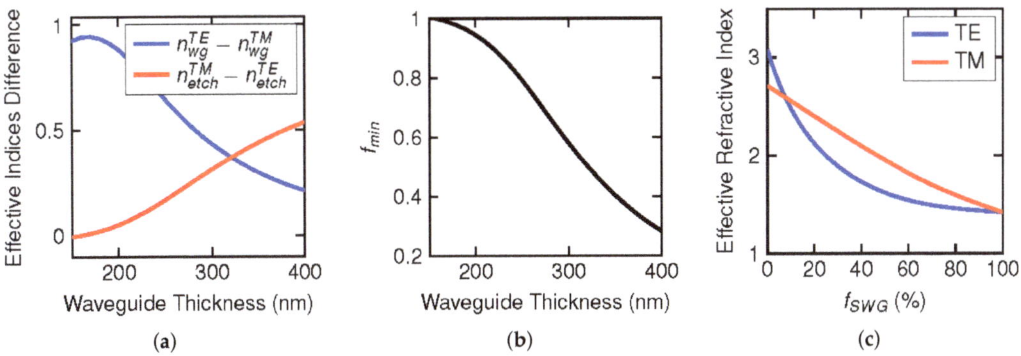

Figure 9. (**a**) Effective index difference and (**b**) minimum attainable value of the filling factor, f_{min}, as a function of the waveguide thickness. Results are given for a stack of SiO_2/Si of $\Lambda = 100$ nm and $f_{SWG} = 40\%$ at $\lambda = 1550$ nm. (**c**) Effective refractive index as a function of the f_{SWG} value for a 320 nm thick slab waveguide with a core made of equivalent material to that of the aforementioned subwavelength structure.

4. Achieving High Optical Performance via Particle Swarm Optimization

The phase-matching condition can be fulfilled for both polarizations using a large set of f_{SWG} values. However, each solution yields a different grating coupler performance. Indeed, the diffraction strength, i.e., the amount of power that is diffracted due to the impedance mismatch of different waveguides, is different for both polarizations. Figure 10a shows the coupling losses between a 320 nm thick Si slab waveguide and a metamaterial slab waveguide comprising the subwavelength structure as a function of the f_{SWG} value. The coupling losses are higher for TE polarization because it presents a more significant impedance mismatch than TM polarization. The difference in the diffraction strength is increased with the fill factor achieving its maximum at $f_{SWG} \approx 85\%$, as shown in Figure 10b. Hence, the optimization of each grating period should be carried out to ensure the maximum performance.

As mentioned in the Introduction, in this work, optimization is based on the meta-heuristic method of PSO. The flow chart of the implemented PSO algorithm is depicted in Figure 11. This begins with the initialization of the particles that comprise the swarm. The swarm consists of N particles, with each particle being M-dimensional. In our case, a particle represents a grating coupler, for which the variables are f_{SWG}, λ^{TE}, and λ^{TM}. The wavelength values are set as variables for greater freedom during optimization. Then, the following steps are repeated I_{max} times. First, the performance of the particles is evaluated using the fitness, which assigns a cost to each particle. If each particle's cost is better (lower) than in the previous iteration, then the pBest (particle's best) is updated. Moreover, among the particles, their positions are compared to the gBest (global best), saving the best particle from all iterations. When the pBests and gBest are updated, if required, the velocity of each particle is updated to obtain their new position.

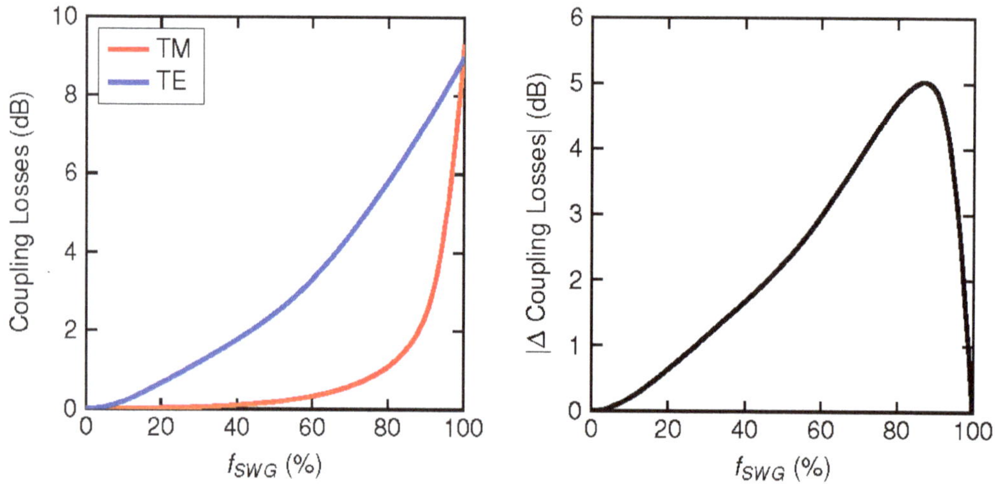

Figure 10. (**a**) Coupling losses and (**b**) increment between 320 nm thick Si and metamaterial waveguides as a function of the subwavelength filling factor for both polarizations. Results are for a stack of SiO_2/Si of $\Lambda_{SWG} = 100$ nm at $\lambda = 1550$ nm.

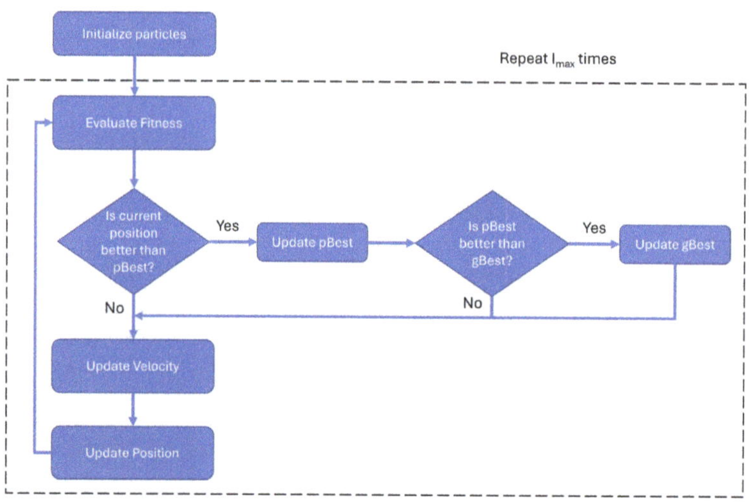

Figure 11. Flow chart of the PSO algorithm.

The velocity of the particles is calculated by using the following expression:

$$\boldsymbol{v}_{i+1}^{(n,m)} = w_i \boldsymbol{v}_i^{(n,m)} + c_1 r_1 \left(\boldsymbol{g}_i^{(m)} - \boldsymbol{x}_i^{(n,m)} \right) + c_2 r_2 \left(\boldsymbol{p}_i^{(n,m)} - \boldsymbol{x}_i^{(n,m)} \right) \tag{12}$$

where w is the inertia weight that prevents the particle from drastically changing its direction, c_1 is the cognitive coefficient that is related to the memory of the previous best position, c_2 is the social coefficient and relates to the neighbors, r is a random number between 0 and 1 following a uniform distribution, $x_i^{(n,m)}$ stands for the current position, and $p_i^{(n,m)}$ and $g_i^{(m)}$ are the particle's best position and the global best position, respectively. The coefficients c_1 and c_2 impose a trade-off between exploration and exploitation, i.e., the ability to explore various regions of the search space or to concentrate the search around a promising area, respectively. For instance, if $c_1 = c_2 > 0$, particles are attracted toward the average of the personal best position and the global best position. On the other hand, $c_2 > c_1$ is more beneficial for unimodal problems, whereas $c_2 < c_1$ is better for multimodal problems. Then, the position is updated as

$$x_{i+1}^{(n,m)} = x_i^{(n,m)} + v_{i+1}^{(n,m)} \tag{13}$$

Every nature-inspired algorithm is not exempt from potential issues. The most common are related to convergence. For instance, the optimal solution can stick to a local minimum, or the particle position goes out of the boundaries. Several solutions have been proposed to tackle these problems [67]. Among them, in this work, we used the following strategies:

- Usually, the positions of particles are initialized to cover the search space uniformly. An efficient initialization method for the particle position is

$$x_0^{(n,m)} = x_{\min} + r(x_{\max} - x_{\min}) \tag{14}$$

- To control the global exploration of particles, velocities are clamped to stay within the boundary constraints. The velocity clamping is defined as

$$v_{i+1}^{(n,m)} = \begin{cases} v_{i+1}^{(n,m)} & \text{if } v_{i+1}^{(n,m)} < v_{\max} \\ v_{\max} & \text{if } v_{i+1}^{(n,m)} \geq v_{\max} \end{cases} \tag{15}$$

  where $v_{\max} = k(v_{ub} - v_{lb})$, with k being a constant between 0 and 1 and v_{up} and v_{lb} the upper and lower boundaries of the velocity, respectively. On the other hand, if the particle is about to go beyond the boundary limits, the velocity is set randomly to set the next position of the particle within the boundaries.
- The inertia weight value is problem-dependent. A common strategy is to start with a large value to facilitate exploration and linearly decrease its value to promote exploitation. Thus,

$$w_i = (w_0 - w_{I_{max}})\frac{I_{max} - i}{I_{max}} + w_{I_{max}} \tag{16}$$

  where $w_0 = 0.9$ and $w_{I_{max}} = 0.4$.
- PSO can find optimal solutions with a small population of between 10 and 30. In this work, we employed 25 particles.
- Particles draw their strength from their cooperative nature and are more effective when nostalgia (c_1) and envy (c_2) coexist in a good balance ($c_1 \approx c_2$).

  Table 1 shows the parameters utilized for all optimization cases.

Table 1. Summary of the PSO parameters used for optimization.

Iterations	# of Particles	c_1	c_2	w_0	$w_{I_{max}}$	k
150	25	2.05	2.05	0.9	0.4	0.1

The fitness of the solutions is assessed by defining a figure of merit (FOM) that includes the following metrics: the coupling efficiency (CE), bandwidth (BW), PDL, and PDL BW. The FOM gives a value that should be minimized in this case. Defining the right

FOM is crucial to obtaining the best solution, and its definition depends on the required specifications. However, one must keep in mind that trade-offs exist among these metrics, and not all can be improved at the same time, known as the Pareto front, in which one parameter or some parameters will not be improved without diminishing others [68].

High coupling, efficiency, and a large bandwidth are desired in a grating coupler. However, there is a trade-off between them (2D Pareto front). Moreover, in a polarization-insensitive grating coupler, the PDL and its bandwidth are included as extra parameters, which extends the Pareto front to four dimensions. The dependence among the four parameters is reflected in Figure 12.

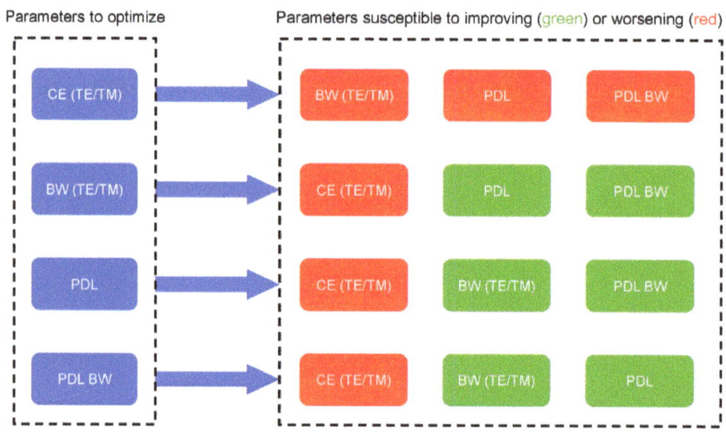

Figure 12. Grating coupler parameters' dependence.

This work considered using grating couplers to cover the telecom C-band (1530–1565 nm) while maximizing the coupling efficiency at 1550 nm. The number of grating periods was fixed at 20. On the other hand, since the perfect coupler does not exist according to Figure 12, three different FOMs were proposed depending on the critical parameter to improve, while the others could be relaxed. The first FOM included all four parameters [Equation (17)] and tried to balance them. The second one was focused on the coupling efficiency, and the PDL and bandwidth constraints were relaxed [Equation (18)]. Finally, the last FOM relaxed the performance at 1550 nm and aimed to cover the entire C-band with PDLs of lower than 1 dB [Equation (19)]. The values of the PDL and CE were calculated in dB and were obtained at 1550 nm, while the bandwidth had nanometer units and was clipped to within the C-band wavelength range (35 nm).

$$FOM = \frac{35}{BW^{PDL}}PDL - \frac{35}{BW^{TE}}CE^{TE} - \frac{35}{BW^{TM}}CE^{TM} \tag{17}$$

$$FOM = -\left(CE^{TE} + CE^{TM}\right)\max(1, PDL) \tag{18}$$

$$FOM = -\left(CE^{TE} + CE^{TM}\right)\left(\frac{35}{BW^{PDL}}\right)^{8} \tag{19}$$

The PSO algorithm was implemented using an in-house script adapted to the RSoft commercial simulation tool [69], which was employed to calculate the optical metrics of the grating couplers using the 2D-FDTD method. Other numerical simulation tools combining the FDTD method with a built-in PSO optimizer could be employed with the proposed PSO settings and workflow.

For the first case, the evolution of the gBest fitness as a function of the iterations is shown in Figure 13a. The cost of the gBest rapidly decreased in the first few iterations. This behavior was attributed to the size of the swarm and the promotion of exploration. Afterward, the cost decreased at a slow pace and converged from the 100th iteration. The spectra for the gBest particle at the beginning and at the end of the optimization are depicted in Figures 13b and 13c, respectively. At the beginning, the PDLs were very low because of the low coupling efficiency for both polarizations (see Figure 12). For the final spectra, the coupling efficiencies were increased from −7.5 dB to −4 dB while keeping the PDL lower than 1 dB across the entire C-band.

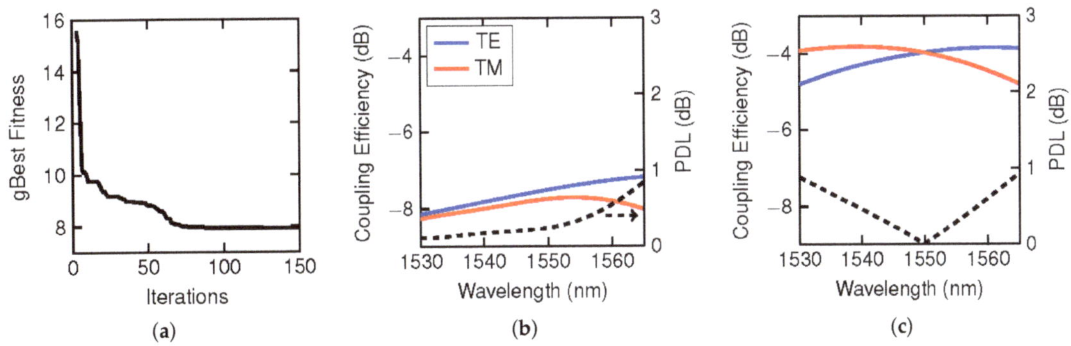

Figure 13. (**a**) Fitness of the gBest as a function of the iterations. (**b**) Initial and (**c**) final spectra of the gBest. Results correspond to FOM #1.

The optimized grating's physical parameters can be found in the Supplementary Materials. Most of the f_{SWG} values were between 35% (35 nm) and 75% (75 nm). Thus, the footprint of the structures was far from the e-beam lithography minimum feature size.

The algorithm converged in the first iterations for the second FOM, as seen in Figure 14a. The initial spectra did not meet the requirement of having less than 1 dB for the PDL at 1550 nm. Moreover, the coupling efficiencies were quite low (see Figure 14b). These values were improved by the end of the optimization, as shown in Figure 14c. The PDL was lower than 1 dB. However, the PDL bandwidth was narrower than in the first case since this metric was not included in the FOM. Regarding the coupling efficiencies, these suffered an increase of up to 5 dB, and values of as high as −2.5 dB were achieved for TE polarization. The geometrical values of the optimized structures can be found in the Supplementary Materials. In this regard, it would be feasible to fabricate these by e-beam lithography.

Finally, for the third analyzed case, the PSO converged similarly to in FOM #1, as seen in Figure 15a. Although the initial spectra met the requirement of having a PDL bandwidth in the C-band with PDLs lower than 1 dB, the coupling efficiencies were relatively low (see Figure 15b). After the optimization, the PDL requirements were still met, but the coupling efficiencies improved by more than 5 dB, as shown in Figure 14c. The physical values corresponding to the last FOM are summarized in the Supplementary Materials. As in the previous designs, the values were within the boundaries, and the physical parameters would be feasible to fabricate.

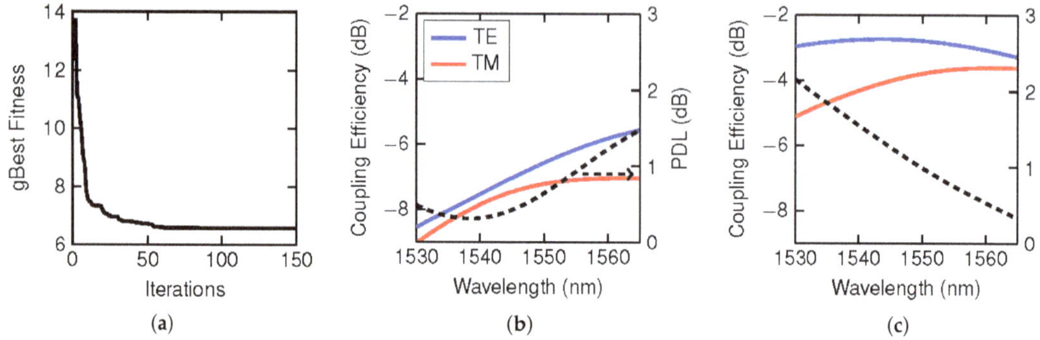

Figure 14. (**a**) Fitness of the gBest as a function of the iterations. (**b**) Initial and (**c**) final spectra of the gBest. Results correspond to FOM #2.

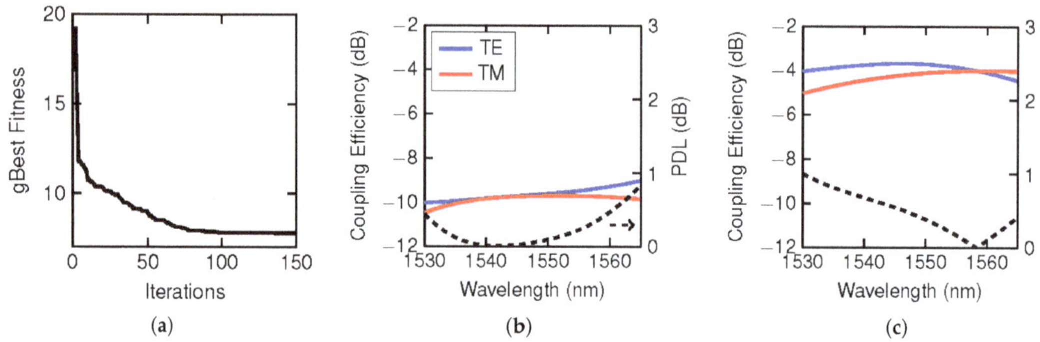

Figure 15. (**a**) Fitness of the gBest as a function of the iterations. (**b**) Initial and (**c**) final spectra of the gBest. Results correspond to FOM #3.

The final metrics of the optimized gratings for the different FOMs are summarized in Table 2. The final geometries of the optimized grating couplers are listed in the Supplementary Materials. The overall best grating performance was achieved with the grating corresponding to FOM #1. The grating of FOM #2 presented a better CE for TE polarization but at the cost of a lower bandwidth and an increase in the PDL. Finally, the grating of FOM #3 covered the entire C-band with a PDL lower than 1 dB but with slightly worse CE values than FOM #1 since this parameter was not evaluated in this FOM.

Table 2. Comparison of the performance of the final gratings. Coupling efficiencies and PDLs are given at $\lambda = 1550$ nm.

	TE		**TM**		**PDL**	
	CE (dB)	**BW (nm)**	**CE (dB)**	**BW (nm)**	**PDL (dB)**	**BW (nm)**
FOM #1	−3.96	35.0	−3.97	35.0	0.01	35.0
FOM #2	−2.79	35.0	−3.78	29.06	0.99	21.3
FOM #3	−3.70	35.0	−4.08	34.4	0.39	35.0

5. Conclusions

In this work, we have demonstrated the potential of polarization-insensitive silicon grating couplers designed using subwavelength metamaterials and the PSO method. The proposed approach achieved efficient coupling with a minimal PDL by leveraging the anisotropic properties of Si/SiO_2 subwavelength structures. Our numerical simulations confirmed that a 100 nm subwavelength pitch ensures UHL behavior in such subwavelength structures by negating the influence of leaky waves. On the other hand, a silicon thickness of 320 nm delivered a broad tunable range for the metamaterial's fill factor. The optimized grating couplers achieved coupling efficiencies exceeding −3 dB and a PDL below 1 dB across the telecom C-band (1530–1565 nm). Three optimization strategies were explored, revealing trade-offs between the coupling efficiency, bandwidth, and PDL through Pareto front analysis. Moreover, our designs are compatible with standard CMOS fabrication processes employed for silicon-based integrated photonics and coupler devices, such as e-beam lithography and reactive ion etching [29].

Therefore, our work establishes a robust framework combining metamaterial engineering with computational optimization, paving the way for advanced photonic devices. The findings are particularly relevant for photonic applications in telecommunications, quantum technologies, and artificial intelligence, where high performance and polarization insensitivity are vital.

Supplementary Materials: The following supporting information can be downloaded at https://www.mdpi.com/article/10.3390/photonics12050428/s1: Table S1: Geometrical grating coupler parameters obtained for FOM #1; Table S2: Geometrical grating coupler parameters obtained for FOM #2; Table S3: Geometrical grating coupler parameters obtained for FOM #3.

Funding: This research was partly funded by the Universitat Politècnica de València, grants PAID-10-23 and PAID-06-23, and partly by grant PID2022-137787OB-I00, funded by MCIN/AEI/10.13039/501100011033.

Institutional Review Board Statement: Not applicable.

Informed Consent Statement: Not applicable.

Data Availability Statement: The original contributions presented in this study are included in the article/Supplementary Material. Further inquiries can be directed to the corresponding author.

Conflicts of Interest: The authors declare no conflicts of interest.

References

1. Marx, V. The big challenges of big data. *Nature* **2013**, *498*, 255–260. [CrossRef] [PubMed]
2. Lord, A.; Soppera, A.; Jacquet, A. The impact of capacity growth in national telecommunications networks. *Philos. Trans. R. Soc. A Math. Phys. Eng. Sci.* **2016**, *374*, 20140431. [CrossRef] [PubMed]
3. Liang, W.; Tadesse, G.A.; Ho, D.; Li, F.; Zaharia, M.; Zhang, C.; Zou, J. Advances, challenges and opportunities in creating data for trustworthy AI. *Nat. Mach. Intell.* **2022**, *4*, 669–677. [CrossRef]
4. Wu, C.J.; Raghavendra, R.; Gupta, U.; Acun, B.; Ardalani, N.; Maeng, K.; Chang, G.; Behram, F.A.; Huang, J.; Bai, C.; et al. Sustainable AI: Environmental Implications, Challenges and Opportunities. *arXiv* **2022**, arXiv:2111.00364.
5. Crawford, K. Generative AI's environmental costs are soaring—And mostly secret. *Nature* **2024**, *626*, 693. [CrossRef]
6. Shastri, B.J.; Tait, A.N.; Ferreira De Lima, T.; Pernice, W.H.P.; Bhaskaran, H.; Wright, C.D.; Prucnal, P.R. Photonics for artificial intelligence and neuromorphic computing. *Nat. Photonics* **2021**, *15*, 102–114. [CrossRef]
7. Winzer, P.J.; Neilson, D.T. From Scaling Disparities to Integrated Parallelism: A Decathlon for a Decade. *J. Light. Technol.* **2017**, *35*, 1099–1115. [CrossRef]
8. Winzer, P.J.; Neilson, D.T.; Chraplyvy, A.R. Fiber-optic transmission and networking: The previous 20 and the next 20 years [Invited]. *Opt. Express* **2018**, *26*, 24190. [CrossRef]
9. Bourzac, K. Fixing AI's energy crisis. *Nature* 2024, *online ahead of print*. [CrossRef]
10. Siew, S.Y.; Li, B.; Gao, F.; Zheng, H.Y.; Zhang, W.; Guo, P.; Xie, S.W.; Song, A.; Dong, B.; Luo, L.W.; et al. Review of Silicon Photonics Technology and Platform Development. *J. Light. Technol.* **2021**, *39*, 4374–4389. [CrossRef]
11. Smit, M.; Williams, K.; Van Der Tol, J. Past, present, and future of InP-based photonic integration. *APL Photonics* **2019**, *4*, 050901. [CrossRef]
12. Blumenthal, D.J.; Heideman, R.; Geuzebroek, D.; Leinse, A.; Roeloffzen, C. Silicon Nitride in Silicon Photonics. *Proc. IEEE* **2018**, *106*, 2209–2231. [CrossRef]

13. Li, N.; Ho, C.P.; Zhu, S.; Fu, Y.H.; Zhu, Y.; Lee, L.Y.T. Aluminium nitride integrated photonics: A review. *Nanophotonics* **2021**, *10*, 2347–2387. [CrossRef]
14. Zhu, D.; Shao, L.; Yu, M.; Cheng, R.; Desiatov, B.; Xin, C.J.; Hu, Y.; Holzgrafe, J.; Ghosh, S.; Shams-Ansari, A.; et al. Integrated photonics on thin-film lithium niobate. *Adv. Opt. Photonics* **2021**, *13*, 242. [CrossRef]
15. Abel, S.; Eltes, F.; Ortmann, J.E.; Messner, A.; Castera, P.; Wagner, T.; Urbonas, D.; Rosa, A.; Gutierrez, A.M.; Tulli, D.; et al. Large Pockels effect in micro- and nanostructured barium titanate integrated on silicon. *Nat. Mater.* **2019**, *18*, 42–47. [CrossRef]
16. Dong, P.; Chen, Y.K.; Duan, G.H.; Neilson, D.T. Silicon photonic devices and integrated circuits. *Nanophotonics* **2014**, *3*, 215–228. [CrossRef]
17. Lipson, M. The revolution of silicon photonics. *Nat. Mater.* **2022**, *21*, 974–975. [CrossRef]
18. Shekhar, S.; Bogaerts, W.; Chrostowski, L.; Bowers, J.E.; Hochberg, M.; Soref, R.; Shastri, B.J. Roadmapping the next generation of silicon photonics. *Nat. Commun.* **2024**, *15*, 751. [CrossRef]
19. Sibson, P.; Kennard, J.E.; Stanisic, S.; Erven, C.; O'Brien, J.L.; Thompson, M.G. Integrated silicon photonics for high-speed quantum key distribution. *Optica* **2017**, *4*, 172. [CrossRef]
20. Moody, G.; Sorger, V.J.; Blumenthal, D.J.; Juodawlkis, P.W.; Loh, W.; Sorace-Agaskar, C.; Jones, A.E.; Balram, K.C.; Matthews, J.C.F.; Laing, A.; et al. 2022 Roadmap on integrated quantum photonics. *J. Phys. Photonics* **2022**, *4*, 012501. [CrossRef]
21. Harris, N.C.; Bunandar, D.; Pant, M.; Steinbrecher, G.R.; Mower, J.; Prabhu, M.; Baehr-Jones, T.; Hochberg, M.; Englund, D. Large-scale quantum photonic circuits in silicon. *Nanophotonics* **2016**, *5*, 456–468. [CrossRef]
22. Xu, B.; Huang, Y.; Fang, Y.; Wang, Z.; Yu, S.; Xu, R. Recent Progress of Neuromorphic Computing Based on Silicon Photonics: Electronic–Photonic Co-Design, Device, and Architecture. *Photonics* **2022**, *9*, 698. [CrossRef]
23. Farmakidis, N.; Dong, B.; Bhaskaran, H. Integrated photonic neuromorphic computing: Opportunities and challenges. *Nat. Rev. Electr. Eng.* **2024**, *1*, 358–373. [CrossRef]
24. El Srouji, L.; Krishnan, A.; Ravichandran, R.; Lee, Y.; On, M.; Xiao, X.; Ben Yoo, S.J. Photonic and optoelectronic neuromorphic computing. *APL Photonics* **2022**, *7*, 051101. [CrossRef]
25. Kutluyarov, R.V.; Zakoyan, A.G.; Voronkov, G.S.; Grakhova, E.P.; Butt, M.A. Neuromorphic Photonics Circuits: Contemporary Review. *Nanomaterials* **2023**, *13*, 3139. [CrossRef]
26. Zhang, X.; Kwon, K.; Henriksson, J.; Luo, J.; Wu, M.C. A large-scale microelectromechanical-systems-based silicon photonics LiDAR. *Nature* **2022**, *603*, 253–258. [CrossRef]
27. Sun, X.; Zhang, L.; Zhang, Q.; Zhang, W. Si Photonics for Practical LiDAR Solutions. *Appl. Sci.* **2019**, *9*, 4225. [CrossRef]
28. Shi, Y.; Zhang, Y.; Wan, Y.; Yu, Y.; Zhang, Y.; Hu, X.; Xiao, X.; Xu, H.; Zhang, L.; Pan, B. Silicon photonics for high-capacity data communications. *Photonics Res.* **2022**, *10*, A106. [CrossRef]
29. Marchetti, R.; Lacava, C.; Carroll, L.; Gradkowski, K.; Minzioni, P. Coupling strategies for silicon photonics integrated chips [Invited]. *Photonics Res.* **2019**, *7*, 201. [CrossRef]
30. Larrea, R.; Gutierrez, A.M.; Griol, A.; Brimont, A.; Sanchis, P. Fiber-to-Chip Spot-Size Converter for Coupling to Silicon Waveguides in the O-Band. *IEEE Photonics Technol. Lett.* **2019**, *31*, 31–34. [CrossRef]
31. Galán, J.V.; Sanchis, P.; Sánchez, G.; Martí, J. Polarization insensitive low-loss coupling technique between SOI waveguides and high mode field diameter single-mode fibers. *Opt. Express* **2007**, *15*, 7058. [CrossRef] [PubMed]
32. Fang, Q.; Liow, T.Y.; Song, J.F.; Tan, C.W.; Yu, M.B.; Lo, G.Q.; Kwong, D.L. Suspended optical fiber-to-waveguide mode size converter for Silicon photonics. *Opt. Express* **2010**, *18*, 7763. [CrossRef]
33. Almeida, V.R.; Panepucci, R.R.; Lipson, M. Nanotaper for compact mode conversion. *Opt. Lett.* **2003**, *28*, 1302. [CrossRef]
34. Taillaert, D.; Laere, F.V.; Ayre, M.; Bogaerts, W.; Thourhout, D.V.; Bienstman, P.; Baets, R. Grating Couplers for Coupling between Optical Fibers and Nanophotonic Waveguides. *Jpn. J. Appl. Phys.* **2006**, *45*, 6071. [CrossRef]
35. Van Laere, F.; Claes, T.; Schrauwen, J.; Scheerlinck, S.; Bogaerts, W.; Taillaert, D.; O'Faolain, L.; Van Thourhout, D.; Baets, R. Compact Focusing Grating Couplers for Silicon-on-Insulator Integrated Circuits. *IEEE Photonics Technol. Lett.* **2007**, *19*, 1919–1921. [CrossRef]
36. Ang, T.W.; Reed, G.T.; Vonsovici, A.; Evans, A.G.R.; Routley, P.R.; Josey, M.R. Highly efficient unibond silicon-on-insulator blazed grating couplers. *Appl. Phys. Lett.* **2000**, *77*, 4214–4216. [CrossRef]
37. Vermeulen, D.; Selvaraja, S.; Verheyen, P.; Lepage, G.; Bogaerts, W.; Absil, P.; Van Thourhout, D.; Roelkens, G. High-efficiency fiber-to-chip grating couplers realized using an advanced CMOS-compatible Silicon-On-Insulator platform. *Opt. Express* **2010**, *18*, 18278. [CrossRef]
38. Zaoui, W.S.; Kunze, A.; Vogel, W.; Berroth, M.; Butschke, J.; Letzkus, F.; Burghartz, J. Bridging the gap between optical fibers and silicon photonic integrated circuits. *Opt. Express* **2014**, *22*, 1277. [CrossRef]
39. Chen, X.; Li, C.; Fung, C.K.Y.; Lo, S.M.G.; Tsang, H.K. Apodized Waveguide Grating Couplers for Efficient Coupling to Optical Fibers. *IEEE Photonics Technol. Lett.* **2010**, *22*, 1156–1158. [CrossRef]
40. Tang, Y.; Wang, Z.; Wosinski, L.; Westergren, U.; He, S. Highly efficient nonuniform grating coupler for silicon-on-insulator nanophotonic circuits. *Opt. Lett.* **2010**, *35*, 1290. [CrossRef]

41. Larrea, R.A.; Gutiérrez, A.M.; Hurtado, J.; Ramírez, J.M.; Garrido, B.; Sanchis, P. Compact focusing grating couplers for silicon horizontal slot waveguides. *Opt. Lett.* **2017**, *42*, 490. [CrossRef] [PubMed]
42. Cheng, L.; Mao, S.; Li, Z.; Han, Y.; Fu, H. Grating Couplers on Silicon Photonics: Design Principles, Emerging Trends and Practical Issues. *Micromachines* **2020**, *11*, 666. [CrossRef] [PubMed]
43. Carroll, L.; Gerace, D.; Cristiani, I.; Menezo, S.; Andreani, L.C. Broad parameter optimization of polarization-diversity 2D grating couplers for silicon photonics. *Opt. Express* **2013**, *21*, 21556. [CrossRef]
44. Pathak, S.; Vanslembrouck, M.; Dumon, P.; Van Thourhout, D.; Bogaerts, W. Compact SOI-based polarization diversity wavelength de-multiplexer circuit using two symmetric AWGs. *Opt. Express* **2012**, *20*, B493. [CrossRef]
45. Bogaerts, W.; Taillaert, D.; Dumon, P.; Van Thourhout, D.; Baets, R.; Pluk, E. A polarization-diversity wavelength duplexer circuit in silicon-on-insulator photonic wires. *Opt. Express* **2007**, *15*, 1567. [CrossRef]
46. Zou, J.; Yu, Y.; Zhang, X. Two-dimensional grating coupler with a low polarization dependent loss of 025 dB covering the C-band. *Opt. Lett.* **2016**, *41*, 4206. [CrossRef]
47. Chen, X.; Tsang, H.K. Polarization-independent grating couplers for silicon-on-insulator nanophotonic waveguides. *Opt. Lett.* **2011**, *36*, 796. [CrossRef]
48. Cheng, Z.; Tsang, H.K. Experimental demonstration of polarization-insensitive air-cladding grating couplers for silicon-on-insulator waveguides. *Opt. Lett.* **2014**, *39*, 2206. [CrossRef]
49. Zhang, B.; Schiller, M.; Al Qubaisi, K.; Onural, D.; Khilo, A.; Naughton, M.J.; Popović, M.A. Polarization-insensitive 1D grating coupler based on a zero-birefringence subwavelength corelet waveguide. *Opt. Lett.* **2022**, *47*, 3167. [CrossRef]
50. Li, W.; Cheng, S.; Yi, Z.; Zhang, H.; Song, Q.; Hao, Z.; Sun, T.; Wu, P.; Zeng, Q.; Raza, R. Advanced optical reinforcement materials based on three-dimensional four-way weaving structure and metasurface technology. *Appl. Phys. Lett.* **2025**, *126*, 033503. [CrossRef]
51. Li, Z.; Cheng, S.; Zhang, H.; Yang, W.; Yi, Z.; Yi, Y.; Wang, J.; Ahmad, S.; Raza, R. Ultrathin broadband terahertz metamaterial based on single-layer nested patterned graphene. *Phys. Lett.* **2025**, *534*, 130262. [CrossRef]
52. Cheng, S.; Li, W.; Zhang, H.; Akhtar, M.N.; Yi, Z.; Zeng, Q.; Ma, C.; Sun, T.; Wu, P.; Ahmad, S. High sensitivity five band tunable metamaterial absorption device based on block like Dirac semimetals. *Opt. Commun.* **2024**, *569*, 130816. [CrossRef]
53. Cheben, P.; Halir, R.; Schmid, J.H.; Atwater, H.A.; Smith, D.R. Subwavelength integrated photonics. *Nature* **2018**, *560*, 565–572. [CrossRef] [PubMed]
54. Halir, R.; Bock, P.J.; Cheben, P.; Ortega-Moñux, A.; Alonso-Ramos, C.; Schmid, J.H.; Lapointe, J.; Xu, D.; Wangüemert-Pérez, J.G.; Molina-Fernández, I.; et al. Waveguide sub-wavelength structures: A review of principles and applications. *Laser Photonics Rev.* **2015**, *9*, 25–49. [CrossRef]
55. Luque-González, J.M.; Sánchez-Postigo, A.; Hadij-ElHouati, A.; Ortega-Moñux, A.; Wangüemert-Pérez, J.G.; Schmid, J.H.; Cheben, P.; Molina-Fernandez, I.; Halir, R. A review of silicon subwavelength gratings: Building break-through devices with anisotropic metamaterials. *Nanophotonics* **2021**, *10*, 2765–2797. [CrossRef]
56. Park, J.; Kim, S.; Nam, D.W.; Chung, H.; Park, C.Y.; Jang, M.S. Free-form optimization of nanophotonic devices: From classical methods to deep learning. *Nanophotonics* **2022**, *11*, 1809–1845. [CrossRef]
57. Banks, A.; Vincent, J.; Anyakoha, C. A review of particle swarm optimization. Part I: Background and development. *Nat. Comput.* **2007**, *6*, 467–484. [CrossRef]
58. Kikuta, H.; Yoshida, H.; Iwata, K. Ability and Limitation of Effective Medium Theory for Subwavelength Gratings. *Opt. Rev.* **1995**, *2*, 92–99. [CrossRef]
59. Yeh, P.; Yariv, A.; Hong, C.S. Electromagnetic propagation in periodic stratified media I General theory. *J. Opt. Soc. Am.* **1977**, *67*, 423. [CrossRef]
60. Gu, C.; Yeh, P. Form birefringence dispersion in periodic layered media. *Opt. Lett.* **1996**, *21*, 504. [CrossRef]
61. Hwang, R.B. *Periodic Structures: Mode-Matching Approach and Applications in Electromagnetic Engineering*; J. Wiley & Sons: Singapore, 2013.
62. Niklasson, G.A.; Granqvist, C.G.; Hunderi, O. Effective medium models for the optical properties of inhomogeneous materials. *Appl. Opt.* **1981**, *20*, 26. [CrossRef] [PubMed]
63. Rytov, S.M. Electromagnetic properties of a finely stratified medium. *Sov. Phys. JETP* **1956**, *2*, 446–475.
64. Bienstman, P. Rigorous and Efficient Modelling of Wavelength Scale Photonic Components. Ph.D. Thesis, Ghent University, Ghent, Belgium, 2001.
65. Chen, W.; Ahmed, H. Fabrication of 5–7 nm wide etched lines in silicon using 100 keV electron-beam lithography and polymethylmethacrylate resist. *Appl. Phys. Lett.* **1993**, *62*, 1499–1501. [CrossRef]
66. Mirza, M.M.; Velha, P.; Ternent, G.; Zhou, H.P.; Docherty, K.E.; Paul, D.J. Silicon nanowire devices with widths below 5 nm. In Proceedings of the 2012 12th IEEE International Conference on Nanotechnology (IEEE-NANO), Birmingham, UK, 20–23 August 2012; pp. 1–4. [CrossRef]
67. Engelbrecht, A.P. *Computational Intelligence: An Introduction*, 1st ed.; Wiley: Hoboken, NJ, USA, 2007. [CrossRef]

68. Hu, X.; Wang, M.; Di Paolo, E. Calculating Complete and Exact Pareto Front for Multiobjective Optimization: A New Deterministic Approach for Discrete Problems. *IEEE Trans. Cybern.* **2013**, *43*, 1088–1101. [CrossRef]
69. Synopsys. *RSoft*; Synopsys: Mississauga, ON, Canada, 2019.

photonics

MDPI

Article

TiN-Only Metasurface Absorber for Solar Energy Harvesting

Hongfu Liu [1], Jijun Li [1,*], Hua Yang [2], Junqiao Wang [3], Boxun Li [4], Han Zhang [5] and Yougen Yi [6,*]

1 School of Physics and Optoelectronic Engineering, Yangtze University, Jingzhou 434023, China; kidvrr@163.com
2 School of Science, Lanzhou University of Technology, Lanzhou 730050, China; hyang@lut.cn
3 School of Physics, Zhengzhou University, Zhengzhou 450001, China; junqiaowang@126.com
4 School of Physics and Optoelectronics, Xiangtan University, Xiangtan 411105, China; lbxcsu@xtu.edu.cn
5 Hubei Feilihua Quartz Glass Co., Ltd., Jingzhou 434000, China; zhanghan@feilihua.net
6 College of Physics, Central South University, Changsha 410083, China
* Correspondence: jjli@yangtzeu.edu.cn (J.L.); yougenyi@csu.edu.cn (Y.Y.); Tel./Fax: +86-0716-8067167 (J.L.); +86-0731-88830995 (Y.Y.)

Abstract: With global energy demand surging and traditional energy resources diminishing, the solar absorber featuring optimized design shows substantial potential in areas like power generation. This study proposes a solar absorber that is insensitive to wide-angle incidence and polarization. It has a cylindrical structure with square holes, which is constructed from titanium nitride (TiN). The calculation results indicate that, for plane waves, the average absorption of this solar absorber across the wavelength range of 300–2500 nm reaches 92.4%. Moreover, its absorption rate of the solar spectrum corresponding to AM1.5 reaches 94.8%. The analysis of the characteristics within the electric and magnetic field profiles indicates that the superior absorption properties arise from a cooperative resonance effect. This effect originates from the interaction among surface plasmon resonance, guided-mode resonance, and cavity resonance. In this study, the geometric parameters of the solar absorber's structure significantly influence its absorption performance. Therefore, we optimized these parameters to obtain the optimal values. Even at a large incident angle, this absorber maintains high absorption performance and shows insensitivity to the polarization angle. The findings expected from this study are likely to be of considerable practical importance within the realm of solar photothermal conversion.

Keywords: titanium nitride; solar absorber; finite element method; surface plasmon resonance; solar photothermal utilization

1. Introduction

New energy is an important direction of current energy development. It is sustainable and capable of offering stable energy support for the long-term progress of human society. As a prime example of new energy, solar energy is not only clean and pollution-free but also inexhaustible, which makes its position in the global energy structure increasingly prominent [1–3]. Solar absorbers are of importance in the exploitation of solar photothermal energy. They effectively take in solar radiation and transform it into heat. During the procedure of photothermal conversion, the surface plasmon resonance of metal materials plays an important role. By optimizing the surface structure of metal materials, different resonances are coupled, thus greatly improving absorption efficiency. Currently, solar absorbers, e.g., metallic nanoparticle-based absorbers, graphene absorbers, and metamaterial absorbers, possess substantial application prospects such as desalination of seawater, irrigation in agriculture, energy resupply in aerospace [4–7]. These advanced configurations leverage nanoscale light-matter interactions to optimize solar energy harvesting efficiency [8–10]. Solar absorbers possess substantial application prospects in areas like desalination of seawater, irrigation in agriculture, as well as energy resupply in aerospace. It is also evidenced that modifications in the polarization angle scarcely influence the absorption performance. The findings expected from this study are likely to be of considerable practical importance within the realm of solar photothermal conversion, effectively covering the visible and infrared light bands, and showing certain polarization-independent characteristics in the full-absorption band [11]. Elsharabasy et al. [12] engineered a broadband metamaterial perfect absorber featuring a Au-TiO_2-Au. This structure consists of concentric arrangements made up of crossed ellipses and quadruple split-ring resonators [12]. Zhang and colleagues [13] developed a broadband solar metamaterial absorber (BSMA). It spans from 280 to 2500 nm, featuring a structure composed of patterned Si_3N_4-TiN, SiO_2, and Ti. The average absorption achieves 97.77%. Many studies show that solar absorbers based on different materials and structural designs all exhibit significant light-absorption performance. Although these metamaterial solar absorbers perform well in absorption performance, and some even achieve high or even perfect absorption, their complex preparation processes and high costs limit their application in large-scale scenarios. Therefore, finding simple materials that can achieve efficient solar energy absorption has become a major challenge. Titanium nitride (TiN), as a newly synthesized artificial metal material, not only has excellent optical properties, enabling it to efficiently absorb solar energy, but also has good thermal conductivity, allowing it to quickly transfer heat energy and achieve effective energy transportation [14–16]. Its melting point attains a value as high as 2950 °C. Additionally, its structure and properties exhibit stability at elevated temperatures. Owing to its cost-effective production, this method has been extensively utilized in the development of solar absorbers [17–21].

Columnar structures are widely used in solar absorbers. Yu et al. designed a broadband solar absorber composed of a TiO_2-TiN nano-elliptical cylinder array, which achieved an average absorption of 95.68% in the wavelength range of 360–1624 nm [22]. Chen et al. proposed a solar absorber with a cylindrical array, which achieved a total solar absorption of 94% in the wavelength range of 300–2500 nm, providing a reliable thermal management solution for concentrated solar power systems [23]. We also focused on solar absorbers with square hole structures. Gao et al. [24] designed a TiN square-ring superstructure absorber. The thickness of the TiN square ring is 25 nm. It achieves an average absorption of 95.69% in the wavelength range of 280–2500 nm, supporting wide-angle incidence, and is insensitive to polarization. Song et al. [25] developed a four-layer metamaterial absorber of Ti/Si_3N_4/Ti-SiO_2, covering the wavelength range of 200–4200 nm with an average absorption of 98.16%. It had the characteristics of wide-angle insensitivity and high manufacturing tolerance. Pan et al. proposed an ultra-broadband absorber based on the TiN rectangular column-ring structure, which had an average absorption of 97.02% in the wavelength range of 300–4962 nm and was suitable for photovoltaic and infrared imaging applications [26]. All the above designs with square holes have good absorption performance. We referred to the above models and considered the manufacturing feasibility [27–29], so we controlled the difference between the radius of the circumscribed circle of the square hole and the radius of the cylinder to be approximately 30 nm. Therefore, we attempted to combine the advantages of the two and add square holes to the columnar structure to see if its performance can be improved. In addition, although TiN is widely used, most of the above structures are about the combinations of multiple materials, and there are relatively few studies using a kind of material just. Constructing the surface structure of an absorber with one kind of material can not only exhibit excellent performance but also avoid problems such as instability caused by the superposition of materials. Therefore, it is necessary to conduct further research on TiN-based solar absorbers.

In the present study, the finite-element method was employed to perform simulations and proposed a solar absorber that uses only titanium nitride (TiN). By discussing the absorption effect of 300–2500 nm plane waves and the absorptivity of the actual solar spectrum, we analyzed its absorption performance of solar radiation. By comparing the electric and magnetic field profiles, changes in local electric field intensity were identified, and the mechanism by which the solar absorber achieves efficient absorption was clarified. Meanwhile, the effect of geometric parameters on the absorber's absorption characteristics was taken into account. Furthermore, the influence of the incident angle and polarization on the spectral absorption patterns was also analyzed. As a result, a thorough evaluation of the solar energy absorption capability of the absorber was conducted.

2. Models and Methods

Considering the significant characteristics of titanium nitride (TiN), including its high hardness, excellent chemical stability, and elevated melting point, TiN was chosen as the material in this study. In terms of micro-nano structures, through extensive literature comparison and analysis, we found that square-columnar and cylindrical solar absorbers each exhibit unique advantages. The square-columnar structure provides higher light-absorption efficiency due to its larger surface area, while the cylindrical structure is favored for its lower manufacturing complexity. Based on these findings, we proposed an innovative design that combines the advantages of these two structures, aiming to achieve better photothermal conversion performance [30]. We conducted a detailed analysis of the impact exerted by diverse structural parameters on the light-absorption efficacy [31]. Regarding the proposed structure, the finite-element approach was employed to optimize the parameters [32–34] with the aim of improving the absorption performance and exploring the maximum possible absorptivity within the required wavelength range. All geometric parameters were optimized based on COMSOL software (COMSOL 6.2).The configuration of the proposed solar absorber is illustrated in Figure 1. After optimization, the parameters are as follows: the height of the cylinder (h_1) is equivalent to the height of the square column(h_2), $h_1 = h_2 = 2000$ nm, the diameter (D) of the cylinder is 200 nm, the period (P) amounts to 300 nm, and the width (L) of the square-shaped column is 130 nm.

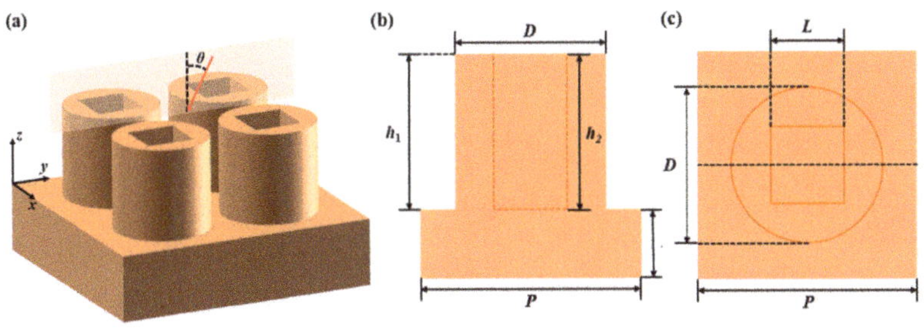

Figure 1. The proposed TiN absorber. (**a**) 3D view, (**b**) x–z view, and (**c**) x–y view.

Material parameters are of great importance in research. The refractive index and the extinction coefficient together form the complex refractive index of the material. The complex refractive index exerts a direct impact on the way light interacts with the material. Consequently, it dictates the energy conversion efficiency of the solar absorber. In this context, the real part (n), which denotes the refractive index of the light-absorbing medium, is closely related to the velocity at which light waves propagate through the absorbing material [35]. It influences the propagation path and the refraction of light within the medium. The imaginary part (k) is governed by the degree of light wave attenuation during its propagation through the absorbing medium, which is directly related to the material's ability to absorb light. As presented in Figure 2, the data from reference [36] were employed in our simulation process.

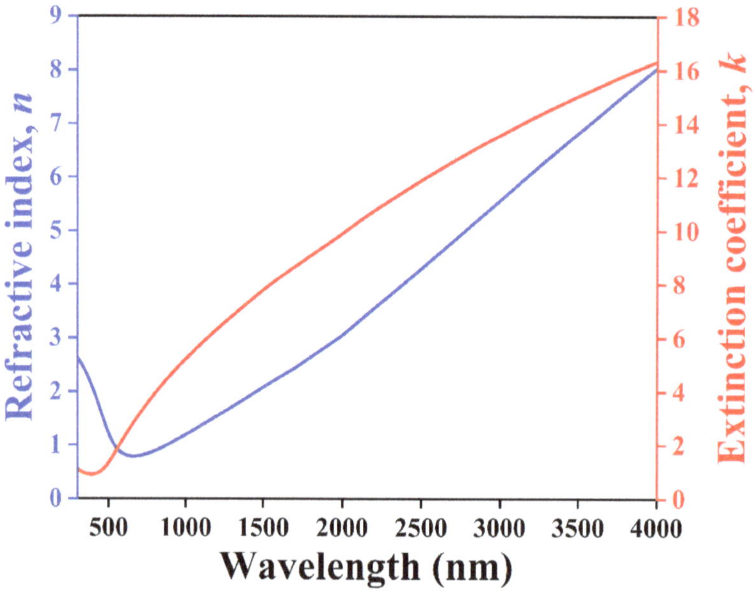

Figure 2. The refractive index and extinction coefficient values for TiN.

In this paper, the finite element method (FEM) is utilized to numerically simulate and analyze the proposed solar absorber [37–39]. This method is widely applied to the calculation of solar absorbers. The unit-cell model of the proposed solar absorption structure is depicted in Figure 3a. Above and below it, there are, respectively, an air layer and a perfect matching layer. To save simulation time, we took one unit cell and performed a parameter-sweep calculation in the 300–2500 nm spectrum range. A normally incident plane wave was adopted as the light source. Due to the fact that the structure is a periodic array, during the specific calculation process, a solitary unit cell featuring Floquet periodic boundary conditions was employed [40,41]. To simulate an infinite surface, periodic boundary conditions were applied to the walls in the x–z and y–z planes, as depicted in Figure 3b. The periodic ports illustrated in Figure 3c are denoted by arrows. The light is incident from the top through the specified periodic port, and a second port is added at the bottom. The upper horizontal port functions as the inlet for the incident light and simultaneously serves to model the reflected light. In the meantime, the lower horizontal port acts as the outlet port and is employed to simulate the transmitted light. In the simulation calculation, the setting of the grid deserves special attention. To a large extent, the grid determines whether the establishment and calculation of this model are reasonable. Since the cylindrical structure model is relatively complex, a relatively detailed meshing was carried out in this area. Looser meshes were adopted in the air and the perfectly matched layer. This approach not only ensures the accuracy of the calculation but also guarantees a reduction in computational costs [42,43]. Figure 3d outlines the grid used in the simulation of this study.

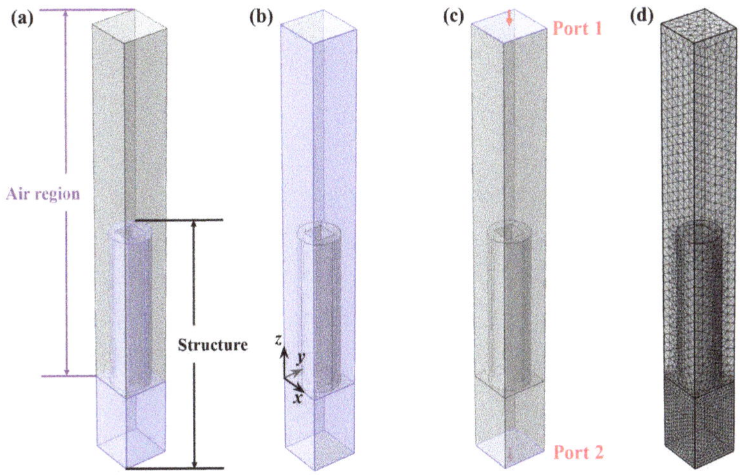

Figure 3. (**a**) The proposed structure, (**b**) Boundary conditions, (**c**) Port 1 and Port 2, (**d**) Mesh.

In the research of solar absorbers, theoretical calculations guide the final results. When light strikes the surface of TiN (which has a complex refractive index $n_2 = n + ik$) from air (where the refractive index n_1 is approximately 1), Maxwell's equations, in combination with the relevant boundary conditions, govern the values of the amplitude coefficients for reflection and transmission. For vertically incident light, the reflection coefficient r is determined by the complex refractive index [44,45]:

$$r = \frac{n_2 - n_1}{n_2 + n_1}. \tag{1}$$

The reflectivity is $R(\lambda)$:

$$R(\lambda) = \frac{(n-1)^2 + k^2}{(n+1)^2 + k^2}. \tag{2}$$

Provided that the TiN substrate has sufficient thickness, the transmittance is close to zero, that is, when $T(\lambda) = 0$, the light is completely reflected or absorbed. Therefore, the absorptivity $A(\lambda)$ [46] is

$$A(\lambda) = 1 - A(\lambda) - T(\lambda). \tag{3}$$

Here, λ represents the wavelength of the incident light beam. To assess the practical solar absorption capability of the solar absorber, we examined its practical absorption characteristics by taking into account the AM1.5 solar irradiance spectrum [24]:

$$I_{abs} = A(\lambda) \cdot I_{AM1.5}(\lambda) \tag{4}$$

$$I_{loss}(\lambda) = I_{AM1.5}(\lambda) - I_{abs}(\lambda). \tag{5}$$

Here, $I_{AM1.5}(\lambda)$ corresponds to the solar spectrum distribution under $AM1.5$ conditions, I_{abs} denotes the absorption spectrum, and $I_{loss}(\lambda)$ stands for the solar spectrum that is lost. According to the computations in Equations (4) and (5), The ratio of the total solar energy absorbed to the energy within the solar radiation spectrum provides the amount of energy that is actually absorbed [25]:

$$\alpha = \frac{\int_{\lambda_2}^{\lambda_1} I_{abs} d\lambda}{\int_{\lambda_2}^{\lambda_1} I_{AM1.5} d\lambda}. \tag{6}$$

Since solar energy is mainly concentrated in the wavelength band of 300–2500 nm, λ_1 = 300 nm and λ_2= 2500 nm.

3. Results and Discussions

Figure 4a shows that the average absorptivity of this solar absorber within the wavelength range of 300–2500 nm is 92.4%. Under the same conditions, the average spectrum absorptions of the planar unstructured titanium nitride (TiN), columnar TiN, and the absorber proposed by us within the same wavelength range are 30.1%, 91.2%, and 92.4%, respectively. In comparison, the absorber proposed by us has an average spectrum absorption more than three times that of the planar unstructured TiN. Since the materials used for them are the same, this is entirely attributed to the structural optimization. After the structural design, a remarkable enhancement occurs. This is attributable to the fact that the periodic array of TiN structures gives rise to a matched impedance, which mitigates solar reflection and substantially boosts solar absorption. Moreover, there exist four characteristic absorption maxima in the vicinity of λ_1 = 400 nm, λ_2= 1100 nm, λ_3= 1450 nm, and λ_4 = 2100 nm. The absorbance values at these wavelengths are 99.2%, 92.4%, 94.8%, and 99.1%, respectively. Additionally, by incorporating the AM1.5 solar radiation spectrum, we examined the actual absorption rate of solar energy by the solar absorber. As illustrated in Figure 4b, solar energy is predominantly distributed within the visible light spectrum, and our absorber demonstrates optimal absorption performance precisely within this wavelength band. At the 2500 nm wavelength band, the absorption of solar energy is nearly negligible. Although the absorption of our absorber declines beyond 2500 nm, this does not impact the overall actual absorption of solar energy. The spectrum absorptivity of the solar absorber reaches 94.8%. It is evident that the energy loss is minimal, with the ratio of energy loss to energy absorption being merely 0.052. This substantiates the superiority of our designed structure and offers compelling evidence for its promising potential in solar energy harvesting.

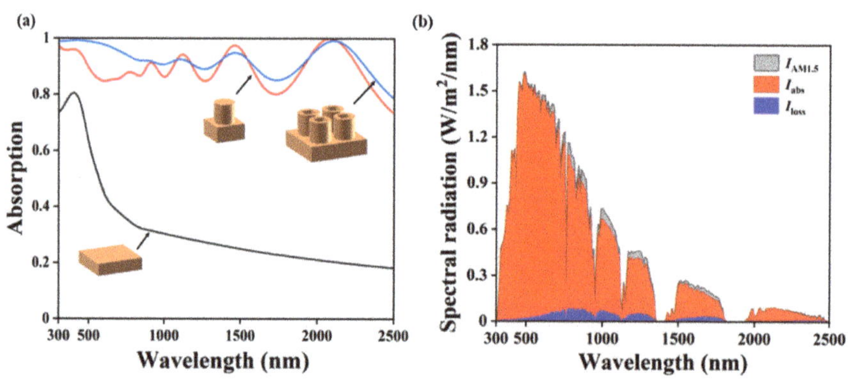

Figure 4. The absorption performance of TiN absorbers (**a**) Planar wave absorption for planar unstructured TiN (black), columnar TiN (red), and the structural model (blue) (**b**) The absorption spectrum of the absorber.

In Table 1, we compare the absorption characteristics of TiN-based devices in prior works [22,23,47,48] with those of the proposed structure in the manuscript. The proposed design in this paper achieves a broad absorption band spanning 300–2500 nm. Reference [23] also reported the same range, while our structure exhibits a higher average absorption of 92.4%. Notably, our structure eliminates complex multilayer configurations (e.g., nanocone arrays in reference [47], cross-based layers in reference [48], or nano-elliptical disks in reference [22]) and employs a simplified single-material TiN architecture. This structural simplicity enhances the fabrication feasibility and provides potential application for broadband absorption.

Table 1. Comparison of our designed absorption device with previous designs.

Reference	Absorption Band	Absorption	Structure
[47]	400–1500 nm	99.6%	TiN nanocone array/Al_2O_3/TiN
[48]	300–900 nm	93%	cross-based TiN/AlN/TiN
[22]	360–1624 nm	95.68%	TiO_2-TiN nano-elliptical disk arrays
[23]	300–2500 nm	88%	TiN cylinder array/SiO_2/TiN
proposed	300–2500 nm	92.4%	TiN

The distribution of the electric field and the magnetic field inside the solar absorber are shown in Figures 5 and 6. The absorber's broadband absorption performance stems from three resonance mechanisms, each supported by distinct electric and magnetic field behaviors [49,50]. Surface plasmon resonance (SPR) is confirmed by the electric field intensity near the cylindrical periphery (Figure 5a–d), which increases with wavelength, driven by free electron oscillations on the TiN surface at shorter wavelengths [51,52]. When the frequency of the incident light matches the collective oscillation frequency of the free electrons on the metallic surface, a strong interaction will occur. This enables the light energy to be effectively coupled into the absorber and converted into the kinetic energy of electrons. Subsequently, the energy is dissipated in the form of heat energy through the processes, thus achieving a high absorption. While the corresponding magnetic field in Figure 6a–d shows localized hotspots near the same edge, spatially overlapping with electric field maxima. This coupling between SPR-induced electric dipoles and magnetic dipoles enhances energy retention at metallic-dielectric interfaces. Guided mode resonance (GMR) manifests in the electric field's standing-wave pattern beyond 1100 nm (Figure 5e–h), where light undergoes multiple reflections and interferences, extending its propagation path [53,54]. Correspondingly, the magnetic field in Figure 6e–h exhibits periodic oscillations along the absorber, aligning with GMR's phase-matched interference. Cavity Resonance (CR) plays a critical role in enhancing the absorber's energy localization, as evidenced by both electric and magnetic field behaviors [48]. Figure 5a–d exhibits stronger field intensity compared to its surface. This contrasts with the weakening field in the cavity's periphery, a hallmark of CR induced energy confinement. This phenomenon arises when the incident wavelength is smaller than the cavity's cutoff wavelength, trapping light within the cavity and amplifying the field through resonant oscillations. Figure 6g,h reflects the influence of cavity resonance (CR) on energy redistribution. At wavelengths below the cavity's cutoff condition, the magnetic field shifts toward the cavity interior due to confined modal profiles, further corroborating CR's role in vertical energy trapping. The absorber's symmetric geometry minimizes radiative losses, directing energy toward thermal dissipation. SPR enhances surface coupling, GMR prolongs light-matter interaction, and CR localizes energy. Together,

SPR, GMR, and CR optimize broadband absorption through correlated electromagnetic field interactions.

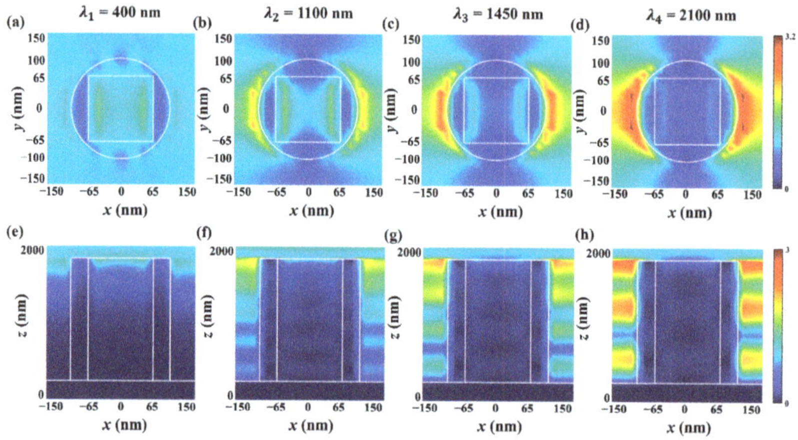

Figure 5. Electric field distributions of the proposed solar absorber at the absorption peaks (**a**–**d**) top view; (**e**–**h**) side view.

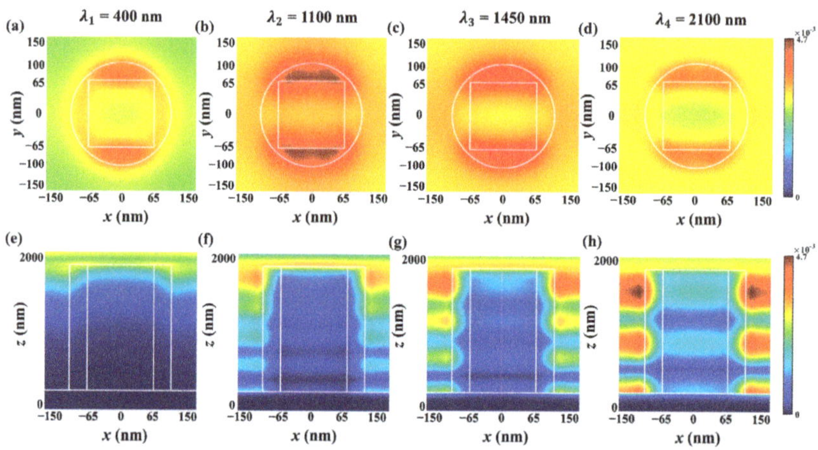

Figure 6. Magnetic field distributions of the proposed solar absorber at the absorption peaks (**a**–**d**) top view; (**e**–**h**) side view.

Moreover, the geometric parameters of the structure of the solar absorber exert a remarkable influence on its absorption capabilities. Therefore, each parameter in this structure was optimized individually to ensure that the final determined structural parameters are optimal. Figure 7a shows that altering the height of the absorber's cylinder significantly impacts the absorption ratio. As the cylinder height increases, the absorption ratio also increases, reaching its peak value when the height reaches 2000 nm. The change in height is capable of significantly altering the light scattering pattern and propagation path. It has an impact on the activation of surface plasmon polaritons and leads to a substantial variation in the region of interaction with the incident light [55,56]. Consequently, it significantly impacts the absorption performance. Furthermore, considering the cylinder's diameter, as illustrated in Figure 7b, the absorption performance reaches its peak when the cylinder's diameter is 200 nm. An appropriate diameter can better excite SPR and GMR phenomena, thereby enhancing absorption [57]. Figure 7c illustrates the impact of the period (P) for the absorber regarding its absorption characteristics. When the period reaches 300 nm, there is a remarkable enhancement in the absorption performance. However, as the period increases, the improvement in absorption performance becomes slow, and there is even no obvious increase. This implies that an augmentation in the period can boost absorption. However, if the period is excessively large, it has a negligible effect on the absorption result. Finally, Figure 7d illustrates the effect of the square column's width on the absorption characteristics. Clearly, as the width of the square column increases, the absorber's performance in absorption improves. Overall, alterations in the geometric parameters of the absorber play a crucial role in determining its performance. The optimized structural parameters provide guidance for the precision of the preparation process, aiming to maximize the efficiency of the solar absorber [58].

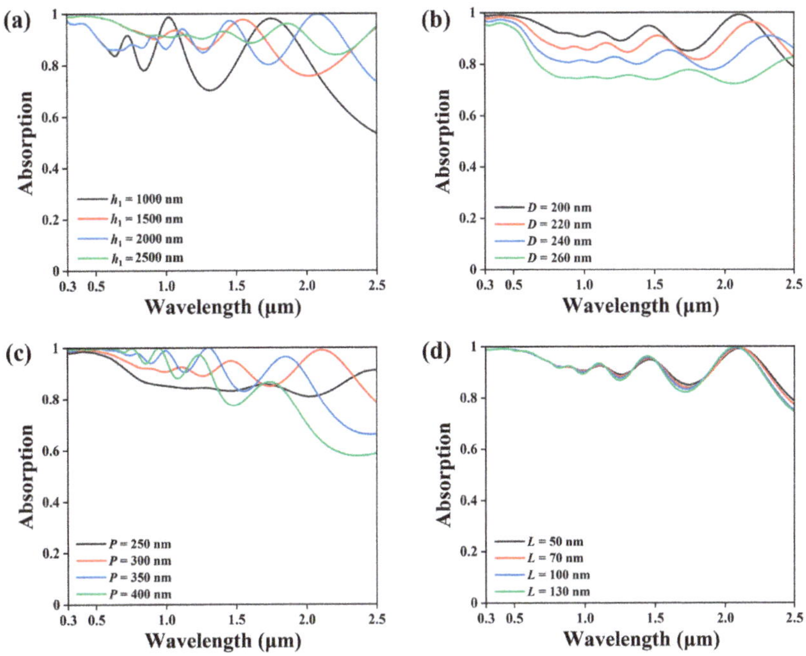

Figure 7. Impact of geometric parameters on absorption properties: (**a**) The height h_1 of the cylinder, (**b**) the diameter D of the cylinder, (**c**) period P, and (**d**) the width L of the square column.

To investigate the relationship between structural dimensions and resonance modes, cylindrical column height (h_1) and square column width (L) were selected as observation parameters. Figures 8a and 8b, respectively, depict the distribution of the electric field diagrams corresponding to the absorption peak at 1840 nm along the red line when different parameters are changed. As can be seen from the figures, at the titanium nitride metal-dielectric interface at the edge of the square hole, the incident light excites the collective oscillation of free electrons, leading to the enhancement of the local electric field and significantly improving the light absorption [59]. As h_1 increases, the effective optical thickness of the waveguide layer increases and the supported guided mode resonance wavelength experiences a red shift. At the same time, it can be observed that the square hole, as a resonant cavity, concentrates the energy within the cavity, causing cavity resonance. In Figure 8b, it can be seen that as L decreases, the local ability of the surface plasmon resonance weakens and the cavity resonance gradually becomes less obvious, resulting in a slight decrease in absorption [60]. In summary, the two parameters jointly optimize the synergistic effect of the guided mode resonance, cavity resonance, and surface plasmon resonance, enabling the absorption rate at 1840 nm to reach its peak.

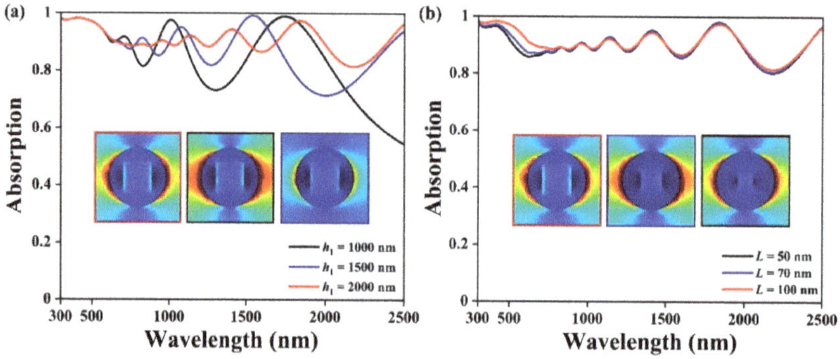

Figure 8. Absorption and electric field distribution: (**a**) The height h_1 of the cylinder and (**b**) the width L of the square column.

Through extensive comparisons with the existing literature [24–26] and meticulous consideration of practical fabrication feasibility [27,28], we optimized structural parameters under mutual constraints. Our design carefully addresses critical manufacturing limitations including aspect ratio and vertical inclination. As demonstrated in references [61,62], which exhibit high absorption performance with identical vertical inclination to our proposed design, and their reported aspect ratios exceed 11. In contrast, our model adopts a more manufacturable aspect ratio of 10, well within the achievable range of standard fabrication processes. Furthermore, we incorporated the inevitable rounding of corners during machining into our simulations. While maintaining a cylinder radius of 100 nm, we ensured a minimum gap of approximately 30 nm between the circumcircle radius of square apertures and cylindrical outer boundaries, resulting in a corresponding square pillar width of 100 nm. Figure 9a presents comparative simulations with and without rounded corners to mimic actual fabrication processes. Analysis confirms negligible variation in absorption efficiency between these configurations, demonstrating our proposed solar absorber's exceptional tolerance to typical manufacturing imperfections. This robust performance under practical fabrication conditions substantiates the design's high technical feasibility for scalable production.

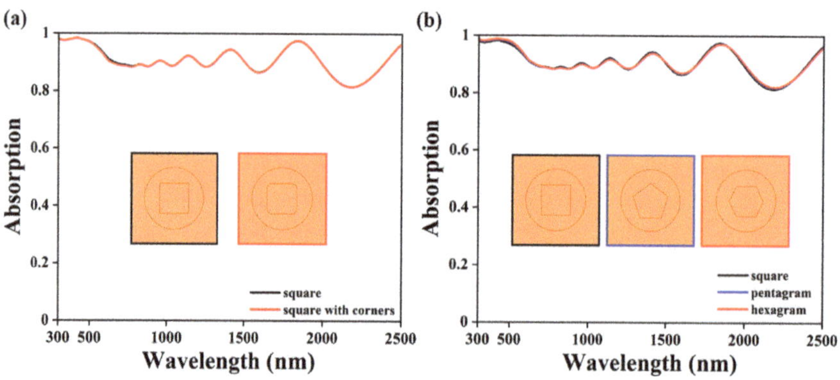

Figure 9. Comparison of (**a**) square with and without corners and (**b**) square versus pentagram and hexagram.

To further validate structural superiority, we conducted a comparative analysis of our square-aperture design with alternative geometric configurations as illustrated in Figure 9b, including pentagram and hexagram patterns. Although these star-shaped structures theoretically offer enhanced surface areas under comparable fabrication precision, experimental results reveal only negligible absorption enhancement relative to our original design. Such marginal performance gains fail to justify the substantially increased fabrication complexity arising from the multi-angle vertices inherent to these intricate geometries [63,64]. Consequently, the square aperture configuration was selected as the optimal architecture, achieving an optimal balance between optical performance and superior manufacturability [65].

As a broadband incoherent light source, the incident angle, polarization state, and spectrum distribution of sunlight directly affect the performance of the absorber [66]. In practical applications, sunlight does not directly irradiate the solar absorber vertically but at different angles [67,68]. Therefore, in the design of the absorber, it is crucial to consider the effect of the incident-light angle on absorption performance [69,70]. The spectral absorption characteristics of both transverse magnetic (TM) waves and transverse electric (TE) waves were analyzed under various incident angle conditions [71–73]. As presented in Figure 10a, the spectrum absorption behavior for TM waves under various incident angles is depicted. It is evident that when the incidence angle reaches 30°, the absorber's average absorption ratio achieves 92.3%. Even when the angle of incidence is relatively large, at 60°, the absorption ratio remains at 77.1%. This maintenance of performance implies that the absorber continues to exhibit a high-level absorption ability at varying angles, signifying the stability of the absorber's performance across different angles. In a similar vein, the alterations in the spectrum absorption characteristics of TE waves at diverse angles were also explored, and the findings are presented in Figure 10b. At an incidence angle of 60°, the average absorption for TE waves is 89.8%. Specifically, the absorber can effectively capture and absorb both TE and TM waves, which gives the absorber high angular sensitivity. Even as the incident angle of the light changes, the alteration in absorption performance remains negligible [74]. Figure 10c further illustrates the effect of the polarization angle on the spectrum absorption characteristics of the solar absorber. Owing to the isotropic properties and high degree of symmetry in the designed solar absorber structure, it demonstrates a high level of resistance to alterations in the polarization angle. This indicates that regardless of changes in the polarization angle of the incident light, the variations in the absorption characteristics of the absorber are barely noticeable [75,76]. This further bolsters its adaptability within practical scenarios, particularly in intricate lighting conditions, allowing it to uphold high efficiency.

Figure 10. (**a**) TM polarization, (**b**) TE polarization, and (**c**) variation in spectrum with polarization angle.

4. Conclusions

In summary, a wide-angle, polarization-independent metamaterial solar absorber, made entirely of TiN, was proposed. Utilizing the finite-element method, the overall solar absorption within the wavelength range of 300–2500 nm was determined to be 92.4% and the absorption under the AM1.5 solar irradiance spectrum reached 94.8%. Analysis of the electric and magnetic field distributions showed that the synergistic effect of SPR, GMR, and CR enhances the light absorption efficiency of the absorber. Additionally, the effects of the geometric parameters of the solar absorber on its absorption characteristics were also investigated. Even at a relatively large incident angle of 60 degrees, the solar absorber still maintains excellent performance and achieves an absorption of 89.8%. This paper plays an exploratory role in the application of TiN materials in solar absorbers and in the optimization of solar absorbers.

Author Contributions: Conceptualization, H.L., J.L., H.Y., and Y.Y.; data curation, H.L., J.L., H.Y., and Y.Y.; formal analysis, J.W., B.L., H.Z., and H.Y.; methodology, J.W., B.L., H.Z., and Y.Y.; resources, Y.Y.; software, J.W., B.L., H.Z., and Y.Y.; data curation, H.Y.; writing—original draft preparation, H.L.; writing—review and editing, H.L., J.L., H.Y., and Y.Y. All authors have read and agreed to the published version of the manuscript.

Funding: The authors are grateful to the support by National Natural Science Foundation of China (No. 52162040).

Institutional Review Board Statement: Not applicable.

Informed Consent Statement: Not applicable.

Data Availability Statement: The data that support the findings of this study are available from the corresponding author upon reasonable request.

Conflicts of Interest: Han Zhang was employed by Hubei Feilihua Quartz Glass Co., Ltd. The remaining authors declare that the research was conducted in the absence of any commercial or financial relationships that could be construed as a potential conflict of interest.

References

1. Xiao, T.X.; Tu, S.; Liang, S.Z.; Guo, R.J.; Tian, T.; Müller-Buschbaum, P. Solar cell-based hybrid energy harvesters towards sustainability. *Opto-Electron. Sci.* **2023**, *2*, 230011. [CrossRef]
2. Zhang, H.; Feng, L.; Wang, F.Y.; Liu, M.Z.; Zhang, Y.Y.; Zhu, J.; Lu, Y.; Xu, T. Janus aramid nanofiber aerogel incorporating plasmonic nanoparticles for high-efficiency interfacial solar steam generation. *Opto-Electron. Adv.* **2023**, *6*, 220061. [CrossRef]
3. Sfirloaga, P.; Bognár, S.; Taranu, B.-O.; Vlazan, P.; Poienar, M.; Merkulov, D.Š. Co- and Sn-Doped $YMnO_3$ Perovskites for Electrocatalytic Water-Splitting and Photocatalytic Pollutant Degradation. *Coatings* **2025**, *15*, 475. [CrossRef]
4. Zhang, D.; Xia, S.; Xu, W.; Zhai, X.; Wang, L. Topological plasmonically induced transparency in a graphene waveguide system. *Phys. Rev. B* **2024**, *109*, 245420. [CrossRef]
5. Xia, S.; Zhang, D.; Zhai, X.; Wang, L.; Wen, S. Phase-controlled topological plasmons in 1D graphene nanoribbon array. *Appl. Phys. Lett.* **2023**, *123*, 101102. [CrossRef]
6. Xia, S.; Zhai, X.; Wang, L.; Wen, S. Plasmonically induced transparency in double-layered graphene nanoribbons. *Photonics Res.* **2018**, *6*, 692–702. [CrossRef]
7. Wang, Y.J.; Liu, Z.M.; Zhou, F.Q.; Yi, Z.; Wang, J.Q. Perfect absorption properties of a near-infrared super-surface perfect absorber based on a multilayer subwavelength array structure. *Phys. Lett. A* **2025**, *540*, 130395. [CrossRef]
8. Sornek, K.; Augustyn-Nadzieja, J.; Rosikoń, I.; Łopusiewicz, R.; Łopusiewicz, M. Status and Development Prospects of Solar-Powered Unmanned Aerial Vehicles—A Literature Review. *Energies* **2025**, *18*, 1924. [CrossRef]
9. Li, W.X.; Cheng, S.B.; Yi, Z.; Zhang, H.F.; Song, Q.J.; Hao, Z.Q.; Sun, T.Y.; Wu, P.H.; Zeng, Q.D.; Raza, R. Advanced optical reinforcement materials based on three-dimensional four-way weaving structure and metasurface technology. *Appl. Phys. Lett.* **2025**, *126*, 033503. [CrossRef]
10. Yang, C.; Luo, M.H.; Ju, X.W.; Hu, J.Y. Ultra-narrow dual-band perfect absorber based on double-slotted silicon nanodisk arrays. *J. Phys. D Appl. Phys.* **2024**, *57*, 345104. [CrossRef]

11. Berka, M.; Mahdjoub, Z.; Afif, B.; Bendaoudi, A. Towards a novel high-performance broadband plasmonic metasurface based nano-absorber in visible light to ultra-violet windows for enhanced and maximizing solar energy harvesting in photovoltaic systems. *Phys. Scr.* **2025**, *100*, 035985. [CrossRef]
12. Elsharabasy, A.; Bakr, M.; Deen, M.J. Wide-angle, wide-band, polarization-insensitive metamaterial absorber for thermal energy harvesting. *Sci. Rep.* **2020**, *10*, 16215. [CrossRef]
13. Zhang, J.; Chen, Y.; Chen, S.; Hou, J.; Yi, Z. A near-perfect solar metamaterial absorber with material substitutability for high-efficiency solar thermal conversion. *Appl. Mater. Today* **2025**, *43*, 102630. [CrossRef]
14. Wen, J.; Li, X.; Chen, W.; Liu, J. Systematical investigation on the solar-thermal conversion performance of TiN plasmonic nanofluids for the direct absorption solar collectors. *Colloids Surf. A Physicochem. Eng. Asp.* **2021**, *624*, 126837. [CrossRef]
15. Wang, L.; Zhu, G.; Wang, M.; Yu, W.; Zeng, J.; Yu, X.; Xie, H.; Li, Q. Dual plasmonic Au/TiN nanofluids for efficient solar photothermal conversion. *Sol. Energy* **2019**, *184*, 240–248. [CrossRef]
16. Zeng, J.; Xuan, Y. Tunable full-spectrum photo-thermal conversion features of magnetic-plasmonic Fe_3O_4/TiN nanofluid. *Nano Energy* **2018**, *51*, 754–763. [CrossRef]
17. Yang, C.; Lin, Q.; Du, W.J.; Wang, L.L.; Liu, G.D. Bi-tunable absorber based on borophene and VO_2 in the optical telecommunication band. *J. Opt. Soc. Am. B* **2022**, *39*, 2969–2974. [CrossRef]
18. Yan, D.Y.; Tang, C.J.; Yi, Z.; Wang, J.Q.; Li, B.X. A fully symmetric solar absorber for thermophotovoltaic power generation. *Phys. Lett. A* **2025**, *542*, 130461. [CrossRef]
19. Li, Y.M.; Tan, C.X.; Hu, J.Y.; Bai, W.D.; Zhang, R.L.; Lin, Q.; Zhang, Y.; Wang, L.L. Ultra-narrow band perfect absorbance induced by magnetic lattice resonances in dielectric dimer metamaterials. *Results Phys.* **2022**, *39*, 105730. [CrossRef]
20. Luo, M.H.; Hu, J.Y.; Li, Y.M.; Bai, W.D.; Zhang, R.L.; Lin, Q.; Wang, L.L. Anapole-assisted ultra-narrow-band lattice resonance in slotted silicon nanodisk arrays. *J. Phys. D Appl. Phys.* **2023**, *56*, 375102. [CrossRef]
21. Mehrabi, S.; Rezaei, M.; Zarifkar, A. Ultra-broadband metamaterial absorber based on cross-shaped TiN resonators. *J. Opt. Soc. Am. A* **2020**, *37*, 697–704. [CrossRef] [PubMed]
22. Yu, P.Q.; Chen, X.F.; Yi, Z.; Tang, Y.J.; Yang, H.; Zhou, Z.G.; Duan, T.; Cheng, S.B.; Zhang, J.G.; Yi, Y.G. A numerical research of wideband solar absorber based on refractory metal from visible to near infrared. *Opt. Mater.* **2019**, *97*, 109400. [CrossRef]
23. Yang, B.; Zou, Y.; Zhou, K.; Liu, H.; Wu, X. TiN-based metasurface absorber for efficient solar energy harvesting. *Int. J. Therm. Sci.* **2023**, *192*, 108428. [CrossRef]
24. Gao, Z.; Yu, S.; Li, Z.; Pan, D.; Xu, Z.; Zhao, T. Ultra-Broadband Spectrally Selective Absorber for Solar Thermal Absorption Based on TiN Square-Ring Meta-Structure. *IEEE Photonics J.* **2023**, *15*, 1–7. [CrossRef]
25. Song, S.; Luo, W. Ultra-wideband solar perfect absorber for photothermal conversion with high manufacturing error tolerance. *Mater. Today Sustain.* **2024**, *27*, 100877. [CrossRef]
26. Pan, Y.; Li, Y.; Chen, F.; Cheng, S.; Yang, W.; Wang, B.; Yi, Z.; Yao, D. Ultra-broadband solar absorber based on TiN metamaterial from visible light to mid-infrared. *J. Opt. Soc. Am. B* **2023**, *40*, 3057–3064. [CrossRef]
27. Zhang, B.W.; Luo, Y.N. Dynamic optical tuning and sensing in L-shaped dirac semimetal-based terahertz metasurfaces. *Phys. Lett. A* **2025**, *541*, 130419. [CrossRef]
28. Li, B.X.; Liu, M.L.; Wen, R.Q.; Wei, Y.; Zeng, L.L.; Deng, C.S. Dynamic control of Fano-like interference in the graphene periodic structure. *J. Phys. D Appl. Phys.* **2023**, *56*, 115104. [CrossRef]
29. Wang, H.Y.; Ma, R.; Liu, G.D.; Wang, L.L.; Lin, Q. Optical force conversion and conveyor belt effect with coupled graphene plasmon waveguide modes. *Opt. Express* **2023**, *31*, 32422. [CrossRef]
30. Chen, S.; Wu, X.H.; Fu, C.J. Active tuning of anisotropic phonon polaritons in natural van der Waals crystals with negative permittivity substrates and its application in energy transport. *Opto-Electron. Sci.* **2024**, *3*, 240002. [CrossRef]
31. Ai, Z.; Liu, H.F.; Cheng, S.B.; Zhang, H.F.; Yi, Z.; Zeng, Q.D.; Wu, P.H.; Zhang, J.G.; Tang, C.J.; Hao, Z.Q. Four peak and high angle tilted insensitive surface plasmon resonance graphene absorber based on circular etching square window. *J. Phys. D Appl. Phys.* **2025**, *58*, 185305. [CrossRef]
32. Chen, Z.Y.; Cheng, S.B.; Zhang, H.F.; Yi, Z.; Tang, B.; Chen, J.; Zhang, J.G.; Tang, C.J. Ultra wideband absorption absorber based on Dirac semimetallic and graphene metamaterials. *Phys. Lett. A* **2024**, *517*, 129675. [CrossRef]
33. Liu, M.L.; Li, B.X.; Zeng, L.L.; Wei, Y.; Wen, R.Q.; Zhang, X.J.; Deng, C.S. Dynamic tunable narrow-band perfect absorber for fiber -optic communication band based on liquid crystal. *J. Phys. D Appl. Phys.* **2023**, *56*, 505102. [CrossRef]
34. Li, Z.T.; Li, X.; Liu, G.D.; Wang, L.L.; Lin, Q. Analytical investigation of unidirectional reflectionless phenomenon near the exceptional points in graphene plasmonic system. *Opt. Express* **2023**, *31*, 30458. [CrossRef] [PubMed]
35. Li, Z.T.; Cheng, S.B.; Zhang, H.F.; Yang, W.X.; Yi, Z.; Yi, Y.G.; Wang, J.Q.; Ahmad, S.; Raza, R. Ultrathin broadband terahertz metamaterial based on single-layer nested patterned graphene. *Phys. Lett. A* **2025**, *534*, 130262. [CrossRef]
36. Wang, J.T.; Tonkaev, P.; Koshelev, K.; Lai, F.; Kruk, S.; Song, Q.; Kivshar, Y.; Panoiu, N.C. Resonantly enhanced second- and third-harmonic generation in dielectric nonlinear metasurfaces. *Opto-Electron. Adv.* **2024**, *7*, 230186. [CrossRef]

37. Zeng, Z.L.; Liu, H.F.; Zhang, H.F.; Cheng, S.B.; Yi, Y.G.; Yi, Z.; Wang, J.Q.; Zhang, J.G. Tunable ultra-sensitive four-band terahertz sensors based on Dirac semimetals. *Photonics Nanostruct. Fundam. Appl.* **2025**, *63*, 101347. [CrossRef]
38. Huang, Z.; Liu, G.D.; Wang, L. Active modulation of quasi-bound state in the continuum based on bulk Dirac semimetals metamaterial. *Appl. Phys. Express* **2022**, *15*, 032006. [CrossRef]
39. Zeng, L.L.; Li, B.X.; Wen, R.Q.; Zhang, X.J. Plasmonic Sensor Based on Multi Fano Resonance in Inverse T Shape Structure for Detection of CO_2 Concentration. *IEEE Photonics J.* **2023**, *15*, 2201805. [CrossRef]
40. Wang, J.Q.; Sun, J.Y.; Sun, S.; Zhang, H.; Wang, Q.Q.; Yang, J.Y.; Mei, Y.W. Numerical simulation of electromagnetically induced transparency in composite metamaterial. *Phys. Scr.* **2025**, *100*, 025512. [CrossRef]
41. Cheng, S.B.; Li, W.X.; Zhang, H.F.; Akhtar, M.N.; Yi, Z.; Zeng, Q.D.; Ma, C.; Sun, T.Y.; Wu, P.H.; Ahmad, S. High sensitivity five band tunable metamaterial absorption device based on block like Dirac semimetals. *Opt. Commun.* **2024**, *569*, 130816. [CrossRef]
42. Wang, Z.; Pan, W.K.; He, Y.; Zhu, Z.; Jin, X.; Liu, M.; Ma, S.; He, Q.; Sun, S.; Zhou, L.; et al. Efficient generation of vectorial terahertz beams using surface-wave excited metasurfaces. *Opto-Electron. Sci.* **2025**, *4*, 240024. [CrossRef]
43. Hu, J.Y.; Tan, C.X.; Bai, W.D.; Li, Y.M.; Lin, Q.; Wang, L.L. Dielectric nanocavity-coupled surface lattice resonances for high-efficiency plasmonic sensing. *J. Phys. D Appl. Phys.* **2022**, *55*, 075105. [CrossRef]
44. Rane, S.; Prabhu, S.; Chowdhury, D.R. Physics and applications of terahertz metagratings. *Opto-Electron. Sci.* **2024**, *3*, 230049. [CrossRef]
45. Li, W.; Cheng, S.; Zhang, H.; Yi, Z.; Tang, B.; Ma, C.; Wu, P.; Zeng, Q.; Raza, R. Multi-functional metasurface: Ultra-wideband/multi-band absorption switching by adjusting guided mode resonance and local surface plasmon resonance effects. *Commun. Theor. Phys.* **2024**, *76*, 065701. [CrossRef]
46. Yu, Z.; Li, M.; Xing, Z.; Gao, H.; Liu, Z.; Pu, S.; Mao, H.; Cai, H.; Ma, Q.; Ren, W.; et al. Genetic algorithm assisted meta-atom design for high-performance metasurface optics. *Opto-Electron. Sci.* **2024**, *3*, 240016. [CrossRef]
47. Huo, D.; Zhang, J.; Wang, Y.; Wang, C.; Su, H.; Zhao, H. Broadband Perfect Absorber Based on TiN-Nanocone Metasurface. *Nanomaterials* **2018**, *8*, 485. [CrossRef]
48. Wang, H.; Chen, Q.; Wen, L.; Song, S.; Hu, X.; Xu, G. Titanium-nitride-based integrated plasmonic absorber/emitter for solar thermophotovoltaic application. *Photonics Res.* **2015**, *3*, 329–334. [CrossRef]
49. Ma, R.; Zhang, L.G.; Zeng, Y.; Liu, G.D.; Wang, L.L.; Lin, Q. Extreme enhancement of optical force via the acoustic graphene plasmon mode. *Opt. Express* **2023**, *31*, 482723. [CrossRef]
50. Yan, X.F.; Lin, Q.; Wang, L.L.; Liu, G.D. Tunable strong plasmon–exciton coupling based modulator employing borophene and deep subwavelength perovskite grating. *J. Phys. D Appl. Phys.* **2023**, *56*, 435106. [CrossRef]
51. Berhe, A.M.; As'ham, K.; Al-Ani, I.; Hattori, H.T.; Miroshnichenko, A.E. Strong coupling and catenary field enhancement in the hybrid plasmonic metamaterial cavity and TMDC monolayers. *Opto-Electron. Adv.* **2024**, *7*, 230181. [CrossRef]
52. im Sande, S.; Deng, Y.D.; Bozhevolnyi, S.I.; Ding, F. Spin-controlled generation of a complete polarization set with randomly-interleaved plasmonic metasurfaces. *Opto-Electron. Adv.* **2024**, *7*, 240076. [CrossRef]
53. Ling, Z.X.; Zeng, Y.; Liu, G.D.; Wang, L.L.; Lin, Q. Unified model for plasmon-induced transparency with direct and indirect coupling in borophene-integrated metamaterials. *Opt. Express* **2022**, *30*, 21966. [CrossRef]
54. Li, W.H.; Chen, J.; Gao, S.Z.; Niu, L.; Wei, J.; Sun, R.; Wei, Y.; Tang, W.; Cui, T.J. An externally perceivable smart leaky-wave antenna based on spoof surface plasmon polaritons. *Opto-Electron. Adv.* **2024**, *7*, 240040. [CrossRef]
55. Li, W.; Yi, Y.; Yang, H.; Cheng, S.; Yang, W.X.; Zhang, H.; Yi, Z.; Yi, Y.; Li, H. Active Tunable Terahertz Band-width Absorber Based on single layer Graphene. *Commun. Theor. Phys.* **2023**, *75*, 045503. [CrossRef]
56. Zeng, Y.; Ling, Z.X.; Liu, G.D.; Wang, L.L.; Lin, Q. Tunable plasmonically induced transparency with giant group delay in gain-assisted graphene metamaterials. *Opt. Express* **2022**, *30*, 455954. [CrossRef] [PubMed]
57. Monrós, G.; Badenes, J.A.; Delgado, C.; Monrós-Andreu, G.; Llusar, M. New Mcconnellite Ceramic Pigment as a Selective Solar Absorber: Effects of Microwave Firing and Rare Earth Doping. *Materials* **2025**, *18*, 1520. [CrossRef] [PubMed]
58. Li, X.; Li, Z.T.; Wang, L.L.; Lin, Q. Unidirectional reflectionless propagation in borophene plasmonic metamaterials. *J. Phys. D Appl. Phys.* **2023**, *57*, 015103. [CrossRef]
59. Long, W.; Li, Y.; Chen, Y.; Chen, Q.; Yu, D. Inverse Design of Wavelength-Selective Film Emitter for Solar Thermal Photovoltaic System. *Photonics* **2025**, *12*, 286. [CrossRef]
60. Long, T.; Zhang, L.; Wang, L.L.; Lin, Q. Tunable narrow transparency windows induced by the coupled quasi-guided modes in borophene plasmonic nanostructure. *J. Phys. D Appl. Phys.* **2022**, *55*, 315101. [CrossRef]
61. Wu, L.J.; Li, Z.W.; Wang, W.J.; Chen, S.L.; Ruan, H. Near-ideal solar absorber with ultra-broadband from UV to MIR. *Results Phys.* **2022**, *40*, 105883. [CrossRef]
62. Feng, Y.; Cao, Y.; Zhang, H.; Yi, Z.; Liu, H.; Wu, X. TiN-based broadband wide-angle solar absorber. *Plasmonics* **2024**, *19*, 963–972. [CrossRef]
63. Hu, J.Y.; Bai, W.D.; Tan, C.X.; Li, Y.M.; Lin, Q.; Wang, L.L. Highly electric field enhancement induced by anapole modes coupling in the hybrid dielectric-metal nanoantenna. *Opt. Commun.* **2022**, *511*, 127987. [CrossRef]

64. Yang, Q.X.; Yu, M.H.; Chen, Z.X.; Ai, S.; Kentsch, U.; Zhou, S.; Jia, Y.; Chen, F.; Liu, H. A novel approach towards robust construction of physical colors on lithium niobate crystal. *Opto-Electron. Adv.* **2025**, *8*, 240193. [CrossRef]
65. Zhang, S.W.; Yang, H.; Tang, C.J.; Yi, Z.; Zhang, J.G.; Wang, J.Q.; Li, B.X. Multiple tunable six-peak graphene absorber for high-performance refractive index sensing. *Phys. B Condens. Matter* **2025**, *708*, 417225. [CrossRef]
66. Ma, R.; Zhang, L.; Liu, G.; Wang, L.; Lin, Q. The total optical force exerted on black phosphorus coated dielectric cylinder pairs enhanced by localized surface plasmon. *J. Appl. Phys.* **2021**, *130*, 113103. [CrossRef]
67. Li, Z.; Song, Q.J.; Jia, L.B.; Yi, Z.; Cheng, S.B.; Wang, J.Q.; Li, B.X. Actively tunable multi-frequency narrowband terahertz absorber using graphene metamaterials. *Opt. Commun.* **2025**, *583*, 131768. [CrossRef]
68. Tan, Z.Q.; Lin, Q.; Du, W.J.; Wang, L.L.; Liu, G.D. Simultaneously enhance electric and magnetic Purcell factor by strong coupling between toroidal dipole quasi-BIC and electric dipole. *J. Appl. Phys.* **2025**, *137*, 033103. [CrossRef]
69. Lv, J.; Ren, Y.; Wang, D.; Wang, J.; Lu, X.; Yu, Y.; Li, W.; Liu, Q.; Xu, X.; Liu, W.; et al. Optical switching with high-Q Fano resonance of all-dielectric metasurface governed by bound states in the continuum. *Opt. Express* **2024**, *32*, 28334–28347. [CrossRef]
70. Li, B.X.; Zeng, L.L.; Wen, R.Q.; Zhang, X.J. Dynamic Regulation of Multiple Fano Resonances Based on Liquid Crystal. *IEEE Photonics J.* **2023**, *15*, 2200506. [CrossRef]
71. Zhang, Y.X.; Lin, Q.; Yan, X.Q.; Wang, L.L.; Liu, G.D. Flat-band Friedrich-Wintgen bound states in the continuum based on borophene metamaterials. *Opt. Express* **2024**, *32*, 10669–10678. [CrossRef] [PubMed]
72. Wang, J.Q.; Yang, J.Y.; Mei, Y.W. Non-radiating anapole state in dielectric nanostructures and metamaterials. *J. Phys. D Appl. Phys.* **2025**, *58*, 203001. [CrossRef]
73. Xiang, T.; Sun, Z.; Wang, L.L.; Lin, Q.; Liu, G.D. Polarization independent perfect absorption of borophene metamaterials operating in the communication band. *Phys. Scr.* **2024**, *99*, 085519. [CrossRef]
74. Gu, X.; Liu, G.D.; Wang, L.L.; Lin, Q. Robust Fano resonance induced by topologically protected interface modes interference at gigahertz. *Appl. Phys. Express* **2022**, *15*, 082004. [CrossRef]
75. Pham, T.S.; Khuyen, B.X.; Lam, V.D.; Chen, L.; Lee, Y. Wide-Angle, Polarization-Independent Broadband Metamaterial Absorber by Using Plasmonic Metasurface-Based Split-Circular Structure. *Photonics* **2025**, *12*, 334. [CrossRef]
76. Shen, Z.; Ni, J. Multi-Resonant Full-Solar-Spectrum Perfect Metamaterial Absorber. *Nanomaterials* **2024**, *14*, 1959. [CrossRef]

photonics

MDPI

Article

A Dual-Band Tunable Electromagnetically Induced Transparency (EIT) Metamaterial Based on Vanadium Dioxide

Lei Zhu [1,2,3,*], Shujie Wang [3], Yun Wang [3], Liang Dong [3], Hailong Li [3], Yiya Wang [3] and Xumin Ding [4,*]

1 School of Computer and Control Engineering, Qiqihar University, Qiqihar 161006, China
2 Heilongjiang Key Laboratory of Big Data Network Security Detection and Analysis, Qiqihar University, Qiqihar 161006, China
3 Communication and Electronics Engineering Institute, Qiqihar University, Qiqihar 161006, China; 2024935290@qqhru.edu.cn (S.W.); 2024935297@qqhru.edu.cn (Y.W.); 01497@qqhru.edu.cn (L.D.); 2023910315@qqhru.edu.cn (H.L.); 2023935726@qqhru.edu.cn (Y.W.)
4 Advanced Microscopy and Instrumentation Research Center, Harbin Institute of Technology, Harbin 150080, China
* Correspondence: zhulei@qqhru.edu.cn (L.Z.); xuminding@hit.edu.cn (X.D.)

Abstract: A dual-band tunable terahertz electromagnetically induced transparency (EIT) metamaterial is introduced. The EIT metamaterial consists of two rectangular split rings, two metal strips, and a patterned vanadium dioxide (VO_2) located at the back. The rectangular split rings serve as the bright resonator to generate two resonance valleys at distinct frequencies. The metal strips act as the dark resonator and are indirectly activated via the coupling influence of the bright resonator. The EIT metamaterial's response mechanism is analyzed via the field effect and the two-particle model, with theoretical fitting results showing strong agreement with the simulation results. Moreover, VO_2's conductivity is altered to dynamically control the EIT effect in both frequency bands. Two transparency windows, with modulation depths of 70% and 75%, are observed as the conductivity of VO_2 decreases. Simultaneously, the simulation results reveal a favorable slow light effect, with group delays reaching 51 ps and 74 ps at the transparency windows. The proposed metamaterial holds considerable promise for future modulator, filter, and slow light device applications.

Keywords: metamaterial; electromagnetically induced transparency (EIT) effect; vanadium dioxide (VO_2)

1. Introduction

The electromagnetically induced transparency (EIT) phenomenon results from destructive interference effects in quantum systems [1] and shows significant advantages in achieving narrow and sharp resonances. It can eliminate materials' medium absorption during electromagnetic wave transmission, thereby creating a narrow transparent window within a broad opaque spectrum. This process significantly enhances nonlinear magnetic susceptibility [2] and is often accompanied by a slow light effect [3]. A more pronounced slow light effect enhances light–matter interaction, thereby improving the performance of optical devices. However, the traditional achievement of the EIT effect requires a stable pump light source and a low-temperature environment as strict test conditions [4], which severely limits practical applications. The development of metamaterials has enabled the successful simulation and realization of the EIT effect, thus no longer relying on traditional strict conditions. This progress provides new development opportunities for multiple potential application fields, including optical switches [5], biosensors [6], and slow light devices [7].

The EIT effect arises from either bright–dark mode or bright–bright mode coupling [8]. Bright modes [9] can be easily activated by incident waves, while dark modes [10] cannot couple directly with the incident waves. It is excited through indirect coupling with the bright mode. Additionally, similar resonance frequencies are observed for bright and dark modes, with differing quality (Q) values. Transparent windows emerge in metamaterials composed of bright and dark resonators through mode coupling. Likewise, the EIT effect arises from bright mode coupling.

Recently, the dynamically tunable EIT metamaterials have garnered significant attention. By incorporating tunable materials into EIT metamaterials, the EIT effect can be actively modulated [11]. Very recently, due to its excellent adjustability, vanadium dioxide (VO_2) has received increasing attention in related fields and has become one of the most notable materials [12]. As the temperature increases, the VO_2 lattice transitions from monoclinic to tetragonal, causing a sharp rise in electrical conductivity. This transformation is complete when the temperature reaches 340 K, marking the shift from an insulator to a metal [13]. This temperature-induced phase transition makes VO_2 an ideal material for tuning the EIT effect in metamaterials. A metamaterial structure of VO_2 as a terahertz modulator in an insulating state was proposed by Wang et al. [14]. The EIT window peak decreases as VO_2 conductivity increases, and the EIT effect gradually disappears. Ning et al. introduced a two-layer metamaterial structure with VO_2 as its substructure [15]. The EIT effect is observed when VO_2 exhibits a metallic phase, whereas the disappearance of the transparency window occurs when VO_2 transitions to an insulating phase, demonstrating that controlling the temperature can tune and switch the EIT effect. However, these studies predominantly focused on tuning a single EIT window, with limited exploration of dual-band EIT tuning. With the increasing demand for multi-band tunable devices, the research on dual-band EIT effects has become an important research direction. Chen et al. reported a dual-frequency EIT effect VO_2 metamaterial with modulation depths of 45.2% and 42.7% and group delays of 3.24 ps and 3.17 ps [16]. However, the relatively low modulation depth and group delay significantly weaken the light signal processing capability, affecting its performance in practical applications. Therefore, utilizing VO_2 to dynamically control the dual-band EIT effect while achieving a high modulation depth and high group delay remains a significant challenge.

In this context, a dynamically tunable EIT metamaterial based on thin VO_2 films has been proposed, aiming to achieve dual-band EIT effects with a higher modulation depth. The top layer of this metamaterial is constructed from two symmetrical rectangular split rings (as the bright mode) and two metal strips (as the dark mode). The coupled bright–dark mode generates transparency windows with high Q values of 102 and 80 at 1.02 THz and 1.15 THz, respectively. Meanwhile, the VO_2 structure is added at the bottom layer of the unit, and its conductivity is changed to achieve the simultaneous dynamic regulation of the dual-band EIT effect. Modulation depths of 70% and 77% and group delays of 41 ps and 74 ps are achieved, respectively, on the two transparent windows. The two-particle model further illustrates the generation mechanism of the EIT effect and calculates the theoretical transmission spectrum, which fits well with the simulation results. The proposed metamaterial holds great potential for making significant contributions to the evolution of modulators, filters, sensors, and slow light devices.

2. Structure Design and the Material Properties of VO_2

Figure 1 illustrates the designed EIT metamaterial, which is deposited on a SiO_2 substrate in a periodic arrangement. The top unit cell is made up of two rectangular split rings and two metal strips, as shown in Figure 1b. Compared with continuous thin films, patterned VO_2 structures exhibit significant advantages in achieving EIT effects, reducing fabrication costs and enhancing the applications' performance. Consequently, we have designed a patterned VO_2 structure as the bottom layer of the unit, as shown in Figure 1c. Furthermore, through simulating the thickness of VO_2 and comprehensively evaluating the EIT effect for the two transparent windows, the parameter t = 0.2 μm is determined. The substrate is made of lossless SiO_2, and the metal layer is gold with a conductivity of 4.56×10^7 S/m. The geometric parameters of the unit cell are as follows: $P_x = P_y$ = 160 μm, a = 90 μm, b = 40 μm, c = 10 μm, d = 13 μm, and g = 16 μm. In this paper, we utilize commercial CST simulation software (Version 2020) to numerically model the metamaterial's structure. Periodic boundary conditions are adopted in the x- and y-directions, while open boundary conditions are applied in the z-direction.

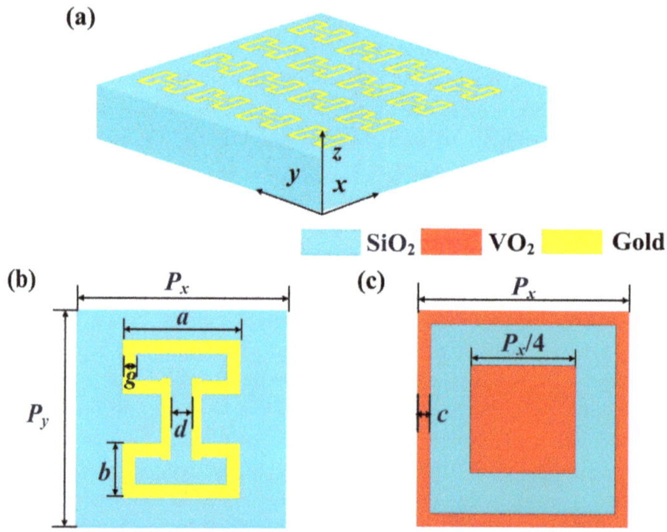

Figure 1. (**a**) Three-dimensional view of the EIT metamaterial array; (**b**) top and (**c**) back view of the EIT metamaterial structure unit.

The Drude model provides an accurate characterization of the terahertz optical properties of VO_2 [14], which is as follows:

$$\varepsilon(\omega) = \varepsilon_\infty - \frac{\omega_p^2(\sigma_{VO_2})}{\omega^2 + j\gamma\omega} \tag{1}$$

$$\omega_p^2(\sigma_{VO_2}) = \frac{\sigma_{VO_2}}{\sigma_0}\omega_p^2(\sigma_0) \tag{2}$$

where ε_∞ is 12, representing the high-frequency dielectric permittivity, and γ (5.75×10^{13} rad/s) represents the collision frequency. Additionally, $\omega_p(\sigma_{VO_2})$ is the plasma frequency at σ_{VO_2}, with $\omega_p(\sigma_0) = 1.4 \times 10^{15}$ rad/s and $\sigma_0 = 3 \times 10^5$ S/m. Generally speaking, the conductivity of VO_2 varies in response to changes in the external temperature. When the external temperature increases, the internal lattice structure of the VO_2 undergoes a transformation, thereby causing a phase transition from an insulator to a metal [14]. This transformation induces changes in physical parameters, such as conductivity and the transmission coefficient. Consequently, the designed metamaterial structure is dynamically modulated by adjusting the electrical conductivity of VO_2.

3. Results

3.1. Mechanism Analysis of the EIT Effect

Fano resonance [17], surface plasmon resonance [18], and guided mode resonance (GMR) [19] are common physical phenomena used to generate sharp spectral features in filtering and sensing applications. For example, Lotfiani et al. proposed a Ge-on-Si self-powered photodetector enhanced by GAM [19]. In contrast, this paper achieves sharp EIT resonances via bright–dark mode near-field coupling, generating two transparent windows with a higher modulation depth and group delay. Figure 2a illustrates the transmission spectra corresponding to the different substructures. Observations indicate that, for the x-polarized incident wave, the rectangular split rings, as the bright modes, can directly couple with the incident wave, resulting in resonance valleys at 1.02 THz and 1.15 THz. In contrast, the metal strips represent the dark mode and are not directly stimulated by the incident wave. When the two substructures are combined, destructive interference occurs between their scattered fields through bright–dark mode coupling, resulting in resonance transmission amplitudes of 0.86 and 0.85 at 1.02 THz and 1.15 THz, respectively. Figure 2b depicts that these two transparent windows are regarded as the EIT effect. At this point, the peak value of the transparent window has not reached 1 because the ohmic loss dominates the total loss [20]. Moreover, the Q factor is the core parameter for measuring the energy storage efficiency of a resonant system, formulated as the ratio of the resonant frequency f_0 to the full width at half maximum (FWHM) Δf [21]. The Q factors of the two transparent windows are 102 and 80. The larger the Q factor, the greater the dispersion and the higher the group delay.

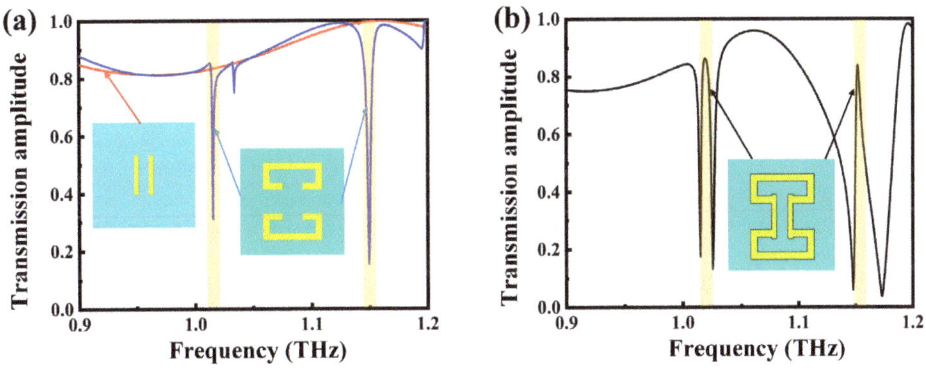

Figure 2. Transmission spectra of (**a**) the two substructures and (**b**) the overall structure of the EIT metamaterial. The yellow shadows represent the resonant regions of 1.02 THz and 1.15 THz.

To further research the underlying mechanisms of the EIT effect, Figure 3a–c present the surface current distributions of the first transparency window. In these figures, the arrow direction in the figure indicates the direction of current flow, and the arrow color changing from blue to red represents the trend of current density increasing from small to large. At 1.015 THz, quasi-circular currents flow in opposite directions on the rectangular split rings, as shown in Figure 3a. The weak coupling to free space increases the radiation loss [22], leading to reduced transmission, while the surface current in the metal strips (dark mode) remains weak. At 1.03 THz, a redistribution of the surface current within the metamaterial structure becomes evident. The current concentrates at the junction between the rectangular split rings and the metal strips, indicating a weakening of the destructive interference among the bright and dark modes. This results in increased radiation loss and further decreases transmission, as presented in Figure 3c. Figure 3b illustrates that the current flows in the opposite direction at 1.02 THz, resulting in destructive regions, as indicated by the red arrow. Destructive interference between the bright and dark modes leads to a strong current distribution throughout the entire metamaterial structure, enhancing transmission and generating the transparency window.

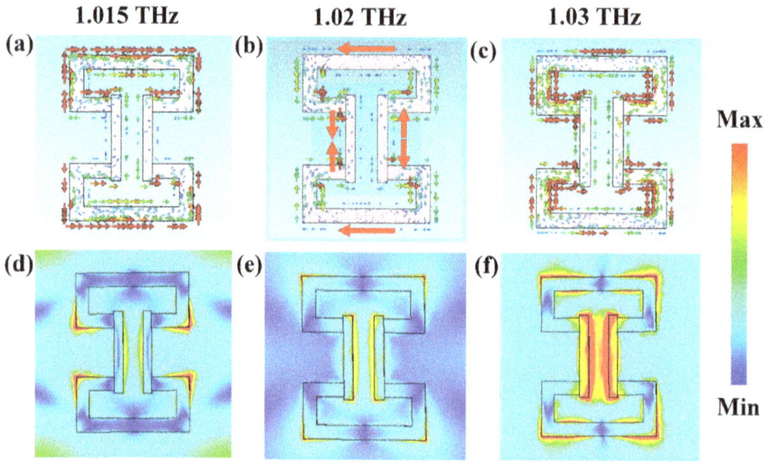

Figure 3. The current and electric field distributions of the EIT metamaterial at (**a**,**d**) 1.015 THz, (**b**,**e**) 1.02 THz, and (**c**,**f**) 1.03 THz. The big red arrow represents the overall trend of the current flow direction.

Figure 3d–f illustrate the electric field distribution at the first transparent window, with red areas indicating regions of stronger electric field intensity. As depicted in Figure 3d, the electric field strength of the rectangular split rings are significantly enhanced at 1.015 THz, leading to increased loss and consequently reduced transmittance. Figure 3f indicates that at 1.03 THz, both the rectangular split rings and the metal strips exhibit further enhancement in electric field strength, resulting in increased losses and a corresponding decrease in transmittance. In contrast, Figure 3e displays the overall structure's electric field distribution at 1.02 THz, which is relatively weak, suggesting lower losses and higher transmittance. It is worth noting that the junctions are the key regions where the interaction between bright and dark resonators occurs. Therefore, the currents and electric field intensities generated by coupling are concentrated in these regions. The formation mechanism of the second transparent window at 1.15 THz is fundamentally similar to that of the first transparent window.

3.2. Parameter Analysis

The EIT effect in metamaterials is influenced by geometric parameters. To optimize the performance of the EIT metamaterial, we systematically simulated and analyzed the impact of various structural parameters (g, d, a, b) on performance, and ultimately determined the optimal parameter combination. As shown in Figure 4a, when g increases from 10 μm to 18 μm, the transmission amplitudes at both transparency windows exhibit slight enhancements, while their resonant frequencies remain nearly constant. Similarly, as d increases from 9 μm to 13 μm, the resonant frequency of the two transparent windows shows no significant change, and the transmission amplitude only slightly increases, as shown in Figure 4b. Furthermore, the first resonance dip frequency is almost unaffected, while the second resonance dip displays a blue shift. Figure 4c shows the changes in the two transparent windows when a increases from 86 μm to 94 μm. The results reveal that neither the transmission amplitude nor the resonant frequency of the two transparent windows exhibits significant changes. When the structural parameter *b* increases from 36 μm to 44 μm, the changes in the transmission amplitude and resonant frequency of the first transparent window can be ignored. However, the second transparent window shows a significant reduction in the transmission amplitude, accompanied by a distinct redshift of the resonant dip frequency, as shown in Figure 4c. To sum up, the change in the structural parameters has a certain influence on the transmission spectra of the two transparent windows. It may even lead to the case that the resonant valley cannot be completely and accurately reproduced at 1.02 THz and 1.15 THz during the simulation process. However, this deviation is almost negligible and has a negligible impact on the final result.

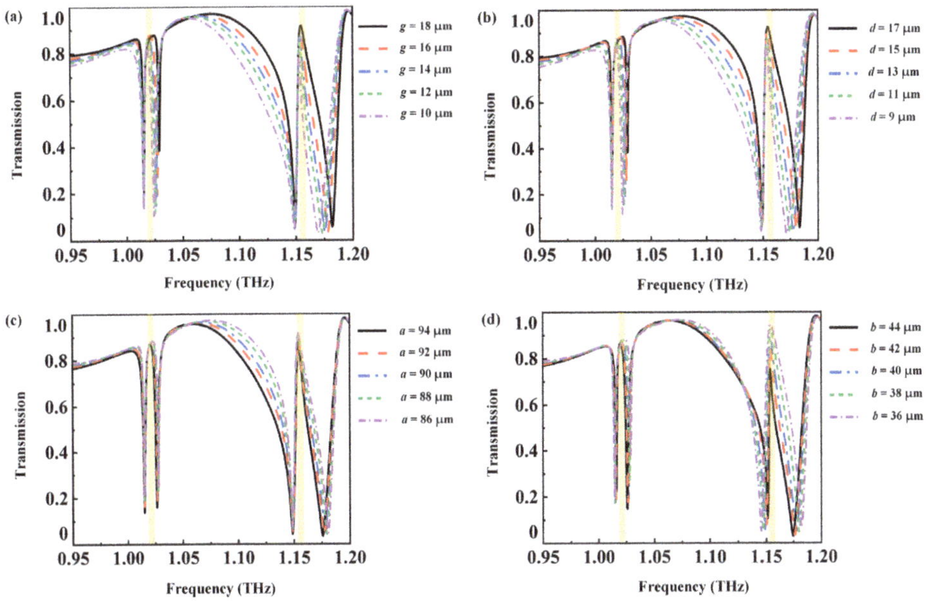

Figure 4. Transmission spectra of EIT metamaterial structures with different values of (**a**) *g*, (**b**) *d*, (**c**) *a*, and (**d**) *b*.

3.3. Dynamic Regulation of VO_2

The tunable EIT effect has broad applications in modulators, filters, sensing, slow light, and other fields [23]. To achieve the tunable EIT effect, VO_2 is incorporated into the bottom layer of the metamaterial structure, enabling dynamic regulation of the EIT effect by modulating the conductivity of VO_2. Figure 5 displays the transmission spectra of the hybrid structure across a range of conductivity of VO_2. It is observed that as the conductivity of VO_2 increases from 10 S/m to 200,000 S/m, the resonance frequencies associated with the two transparency windows are nearly unchanged. However, notable variations are observed in the resonance strengths of the EIT effects. As conductivity increases from 10 S/m to 200,000 S/m, the strengths of the EIT effects gradually decrease and eventually disappear. This is attributed to a reduction in the destructive interference between the bright and dark modes as the conductivity of VO_2 increases. Furthermore, the terahertz metasurface based on VO_2 proposed by Chen et al. can increase the electrical conductivity of VO_2 from 0.02 S/m to 250,000 S/m by raising the temperature from 40 °C to approximately 110 °C [24], further verifying the feasibility of thermal tuning for the regulation of VO_2 within the terahertz range. The modulation depth of the transparency window is determined using the formula $\rho_T = (T_{on} - T_{off})/T_{on}$. T_{on} represents transmission when VO_2 is insulating (10 S/m), while T_{off} denotes transmission in the metallic state (200,000 S/m) of VO_2. Although our modulation depth has not yet reached the theoretical limit, it is already at higher levels of 70% and 75%.

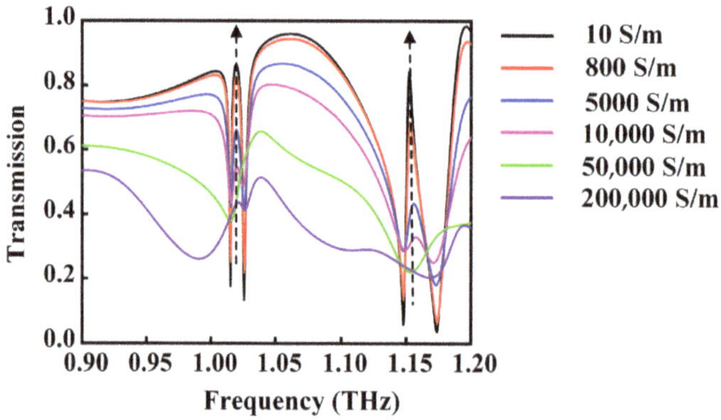

Figure 5. Transmission spectra of EIT metamaterial structures with different levels conductivity of VO_2.

To highlight the modulation depth of the introduced metamaterial, the modulation properties of various tunable metamaterials are compared, as shown in Table 1. Chen et al. proposed a VO_2-based terahertz metamaterial with dual-band EIT properties, achieving modulation depths of 46% and 45% for the two transparency windows [25]. They also presented an EIT metamaterial that exhibits multi-resonance, polarization insensitivity, and dynamic adjustability, with modulation depths of 62.5% and 65% for the two transparency windows [26]. Xu et al. introduced a terahertz metamaterial based on $MoTe_2$, which achieves a modulation depth of 77% at the transparency window but supports only a single band of EIT effect [27]. Wang et al. realized an EIT effect, achieving a maximum modulation depth of 80% and a Q factor of 544 in a single band by adjusting the Fermi level of graphene [28]. Although the EIT modulation depth of graphene and $MoTe_2$ is slightly higher than in this study, their single-band EIT effect limits their application in multi-frequency communication systems. Moreover, $MoTe_2$ has a limited conductivity tuning range, restricting modulation depth enhancement. While graphene offers strong tunability, it requires complex electrocontrol to adjust the Fermi level, increasing the complexity and cost. In contrast, the proposed dual-band EIT metamaterial not only enhances flexibility for multi-frequency applications but also achieves efficient and rapid modulation through a broad conductivity tuning range and the fast phase transition properties of VO_2

Table 1. Comparisons of the properties of different tunable metamaterials.

Reference	Active Material	Center Frequency of the EIT Window	Modulation Depth
Ref. [25]	VO_2	1.23 THz, 1.41 THz	46%, 45%
Ref. [26]	VO_2	0.90 THz, 1.06 THz	62.5%, 65%
Ref. [27]	$MoTe_2$	1.23 THz	77%
Ref. [28]	Graphene	220.6 THz	80%
Proposed structure	VO_2	1.02 THz, 1.15 THz	75%, 70%

3.4. The "Two-Particle" Model

The "two-particle" model serves as a classical framework to characterize the EIT effect and quantitatively analyze its occurrence in metamaterials [29]. To clarify the coupling mechanism underlying the EIT effect in the metamaterial system, the "two-particle" model was employed. In this model, the bright mode structure and the dark mode structure are represented as two particles, and the system of coupled particles satisfies the following coupling equation [29]

$$\begin{aligned}\ddot{x}_1(t) + \gamma_1\dot{x}_1(t) + \omega_0^2 x_1(t) + k^2 x_2(t) = qE_0 \\ \ddot{x}_2(t) + \gamma_2\dot{x}_2(t) + (\omega_0 + \delta)^2 x_2(t) + k^2 x_1(t) = 0\end{aligned} \quad (3)$$

where x_1 and γ_1 denote the amplitude and loss of the bright particle, respectively; x_2 and γ_2 denote those of the dark particle, respectively; δ is the detuning of the resonance frequencies between particles; k characterizes the coupling strength between the bright and the dark particles; and q represents the coupling intensity between the bright particle and the incident field. On the basis of Equation (3), the transmission coefficient for the EIT metamaterial can be determined in the following manner

$$|T| = \left| \frac{4\sqrt{x_{eff}+1}}{\left(\sqrt{x_{eff}+1}+1\right)^2 e^{j\frac{2\pi d}{\lambda_0}\sqrt{x_{eff}+1}} - \left(\sqrt{x_{eff}+1}-1\right)^2 e^{-j\frac{2\pi d}{\lambda_0}\sqrt{x_{eff}+1}}} \right| \quad (4)$$

where d represents the metamaterial's thickness, λ_0 represents the vacuum wavelength, and x_{eff} denotes the equivalent magnetic susceptibility of metamaterial, which can be evaluated with the following formula [29]:

$$x_{eff} = \frac{p}{\varepsilon_0 E_0} = \frac{q^2}{\varepsilon_0} \cdot \frac{\left[\omega^2 - (\omega_0+\delta)^2 + i\gamma_2\omega\right]}{k^4 - \left[\omega^2 - (\omega_0+\delta)^2 - i\gamma_2\omega\right] \cdot (\omega^2 - \omega_0^2 + i\gamma_1\omega)} \quad (5)$$

On the basis of the previous discussion, the theoretical results of the EIT effect with different VO_2 conductivities in the metamaterial are calculated, which are basically consistent with the simulation results. To further analyze the relationship between the fitting parameters and the conductivity, Figure 6 presents the fitting parameters (γ_1, γ_2, k) of different conductivities for the first transparent window. Among them, γ_1 and γ_2 are the losses of the bright mode and the dark mode, respectively, and k represents the coupling strength between the bright mode and the dark mode. Since the bright mode can be directly excited by the incident wave, its loss γ_1 is relatively large and remains almost unchanged with the increase in the electrical conductivity. However, the dark mode cannot be directly excited by the incident wave, the loss γ_2 is relatively small, and it will slightly increase with the increase in electrical conductivity. Meanwhile, with the increase in conductivity, the coupling strength k remains almost constant. In conclusion, as the conductivity increases from 10 S/m to 200,000 S/m, γ_1 remains almost unchanged, while γ_2 gradually increases. This indicates that the conductivity mainly affects the value of γ_1.

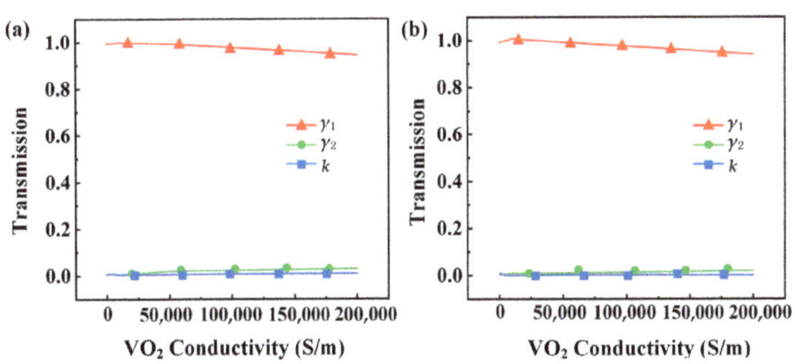

Figure 6. The relationship graph between γ_1, γ_2, and k and the conductivity of VO_2 at (**a**) 1.02 THz and (**b**) 1.15 THz.

3.5. Slow Light Effect

The pronounced phase dispersion in the transparency window is a prominent characteristic of the EIT effect. To calculate the group delay and analyze the characteristics of the EIT transparent window in this metamaterial, we adopted the following formula [30]

$$t_g = -\frac{d\psi}{d\omega} \tag{6}$$

where the phase shift of the transparent window is represented by ψ, and the resonant frequency is represented by ω.

Figure 7 illustrates the relationship between the conductivity of VO_2 and the group delay at 1.02 THz and 1.15 THz. Specifically, the group delays at both transparent windows exhibit a pronounced decrease as the conductivity of VO_2 rises from 10 S/m to 50,000 S/m, thereby gradually weakening the slow light effect. When the conductivity of VO_2 increases from 50,000 S/m to 200,000 S/m, the group delay only shows a slight increase. Obviously, the maximum group delay occurred at the two transparent windows of the EIT effect, reaching 51 ps and 74 ps. Compared with Refs. [31–33], the metamaterial designed in this work achieves higher group delays at both transparency windows simultaneously, with values surpassing those in Ref. [31] (9.98 ps and 6.23 ps), and Ref. [32] (0.34 ps and 0.62 ps). Although the group delay of the metamaterial presented here is lower than that in Ref. [33] (117.21 ps), it enables the dual-band slow light effect, which holds promise for multi-band or broad-band slow light applications. Moreover, modifying the conductivity of VO_2 allows for the active adjustment of the group delay. As the conductivity increases, the slow light effect gradually diminishes. Thus, the presented EIT metamaterial provides a new method for designing flexible and tunable slow light devices, opening up exciting possibilities for future applications.

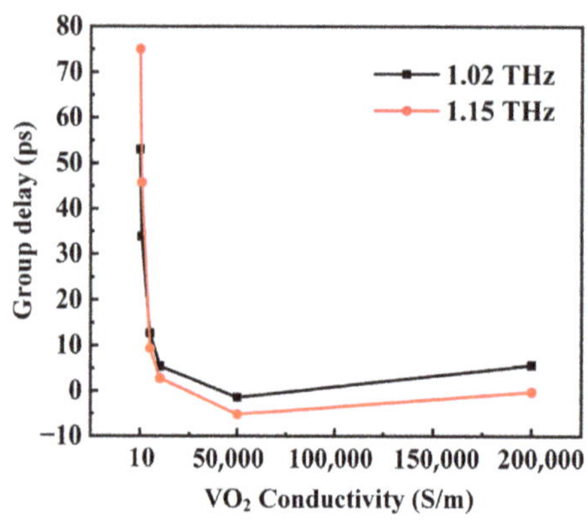

Figure 7. Relationship graph between the conductivity of VO_2 and group delay at 1.02 and 1.15 THz.

4. Discussion

Owing to the experimental constraints, we have currently verified the feasibility of dual-band tunable EIT metamaterials based on VO_2 through simulation. If possible, in the future, the processing and manufacturing of this metamaterial will be carried out in accordance with the following steps. Firstly, a gold film is deposited on the SiO_2 dielectric substrate using radio frequency sputtering. Secondly, the photoresist film is spin-coated and the top layer of the metal structure is patterned using ultraviolet lithography technology. Thirdly, the VO_2 film is deposited at the bottom layer by the pulsed laser deposition method, and the same method is used for spin-coating and patterning. Finally, the VO_2 film is etched by the reactive ion etching method to form two rectangular split ring structures at the bottom layer [34]. However, in actual manufacturing and testing, the manufacturing tolerances of the terahertz structure's nanolithography [35], the uniformity of the VO_2 film's deposition [36], the accuracy of electrical conductivity regulation [37], the loss of the SiO_2 substrate [38], and the non-ideality of the VO_2 phase transition's sharpness [39] may affect the performance of metamaterials. In addition, the influence of the loss of the SiO_2 substrate on the transmission peak, Q factor, and modulation depth cannot be ignored.

To further optimize the material design, on the one hand, in the future, we plan to introduce advanced computing technologies such as artificial intelligence and machine learning to predict the interplay under different combinations of geometry and tunable material properties [40,41]. Meanwhile, we will explore methods for optimizing the phase transition characteristics of VO_2 to improve the design's efficiency. On the other hand, the modulation depth of the metamaterials can be further enhanced through a multi-layer structure [42], material optimization [43], and dynamic tuning [44].

According to the discussion above, the dual-band EIT metamaterial based on VO_2 that we propose, with its tunability, high Q factor, high modulation depth, and slow light characteristics, holds substantial application prospects for fields like terahertz communication, high-precision sensing, and intelligent cloaking. It can achieve functions such as multi-band signal control, high-sensitivity parameter detection, and targeted interference shielding. Among them, the EIT effect limits the light field by enhancing light–matter interaction, improving the sensors' sensitivity and spectral clarity, offering a unique advantage in high-precision sensing. In comparison, although other nanoplasmonic or resonant sensors are applied in biological field, such as the infrared photodetector combining PbS quantum dots with silver nanoparticles proposed by Izadpour et al. [45], and the PIN photodetector based on plasmonic nanostructures designed by Lotfiani et al. [46], they are still inferior to EIT-based sensors in terms of spectral resolution and sensitivity to environmental changes.

5. Conclusions

In summary, a dual-band tunable EIT metamaterial based on VO_2 is designed in this paper. The top layer structure of the designed metamaterial consists of rectangular split rings and metallic strips, with the dual-band EIT effect is generated through the coupling of bright modes (rectangular split rings) and dark modes (metallic strips). By introducing VO_2 into the bottom layer of the metamaterial structure and tuning its conductivity, dynamic regulation of the dual-band EIT effect is manifested. The maximum modulation depths of 70% and 75% are achieved at 1.02 THz and 1.15 THz, along with higher Q values of 102 and 80. The two-particle model further analyzed the physical and modulation mechanisms of the designed metamaterials. The good fit between the calculated and simulated transmission spectra verified the reliability of the proposed metamaterial. Furthermore, by taking advantage of the tunability of VO_2, the slow light effect is dynamically controlled at two frequency bands to achieve group delays of 41 ps and 74 ps. Therefore, the designed dual-band tunable EIT metamaterials have significant potential in the progress of modulators and slow light devices with a high modulation depth.

Author Contributions: Conceptualization, L.Z.; software, H.L.; validation, Y.W. (Yiya Wang); formal analysis, Y.W. (Yiya Wang); investigation, Y.W. (Yiya Wang); writing—original draft preparation, S.W. and Y.W. (Yun Wang); writing—review and editing, L.Z. and L.D.; supervision, L.Z., L.D. and X.D.; project administration, L.Z.; funding acquisition, L.Z. and X.D. All authors have read and agreed to the published version of the manuscript.

Funding: This work was supported by the Fundamental Research Funds of Heilongjiang Provincial Universities of China (Grant No. 145409208), the Natural Science Foundation of Heilongjiang Province (Grant No. LH2022F053), the National Natural Science Foundation of China (Grant No. 62275063,

62171153), and the Postdoctoral Research Fund Project of Heilongjiang Province of China (Grant No. LBH-Q21195).

Institutional Review Board Statement: Not applicable.

Informed Consent Statement: Not applicable.

Data Availability Statement: The original contributions presented in this study are included in the article. Further inquiries can be directed to the corresponding author.

Conflicts of Interest: The authors declare no conflicts of interest.

Abbreviations

The following abbreviations are used in this manuscript.

EIT	Electromagnetically induced transparency
VO2	Vanadium dioxide
Q	Quality
GMR	Guided mode resonance
FWHM	Full width at half maximum

References

1. Liu, S.B.; Ning, R.X.; Zhu, J.L.; Zhang, Y.F. Multiband electromagnetically induced transparency-like on metamaterials. *Opt. Commun.* **2025**, *574*, 131065. [CrossRef]
2. Zhang, X.L.; Zhao, Z.Y.; Ako, R.T.; Sriram, S.; Zhao, X.; Liu, H.X.; Bu, H.J. Moire photonic superlattice-induced transparency at commensurate angle in a terahertz metasurface composed of triple layer square lattices. *Appl. Phys. Lett.* **2024**, *125*, 141701. [CrossRef]
3. Zhang, B.H.; Dong, A.L.; Wang, J.R.; Qin, M.; Liu, J.Q.; Huang, W.; Xiao, S.Y.; Li, H.J. Strong coupling and electromagnetically induced transparency in multiple-BIC-driven metasurfaces. *Nano Lett.* **2025**, *25*, 4568–4575. [CrossRef] [PubMed]
4. Cai, W.J.; Xiao, B.G.; Yu, J.B.; Xiao, L.H. A compact graphene metamaterial based on electromagnetically induced transparency effect. *Opt. Commun.* **2020**, *475*, 126266. [CrossRef]
5. Zhu, J.; Xiong, J.Y. Tunable terahertz graphene metamaterial optical switches and sensors based on plasma-induced transparency. *Measurement* **2023**, *220*, 113302. [CrossRef]
6. Wang, J.Q.; Sun, J.Y.; Mei, Y.W. Electric field dropping effect enhanced extraordinary sensitivity of THz electromagnetically induced transparency metamaterial. *IEEE Sens. J.* **2024**, *24*, 7807–7815.
7. Vaswani, L.K.; Chouhan, B.S.; Panwar, A.; Kumar, G. Electromagnetically induced transparency in a strongly coupled orthogonal polarization-insensitive planar terahertz metamaterial. *J. Phys. D Appl. Phys.* **2024**, *57*, 075103. [CrossRef]
8. Chen, M.M.; Yang, X.X. Frequency shift in tunable electromagnetically induced transparency-like metamaterials. *Plasmonics* **2024**, *19*, 1487–1498. [CrossRef]
9. Zhao, S.X.; Gao, L.; Zhang, W. Chirality-selective electromagnetically induced transparency in a dielectric metasurface based on chirality transfer between bright and dark modes. *Appl. Phys. Lett.* **2023**, *123*, 151701. [CrossRef]
10. Shu, C.; Mei, J.S.; Sun, H.Y.; Chen, L.Y.; Sun, Y. Tunable electromagnetically induced transparency metamaterial utilizing bright-dark mode coupling between electric and toroidal resonances. *Opt. Quant. Electron.* **2024**, *56*, 43. [CrossRef]
11. Zhu, L.; Dong, L. Electromagnetically induced transparency metamaterials: Theories, designs and applications. *J. Phys. D Appl. Phys.* **2022**, *55*, 263003. [CrossRef]
12. Zhu, L.; Li, H.; Dong, L.; Zhou, W.J.; Rong, M.X.; Zhang, X.Z.; Guo, J. Dual-band electromagnetically induced transparency (EIT) terahertz metamaterial sensor. *Opt. Mater. Express* **2021**, *11*, 2109. [CrossRef]
13. Long, L.S.; Taylor, S.; Wang, L.P. Enhanced infrared emission by thermally switching the excitation of magnetic polariton with scalable microstructured VO_2 metasurfaces. *ACS Photoincs* **2020**, *7*, 2219–2227. [CrossRef]
14. Wang, T.L.; Zhang, H.Y.; Zhang, Y.; Zhang, Y.P.; Cao, M.Y. Tunable bifunctional terahertz metamaterial device based on Dirac semimetals and vanadium dioxide. *Opt. Express* **2020**, *28*, 17434–17448. [CrossRef] [PubMed]
15. Ning, R.X.; Xiao, Z.Q.; Chen, Z.H.; Huang, W. Temperature tunable like-electromagnetically induced transparency in metasurface with vanadium dioxide. *Eur. Phys. J. Plus* **2021**, *136*, 751. [CrossRef]
16. Chen, M.M.; Yang, X.X.; Shu, F.Z. Dual-band tunable electromagnetically induced transparency in vanadium dioxide-based miniaturized terahertz metasurfaces. *Mater. Res. Bull.* **2024**, *180*, 113000. [CrossRef]
17. Salama, N.A.; Alexeree, S.M.; Obayya, S.S.A.; Swillam, M.A. Silicon-based double fano resonances photonic integrated gas sensor. *Sci. Rep.* **2024**, *14*, 24811. [CrossRef] [PubMed]

18. Hossain, N.I.; Nayemuzzaman, S.; Tabassum, S. Dual localized surface plasmon resonance sensors on a single optical fiber. *IEEE Sens. Lett.* **2023**, *7*, 9. [CrossRef]
19. Lotfiani, A.; Jahromi, H.D. Guided-mode resonance enhanced Ge-on-Si self-powered surface illuminated photodetector for ultrahigh-speed optical communication systems. *IEEE Sens. J.* **2024**, *24*, 40669–40677. [CrossRef]
20. Yang, T.; Liu, X.M.; Zhou, J. Terahertz polarization conversion in an electromagnetically induced transparency (EIT)-like metamaterial. *Ann. Phys.* **2021**, *533*, 2000528. [CrossRef]
21. Cheng, Y.; Cai, S.J.; Li, J.J.; Gong, H.Y.; Liu, Z.Q.; Chen, J.; Fu, G.L.; Liu, X.S.; Pan, P.P.; Liu, G.Q. High *Q* transparency, strong third harmonic generation, and giant nonlinear chirality driven by toroidal dipole-quasi-BIC. *Appl. Phys. Lett.* **2024**, *125*, 181702. [CrossRef]
22. Hu, J.; Lang, T.; Hong, Z.; Shen, C.Y.; Shi, G.H. Comparison of electromagnetically induced transparency performance in metallic and all-dielectric metamaterials. *J. Lightwave Technol.* **2018**, *10*, 1109. [CrossRef]
23. Makhloufi, M.; Salah, H. Graphite based metamaterial for high frequency applications. *J. Magn. Magn. Mater.* **2022**, *557*, 169410. [CrossRef]
24. Chen, M.M.; Yang, X.X. Polarization-insensitive electromagnetically induced transparency and its sensing performance based on spoof localized surface plasmons in vanadium dioxide-based terahertz metasurfaces. *Phys. Chem. Chem. Phys.* **2023**, *25*, 21074–21081. [CrossRef]
25. Chen, M.M.; Xiao, Z.Y.; Lv, F.; Cui, Z.T.; Xu, Q.D. Dynamically tunable dual-band electromagnetically induced transparency-like in terahertz metamaterial. *Opt. Mater.* **2020**, *107*, 110060. [CrossRef]
26. Chen, M.M.; Xiao, Z.Y.; Lu, X.J.; Lv, F.; Cui, Z.T. Dynamically tunable multi-resonance and polarization-insensitive electromagnetically induced transparency-like based on vanadium dioxide film. *Opt. Mater.* **2020**, *102*, 109811. [CrossRef]
27. Xu, X.; Lou, J.; Gao, M.X.; Wu, S.Y.; Fang, G.Y.; Huang, Y.D. Ultrafast modulation of THz waves based on $MoTe_2$-covered metasurface. *Sensors* **2023**, *23*, 1174. [CrossRef]
28. Wang, Z.C.; Huang, H.H.; Zhang, H.; Zhao, W.R. Controllable dual resonances of Fano and EIT in a graphene-loaded all-dielectric GaAs metasurface and its sensing and slow-light applications. *J. Phys. D Appl. Phys.* **2024**, *48*, 485101. [CrossRef]
29. Zhu, L.; Zhang, C.; Dong, L.; Rong, M.X.; Gong, J.Y.; Meng, F.Y. Inverse design of electromagnetically induced transparency (EIT) metasurface based on deep convolutional generative adversarial network. *Phys. Scr.* **2023**, *10*, 105501. [CrossRef]
30. Meng, Y.W.; Wang, Z.B.; Fang, S.J.; Liu, H.M. Reconfigurable negative group delay circuit with tunable group delay flatness. *Int. J. RF Microw. Comput. Aided Eng.* **2022**, *32*, e23159. [CrossRef]
31. Sun, Z.S.; Wang, X.; Wang, J.L.; Fan, B.; Zhang, Y.; Feng, Y. Sensing and slow light properties of dual-band terahertz metamaterials based on electromagnetically induced transparency-like. *Acta. Phys. Sinch. Ed.* **2022**, *71*, 138101. [CrossRef]
32. Zhong, X.; Wu, T.S.; Liu, Z.H.; Yang, D.; Yang, Z.N.; Liu, R.; Liu, Y.; Wang, J.Y. Dual tunable electromagnetically induced transparency based on a grating-assisted double-layer graphene hybrid structure at terahertz frequencies. *Nanomaterials* **2022**, *12*, 3853. [CrossRef]
33. Cui, Y.; Wang, X.S.; Ren, B.; Jiang, Y.Y. Polarization-controlled electromagnetically induced transparency analogue based on multipolar resonances. *Opt. Mater. Express* **2022**, *12*, 3738–3748. [CrossRef]
34. Chen, B.W.; Yang, S.X.; Chen, J.; Wu, J.B.; Chen, K.; Li, W.L.; Tan, Y.H.; Wang, Z.S.; Qiu, H.S.; Fan, K.B.; et al. Directional terahertz holography with thermally active Janus metasurface. *Light Sci. Appl.* **2023**, *12*, 136. [CrossRef]
35. Chen, Z.; Li, Y.; Man, Z.W.; Tang, A.W. Research progress of quantum dot photolithography patterning and direct photolithography application. *Nano Res.* **2024**, *17*, 10386–10411. [CrossRef]
36. Zhou, X.C.; Wu, Y.; Yan, F.B.; Zhang, T.Z.; Ke, X.Y.; Meng, K.K.; Xu, X.G.; Li, Z.P.; Miao, J.; Chen, J.K.; et al. Revealing the high sensitivity in the metal toinsulator transition properties of the pulsed laser deposited VO_2 thin films. *Ceram. Int.* **2021**, *47*, 25574–25579. [CrossRef]
37. Lian, H.C.; Qi, L.H.; Luo, J.; Zhang, R.R. Drop-on-demand printing of edge-enhanced and conductive graphene twin-lines by coalescence regulation and multi-layers overwriting. *2D Mater.* **2021**, *8*, 035004. [CrossRef]
38. Luo, K.; Wang, H.; Li, E.Z.; Tang, B.; Yuan, Y. Polyphenylene oxide/SiO_2 composites: Towards ultra-low dielectric loss and high thermal conductivity of microwave substrates. *Appl. Surf. Sci.* **2024**, *672*, 160809. [CrossRef]
39. Hong, S.C.; Lee, M.; Lee, M.W.; Kim, D. Sharp phase transition by the enhanced lattice stability of low-temperature phase of Cr-doped VO_2. *Bull. Korean Chem. Soc.* **2021**, *42*, 1232–1238. [CrossRef]
40. Jahromi, H.D.; Hamedi, S. Artiffcial intelligence approach for calculating electronic and optical properties of nanocomposites. *Mater. Res. Bull.* **2021**, *141*, 111371. [CrossRef]
41. Hamedi, S.; Jahromi, H.D.; Lotfiani, A. Artificial intelligence-aided nanoplasmonic biosensor modeling. *Eng. Appl. Artif. Intel.* **2023**, *118*, 105646. [CrossRef]
42. Liu, C.J.; Sang, C.L.; Wu, X.S.; Cai, J.L.; Wang, J.W. Grating double-slot micro-ring resonator for sensing. *Opt. Commun.* **2021**, *499*, 127280. [CrossRef]

43. Esmail, B.A.F.; Koziel, S.; Szczepanski, S.; Majid, H.A. Overview of approaches for compensating inherent metamaterials losses. *IEEE Access* **2022**, *10*, 67058–67080. [CrossRef]
44. Guo, Y.Y.; Huo, Y.P.; Niu, Q.Q.; He, Q.; Hao, X.X. Band-stop filter based on tunable Fano resonance and electromagnetically induced transparency in metal-dielectric-metal waveguide coupling systems. *Phys. Scripta* **2021**, *96*, 015507. [CrossRef]
45. Izadpour, A.R.; Jahromi, H.D.; Sheikhi, M.H. Plasmonic enhancement of colloidal quantum dot infrared photodetector photosensitivity. *IEEE J. Quantum Electron.* **2018**, *3*, 1–7. [CrossRef]
46. Lotfiani, A.; Jahromi, H.D.; Samaneh, H. Silicon-Based Photovoltaic-nanoplasmonic biosensor with enhanced limit of detection and minimum detectable power. *J. Lightwave Technol.* **2022**, *4*, 1231–1237. [CrossRef]

hv photonics

MDPI

Article

Topological Rainbow Trapping with Expanded Bandwidth in Valley Photonic Crystals

Sayed El. Soliman [1,*], Israa Abood [2], Naglaa Abdel All [3] and Chii-Chang Chen [4,*]

1 Physics Department, Faculty of Science, Assiut University, Assiut 71516, Egypt
2 College of Physics and Optoelectronic Engineering, Shenzhen University, Shenzhen 518060, China; israa_abood@yahoo.com
3 Department of Physics, College of Science, Imam Mohammad Ibn Saud Islamic University, Riyadh 11432, Saudi Arabia; nfabdelall@imamu.edu.sa
4 Department of Optics and Photonics, National Central University, Taoyuan City 320, Taiwan
* Correspondence: selshahat@aun.edu.eg (S.E.S.); trich@dop.ncu.edu.tw (C.-C.C.)

Abstract: We introduce a novel approach to achieve broadband rainbow trapping in a 2D photonic crystal (PC) platform. By exploiting the concept of valley PCs, we engineer a structure that supports robust topological edge states. A carefully designed rotational angle gradient along the edge state path induces frequency-dependent light localization, forming a topological rainbow with a significantly expanded bandwidth. This phenomenon of topological rainbow trapping is attributed to the interplay between valley-dependent topological edge states and the engineered rotational angle gradient. To further enhance light localization and broaden the trapping spectrum, we incorporate a graded radius profile in the bottom row of dielectric columns. Through a combination of rotational angle modulation and radius grading, we successfully realize broadband rainbow trapping with enhanced light localization. Our findings reveal a broad trapping bandwidth spanning from $0.8314c/a$ to $0.9205c/a$, showcasing the potential of this approach for applications in optical frequency filtering, sensing, and information processing.

Keywords: photonic crystals; topological states; broadband; rainbow tapping

1. Introduction

The manipulation and control of light at the nanoscale level has been a cornerstone of scientific and technological advancement [1–3]. Photonic crystals (PCs), periodic structures that modulate light propagation, have emerged as a promising platform for achieving unprecedented control over light flow [4–11]. In recent years, exploring topological phenomena in photonic systems has opened up new avenues for designing innovative optical devices with unprecedented functionalities [12–15]. Among these, valley topological PCs have garnered significant attention due to their ability to support topologically protected edge states, characterized by their robustness to defects and disorder [16–20]. These properties hold immense potential for applications in various fields, including optical communication, sensing, and information processing.

In recent years, there has been a growing interest in developing methods for controlling the frequency-dependent behavior of light within PC structures. This has led to exploring strategies for achieving broadband light manipulation, essential for various applications, including optical communication, sensing, and imaging. One particularly intriguing phenomenon is the creation of a "topological rainbow", where light of different frequencies is localized at distinct spatial positions within a PC structure [21–26]. This phenomenon is attributed to the interplay between the valley-dependent topology of the system and the engineered structural gradients.

The core of the design is a two-dimensional (2D) PC platform. Leveraging a square lattice structure with judiciously designed circular dielectric columns, we induce topological phase transitions by manipulating rotational symmetry with varying radii arranged in a specific pattern by carefully adjusting the rotational angle gradient of these columns along the edge state path. This gradient induces a frequency-dependent modulation of light propagation, resulting in the localization of different frequencies at specific points within the structure, forming a topological rainbow. This phenomenon arises from the interplay between the valley-dependent topological edge states (TESs) and the engineered rotational angle gradient.

Moreover, introducing a graded radius profile to the dielectric columns further enhances light localization and broadens the trapping spectrum. This approach offers several advantages over existing methods. Firstly, the utilization of valley topological PC ensures the robustness of the TESs, making the system resilient to imperfections and disorder. Secondly, the combination of rotational angle modulation and radius grading allows for a significant enhancement in light localization and a broadening of the trapping spectrum. Consequently, we achieve a broadband rainbow trapping with a bandwidth spanning from $0.8314c/a$ to $0.9205c/a$, surpassing the limitations of previous approaches. This study presents a groundbreaking advance in the field of PCs by demonstrating broadband rainbow trapping through the integration of valley topological photonics and carefully engineered structural parameters. These findings pave the way for developing novel photonic devices with enhanced functionalities and open up new possibilities for applications in optical communication, sensing, and information processing.

2. Valley Topological PC and Photonic Band Structures

A 2D PC based on a square lattice was designed, consisting of four circular dielectric columns embedded in an air background. The rods have radii $r_1 = 0.05a$ and $r_2 = 0.1a$, where $a = 1000$ nm is the lattice constant. Each rod is positioned at a distance $l = 0.20a$ from the unit cell (UC) center that supports symmetry manipulation. The dielectric rods have a refractive index of $n = 3.46$, which is representative of silicon in the infrared regime. Figure 1a shows the three UC configurations used throughout the study, each representing a distinct symmetry or perturbation. To analyze the photonic band structure, the finite element method (FEM) was employed via COMSOL Multiphysics (version 6.2), applying periodic boundary conditions and computing along the high-symmetry path Γ–X–M–Γ of the first Brillouin zone (BZ). The focus was on transverse magnetic (TM) modes with non-zero components E_z, H_x, and H_y. For TM polarization, the governing Maxwell equation for electromagnetic wave propagation in the PC is given by [5]:

$$\nabla \times \left[\frac{1}{\varepsilon(r)}\nabla \times E_z(r)\right] = \frac{\omega^2}{c^2}E_z(r), \tag{1}$$

where $\varepsilon(r)$ is the position-dependent permittivity, ω is the angular frequency, and c is the speed of light in the vacuum. The corresponding magnetic field components H_x, and H_y are derived from E_z using Faraday's law: $H = [i/(\mu_0\omega)]\nabla \times E$, assuming vacuum permeability μ_0. Due to the periodicity of the dielectric structure, the out-of-plane electric field component $E_z(r)$ can be expanded using Bloch's theorem:

$$E_z(r) = u_{nk}(r)e^{ik.r}, \tag{2}$$

where u_{nk} is the periodic part of the Bloch function for the nth band. This periodic function can be further expanded in a Fourier series over reciprocal lattice vectors:

$$u_{nk}(r) = \sum_G E_{kn}(G)e^{iG.r}, \quad (3)$$

Substituting this into Maxwell's equations and applying orthogonality of plane waves leads to the eigenvalue problem [27]:

$$\sum_{G'} \kappa(G - G')|k + G'|^2 E_{kn}(G') = \frac{\omega_{kn}^2}{c^2} E_{kn}(G), \quad (4)$$

where $\kappa(G)$ is the Fourier component of $1/\varepsilon(r)$. By solving this eigenvalue problem using FEM with periodic boundary conditions, the eigenfrequency ω_{kn} and Fourier coefficients $E_{kn}(G)$ of the corresponding eigenmode can be obtained, leading to the photonic band structures, as shown in Figure 1b–d. The formation of topological phases in PCs fundamentally relies on symmetry breaking, which modifies the degeneracies and field distributions of photonic modes [28–31]. Two primary methods of symmetry breaking are commonly used: (i) shifting the positions of adjacent rods to break inversion symmetry, or (ii) rotating dielectric rods to perturb mirror and rotational symmetries. In this study, we adopt the latter method by introducing a rotational angle θ for the rods, enabling controlled topological transitions within the same lattice geometry.

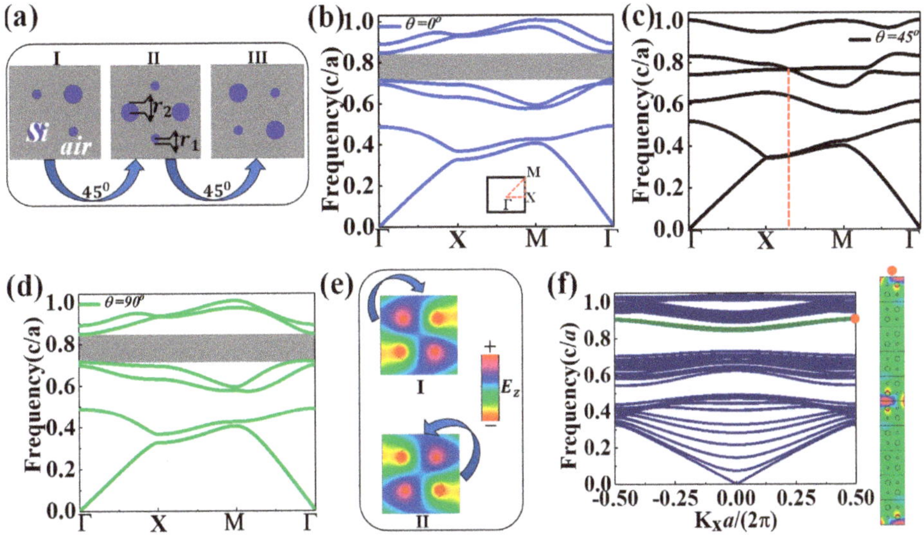

Figure 1. (**a**) The three possible unit cells (UCs) that contributed to the designed structure at (I) of $\theta = 0°$, (II) $\theta = 45°$, and (III) $\theta = 90°$, (**b–d**) the band structures of I, II, and III UCs, respectively, (**e**) the electric field E_z distribution for band four at the frequency of $0.6944c/a$ at $\theta = 0°$ and $\theta = 90°$, (**f**) the projected band structure of the supercell with forming topological edge mode inside the bandgap (green). The red dot marks the point $k_x = 1$, at which the electric field distribution (right panel) is evaluated, demonstrating strong localization along the domain wall.

To understand the topological origin of the valley-protected edge states in the designed PC, we analyze the evolution of the photonic band structure and field characteristics associated with the three rotational configurations of UCs, illustrated in Figure 1a. In the unrotated configuration $\theta = 0°$ (Figure 1b), the structure preserves both mirror and inversion symmetry. This high-symmetry state results in multiple clear photonic bandgaps for TM modes and supports degenerate modes at certain high-symmetry points, leading to topologically trivial bands. At intermediate rotation $\theta = 45°$ (Figure 1c), symmetry is partially broken, lifting degeneracies and inducing accidental Dirac points, and linear band crossings away from high-symmetry points (highlighted by the red dashed line) [32,33]. These crossings arise due to the hybridization of photonic modes carrying opposite orbital angular momentum, signaling a critical point in a topological phase transition [34,35]. The Berry curvature becomes non-zero near these valleys, and multiple peaks appear, implying potential non-trivial topological indices [34,36]. Upon further rotation to $\theta = 90°$ (Figure 1d), symmetry breaking is maximized again but with an inverted topological character. The bandgap reopens with opposite Berry curvature signs compared to $\theta = 0°$, completing the topological phase transition. The effective Hamiltonian governing this valley Hall transition is: [37–39] $H_{eff}(k) = v_D(\tau k_x \sigma_x + k_y \sigma_y) + \lambda(\theta)\sigma_z$, where v_D is the Dirac velocity, τ is the valley index for the two inequivalent momentum valleys, and σ_{xyz} are the Pauli matrices representing the pseudospin degrees of freedom. The term $\lambda(\theta)$ is an external perturbative parameter determined by the rotational angle θ. It reflects the degree of inversion and mirror symmetry breaking introduced by rod rotation [40,41]. The Berry curvature is defined via the Berry connection $\Omega_n(k) = \nabla_k \times \mathrm{A}_n(k)$, with $\mathrm{A}_n(k) = i\langle u_{nk}|\nabla_k|u_{nk}\rangle$. In inversion symmetry-broken PCs, this curvature becomes sharply localized near valley centers [42]. Integrating over the Brillouin zone (BZ) gives the valley Chern number $C = \frac{1}{2\pi}\int_{BZ} \Omega_n(k) d^2k$. The term $\lambda(\theta)$ is an external perturbative parameter determined by the rotational angle θ. The continuous variation $\lambda(\theta)$ along the interface results in a graded energy landscape for TESs, which translates into frequency-selective localization. The field distribution maps of the designed valley PC at two symmetry-broken configurations $\theta = 0°$, and $\theta = 90°$ were constructed to reflect the vortex behavior of the out-of-plane electric field E_z, which was directly extracted at the Dirac point frequency $0.6944c/a$ (Figure 1e). At $\theta = 0°$, E_z exhibits a clockwise vortex, corresponding to a negative Berry curvature concentrated at the upper valleys and a positive Berry curvature at the lower valleys. In contrast, at $\theta = 90°$, the phase vortex is counterclockwise, yielding an opposite Berry curvature distribution with positive curvature at the upper valleys. This reversal of Berry curvature is a direct consequence of the valley pseudospin switching induced by symmetry manipulation [26]. These antisymmetric Berry curvature distributions are centered at the valley points of the Brillouin zone and serve as the real-space indicator of the valley Hall topological phase transition. The sign reversal leads to a valley Chern number difference of $\Delta C = 1$ between $\theta = 0°$ and $\theta = 90°$, ensuring the existence of a robust topological edge mode along their interface, consistent with the bulk-boundary correspondence [13,43–45]. Figure 1f shows the projected band structure of a supercell composed of UCs with $\theta = 0°$ and $\theta = 90°$, forming a domain wall. A topological edge mode clearly appears inside the bandgap (highlighted in green). The electric field distribution at $k_x = 1$ (red dot) shows strong localization at the interface (right panel), with negligible leakage into the bulk [46]. This is a hallmark of valley Hall topological edge states, confirmed by prior studies [47]. These states are immune to backscattering and remain stable even in the presence of structural disorder, making them suitable for broadband slow-light applications and frequency-dependent light localization.

The capacity to precisely control and localize light within PC structures is paramount for advancing optical technologies. By strategically manipulating the rotational angles of the four circular dielectric columns in the valley topological PC design, we introduce a novel method to achieve highly customizable and robust light confinement. This approach gradually modulates the rotational angle for the four rods (orange color) along the edge state path, as shown in Figure 2a, which subtly alters the TESs by modifying their group velocities, implementing a graded interface to achieve broadband light manipulation. To understand this effect, we analyze the evolution of the TES dispersion curves shown in Figure 2b. As the rotational angle increases, the slope of the TES dispersion curve progressively decreases, indicating a reduction in the group velocity. This modulation is crucial for inducing spatial light localization, as lower group velocities correspond to enhanced energy confinement. The group velocity is defined as:$v_g = d\omega/dk$, progressively decreases with the angle gradient. This modulation is crucial for inducing spatial light localization, as lower group velocities correspond to enhanced energy confinement. Figure 2b illustrates this behavior, showing the evolution of the dispersion curves for the TESs as the rotational angle increases from $\theta = 0°$ to $\theta = 40°$. The band structure transitions from a positive to an opposing group velocity regime, as depicted in Figure 2c. The group velocity ultimately approaches zero for certain frequency components, forming a localized "slow light" region. This localized region is the fundamental mechanism enabling broadband rainbow trapping. The resulting frequency-dependent control over light propagation forms a topological rainbow, wherein distinct light frequencies are located at specific spatial positions within the structure. This phenomenon is a direct consequence of the interplay between the valley-dependent nature of the TESs and the meticulously engineered rotational angle gradient. The physical origin of this effect lies in the symmetry perturbation induced by the rotational angle gradient. By varying the rotational angle, the inversion symmetry between neighboring unit cells is progressively broken. This perturbation modifies the local dispersion properties, causing different frequency components to propagate at distinct velocities. As a result, different frequencies experience varying delays, effectively leading to spatial separation across the graded structure, which is the signature behavior of the rainbow trapping effect. We have demonstrated an effective method for controlling light propagation and localization within a PC structure by harnessing the intricate relationship between rotational angle, group velocity, and frequency. This approach holds significant potential for applications in optical frequency filtering, wavelength division multiplexing, and other photonic devices.

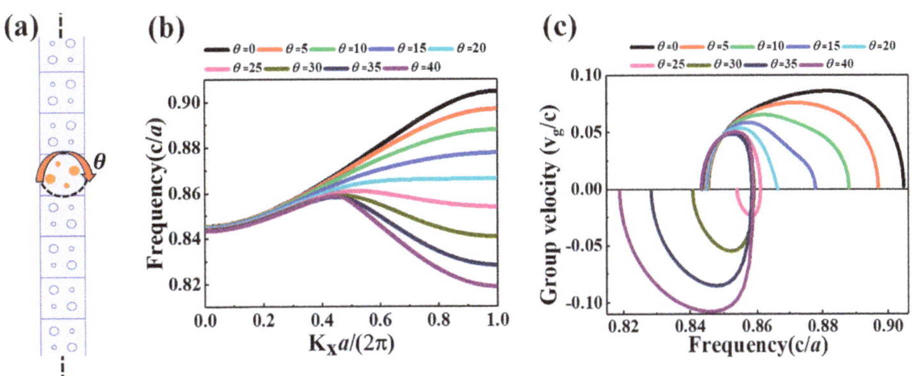

Figure 2. (**a**) The designed supercell structure with a graded rotational angle θ applied to the four dielectric rods (orange color) along the edge interface, (**b**) dispersion relations of the guided TESs at various values from $\theta = 0°$ to $\theta = 40°$, (**c**) corresponding group velocities v_g vs. frequency, showing the transition from positive to negative slopes, indicative of slow-light behavior and potential localization due to the graded symmetry breaking.

To further enhance light localization and broaden the rainbow trapping bandwidth, a radial gradient was introduced to the dielectric rods along the propagation path. As illustrated in Figure 3a, the bottom-row rod r_d (red) gradually chirped in radius from $0.25a$ to $0.29a$. The bottom row does not form a full topological interface; the resonant frequency gradient induced by refractive index modulation that supports gradual frequency localization effectively creates a quasi-1D slow-light path that enables strong localization via slow-light effects. This modulation increases the local dielectric density, effectively raising the refractive index along the direction of light propagation. The adjacent bulk region with non-trivial topology still provides a bandgap barrier, allowing light to remain confined along this quasi-guided path, as shown in Figure 3b, which displays the dispersion relations for different values of r_d. The progressive downward shift in the dispersion curves with increasing r_d confirms the redshift in modal frequency induced by the structural chirp. Moreover, the structure was engineered such that the effective operating point approaches the band edge, leading to a flattening of the dispersion curves. This behavior is clearly illustrated in Figure 3c, where the group velocity v_g decreases significantly for increasing rod radii. The resulting slow light regime stems from the constructive interference of multiple scattering events within the spatially graded medium. The prolonged light–matter interaction enabled by this gradient design opens pathways for enhanced optical confinement and spectral control. Our simulations are intentionally performed in 2D, assuming infinite structural thickness. While in practical three-dimensional structures, parameters such as rod depth indeed influence resonant frequencies and rainbow trapping behavior (see, e.g., Refs. [48,49]), the current design specifically isolates the effects arising solely from the rotational angle gradient and graded radius profile. Investigating additional frequency-tuning mechanisms via structural depth variation potentially extends and further optimizes these trapping phenomena.

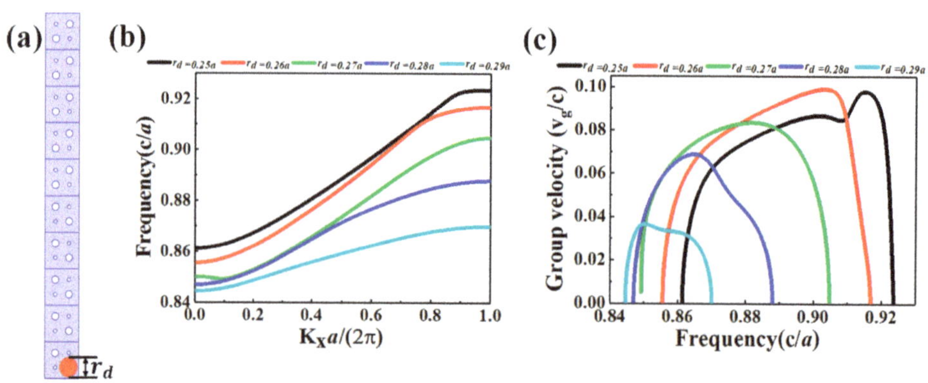

Figure 3. (**a**) Supercell structure incorporating a chirped dielectric rod (red) with a radius r_d gradually increasing from 0.25a to 0.29a along the propagation direction, (**b**) TES dispersion curves showing a frequency redshift as r_d increases, (**c**) corresponding group velocities of the guided modes (TESs) versus frequency, demonstrating dispersion flattening and slow light behavior as the rod radius grows.

3. Rainbow Trapping Structures

To engineer a rainbow-trapping structure, we designed a valley topological PC with a gradually modulated rotational angle along its edge state path. This was achieved by coupling supercells together, as illustrated in Figure 4a, where the four orange rods define the rotational angle. The graded structure spans 21-unit cells along the x-direction, with the rotational angle of the orange rods increasing from 0° to 40° degrees in 2-degree increments to span the topological bandgap without crossing the Dirac point at 45°, ensuring stable mode confinement and smooth field evolution over 21 units in the propagation direction. The upper unit cells remained constant throughout. To characterize the rainbow trapping phenomenon, FEM simulations were performed using COMSOL Multiphysics. A plane wave broadband source was introduced at the left boundary while scattering boundary conditions were applied to all other sides, as depicted in Figure 4a. This configuration allowed for a comprehensive analysis of the structure's response to multifrequency light input and the electric field distribution visualization. The origin of rainbow trapping lies in the structure's dispersive nature. When a spatial gradient in the rotational angle is introduced along the propagation direction (from $\theta = 0^\circ$ to $\theta = 90^\circ$ across 21 units), each interface segment locally supports a slightly different TES with a distinct dispersion relation. This results in a position-dependent group velocity, as shown in Figure 2b,c. Specifically, the slope of the TES dispersion curve ($v_g = d\omega/dk$) flattens progressively with increasing θ, indicating a reduction in the group velocity. The slower light is more confined, and near-zero group velocities correspond to frequency components being effectively "trapped" at specific spatial positions. This behavior is physically analogous to an adiabatic modulation of the local photonic potential, where each unit cell along the gradient acts as a locally perturbed resonator with a slightly shifted resonance [50]. This detuning along the interface forms a graded photonic landscape, where light experiences increasing delay as it propagates, leading to the rainbow trapping effect. Importantly, this is achieved without altering the global topology of the structure, thereby maintaining topological protection. Accordingly, by incident broadband plane wave comprises a spectrum of frequencies, and due to the structure's properties, these frequencies interact differently, leading to distinct propagation delays. This differential delay results in the spatial separation of frequencies, or rainbow trapping, as different frequency components become localized at specific positions within the structure. The engineered rotational angle gradient plays a crucial role in this process. Inducing a frequency-dependent modulation of light propagation enhances light localization and broadens the trapping spectrum. This interplay between valley-dependent TESs and the rotational angle gradient creates a topological rainbow formed, where each frequency component is spatially separated. Our design achieved broadband rainbow trapping with a bandwidth spanning from $0.8314c/a$ to $0.8960c/a$, significantly surpassing the capabilities of previous approaches. This is evident in the electric field distribution shown in Figure 4b, where the spatial separation of frequencies across the structure is visible. Although the demonstrated design inherently localizes energy into a relatively narrow spatial region at the topological interface, such strong localization is highly beneficial for specific advanced applications, such as integrated nanolasers [51], optical multiplexers [52], high-Q resonant cavities, and sensors [53,54]. The barrier regions (bulk structures above and below the interface) are critical to maintaining robust topological confinement by preventing scattering or leakage.

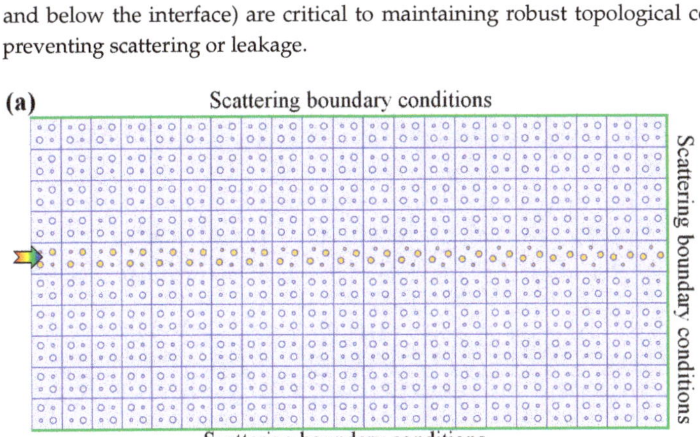

Figure 4. *Cont.*

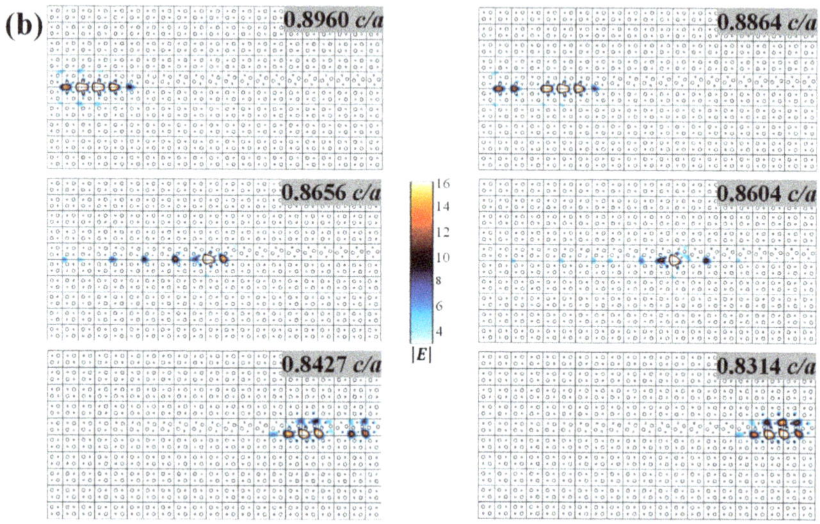

Figure 4. (**a**) Schematic diagram of valley topological PC with a gradually modulated rotational angle from 0° to 40° degrees in 2-degree increments along its edge state path. A plane wave broadband source was introduced at the left boundary, while scattering boundary conditions were applied to all other sides, (**b**) the electric field distribution $|E|$ demonstrated broadband rainbow trapping with a bandwidth spanning from $0.8314c/a$ to $0.8960c/a$.

A radial gradient was introduced to the dielectric columns in the bottom row of the structure to achieve broadband and spatially localized topological rainbow trapping. This graded configuration consisted of 21 units along the *x*-axis, with the column radius (r_d) increasing linearly from 0.25*a* to 0.29*a* in increments of 0.002*a* (Figure 5a). Eigenfrequency simulations were performed using the Wave Optics module in COMSOL Multiphysics to characterize the resonant frequencies and the field within this gradient structure. A 2D cut line was defined along the arc length of the dielectric column gradient (from 0*a* to 21*a*) as shown by black and green arrows in Figure 5a, allowing for analysis of the electric field amplitude as a function of both radii (r_d) and position along the arc length. This analysis revealed a systematic shift in the maximum electric field amplitude position towards larger arc lengths with decreasing frequency, confirming the occurrence of rainbow trapping (Figure 5b). Further validation of the rainbow trapping effect was obtained by simulating the structure's response to a broadband plane wave source. The resulting electric field distribution (Figure 5c) clearly demonstrated the localization of different frequency components within specific spatial regions, spanning a bandwidth from $0.8657c/a$ to $0.9205c/a$. This trapping occurs at frequencies where the group velocity of the slow light mode approaches zero, facilitated by the refractive index gradient induced by the structure. This study demonstrates the realization of broadband and localized topological rainbow trapping, surpassing the limitations of previous approaches. The ability to manipulate and control light frequencies within a compact structure holds promise for various applications. The precise control over frequency localization can enable the development of highly selective optical filters with tailored spectral responses. The sensitivity of the trapped frequencies to environmental changes can be exploited for sensing applications, such as refractive index sensing and biochemical detection. By harnessing the principles of topological photonics and gradient engineering, this work contributes to the advancement of light manipulation and control, with the potential for significant impact across multiple scientific and technological domains [55,56]. While the broadband rainbow trapping presented here is primarily enabled through a spatial gradient in rod radius that tunes local resonances, we note that inter-rod separation can also influence coupling strength and dispersion. However, based on the presented design and trapping mechanism, radius variation is more effective in producing distinct and spatially localized field concentrations.

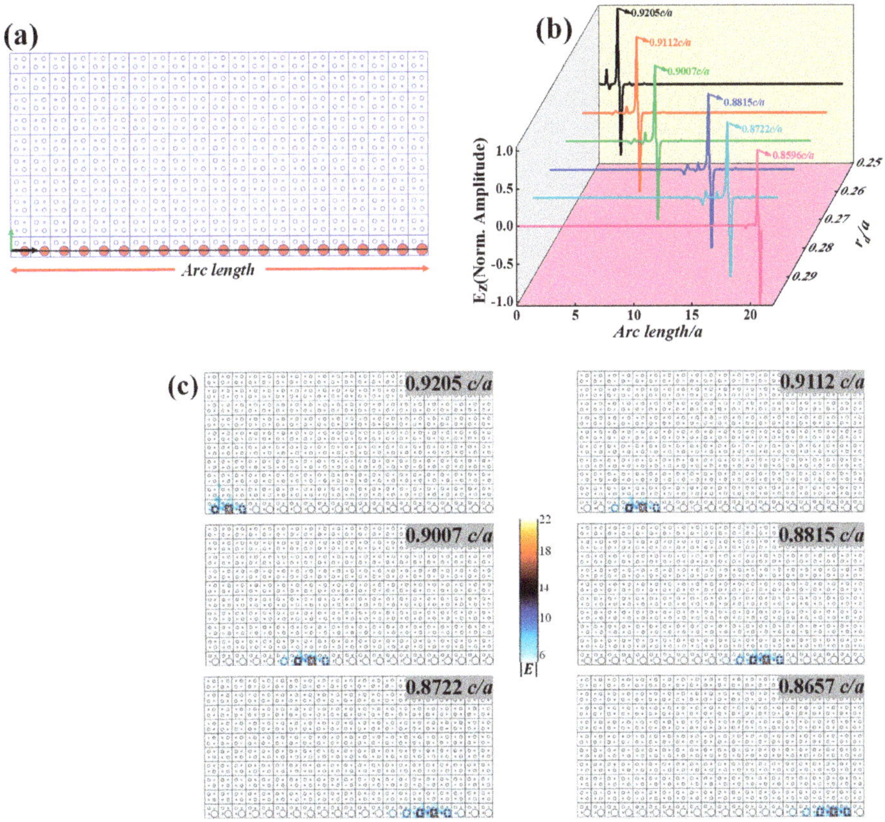

Figure 5. (**a**) Schematic diagram of the proposed PC heterostructure with the designated gradient r_d from $0.25a$ to $0.29a$ by an increment of $0.002a$, (**b**) the normalized electric field amplitude with changing r_d and arc length along the propagation direction of $21a$, (**c**) the electric field distribution $|E|$ to realize broadband rainbow trapping with a bandwidth spanning from $0.8657c/a$ to $0.9205c/a$.

To further examine the influence of radius variation, we conducted additional simulations where the radius gradient was confined to three distinct subranges, each targeting specific modal regimes. As shown in Figure 6, the monopole region ($0.05a$–$0.08a$, step $0.0015a$) supports simple localized field profiles (Figure 6a), while the dipole range ($0.12a$–$0.17a$, step $0.0025a$) leads to modes with increased spatial symmetry (Figure 6b). The quadrupole regime ($0.20a$–$0.24a$, step $0.002a$) enables higher-order field confinement (Figure 6c). These results confirm the ability of the proposed design to achieve spatially resolved rainbow trapping across different frequency bands by selectively exciting monopole-, dipole-, and quadrupole-like edge states. Compared to spacing-based modulation, this radius-tuning strategy proves more robust, controllable, and compatible with broadband operation. The modal localization patterns observed here complement the multipolar coexistence region illustrated in Figure 5c, highlighting the flexibility of the design to enable diverse rainbow trapping responses.

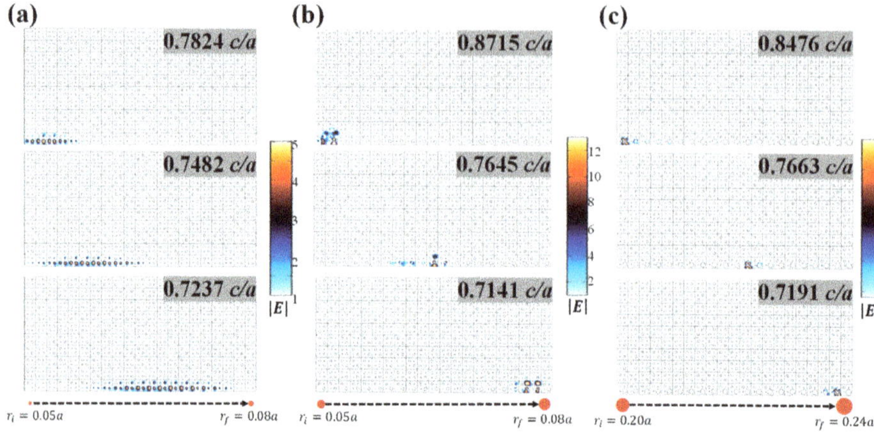

Figure 6. Electric field distributions $|E|$ showing rainbow trapping achieved by three different graded radius profiles, (**a**) from $r_i = 0.05a$ to $r_f = 0.08a$ with step $0.0015a$, exciting monopole-like modes, (**b**) from $r_i = 0.12a$ to $r_f = 0.17a$ with step $0.0025a$, exciting dipole-like modes, (**c**) from $r_i = 0.20a$ to $r_f = 0.24a$ with step $0.002a$, supporting quadrupole-like field patterns. These localized modes contribute to broadband rainbow trapping, each type operating within a different frequency range.

The proposed approach combines insights from previous topological rainbow trapping designs while introducing a novel mechanism that leverages the interplay between topological phase transitions and rotational angle gradients. Previous studies have demonstrated broadband topological rainbow trapping using different strategies. For example, in [57], a dual-mode rainbow was obtained via a sandwiched heterostructure structure with a combined bandwidth of $\sim 0.038c/a$. Meanwhile, in [24], a cavity-coupled TES was used, achieving a normalized bandwidth of approximately $\sim 0.045c/a$ (from $0.295c/a$ to $0.340c/a$). However, both methods rely on dual-mode interference or coupled cavities and are either limited in flexibility or involve complex spatial layouts. In contrast, our structure achieves a significantly broader normalized bandwidth of $\sim 0.089c/a$ (from $0.8314c/a$ to $0.9205c/a$), solely through continuous modulation of rotation angle and rod radius in a single unified interface. This results in a simpler and more compact design, free of external cavities or mode conversion schemes, while preserving topological protection. Additionally, this method allows tunable excitation of different modal families (monopole, dipole, quadrupole), offering a versatile platform for integrated slow-light and light-trapping applications. Thus, the approach not only outperforms previous structures in terms of bandwidth but also demonstrates higher design efficiency and robustness through a single-phase-engineered interface.

4. Verification of Robustness

We introduced controlled disorder into the system to assess the robustness of our proposed rainbow-trapping design against structural imperfections. Specifically, random displacements and omissions of rods near the edge state path were implemented, as shown in Figure 7a. Disordered rods were shifted by $\pm 0.05a$ in the x or y directions and, in some cases, entirely removed. Remarkably, electric field distributions $|E|$ at the targeted frequencies ($0.8960c/a$ and $0.8314c/a$) remained highly localized and largely unaffected by the introduced disorder, demonstrating the intrinsic robustness of the topological rainbow trapping effect (Figure 7b). While minor perturbations in the field distribution were observed near the defect sites at intermediate frequencies (e.g., $0.8687c/a$ (II)), the overall rainbow trapping phenomenon persisted. This resilience to disorder can be attributed to the topological protection of the edge states, which renders the system less susceptible to localized perturbations. The robustness of the TESs can be further understood by analyzing the system's Berry curvature. In the presence of a rotational angle gradient, the Berry curvature becomes asymmetric, resulting in a net accumulation of geometric phases across the structure. This non-trivial curvature leads to non-reciprocal light propagation along the interface, reinforcing the system's robustness against structural imperfections and disorder. As demonstrated in Figure 7b, even with intentional rod displacement and omissions, the localized field distribution remains robust, confirming the topological protection of the TESs. The observed robustness underscores the potential of this design for practical applications where fabrication imperfections are inevitable. Moreover, the system's sensitivity to local perturbations suggests a potential avenue for tuning the rainbow trapping characteristics through controlled defect engineering, offering opportunities for dynamic control and modulation of the trapped light.

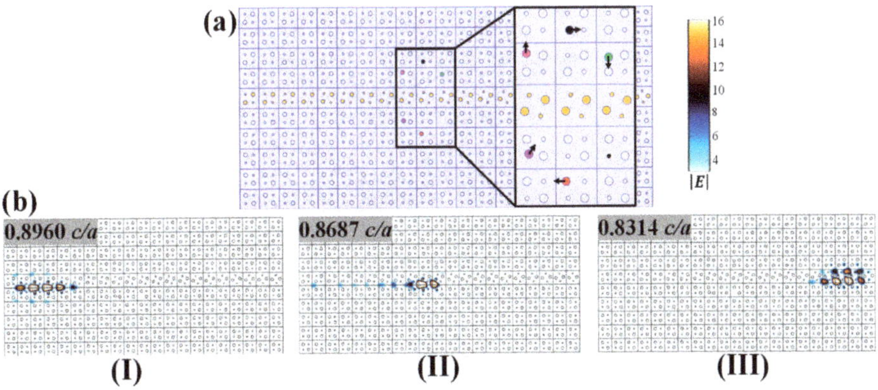

Figure 7. (**a**) Schematic of the disordered structure. Random perturbations were introduced to the Si rods, including $\pm 0.05a$ displacements in the x-direction (black and red rods), $\pm 0.05a$ displacements in the y-direction (pink and mauve rods), and the removal of one rod (black pattern). Inset: magnified view of the disordered Si rods, (**b**) electric field distribution $|E|$ demonstrates the robustness of topological rainbow trapping in the presence of structural disorder.

The proposed PC heterostructure can be fabricated using standard techniques widely employed in photonic device manufacturing [54,58–61]. Techniques such as electron beam lithography (EBL) and electrochemical etching have proven highly effective for realizing 2D PCs with well-defined rod patterns. In the EBL process, a patterned resist layer is defined on a substrate using electron beam exposure, followed by developing and lift-off procedures to produce the desired rod geometry [62,63]. This approach enables precise control over rod dimensions and orientation, which is essential for realizing the rotational gradient required in the proposed design [64–66]. Electrochemical etching offers an efficient method to fabricate large-area PCs using silicon and hydrofluoric acid, providing excellent control over spatial periodicity [67,68]. Meanwhile, direct laser writing (DLW) provides a flexible and accessible alternative for prototyping, allowing on-demand patterning of complex topological structures.

To ensure fabrication accuracy and performance validation, standard characterization techniques can be employed. Scanning electron microscopy can be used to confirm geometric fidelity, while near-field scanning optical microscopy enables spatial mapping of the localized electric fields. Additionally, far-field broadband sources can be used to monitor spectral-spatial separation of light, serving as direct experimental confirmation of the rainbow trapping phenomenon. Robustness to fabrication imperfections can also be evaluated by intentionally introducing rod displacement or defects and observing field confinement behavior.

These fabrication and characterization techniques are well-established in integrated photonics and provide a feasible path to experimentally realize and validate the proposed structure. Thus, the design aligns well with existing fabrication and measurement capabilities and can be integrated into practical photonic devices for sensing, filtering, and light routing.

5. Conclusions

In conclusion, we proposed and demonstrated topological rainbow trapping in two-dimensional valley PCs. The structure possesses a square lattice with circular dielectric columns of varying radii. We achieve robust localization of different frequency components at specific spatial positions by introducing a rotational angle gradient along the edge of a supercell composed of unit cells with distinct topological phases. This phenomenon, termed topological rainbow trapping, is attributed to the interplay between valley-dependent TESs and the engineered rotational angle gradient. Our design exhibits a wide trapping bandwidth and robust light localization, surpassing conventional rainbow trapping methods. This work paves the way for novel photonic devices with enhanced functionality and robustness.

Author Contributions: S.E.S.: Conceptualization, Methodology, Software, Writing—Original draft preparation, Writing—Reviewing and editing. I.A.: Data curation, Software, Visualization, Investigation. N.A.A.: Software, Writing—Reviewing and editing C.-C.C.: Supervision, Funding acquisition, Writing—Reviewing and editing. All authors have read and agreed to the published version of the manuscript.

Funding: This work was granted by the National Science and Technology Council, Taiwan (Grant Nos. NSTC 112-2221-E-008-089 and NSTC 113-2811-E-008-006).

Institutional Review Board Statement: Not applicable.

Informed Consent Statement: Not applicable.

Data Availability Statement: Data will be made available on request.

Conflicts of Interest: The authors declare that they have no known competing financial interests or personal relationships that could have appeared to influence the work reported in this paper.

References

1. Li, Z.-Y. Optics and photonics at nanoscale: Principles and perspectives. *Europhys. Lett.* **2015**, *110*, 14001. [CrossRef]
2. You, J.; Ma, Q.; Zhang, L.; Liu, C.; Zhang, J.; Liu, S.; Cui, T. Electromagnetic metamaterials: From classical to quantum. *Electromagn. Sci.* **2023**, *1*, 1–33. [CrossRef]
3. Ai, X.-C.; Pan, S.-C.; Wang, Y.-H.; Tian, S.-C. Optimization Design of Photonic-Crystal Surface-Emitting Lasers: Toward Large Bandwidth and Single-Lane 200G Optical Transmission. *Prog. Electromagn. Res.* **2024**, *180*, 89–101. [CrossRef]
4. Firouzjaei, A.S.; Afghahi, S.S.; Valmoozi, A.-A.E. Emerging Trends, Applications, and Fabrication Techniques in Photonic Crystal Technology. In *Recent Advances and Trends in Photonic Crystal Technology*; IntechOpen: London, UK, 2024.
5. Joannopoulos, J.D.; Johnson, S.G.; Winn, J.N.; Meade, R.D. *Photonic Crystals: Molding the Flow of Light*, 2nd ed.; Princeton Univ. Press: Princeton, NJ, USA, 2008.
6. Hou, C.-H.; Tseng, S.-Z.; Chan, C.-H.; Chen, T.-J.; Chien, H.-T.; Hsiao, F.-L.; Chiu, H.-K.; Lee, C.-C.; Tsai, Y.-L.; Chen, C.-C. Output power enhancement of light-emitting diodes via two-dimensional hole arrays generated by a monolayer of microspheres. *Appl. Phys. Lett.* **2009**, *95*, 133105. [CrossRef]
7. Cheng, Y.C.; Cai, D.P.; Chen, C.C.; Chan, C.H.; Lee, C.C.; Tsai, Y.L. Photonic Crystal Cavity with Double Heterostructure in GaN Bulk. *IEEE Photonics J.* **2013**, *5*, 2202606. [CrossRef]

8. Hsiao, F.-L.; Ni, C.-Y.; Tsai, Y.-P.; Chiang, T.-W.; Yang, Y.-T.; Fan, C.-J.; Chang, H.-M.; Chen, C.-C.; Lee, H.-F.; Lin, B.-S.; et al. Design of Waveguide Polarization Convertor Based on Asymmetric 1D Photonic Crystals. *Nanomaterials* **2022**, *12*, 2454. [CrossRef] [PubMed]
9. Liu, L.-Y.; Huang, H.-C.; Chen, C.-W.; Hsiao, F.-L.; Cheng, Y.-C.; Chen, C.-C. Design of Reflective Polarization Rotator in Silicon Waveguide. *Nanomaterials* **2022**, *12*, 3694. [CrossRef] [PubMed]
10. Lu, J.-H.; Cai, D.-P.; Tsai, Y.-L.; Chen, C.-C.; Lin, C.-E.; Yen, T.-J. Genetic algorithms optimization of photonic crystal fibers for half diffraction angle reduction of output beam. *Opt. Express* **2014**, *22*, 22590–22597. [CrossRef]
11. Chen, C.-C. Design of ultra-short polarization convertor with enhanced birefringence by photonic crystals. *Results Phys.* **2021**, *24*, 104138. [CrossRef]
12. Price, H.; Chong, Y.; Khanikaev, A.; Schomerus, H.; Maczewsky, L.J.; Kremer, M.; Heinrich, M.; Szameit, A.; Zilberberg, O.; Yang, Y. Roadmap on topological photonics. *J. Phys. Photonics* **2022**, *4*, 032501. [CrossRef]
13. Lu, L.; Joannopoulos, J.D.; Soljačić, M. Topological photonics. *Nat. Photonics* **2014**, *8*, 821–829. [CrossRef]
14. Ota, Y.; Takata, K.; Ozawa, T.; Amo, A.; Jia, Z.; Kante, B.; Notomi, M.; Arakawa, Y.; Iwamoto, S. Active topological photonics. *Nanophotonics* **2020**, *9*, 547–567. [CrossRef]
15. Ozawa, T.; Price, H.M.; Amo, A.; Goldman, N.; Hafezi, M.; Lu, L.; Rechtsman, M.C.; Schuster, D.; Simon, J.; Zilberberg, O.; et al. Topological photonics. *Rev. Mod. Phys.* **2019**, *91*, 015006. [CrossRef]
16. Liu, J.-W.; Shi, F.-L.; He, X.-T.; Tang, G.-J.; Chen, W.-J.; Chen, X.-D.; Dong, J.-W. Valley photonic crystals. *Adv. Phys. X* **2021**, *6*, 1905546. [CrossRef]
17. Wang, Y.; Wang, H.-X.; Liang, L.; Zhu, W.; Fan, L.; Lin, Z.-K.; Li, F.; Zhang, X.; Luan, P.-G.; Poo, Y.; et al. Hybrid topological photonic crystals. *Nat. Commun.* **2023**, *14*, 4457. [CrossRef]
18. Zeng, Y.; Chattopadhyay, U.; Zhu, B.; Qiang, B.; Li, J.; Jin, Y.; Li, L.; Davies, A.G.; Linfield, E.H.; Zhang, B.; et al. Electrically pumped topological laser with valley edge modes. *Nature* **2020**, *578*, 246–250. [CrossRef]
19. Liu, J.-W.; Liu, G.-G.; Zhang, B. Three-Dimensional Topological Photonic Crystals. *Prog. Electromagn. Res.* **2024**, *181*, 99–112. [CrossRef]
20. Zheng, J.; Guo, Z.; Sun, Y.; Jiang, H.; Li, Y.; Chen, H. Topological Edge Modes in One-Dimensional Photonic Artificial Structures. *Prog. Electromagn. Res.* **2023**, *177*, 1–20. [CrossRef]
21. Elshahat, S.; Wang, C.; Zhang, H.; Lu, C. Perspective on the topological rainbow. *Appl. Phys. Lett.* **2021**, *119*, 230505. [CrossRef]
22. Lu, C.; Sun, Y.-Z.; Wang, C.; Zhang, H.; Zhao, W.; Hu, X.; Xiao, M.; Ding, W.; Liu, Y.-C.; Chan, C.T. On-chip nanophotonic topological rainbow. *Nat. Commun.* **2022**, *13*, 2586. [CrossRef]
23. Xu, Q.; Peng, Y.; Shi, A.; Peng, P.; Liu, J. Dual-band topological rainbows in Penrose-triangle photonic crystals. *J. Opt. Soc. Am. A* **2024**, *41*, 366–370. [CrossRef] [PubMed]
24. AbdelAll, N.; Almokhtar, M.; Khouqeer, G.; Soliman, S.E. Realizing the topological rainbow based on cavity-coupled topological edge state. *Opt. Laser Technol.* **2024**, *171*, 110351. [CrossRef]
25. Elshahat, S.; Esmail, M.S.M.; Yuan, H.; Feng, S.; Lu, C. Broadband Multiple Topological Rainbows. *Ann. Der Phys.* **2022**, *534*, 2200137. [CrossRef]
26. Elshahat, S.; Zhang, H.; Lu, C. Topological rainbow based on coupling of topological waveguide and cavity. *Opt. Express* **2023**, *31*, 20187–20199. [CrossRef]
27. Sakoda, K. *Optical Properties of Photonic Crystals*; Springer: Berlin/Heidelberg, Germany, 2001.
28. Liang, L.; Zhou, X.; Hu, J.-H.; Wang, H.-X.; Jiang, J.-H.; Hou, B. Rainbow trapping based on higher-order topological corner modes. *Opt. Lett.* **2022**, *47*, 1454–1457. [CrossRef]
29. Ezawa, M. Higher-order topological insulators and semimetals on the breathing kagome and pyrochlore lattices. *Phys. Rev. Lett.* **2018**, *120*, 026801. [CrossRef]
30. Lera, N.; Torrent, D.; San-Jose, P.; Christensen, J.; Alvarez, J.V. Valley Hall phases in kagome lattices. *Phys. Rev. B* **2019**, *99*, 134102. [CrossRef]
31. Shen, S.-l.; Li, C.; Wu, J.-F. Investigation of corner states in second-order photonic topological insulator. *Opt. Express* **2021**, *29*, 24045–24055. [CrossRef]
32. Kim, K.-H.; Om, K.-K. Multiband Photonic Topological Valley-Hall Edge Modes and Second-Order Corner States in Square Lattices. *Adv. Opt. Mater.* **2021**, *9*, 2001865. [CrossRef]
33. Wang, X.; Zhao, W.; Elshahat, S.; Lu, C. Topological rainbow trapping based on gradual valley photonic crystals. *Front. Phys.* **2023**, *11*, 1141997. [CrossRef]
34. Xi, X.; Ye, K.-P.; Wu, R.-X. Topological photonic crystal of large valley Chern numbers. *Photonics Res.* **2020**, *8*, B1–B7. [CrossRef]
35. Xiong, L.; Liu, Y.; Zhang, Y.; Zheng, Y.; Jiang, X. Topological Properties of a Two-Dimensional Photonic Square Lattice without C_4 and $M_{x(y)}$ Symmetries. *ACS Photonics* **2022**, *9*, 2448–2454. [CrossRef]
36. Skirlo, S.A.; Lu, L.; Soljačić, M. Multimode One-Way Waveguides of Large Chern Numbers. *Phys. Rev. Lett.* **2014**, *113*, 113904. [CrossRef] [PubMed]

37. Wu, X.; Meng, Y.; Tian, J.; Huang, Y.; Xiang, H.; Han, D.; Wen, W. Direct observation of valley-polarized topological edge states in designer surface plasmon crystals. *Nat. Commun.* **2017**, *8*, 1–9. [CrossRef] [PubMed]
38. Mei, J.; Wu, Y.; Chan, C.T.; Zhang, Z.-Q. First-principles study of Dirac and Dirac-like cones in phononic and photonic crystals. *Phys. Rev. B* **2012**, *86*, 035141. [CrossRef]
39. Tang, G.-J.; Chen, X.-D.; Shi, F.-L.; Liu, J.-W.; Chen, M.; Dong, J.-W. Frequency range dependent topological phases and photonic detouring in valley photonic crystals. *Phys. Rev. B* **2020**, *102*, 174202. [CrossRef]
40. Chen, X.-D.; Zhao, F.-L.; Chen, M.; Dong, J.-W. Valley-contrasting physics in all-dielectric photonic crystals: Orbital angular momentum and topological propagation. *Phys. Rev. B* **2017**, *96*, 020202. [CrossRef]
41. Zhang, H.; Xie, R.; Tao, X.; Gao, J. Topological valley-locked waveguides with C4 impurity. *Nanophotonics* **2024**, *13*, 3727–3736. [CrossRef]
42. Tang, G.-J.; He, X.-T.; Shi, F.-L.; Liu, J.-W.; Chen, X.-D.; Dong, J.-W. Topological Photonic Crystals: Physics, Designs, and Applications. *Laser Photonics Rev.* **2022**, *16*, 2100300. [CrossRef]
43. Ma, T.; Shvets, G. All-Si valley-Hall photonic topological insulator. *New J. Phys.* **2016**, *18*, 025012. [CrossRef]
44. Gong, Y.; Wong, S.; Bennett, A.J.; Huffaker, D.L.; Oh, S.S. Topological Insulator Laser Using Valley-Hall Photonic Crystals. *ACS Photonics* **2020**, *7*, 2089–2097. [CrossRef]
45. He, L.; Addison, Z.; Mele, E.J.; Zhen, B. Quadrupole topological photonic crystals. *Nat. Commun.* **2020**, *11*, 3119. [CrossRef] [PubMed]
46. Li, Y.-Z.; Zhang, Z.; Chen, H.; Gao, F. Polarization-wavelength locked plasmonic topological states. *Polarization* **2023**, *2023*, 9–25. [CrossRef]
47. Noh, J.; Huang, S.; Chen, K.P.; Rechtsman, M.C. Observation of photonic topological valley Hall edge states. *Phys. Rev. Lett.* **2018**, *120*, 063902. [CrossRef]
48. Chaplain, G.J.; De Ponti, J.M.; Aguzzi, G.; Colombi, A.; Craster, R.V. Topological Rainbow Trapping for Elastic Energy Harvesting in Graded Su-Schrieffer-Heeger Systems. *Phys. Rev. Appl.* **2020**, *14*, 054035. [CrossRef]
49. Wang, G.; Wei, Y.; Chen, Z.; Lim, C.W. Controllable subwavelength topological rainbow trapping in water-filling acoustic metamaterials. *Appl. Acoust.* **2023**, *207*, 109366. [CrossRef]
50. Zhu, L.; Liu, N. Multimode resonator technique in antennas: A review. *Electromagn. Sci.* **2023**, *1*, 1–17. [CrossRef]
51. Tian, F.; Wang, Y.; Huang, W.; Fang, X.; Guo, S.; Zhou, T. Ultra-compact topological photonic crystal rainbow nanolasers operating in the 1550 nm telecom band with wavelength-scale mode volumes. *arXiv* **2024**, arXiv:2411.11009.
52. AbdelAll, N.; Almokhtar, M.; Khouqeer, G.; Esmail, M.S.M.; Abood, I.; El Soliman, S. Multifunctional topological photonic crystal device for multichannel frequency routing and highly sensitive refractive index sensing. *Phys. Scr.* **2024**, *99*, 055539. [CrossRef]
53. Soliman, S.E.; Abood, I.; Lu, C. Robust multi-mode rainbow trapping with ultra-high-Q Fano resonances. *Opt. Express* **2024**, *32*, 1010–1019. [CrossRef]
54. Abood, I.; El Soliman, S.; Wenlong, H.; Ouyang, Z. Multi-Fano resonances by TCS and resonance-enhanced TES in hybrid photonic crystals for ultracompact sensing. *Opt. Laser Technol.* **2025**, *180*, 111513. [CrossRef]
55. Luo, Z.-Y.; Zhang, T.; Ye, Y.-T.; Wang, Y.-F.; Yu, C.-C.; Luo, Z.-C.; Zhang, Y.-J.; Xu, M.-C.; Sanders, B.C.; Wang, H.; et al. On-Demand Photon Storage and Retrieval with a Solid-State Photon Molecule at Room Temperature. *Electromagn. Sci.* **2024**, *2*, 1–10. [CrossRef]
56. Ma, K. Regulation and control of electromagnetic field in radio-frequency circuits and systems. *Electromagn. Sci.* **2023**, *1*, 1–28. [CrossRef]
57. Li, M.X.; Wang, Y.K.; Lu, M.J.; Sang, T. Dual-mode of topological rainbow in gradual photonic heterostructures. *J. Phys. D Appl. Phys.* **2022**, *55*, 095103. [CrossRef]
58. Asghari, M.; Krishnamoorthy, A.V. Silicon photonics: Energy-efficient communication. *Nat. Photonics* **2011**, *5*, 268–270. [CrossRef]
59. Loncar, M.; Doll, T.; Vuckovic, J.; Scherer, A. Design and fabrication of silicon photonic crystal optical waveguides. *J. Light. Technol.* **2000**, *18*, 1402–1411. [CrossRef]
60. Feng, J.; Chen, Y.; Blair, J.; Kurt, H.; Hao, R.; Citrin, D.S.; Summers, C.J.; Zhou, Z. Fabrication of annular photonic crystals by atomic layer deposition and sacrificial etching. *J. Vac. Sci. Technol. B* **2009**, *27*, 568–572. [CrossRef]
61. Baba, T.; Matsuzaki, T. Fabrication and Photoluminescence Studies of GaInAsP/InP 2-Dimensional Photonic Crystals. *Jpn. J. Appl. Phys.* **1996**, *35*, 1348–1352. [CrossRef]
62. Tavousi, A.; Rakhshani, M.R.; Mansouri-Birjandi, M.A. High sensitivity label-free refractometer based biosensor applicable to glycated hemoglobin detection in human blood using all-circular photonic crystal ring resonators. *Opt. Commun.* **2018**, *429*, 166–174. [CrossRef]
63. Xu, W.; Li, Y. The effect of anisotropy on light extraction of organic light-emitting diodes with photonic crystal structure. *J. Nanomater.* **2013**, *2013*, 7. [CrossRef]
64. Xu, X.; Zhang, D. The research and progress of micro-fabrication technologies of two-dimensional photonic crystal. *Chin. Sci. Bull.* **2007**, *52*, 865–876. [CrossRef]

65. Chong, T.C.; Hong, M.H.; Shi, L.P. Laser precision engineering: From microfabrication to nanoprocessing. *Laser Photonics Rev.* **2010**, *4*, 123–143. [CrossRef]
66. Trifonov, T.; Marsal, L.F.; Rodriguez, A.; Pallares, J.; Alcubilla, R. Two-dimensional photonic crystals of rods with a dielectric cladding. In Proceedings of the Microtechnologies for the New Millennium 2005, Sevilla, Spain, 9–11 May 2005; Volume 5840, pp. 746–757.
67. Lehmann, V.; Foll, H. Formation Mechanism and Properties of Electrochemically Etched Trenches in n-Type Silicon. *J. Electrochem. Soc.* **1990**, *137*, 653–659. [CrossRef]
68. Gruning, U.; Lehmann, V.; Ottow, S.; Busch, K. Macroporous silicon with a complete two-dimensional photonic band gap centered at 5 μm. *Appl. Phys. Lett.* **1996**, *68*, 747–749. [CrossRef]

photonics

MDPI

Article

Spectral Tuning and Angular–Gap Interrogation of Terahertz Spoof Surface Plasmon Resonances Excited on Rectangular Subwavelength Grating Using Attenuated Total Reflection in Otto Configuration

Oleg Kameshkov [1], Vasily Gerasimov [2,3,*], Boris Goldenberg [2,4] and Vladimir Nazmov [2,5]

1 Research School of Physics, The Australian National University, Canberra, ACT 2601, Australia; oleg.kameshkov@anu.edu.au
2 Budker Institute of Nuclear Physics, Siberian Branch of RAS, Novosibirsk 630090, Russia
3 Department of Physics, Novosibirsk State University, Novosibirsk 630090, Russia
4 Synchrotron Radiation Facility SKIF, 1, pr. Nikolsky, Kol'tsovo 630559, Russia
5 Institute of Solid State Chemistry and Mechanochemistry, Siberian Branch of RAS, Novosibirsk 630090, Russia
* Correspondence: v.v.gerasimov@inp.nsk.su

Abstract: In this paper, we experimentally investigated the excitation of spoof surface plasmon polaritons (SSPPs) supported by a 1D subwavelength grating with a rectangular profile in the terahertz (THz) frequency range. Using the attenuated total reflection technique and the THz radiation of the Novosibirsk free electron laser, we carried out detailed studies of both angular and gap spectra at several wavelengths. A shallow grating supporting a fundamental mode was fabricated by means of multibeam X-ray lithography and used as a test sample. The results indicated that we achieved 1-THz tunability of resonance in the frequency range from 1.51 to 2.54 THz on a single grating, which cannot be obtained with active tunable metamaterials. The Q factors of the resonances in the angular spectra were within the range of 19.4–37.6, while the resonances of the gap spectra had a Q factor lying within the 1.17–2.03 range. The gap adjustment capability of the setup shown in the work has great potential in modulation of the absorption efficiency, whereas the angular tuning and recording data from each point of the grating will enable real-time monitoring of changes in the surrounding medium. All of this is highly important for enhanced terahertz real-time absorption spectroscopy and imaging.

Keywords: attenuated total reflection; Otto configuration; spoof surface plasmon resonance; monochromatic radiation source; frequency tunability; angular and gap interrogation

1. Introduction

Surface plasmon polaritons (SPPs) are TM-polarised waves exponentially bound to a metal–dielectric interface and propagating along it. These kinds of waves have found widespread use in sensor and nonlinear applications in the visible and infrared spectral ranges due to their capability to field enhancement, light concentration, and sensitivity to minute changes of analyte [1–3]. However, the promising technological potential of SPPs on metals cannot be directly transferred to the terahertz (THz) frequency range (0.1–10 THz). Since surface waves arise from a collective oscillation of electrons, their existence is related to the negative dielectric permittivity of metals. Being an evanescent wave, SPPs penetrate into the metal to a distance referred to as the skin depth. In the optical range, a skin layer depth of about 15–30 nm can be observed with a similar penetration depth into the dielectric medium [4], whereas in the THz range the dielectric permittivity of metals increases significantly, leading to a slight excess of these values in metals (30–50 nm depth at 2 THz =150 μm) and a gigantic penetration depth into the space above them (the experimental value for gold is $\approx 5\lambda \approx 750$ μm) [5]. Thus, THz SPPs have low confinement near the surface and weak coupling (the proportion of the SPP energy carried in metals is much less than that in air), which results in their emitting into bulk waves at any slight roughness or optical inhomogeneities of the surface [6,7] and makes them virtually useless in the context of the above applications.

The coupling and confinement problems can be overcome with three effective approaches. First of all, it is possible to replace a metal material with semiconductors [8] or carbon materials, such as graphene [9]. At low frequencies, metals behave like a nearly perfect conductor due to the high density of electrons, while doped semiconductors or carbon nanoparticles can be considered as metallic systems, with the electron density significantly lower than that of metals. Moreover, their properties can be tuned via the doping density [10]. Another way to reduce the delocalization is to use metals with a thin dielectric coating. Such a multilayer system can be seen as a dielectric film waveguide, which has only the TM_0 mode if the film thickness is in the deep subwavelength regime. This mode corresponds to surface plasmons, and using this method can increase their confinement by an order compared to a bare metal [6,11]. The third approach is related to subwavelength textured periodic structures on metal surfaces. In this case, the patterned metal supports surface modes, and its average response corresponds to that of an effective metal with a larger skin depth and lower dielectric permittivity. The excited surface waves on such metamaterials imitate well the properties of classical surface waves, having yet completely different dispersion relations, which depend on the geometry and material. Due to this similarity, they are called spoof surface plasmon polaritons (SSPPs) [12].

The high flexibility in manipulating and changing the properties of the SSPPs not only gave rise to various applications, but also made it possible to transfer some field concepts; for example, localised surface plasmon polaritons, from the visible range to the terahertz frequency one [13]. The experimental configurations for the implementation of the SSPP potentialities, mainly in sensing, include 1D subwavelength metal gratings, on which the SSPP resonance is excited with an attenuated total reflection setup. The first experimental demonstration of refractive index sensing of various fluids with this scheme was given in 2013. In 2014, Ng with his colleagues used the scattering edge coupling configuration to extract broadband dispersion data [14,15]. Since then, various scientists have conducted extensive numerical studies on the influence of various parameters on SSPP resonances and sensor capabilities of 1D subwavelength gratings in this scheme [16–22]. For instance, Yao H. and Zhong S. investigated the properties of high-order modes excited on subwavelength gratings [16], whereas Chen L. et al. demonstrated an analysis of their groove shape [19]. In another work [18], it was shown that by measuring the angular spectra, one can attain a high resolution for refractometry of substances. Further experimental works related to excitation of fundamental SSPP modes appeared in 2019 and 2020, when Huang et al. demonstrated the modulation of resonances via gap variation and realised a liquid sensor with direct phase readout capacity [23,24].

There have only been a few experimental works on the generation of spoof surface plasmon resonances on a subwavelength grating by the Otto method, where the time-domain spectroscopy (TDS) technique with broadband radiation sources was used. In the reflection spectrum, a resonant frequency corresponding to the resonance on the grating was observed, and refractometry of the substance surrounding the grating was performed based on the frequency shift. Despite the TDS method advantages, such as the possibility of simultaneous reconstruction of amplitude and phase spectra, there can be some disadvantages for SSPP sensing: (1) low spectral brightness (especially in the high-frequency region > 2 THz of the THz range for most TDS spectrometers), which complicates study of substances with absorption lines in this range; (2) numerical errors arising during spectrum reconstruction, which reduce the spectral resolution of the surface plasmon resonance (SPR) method; (3) not the entire spectral range of the source can be effectively used because of the subwavelength nature of the grating in relation to the wavelength; (4) excitation of SSPPs requires inducing THz waves with various incident angles at a certain frequency, which could be a challenge with the TDS method because of the need to adjust the optical path [25].

Recently, we made the first attempt to experimentally observe a plasmon resonance in the angular spectrum on a shallow subwavelength grating using a cylindrical silicon prism and monochromatic laser radiation [26]. The results showed poor resonance quality because of (1) the necessity to work near the critical angle (due to the high refractive index of silicon), which led to appearance of interference effects due to the leakage of some radiation from the prism; (2) the focusing properties of the cylindrical prism, which increased the dispersion of the incidence angles of radiation on the prism–air-grating interface; (3) a non-optimal system for recording the reflected signal. In this paper, we modify the experimental setup, taking into account the shortcomings of the first experiments. The new setup allowed us to suppress most of the side effects mentioned in the previous work and carry out detailed studies of both the angular and gap spectra at several wavelengths. A shallow grating supporting a fundamental mode was fabricated by means of multibeam X-ray lithography and used as a test sample.

2. Theoretical Description

Many scientists have proposed descriptions of spoof surface plasmon polaritons excited on 1D subwavelength gratings using an effective medium and the mode matching technique [27–32]. We will use the theory written by Rusina et al. [29] for such structures with losses. In the framework of this consideration, a subwavelength grating (see Figure 1a) with period p, groove width w, and depth d satisfying the conditions $p,\ w \ll \lambda$ and $w \ll d$ can be described as a three-layer medium consisting of a homogeneous anisotropic layer with height d placed between metal and dielectric layers. The dispersion relation of SSPPs is as follows:

$$k_{SSPP} = \left(\varepsilon_d k_0^2 + \left(\frac{w\varepsilon_d}{p\varepsilon_g}\right)^2 \cdot k_g^2 tan^2(k_g d)\right)^{\frac{1}{2}}, \tag{1}$$

where $k_g = k_0\sqrt{\varepsilon_g}\left(1 + \frac{l_s(i+1)}{w}\right)^{\frac{1}{2}}$ is the wave vector of the wave propagating in the groove; ε_g and ε_d is the dielectric permittivity of substance inside and outside grooves, respectively; $k_0 = \frac{2\pi}{\lambda}$ is the wavenumber of the wave in free space. This model takes into account the finite conductivity of the metal ε_m via the skin depth $l_s = \left(k_0 Re\sqrt{-\varepsilon_m}\right)^{-1}$.

The typical dispersion relation is shown in Figure 1b. It exhibits high dependence on the width, depth, and period of the grooves, which provides flexibility in designing gratings for various applications. Among all these parameters, the depth has the greatest influence on the asymptotic frequency of spoof SPPs and excitation of the higher spoof plasmon modes, which can be seen as follows. If the higher-order mode exists, then the dispersion curve will intersect the light line, that is, $k_{SSPP} = \sqrt{\varepsilon_d}k_0$. This condition corresponds to $tan(k_g d) = 0$ and $k_g d = m\pi$, where $m = 0,\ 1,\ 2, \ldots$ is the order of the spoof plasmon. The wavevector of the surface wave propagating along the corrugated surface is limited by the first Brillouin zone of $k_g < \frac{\pi}{p}$. Thus, clearly, modes with $m \geq 1$ begin to be supported when $d \geq mp$ [31].

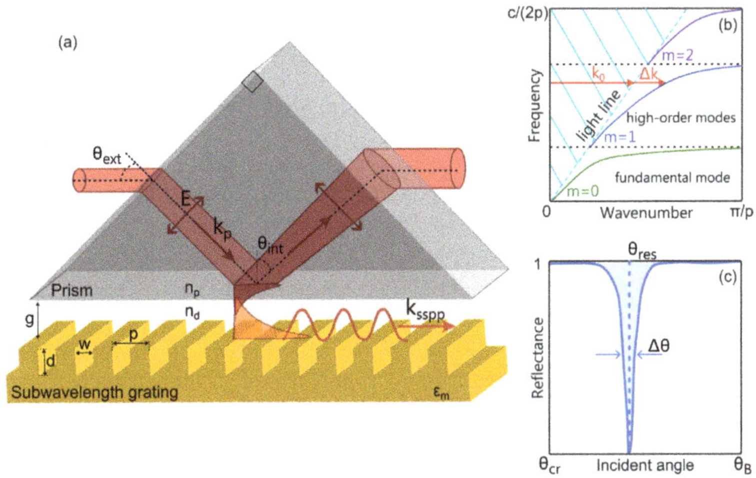

Figure 1. (**a**) Attenuated total reflection scheme in Otto configuration; (**b**) Typical dispersion curve. X-axis is limited by Brillouin zone, while Y-axis is restricted by condition of grating period being twice as small as wavelength, which is necessary for considering grating as subwavelength one; (**c**) Typical reflectance spectrum.

To excite a spoof plasmon, the incident wave must have a similar frequency and wavevector. However, for a given frequency, because of differing dispersion relations, the wavevector of a free-space wave is smaller than that of the SSPPs. This mismatch can be overcome with the attenuated total reflection technique. In this approach, a plane wave incident at the angle θ_{ext} onto a prism positioned against the subwavelength grating in the Otto configuration (see Figure 1a) creates an evanescent field from the prism surface at angles θ_{int} greater than θ_{cr}:

$$\theta_{int} > \theta_{cr} = arcsin(n_d/n_p), \tag{2}$$

where n_p is the refractive index of the prism.

This field satisfies the spoof plasmon wave vector matching condition when

$$k_{SSPP} = k_0 n_p sin(\theta_{int}). \tag{3}$$

The coupling of the spoof plasmon affects not only the matching condition but also the gap between the prism and the subwavelength grating. In the first approximation, excitation of the mode appears when the evanescent wave from the prism overlaps with the mode of the subwavelength grating. This process is related to the penetration depths

$$l_{sspp} = \frac{1}{\sqrt{k_{SSPP}^2 - \varepsilon_d k_0^2}} \tag{4}$$

of the spoof plasmon waves and

$$l_p = \frac{1}{\sqrt{k_0^2 n_p^2 sin(\theta_{int}) - \varepsilon_d k_0^2}} \tag{5}$$

of the evanescent wave from the prism. As can be seen from Equation (4), l_{sspp} is smaller in the region above the surface for the higher operation frequencies due to the larger deviation of the parallel component of wavevector from the light line. In turn, l_p (see Equation (5)) becomes smaller for prisms with higher refractive indices, which leads to higher confinement. The coupling process can be accompanied by side effects. If the prism is in close proximity to the grating, this significantly enhances the scattering and reflection between the prism and grating surfaces, which reduces the coupling efficiency, especially for strongly localised modes. On the contrary, a larger gap will result in a small overlap of the waves and poor coupling.

The coupling process can be explained precisely by the interference of the incident wave and reflected beam from the grating [33,34]. To our knowledge, this problem has not yet been analytically solved for the spoof surface plasmon polaritons excited on sub-wavelength gratings. However, we can draw on theoretical results developed for classical surface plasmon polaritons, assuming that the mechanisms for SSPPs are similar.

In this context, the transfer of the evanescent wave to the SSPPs is associated with losses in the system under consideration. On one hand, the resonance in the spectrum (see Figure 1c) can be described using a complex wavenumber, as follows [35–37]:

$$q_{res} = q'_{res} + iq''_{res} = k_0 n_p sin(\theta_{res}) + ik_0 n_p cos(\theta_{res}) \cdot \Delta\theta_{res}, \quad (6)$$

where q'_{res} corresponds to the angle of minimum reflectivity, i.e., the phase matching condition, and q''_{res} is related to the losses and determines the halfwidth of the resonance $\Delta\theta_{res}$.

On the other hand, the shape of the resonance is entirely determined by the characteristic of the spoof surface plasmon polaritons. The imaginary part of the wavenumber represents the damping rate of the SSPP, which consists of two damping mechanisms:

- Internal damping (q_i), associated with absorption in the metallic grating and proportional to the imaginary part of the effective dielectric constant.
- Radiation damping (q_r), arising from the emission of SSPPs into the prism, which depends on the gap between the prism and the grating.

It is expected that the critical coupling condition (corresponding to a minimum in the reflection spectrum) is achieved when the internal losses equal radiation losses that is $q_i = q_r$ [35,36]. Furthermore, the influence of losses on the optimal gap can be understood in terms of SSPP skin depth into the grating. Lower internal losses result in a smaller skin depth and stronger localization of the spoof plasmon, thereby increasing the coupling between the SSPP and the incident wave. Consequently, a system with lower losses will exhibit a larger optimal gap.

3. Numerical Simulation

The above theoretical model of SSPP on the grating is limited by the lack of consideration of diffraction of higher modes and interaction of fields arising above the grating grooves. These effects can be taken into account through either more rigorous, complicated theories [30] or a numerical simulation chosen as a basis. We performed numerical simulations using the finite difference method with the Radio Frequencies module of the commercially available COMSOL Multiphysics Software (version 6.0) [38]. The COMSOL eigenmode solver was applied to finding numerical dispersion curves, whereas the analysis of the reflection spectra was carried out with the frequency domain solver.

The numerical scheme for searching for eigenmodes is shown in Figure 2a. We calculated the dispersion curves of a 1D subwavelength grating with rectangular grooves infinite in the y-direction, which enabled us to use a 2D Floquet cell during the analysis. The right and left boundaries of the cell were limited by periodic boundary conditions, while the upper side was closed by a scattering boundary condition (SBC) with a perfect matched layer (PML) domain, and the lower one was defined as a perfect electric conductor (PEC). The SBC and the PEC almost did not affect the result of the simulation because of the high absorption of the grating and PML domains.

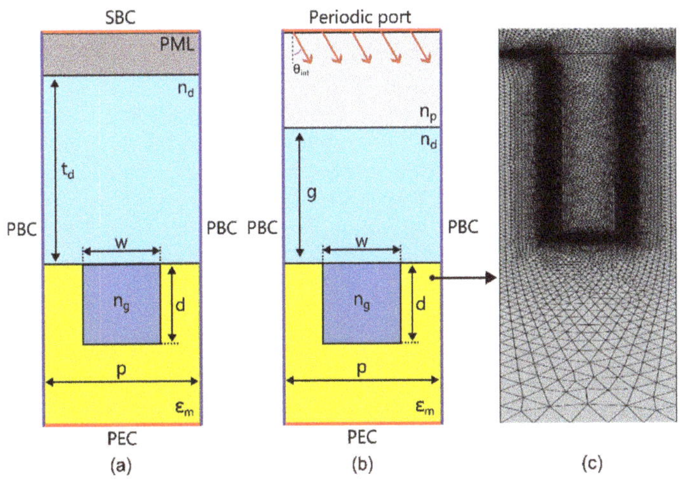

Figure 2. Numerical schemes for calculation of (**a**) eigenmodes, (**b**) reflection spectra, and (**c**) meshing around grating domain.

The x-component of the wavevector directed across the grooves varies, and an eigenmode is excited when k_x coincides with the propagation constant of the spoof plasmon supported by the structure. The variation range was limited by the first Brillouin zone, that is, $|k_x| \leq \frac{\pi}{p}$. The distance between the upper surface of the groove and the PML domain was large enough (greater than the penetration depth of the evanescent wave) to avoid the influence of the latter on the SSPP modes. The number of modes to be found was defined as $M = 1 + \left\lfloor \frac{d}{p} \right\rfloor$, while the initial frequency for search resulted from solution to Equation (1). "Larger real part" was chosen as "the search method around shift" parameter.

The calculation of the reflection spectra was carried out using the numerical configuration consisting of the prism domain, dielectric gap, and grating shown in Figure 2b. As in the previous case, the left and right boundaries were periodic Floquet boundary conditions, but the wavevector magnitude was governed by the periodic port applied to the upper boundary. This boundary must not interact with the evanescent surface plasmon modes; thus, the prism domain size was large enough to allow waves from the port to propagate and reflect inside it. The remaining side was the PEC, which did not affect the response but made it possible to close the cell. The port excited a plane TM wave of $\begin{pmatrix} H_x & H_y & H_z \end{pmatrix} = \begin{pmatrix} 0 & 0 & 1 \end{pmatrix}$. The electric field component was solved for the "in-plane vector." The simulations were conducted at various angles θ_{int}, frequencies, and gaps g. In both simulations, the permittivity of the metal was modeled with the Drude model

$$\varepsilon_m = 1 - \frac{\omega_p^2}{\omega^2 + i\omega\omega_\tau} = 1 - \frac{\omega_p^2\omega_\tau^2}{\omega^4 + \omega^2\omega_\tau^2} + i\frac{\omega_p^2\omega_\tau}{\omega\left(\omega^2 + \omega_\tau^2\right)}, \quad (7)$$

where ω_τ is the electron-collision frequency, and ω_p is the plasma frequency. The parameters of gold were $\omega_\tau = 4.05 \cdot 10^{13}$ rad/s and $\omega_p = 1.37 \cdot 10^{16}$ rad/s [39]. For account of the metal loss and tiny skin depth, the mesh size at the metal–air interface was set to 100 nm and 300 nm as the minimum and maximum element sizes, respectively. The correctness of the definition of the mesh size around the metal–dielectric boundary significantly impacts the value of the optimal gap obtained in the simulation. The rest parameters of meshing were set to "extremely fine", and the mesh was built automatically. The typical meshing around the metal–dielectric interface is shown in Figure 2c.

4. Fabrication Process

The process of rectangular grating fabrication consisted of two stages. At the first stage, we fabricated a rectangular grating using multibeam X-ray lithography in a manner similar to that described in the previous papers [40,41]. The gratings were embedded in a 50 × 50 × 10 mm polished glass substrate. The patterned area was 22 × 20 mm. The lithographic process used synchrotron radiation with energy in the range of 4–12 keV from the VEPP-3 electron storage ring, transmitted to the LIGA station of the Siberian Center of Synchrotron and Terahertz Radiation of Budker Institute of Nuclear Physics SB RAS [42].

The fabrication process was as follows. After wet cleaning in an aqueous solution of sulfuric acid, SU-8–100 photoresist layer about 50 μm thick was deposited by the centrifugation method and then dried on a hotplate. Then the photoresist layer was exposed to synchrotron radiation through a slit diaphragm with step-by-step motion of the substrate [43]. The exposure time depended on the required depth of the grating. This approach enabled quickly changing the design parameters through variation of the slit width and scanning speed. Once the photoresist layer had hardened on the hotplate at a temperature of 95 °C, liquid development of the photoresist layer was carried out in the PGMEA solvent.

At the second stage, a thin metal film was deposited on the patterned substrate using magnetron sputtering. The grating was covered with a 0.3 μm-thick gold (Au) layer with a 10 nm thick chromium sublayer, which was necessary to improve the adhesion of Au to the resist. Due to the small thickness of the Au layer, the geometric dimensions of the grating remained the same as before the sputtering. The choice of gold was due to its chemical inertness to the atmosphere and high conductivity. A fragment of the microstructure with a height and width of 14.5 ± 0.2 μm and period of 40 ± 0.1 μm is shown in the inserts of Figure 3.

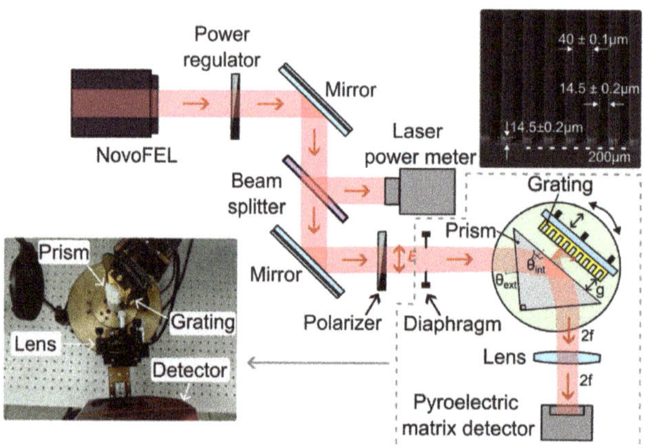

Figure 3. Scheme of experimental setup with photo inserts of its part and subwavelength grating after magnetron sputtering.

5. Experimental Scheme

To excite SSPPs we built the experimental scheme shown in Figure 3. The Novosibirsk free electron laser (NovoFEL) [44] was used as a source of monochromatic terahertz radiation. The power of the incident *p*-polarised beam (E ⊥ grating grooves) was controlled using a metal wire polarizer, whereas its direction was governed by a system of mirrors. The beam splitter directed part of the beam energy to the laser power meter, which enabled detecting the current reference value of the power and tracking changes in the power during the experiment. The other part of the wave passed through the lithographic polarizer to be filtered from the parasitic s-polarised background (E | | grating grooves).

Then the beam impinged on the ZEONEX rectangular prism ($n_p = 1.531$) [45], transparent in both the THz and visible ranges. To adjust the 20-mm beam to the small size of the prism (≈7 mm), we used the 7-mm iris diaphragm placed near the prism. The incident angle exceeded the total internal reflection angle, resulting in the appearance of the evanescent wave interacting with the subwavelength grating. In our case, the relation between θ_{ext} and θ_{int} can be calculated with Snell's law as

$$\theta_{int} = 45^\circ - asin\left(sin(45^\circ - \theta_{ext})\frac{1}{n_p}\right) \quad (8)$$

in the assumption that the prism is surrounded by air. Therefore, the experimental angles were $\theta_{ext} > 38.56^\circ$ and $\theta_{int} > 40.8^\circ$. The beam reflected from the interface (prism–air–grating) was recorded using a TPX lens (f = 50 mm) in the $2f$-$2f$ configuration and a pyroelectric matrix detector (Pyrocam-IV) with a receiving area of 25 × 25 mm^2 (320 × 320 pixels).

The subwavelength grating sample was attached to the optic mount at the distance g from the prism surface along the vertical axis. The motorised rotation stage (Standa) controlled the angular position of the prism with increments of 0.01°, while the vertical motorised stage (Standa) moved the gap with a step of 83 nm. The mechanically adjusted goniometer managed the parallel position of the prism surface and the subwavelength grating. The whole setup was adjusted using the red beam of the diode laser, combined with the THz beam.

6. Simulation Results

6.1. Dispersion Curve

The groove depth of our sample is 14.5 ± 0.2 μm, which is smaller than its period of 40 ± 0.1 μm, and thus it can support only the fundamental mode. The dispersion curves obtained with the simulation and Equation (1) are shown in Figure 4. As can be seen, a significant deviation of the analytical theory from the numerical results can be observed in the case of a shallow grating (w~d), as follows from the Rusina theory limitation (see the theory and numerical sections).

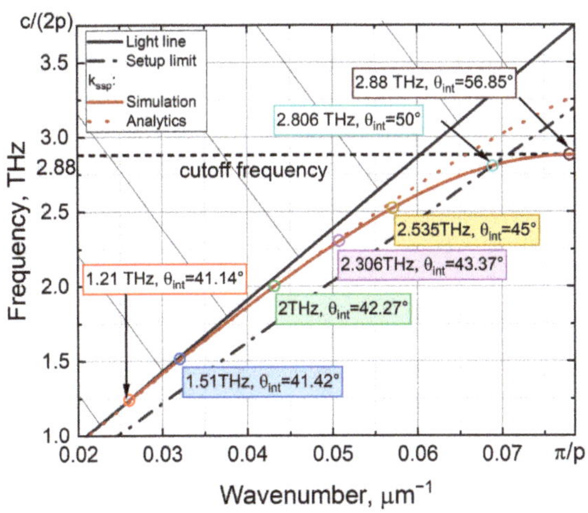

Figure 4. SSPP dispersion curve obtained with numerical simulation and analytical equation. Highlighted points with filled rectangles reflect positions of experimental resonances, while limiting points of our setup and grating are shown in rectangles without color.

We conducted experiments in the vicinity of several frequencies and angles (see Figure 4). Their choice was determined by the atmospheric frequency transparency windows (in the absence of strong water vapor absorption lines) and the angles that can be adjusted with our in-house setup. The NovoFEL frequency can be tuned over a wide range (0.75–38 THz), but there are well-established operating regimes. Some of them correspond to 1.51 THz, 2 THz, 2.306 THz, and 2.535 THz. The range of angles covered by our setup is limited by both the prism size and the diameter of the incident beam. Due to this, the angle has to be between 40.8 (the ATR critical angle) and 50 degrees. The limited angular range has narrowed the operating frequency range to 1.21–2.806 THz. Both the frequencies and angles at which we work satisfy the matching condition (Equation (3)) and limitations of the setup.

6.2. Gap and Frequency Variations

The gap between the grating and the prism base affects the efficiency of coupling of the evanescent wave into spoof surface plasmon polaritons. Through several simulations, we obtained gap–angular maps for the chosen excitation frequencies. One of the maps is shown in Figure 5a. By comparing the dip values of the reflectivity at the SPR points for different gap distances, we could find the optimal coupling distance. The optimal gap distance varies with the position on the dispersion curve, and the larger optimal gap corresponds to the closest position to the light line.

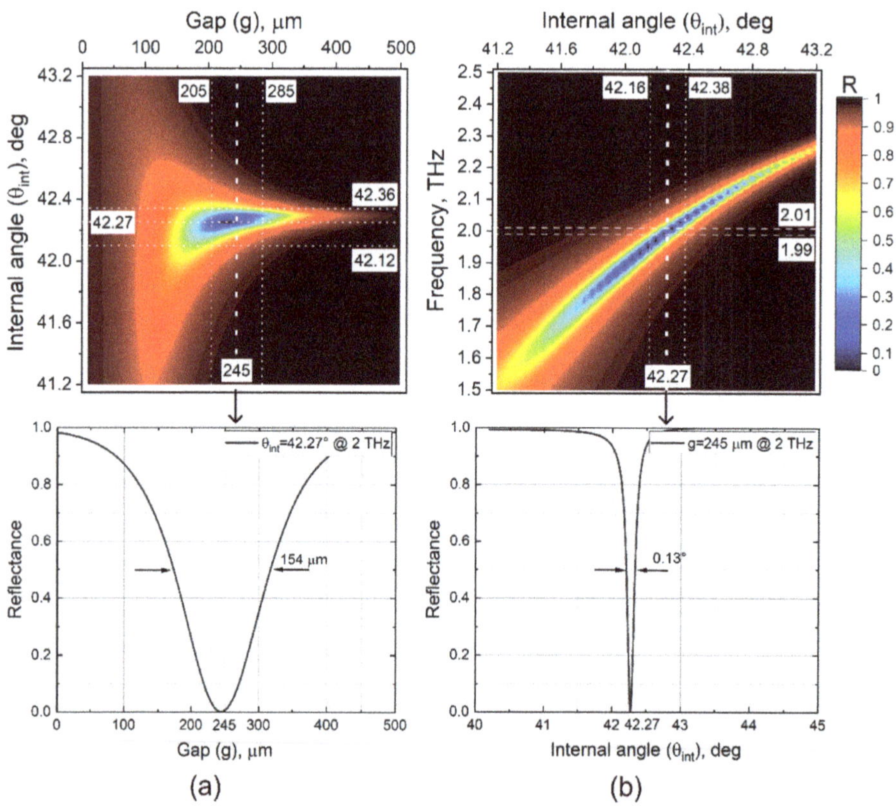

Figure 5. (**a**) Gap–angular map (λ = 150 μm) with inset of gap spectrum with optimal parameters of SPP coupling below it; (**b**) Frequency–angular map (g = 245 μm) with inset of angular spectrum with optimal parameters of SPP coupling below it.

Although for many applications it is preferable to have a highly confined field, it requires more sophisticated experimental solutions in controlling the gap distance. In our setup, this complex task is not fully solved, which leads to the gap dispersion and zero gap error. Because of tilts, the gap distance varies for different points of the grating surface. This effect can be called gap dispersion. As a result, the SSPPs interact non-uniformly with the evanescent wave and are excited with different efficiencies. Another challenge is the difficulty of experimental determination of the zero position of the gap because of the mechanical adjustment prior to the experiment. The inaccuracy of the zero setting could be about 0–40 μm, while the tilt of the prism base could be ~0.2–0.3°, which leads to a gap dispersion of ~25–40 μm (see Figure 5a). All these effects can be eliminated by increasing the optimal gap distances in the experimental measurements and choosing the excitation points on the dispersion curves closer to the light line.

In the experiment, in addition to the gap dispersion, we have the spectral width of the laser. The NovoFEL radiation usually has a spectral width of ~1% of the central frequency [46]. For instance, if we have an excitation frequency of 2 THz, then the spectral width will be 0.02 THz. In addition, the beam incident on the prism has diffraction divergence. This means that the light will fall onto the prism at different angles, which leads to different efficiencies of excitations of SSPPs related to different points on the dispersion curves. This effect can be reflected with the frequency–angular map shown in Figure 5b.

To estimate the contribution of these effects, we will consider excitation of a spoof plasmon on the gold grating at 2 THz (150 μm). At this frequency, we have an optimal gap and an optimal angle, equal to 245 μm and 42.27°, respectively. The spectral width of the incident beam affects the optimal-coupling incident-angle broadening. Since we have a diaphragm diameter of 7 mm in the experiment, the diffraction angular spread will be $\frac{\sim\lambda}{d} = \pm 1.2^\circ$. However, due to the relatively narrow spectral width of the beam, only angles lying in the range of $\pm 0.11^\circ$ from the optimal one will contribute to the excitation process, as shown in Figure 5b. The angular components that do not interact with the grating will reduce the energy transferred into the spoof plasmons in the experiment. The gap dispersion enables more efficient coupling of different angular and frequency components of the beam, resulting in an almost twofold expansion of the angular spectrum width from 0.13° to 0.22°. The gap spectrum remains virtually unchanged due to its large width.

6.3. Field Confinement and Optimal Gap

The confinement of the electromagnetic field above the grooves is dependent on the position of the excited mode on the dispersion curve. According to Equation (4), the closer the wavenumber of the spoof plasmon to the Brillouin zone boundary, the better its confinement. We estimated the penetration depth of the spoof surface plasmon using Equation (4) and the numerical dispersion curve, as shown in Figure 6a. The value of l_{sspp} decreases relatively fast from 780 μm (≈3 λ) in the vicinity of the light line to 20 μm (≈0.2 λ) at the Brillouin zone boundary. Excitation of SSPPs occurs when the wave vector matching condition is met (see Equation (3)), which looks like the intersection of l_p (see Equation (5)) with l_{sspp}.

However, the optimal gap is not equal to the SSPP penetration depth and cannot be fully described as an overlap of two evanescent waves when $l_p = l_{sspp}$. It is associated with the appearance of additional reflection and interference effects, which result in a complicated interaction of the SSPP waves and the evanescent wave of the prism in the gap between the prism and the subwavelength grating. We conducted several simulations to find the optimal gaps at various wavelengths. The obtained dependence of the optimal gap on the wavelength is plotted in Figure 6b. To approximate it, we used the equation

$$g = g_0 + \frac{\Delta g}{\Delta \lambda}(\lambda - \lambda_0), \tag{9}$$

where g_0 is the minimum gap corresponding to the cutoff wavelength, λ_0 is the cutoff wavelength, and $\frac{\Delta g}{\Delta \lambda}$ is the rate of gap variation with change in the wavelength. This approximation is valid up to the proximity to the light line, where transition effects occur and enable quick determination of the optimal coupling gap distance.

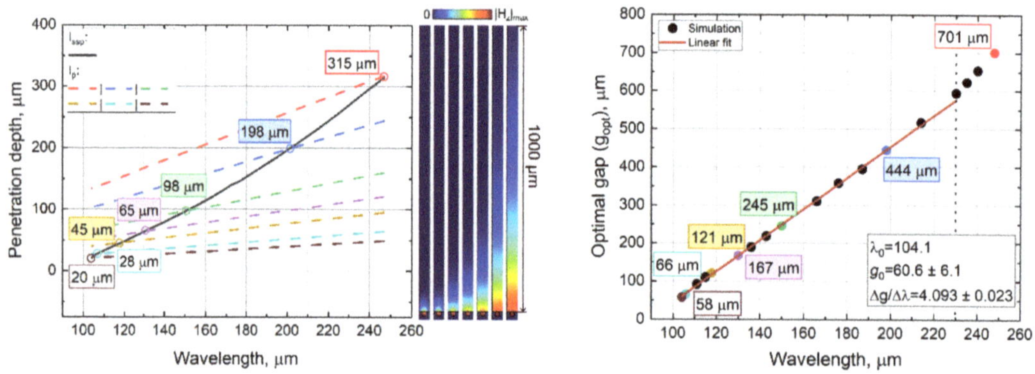

(**a**) (**b**)

Figure 6. (**a**) Penetration depth of electric field above grating and (**b**) optimal coupling gap. Values in coloured circles correspond to those in circles of similar colours in Figure 4.

6.4. Losses and Optimal Gap

As mentioned in the theoretical section, the gap distance between the prism and the grating is influenced by losses in the system. To demonstrate this, we can consider two ways to increase internal losses: either by increasing the metal losses ($\text{Im}\{\varepsilon_m\}$) or by increasing the losses in surrounding media ($\text{Im}\{n_d\}$). We chose the former approach and increased the imaginary part of the metal permittivity by scaling ω_τ by a factor of X. As can be seen from Equation (7), this parameter has only a minor effect on the real part of the metal permittivity at high values of ω_τ, while significantly increasing the imaginary part. Radiative losses have a much more complex and diverse nature, which does not allow us to estimate them for our task. Since the resonance parameters depend on the sum of internal and radiative losses, and we cannot separate them from each other in the experiment, then here and further we will be able to increase the total losses of the SSPPs only by increasing the internal losses by the way mentioned above.

This approach allows us to qualitatively assess the influence of internal losses on the resonance behaviour. Figure 7a,b show the dependence of internal losses on optimal gaps at a wavelength of 198 µm. As the losses increase, the optimal gap decreases, while the FWHM of the gap resonance remains nearly constant at approximately 242 µm. In contrast, the angular spectrum exhibits the opposite trend (see Figure 7c). The optimal angle of the resonance continues to satisfy the phase matching condition, although the angular resonance becomes significantly broader.

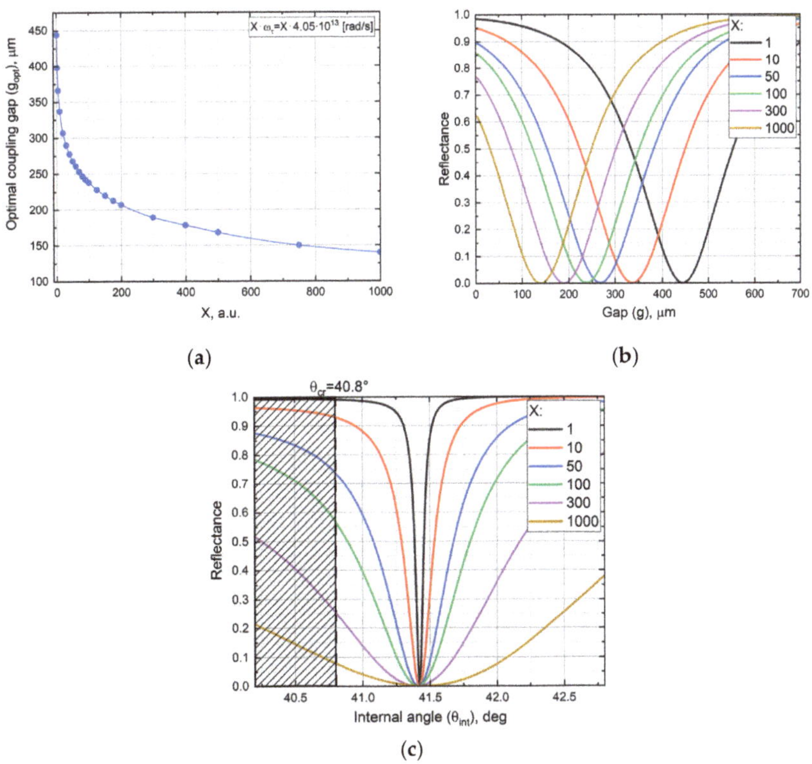

Figure 7. (**a**) Dependence of the optimal coupling gap on factor X at λ = 198 µm; (**b**) Gap reflectance spectra as a function of factor X at λ = 198 µm; (**c**) Angular reflectance spectra as a function of factor X at at λ = 198 µm.

7. Experimental Results

7.1. Measurement Procedure

To find the optimal conditions of SSPP excitation in the experiment, we set up the beam incidence angle according to the simulation results, and then we varied the gap distance. However, the experimental optimal angle usually differs from the simulation one; therefore, we slightly varied it in the vicinity of the first predicted angle and repeated the procedure with the gap until we were able to determine the absolute minimum dip in the reflection spectra.

In the experiments, we recorded the 2D distribution of the beam reflected from the prism base in the ATR regime (see Section 5). This method of beam recording can be applied in the future to the analysis of the local characteristics of the analyte at each point of the grating in the same manner as in our previous article on the excitation of surface plasmons on InSb [47]. In this article, we processed the recorded data in the following way:

1. The beam signal normalised to the reference one was averaged over the detector matrix to yield the integral response of the grating.
2. The obtained angular spectra at various spectral frequencies were then normalised to the attenuated total reflection angular spectrum without the grating in the vicinity of the prism to account for the interaction of the incident beam with the prism under different conditions.

3. To obtain the reflectance, we divided the entire angular and gap spectrum by its maximum value reached at large air gaps and angles far from the resonance, correspondingly.
4. The measured intensity errors were about ±5%, which corresponded to the NovoFEL average power instability.

Figure 8 shows an example of reflected beam images obtained at the resonant angle and wavelength for different gaps. It is evident that at g = 150 μm, the greatest attenuation of the beam is observed, which corresponds to the surface plasmon resonance. The distortions observed in the beam profile in the resonance region indicate re-radiation of the SSPPs generated on the grating and their interference with the reflected beam. At large gaps (g ≈ 300 μm), in the absence of bulk wave transformation into SSPPs, it is possible to observe the reflected beam profile under conditions of total internal reflection, which corresponds to the profile of the beam incident on the air–prism boundary. The white dotted line highlights the region over which the signal was averaged for all frames in this series.

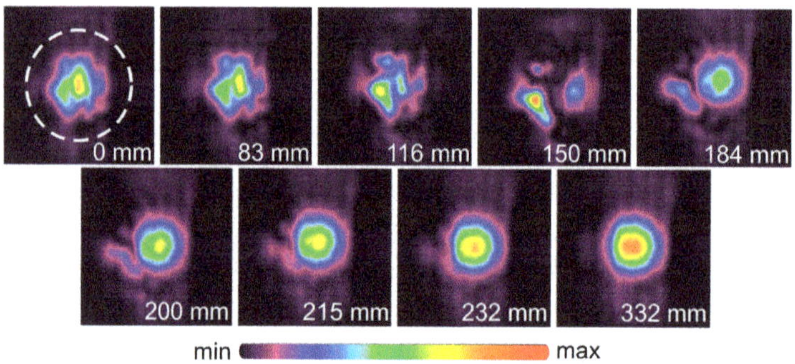

Figure 8. Reflected beam images for subwavelength grating with different gaps between prism and grating. Radiation wavelength λ = 130 μm (2.3 THz); internal angle of incidence θ_{int} = 43.7 degrees. White dotted line: region of averaging of reflected beam signal.

7.2. Angular and Gap Spectra

The angular spectra are shown in Figure 9a. It is immediately clear that the experimental dip width was significantly larger than the simulated one, while the dip depth in the experiments was smaller than that in the simulation. For example, (see Section 6.2), the angular width is estimated at about 0.22° at 150 μm, while the experiment gives 1.2° at 150 μm, which is 5.5 times larger.

This shows that the spectral and angular widths of the incident beam have less impact on the experimental results than the broadening caused by absorption and diffraction due to the grating surface roughness, dispersion of the grating parameters, and the finite size of the grating. The assumption about the significant effect of the sources of the experimental angular broadening is confirmed by the increase in the depth and the decrease in the width of the dip with the growth of the radiation wavelength (see Figure 9a,b).

In addition to this, it can be noted that the experimental dip, corresponded to 198 μm (black line) slightly outside of the ATR regime. This is due to the higher losses in experiments and the relatively close proximity of the resonance to the critical angle. A similar effect is observed in the simulation results when the metal losses are increased (see Figure 7c).

The gap spectra are demonstrated in Figure 9b. The depth values in these spectra are almost similar to those we obtained in the angular spectra at the corresponding wavelengths (see δR_θ and δR_g in Figure 9c). The non-zero depth of the experimental resonance is attributed to the angular width of the beam and indicates that only a portion of the beam transforms into the SSPPs (see Section 6.2). The angular components of the beam that are not involved in the excitation process cannot be excluded from consideration, as the reflected beam is normalised with respect to the entire incident beam. The resonance depth can be increased, for example, by reducing the angular width of the incident beam. The dip width suffers less from the broadening effects than the angular spectra do, and their values are closer to the simulation results.

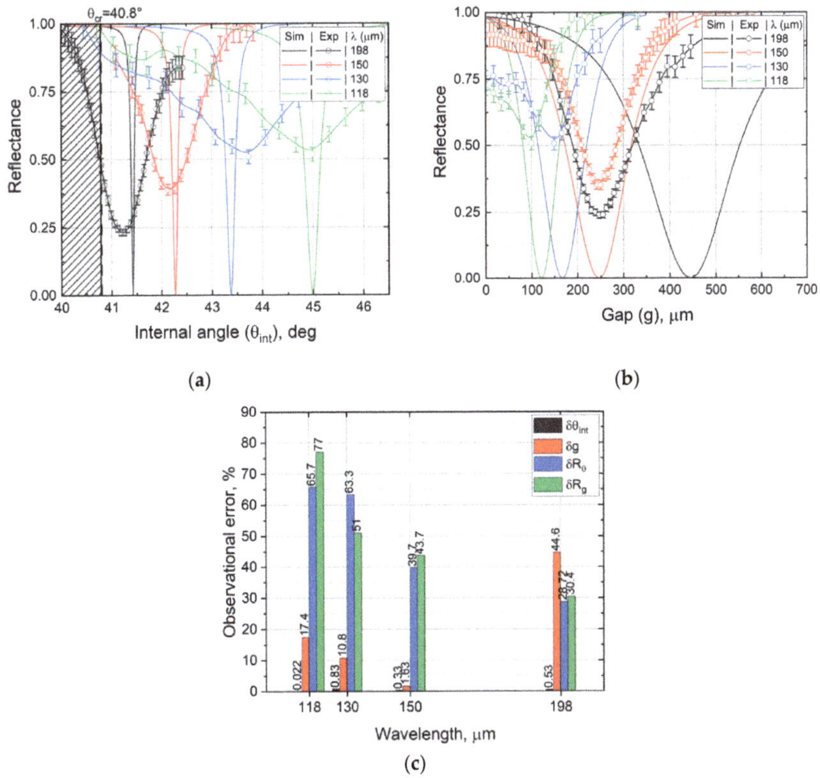

Figure 9. Experimental results: (**a**) angular spectrum; (**b**) gap spectrum; (**c**) observation errors.

7.3. Optimal Incident Angle and Gap Distance

The optimal excitation angle and gap obtained experimentally and numerically are summarised in Table 1. In most cases, the experimental angles and gaps are close to the simulation ones. The observation error in the determination of the angular position of the resonance $\delta\theta_{int}$ does not exceed 1%, while the experimental gap position δg coincides with the simulation one with an accuracy of less than 17.4% (see Figure 9c).

As discussed earlier, the optimal angle is primarily determined by the phase matching conditions and its position can be affected by variation in structural parameters and the system misalignment. This explains the relatively low angular error observed in the experiment. In contrast, the optimal gap distance is governed by a combination of internal and radiation losses, which depend on various interrelated factors such as roughness of the surface, plasmon localisation, metal absorption, boundary effects due to the finite size of the grating, and even zero gap position errors. As a result, a relatively large deviation in the gap is acceptable. In absolute values, the gap error is less than 21 μm.

Table 1. Optimal SSPP excitation incident angles (θ_{int}) and gap distances (g_{opt}).

Wavelength, μm	Frequency, THz	Simulation		Experiment	
		Gap, μm	Angle, deg	Gap, μm	Angle, deg
198 ± 0.51	1.51 ± 0.008	444 ± 0.5	41.42 ± 0.005	245.8 ± 5	41.2 ± 0.1
150 ± 0.85	2 ± 0.01	245 ± 0.5	42.27 ± 0.005	249 ± 5	42.13 ± 0.1
130 ± 0.67	2.306 ± 0.012	167 ± 0.5	43.37 ± 0.005	149 ± 5	43.73 ± 0.1
118 ± 0.34	2.535 ± 0.013	121 ± 0.5	44.99 ± 0.005	100 ± 5	44.98 ± 0.1

There exists only one large shift from the value predicted by the simulation, observed in several experimental series at a wavelength of 198 μm. In this case, the measured gap distance was nearly half that predicted by the simulation. This shift is related to the fact that, at this frequency, the resonance lies close to the light line, indicating strong coupling between the SSPP and the incident wave. Due to weak localisation, the scattering of THz SSPPs on the roughness occurs more intensely [48], which leads to the interference of waves in the gap with diffracted ones, resulting in the non-uniformity of the FEL beam intensity profile at the surface plasmon resonance (see Figure 8). As a result, the optimal coupling gap distance shifted to the lower values in the experiment. This effect can be replicated in the simulation by introducing additional losses into the system. For example, a similar critical coupling condition is observed in the simulation when X=85, as shown in Figure 7a.

Despite the fact that we used only a discrete number of frequencies in the experiment, the results indicate that we can continuously switch the resonant frequency in at least the 1-THz range from 1.51 to 2.54 THz and theoretically can increase the value up to 1.67 THz using a single grating. The obtained frequency tuning is significant compared to the other strategy of resonance switching. For instance, the active tunable metamaterials allow only achieving about 3–5% of frequency tuning from the initial position of resonance, which is typically less than 0.3 THz [49,50]. Table 2 summarises the typical spectral tunable ranges observed experimentally for both active and angle-dependent terahertz metamaterials.

Table 2. Typical tunable range of terahertz metamaterials.

Approach	Spectral Tuning Range, THz	Tunable Range, GHz	Reference
Angle-dependent metasurface	0.139–0.1498	10.8	[51]
Angle-dependent metagrating	0.5–0.55	50	[52]
MEMS, thermal tuning	0.32–0. 43	110	[53]
MEMS, mechanical tuning	2.12–2.28	160	[54]
Liquid crystal, electrical tuning	0.75–1	250	[55]
Metal grating with angular dispersion	1.51–2.54	1030	This work

7.4. Quality Factor

The Q factor for the gap and angular resonances can be defined as $Q_g = \frac{g_{opt}}{\Delta g}$ and $Q_\theta = \frac{\theta_{opt}}{\Delta\theta}$, correspondingly, where g_{opt} and θ_{opt} are the optimal gap and angular resonance positions, and Δg and $\Delta\theta$ are the full widths at half maximum of the resonances. Due to the broadening and position errors described in the previous sections, Q_g in the experiments turned out to be of the same order as in the simulation, while the experimental Q_θ was one order of magnitude smaller than the simulation one, as shown in Figure 10. The Q factor of the resonances in the angular spectra was within the range of 19.4–37.6, while the resonances of the gap spectra had the Q factor lying within the 1.17–2.03 range. Both the gap and angular quality factors grew with an increase in the resonance wavelength and achieved their maximum values at a wavelength of 198 μm. As mentioned in the previous section, this wavelength dependency is clear because of the growing effects of the gap dispersion, absorption, and diffraction on the propagation of the spoof surface plasmon waves along the grating at smaller wavelengths. The maximum angular quality factor $Q_\theta = 37.6 \pm 6.7$ at 1.51 THz(198 μm) is comparable to a value of 43.3 at 0.738 THz, which was obtained with frequency spectrum and phase measurements by Huang [24], but is lower than a Q factor of 170 at 1.71 THz, obtained by the similar phase method by Binghao Ng [14].

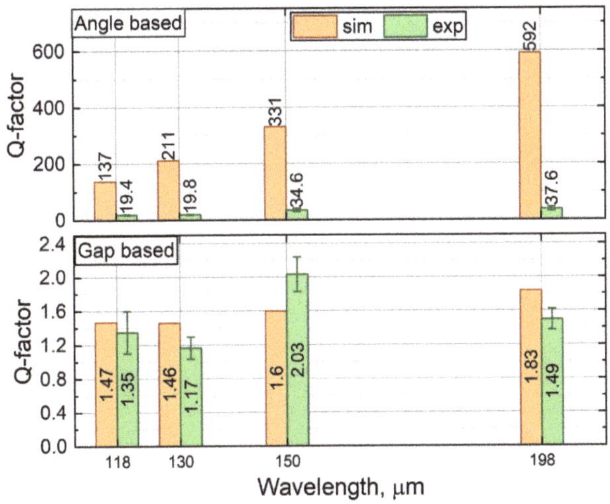

Figure 10. Q factors obtained experimentally and numerically.

8. Discussion

The experimental excitation of spoof surface plasmons on a 1D subwavelength grating in the attenuated total reflection scheme is a new topic. The first experimental demonstration was carried out in 2013 by Binghao Ng [14], repeated in 2019–2020 by Huang [23,24]. Moreover, they showed the sensor capabilities of this type of metamaterial using various liquids. The authors were not limited to that and demonstrated the possibility of improving the quality of resonances through phase measurements and coupling gap variation. Although all these measurements were conducted using a time domain spectrometer, and the grating responses were recorded over a wide frequency range, the work was performed in one frequency resonance regime. The authors did not study, for example, the analyte–grating interaction at various frequencies and were not able to unleash the full potential of the TDS and this metamaterial.

Compared to the above works, we operated in the frequency domain and realised the attenuated total reflection technique with angular interrogation to collect data at various frequencies and angles. This allowed us to change the resonance of the subwavelength grating in the dispersion curve through variation of the incidence angle and variation and implement passive tuning of the resonance. We were not able to collect phase information, which would significantly improve the resonance quality and enable work with highly absorbent substances. However, as shown in our article, the angular spectra with a tunable quasi-monochromatic source of radiation might be a good alternative, without the necessity of using a complicated TDS technique, which has various experimental limitations and uncertainties. It is especially promising in connection with the development of compact frequency-tunable THz radiation sources, such as quantum cascade lasers and frequency multipliers based on avalanche diodes [56].

Moreover, like other authors, we presented gap variation to enable modulation of the coupling efficiency. It is highly important for the adjustment of the field enhancement when the refractive index of the surrounding medium changes or during layer scanning, for instance. Another difference is that we recorded our data with a matrix detector. This new feature of our setup opens up new ways to analyse local interactions of SSPPs with the surrounding media.

Although our experimental results are in good agreement with the theory, the capability of our setup is still limited. (1) We still cannot provide precise control of the zero gap distance. (2) The limited prism size and the large beam diameter do not allow us to tune the full available frequency region across the entire dispersion curve and reduce the diffracted components of the incident beam that do not contribute to the formation of the spoof plasmon resonance. (3) The system of phase measurements has not been implemented yet. We will continue developing the setup to improve the quality of the resonances and expand its potential applications.

The attenuated total reflection scheme in the Otto configuration combined with the subwavelength grating can become a platform for the terahertz real-time enhanced spectroscopy and imaging since it provides full control over spoof plasmon resonances. However, the interaction of spoof plasmon resonances with liquid analytes is unclear, and the problem of calibration of resonances over a wide frequency range to extract parameters of the analyte has not been solved. In addition, excitation and analysis of higher SSPP modes have not been realised experimentally yet. All of these might be topics for further research.

9. Conclusions

We demonstrated excitation of the fundamental spoof surface plasmon mode on a 1D subwavelength rectangular profile grating in the terahertz frequency range. Using the attenuated total internal reflection technique with angular interrogation in the Otto configuration in the frequency domain, we successfully carried out experiments and measured spectra at different frequencies by tuning the resonance from 1.51 to 2.54 THz, which corresponds to 1-THz tunability, which cannot be achieved with active tunable metamaterials. The obtained Q factors of the angular resonances lie within the range of 19.4–37.6, reaching the maximum at a wavelength of 198 μm. These values are comparable to those other authors obtained earlier using similar structures. By changing the gap, we were able to control the efficiency of coupling of the prism evanescent wave and the spoof surface plasmon wave, which leads to modulation of the absorption characteristics. In addition, a new recording system realised in our in-house setup makes it possible to monitor the local response of the grating, which will enable real-time tracking of changes in the surrounding medium. An experimental realisation of all capabilities of this metamaterial at once was shown for the first time. All of this is highly important for enhanced terahertz real-time absorption spectroscopy and imaging, and the presented results bring us closer to the implementation of these techniques.

Author Contributions: This work was realised through the collaboration of all authors. O.K.: contributing to the main results, original draft preparation; V.G.: reviewing and editing; experimental investigation. B.G.: sample preparation, resources V.N.: sample preparation, resources. All authors have read and agreed to the published version of the manuscript.

Funding: This research received no external funding.

Data Availability Statement: The raw data supporting the conclusions of this article will be made available by the authors on request.

Acknowledgments: The work was carried out at the shared research facility Siberian Center for Synchrotron and Terahertz Radiation on the basis of the Novosibirsk Free Electron Laser at Budker Institute of Nuclear Physics SB RAS.

Conflicts of Interest: The authors declare no conflicts of interest.

Abbreviations

The following abbreviations are used in this manuscript:

SSPP	spoof surface plasmon polariton
SPP	surface plasmon polaritons
THz	terahertz
ATR	attenuated total reflection
NovoFEL	Novosibirsk free electron laser

References

1. Zayats, A.V.; Smolyaninov, I.I.; Maradudin, A.A. Nano-Optics of Surface Plasmon Polaritons. *Phys. Rep.* **2005**, *408*, 131–314. [CrossRef]
2. Shalabney, A.; Abdulhalim, I. Sensitivity-enhancement Methods for Surface Plasmon Sensors. *Laser Photonics Rev.* **2011**, *5*, 571–606. [CrossRef]
3. Maier, S.A. *Plasmonics: Fundamentals and Applications*; Springer: New York, NY, USA, 2007; ISBN 978-0-387-33150-8.
4. Knyazev, B.A.; Kuzmin, A.V. Surface electromagnetic waves: From visible range to microwaves. *Vestn. NGU Ser. Fiz.* **2007**, *2*, 108–122. [CrossRef]
5. Kukotenko, V.D.; Gerasimov, V.V.; Khasanov, I.S.; Lemzyakov, A.G.; Ivanov, A.I.; Antonova, I.V. Experimental Study of Terahertz Surface Wave Field on Metallic and Composite Graphene Films. In Proceedings of the 2024 IEEE 25th International Conference of Young Professionals in Electron Devices and Materials (EDM), Altai, Russian, 28 June–2 July 2024; pp. 1070–1073. [CrossRef]
6. Gerasimov, V.V.; Knyazev, B.A.; Lemzyakov, A.G.; Nikitin, A.K.; Zhizhin, G.N. Growth of Terahertz Surface Plasmon Propagation Length Due to Thin-Layer Dielectric Coating. *J. Opt. Soc. Am. B* **2016**, *33*, 2196. [CrossRef]
7. Gerasimov, V.V.; Nikitin, A.K.; Lemzyakov, A.G.; Azarov, I.A.; Kotelnikov, I.A. Obtaining the Effective Dielectric Permittivity of a Conducting Surface in the Terahertz Range via the Characteristics of Surface Plasmon Polaritons. *Appl. Sci.* **2023**, *13*, 7898. [CrossRef]
8. Isaac, T.H.; Barnes, W.L.; Hendry, E. Determining the Terahertz Optical Properties of Subwavelength Films Using Semiconductor Surface Plasmons. *Appl. Phys. Lett.* **2008**, *93*, 241115. [CrossRef]
9. Gerasimov, V.V.; Khasanov, I.S.; Kukotenko, V.D.; Lemzyakov, A.G.; Ivanov, A.I.; Antonova, I.V.; Cherevko, A.G. Terahertz Surface Plasmon Refractometry of Composite Graphene Nanoparticle Films. *IEEE Trans. THZ Sci. Technol.* **2025**, *15*, 61–68. [CrossRef]
10. Rivas, J.G.; Kuttge, M.; Bolivar, P.H.; Kurz, H.; Sánchez-Gil, J.A. Propagation of Surface Plasmon Polaritons on Semiconductor Gratings. *Phys. Rev. Lett.* **2004**, *93*, 256804. [CrossRef]
11. Gong, M.; Jeon, T.-I.; Grischkowsky, D. THz Surface Wave Collapse on Coated Metal Surfaces. *Opt. Express* **2009**, *17*, 17088. [CrossRef]
12. Garcia-Vidal, F.J.; Fernández-Domínguez, A.I.; Martin-Moreno, L.; Zhang, H.C.; Tang, W.; Peng, R.; Cui, T.J. Spoof Surface Plasmon Photonics. *Rev. Mod. Phys.* **2022**, *94*, 025004. [CrossRef]
13. Gao, Z.; Wu, L.; Gao, F.; Luo, Y.; Zhang, B. Spoof Plasmonics: From Metamaterial Concept to Topological Description. *Adv. Mater.* **2018**, *30*, 1706683. [CrossRef]
14. Ng, B.; Wu, J.; Hanham, S.M.; Fernández-Domínguez, A.I.; Klein, N.; Liew, Y.F.; Breese, M.B.H.; Hong, M.; Maier, S.A. Spoof Plasmon Surfaces: A Novel Platform for THz Sensing. *Adv. Opt. Mater.* **2013**, *1*, 543–548. [CrossRef]
15. Ng, B.; Hanham, S.M.; Wu, J.; Fernández-Domínguez, A.I.; Klein, N.; Liew, Y.F.; Breese, M.B.H.; Hong, M.; Maier, S.A. Broadband Terahertz Sensing on Spoof Plasmon Surfaces. *ACS Photonics* **2014**, *1*, 1059–1067. [CrossRef]
16. Yao, H.; Zhong, S. High-Mode Spoof SPP of Periodic Metal Grooves for Ultra-Sensitive Terahertz Sensing. *Opt. Express* **2014**, *22*, 25149. [CrossRef] [PubMed]
17. Yao, H.; Zhong, S.; Tu, W. Performance Analysis of Higher Mode Spoof Surface Plasmon Polariton for Terahertz Sensing. *J. Appl. Phys.* **2015**, *117*, 133104. [CrossRef]
18. Zhang, Y.; Hong, Z.; Han, Z. Spoof Plasmon Resonance with 1D Periodic Grooves for Terahertz Refractive Index Sensing. *Opt. Commun.* **2015**, *340*, 102–106. [CrossRef]
19. Chen, L.; Yin, H.; Chen, L.; Zhu, Y. Ultra-Sensitive Fluid Fill Height Sensing Based on Spoof Surface Plasmon Polaritons. *J. Electromagn. Waves Appl.* **2018**, *32*, 471–482. [CrossRef]

20. Zhao, R.; Lu, G.; Yin, H.; Liang, J.; Zeng, D.; Xiao, H. Terahertz Sensor Study Based on Spoof Surface Plasmon Polaritons. *Int. J. Antennas Propag.* **2020**, *2020*, 2504626. [CrossRef]
21. Chen, X.; Xiao, H.; Lu, G.; Zhao, R. Refractive Index Sensing Based on Terahertz Spoof Surface Plasmon Polariton Structure. *J. Phys. Conf. Ser.* **2020**, *1617*, 012008. [CrossRef]
22. Yao, H.; Zhang, W.; Liu, W.; Mei, H. Resolved Terahertz Spectroscopy of Tiny Molecules Employing Tunable Spoof Plasmons in an Otto Prism Configuration. *J. Opt.* **2022**, *24*, 045301. [CrossRef]
23. Huang, Y.; Zhong, S.; Shi, T.; Shen, Y.-C.; Cui, D. HR-Si Prism Coupled Tightly Confined Spoof Surface Plasmon Polaritons Mode for Terahertz Sensing. *Opt. Express* **2019**, *27*, 34067. [CrossRef] [PubMed]
24. Huang, Y.; Zhong, S.; Shi, T.; Shen, Y.; Cui, D. Terahertz Plasmonic Phase-Jump Manipulator for Liquid Sensing. *Nanophotonics* **2020**, *9*, 3011–3021. [CrossRef]
25. Huang, Y.; Singh, R.; Xie, L.; Ying, Y. Attenuated Total Reflection for Terahertz Modulation, Sensing, Spectroscopy and Imaging Applications: A Review. *Appl. Sci.* **2020**, *10*, 4688. [CrossRef]
26. Kameshkov, O.E.; Gerasimov, V.V. Excitation of Terahertz Plasmonic Resonance on a Subwavelength Metallic Grating in an Attenuated Total Reflection Scheme Using a Monochromatic Radiation Source. *J. Surf. Investig.* **2023**, *17*, 1463–1471. [CrossRef]
27. Garcia-Vidal, F.J.; Martín-Moreno, L.; Pendry, J.B. Surfaces with Holes in Them: New Plasmonic Metamaterials. *J. Opt. A Pure Appl. Opt.* **2005**, *7*, S97–S101. [CrossRef]
28. Shen, L.; Chen, X.; Yang, T.-J. Terahertz Surface Plasmon Polaritons on Periodically Corrugated Metal Surfaces. *Opt. Express* **2008**, *16*, 3326. [CrossRef] [PubMed]
29. Rusina, A.; Durach, M.; Stockman, M.I. Theory of Spoof Plasmons in Real Metals. *Appl. Phys. A* **2010**, *100*, 375–378. [CrossRef]
30. Erementchouk, M.; Joy, S.R.; Mazumder, P. Electrodynamics of Spoof Plasmons in Periodically Corrugated Waveguides. *Proc. R. Soc. A* **2016**, *472*, 20160616. [CrossRef] [PubMed]
31. Jiang, T.; Shen, L.; Zhang, X.; Ran, L.-X. High-order modes of spoof surface plasmon polaritons on periodically corrugated metal surfaces. *PIER M* **2009**, *8*, 91–102. [CrossRef]
32. Bulgakova, V.V.; Gerasimov, V.V.; Goldenberg, B.G.; Lemzyakov, A.G.; Malkin, A.M. Study of Terahertz Spoof Surface Plasmons on Subwavelength Gratings with Dielectric Substance in Grooves. *Procedia Eng.* **2017**, *201*, 14–23. [CrossRef]
33. Khasanov, I.S.; Gerasimov, V.V.; Kameshkov, O.E.; Nikitin, A.K.; Kassandrov, V.V. Observation of Surface Plasmon Resonance in Monochromatic Terahertz Radiation on Indium Antimonide. *J. Surf. Investig.* **2023**, *17*, 1052–1059. [CrossRef]
34. Dominique, B.; Otto, A. Excitations of Surface Plasmon Polaritons by Attenuated Total Reflection, Revisited. *Riv. Nuovo C.* **2013**, *36*, 173–209. [CrossRef]
35. Otto, A. Excitation of Nonradiative Surface Plasma Waves in Silver by the Method of Frustrated Total Reflection. *Z. Phys.* **1968**, *216*, 398–410. [CrossRef]
36. Mason, D.R.; Menabde, S.G.; Park, N. Unusual Otto Excitation Dynamics and Enhanced Coupling of Light to TE Plasmons in Graphene. *Opt. Express* **2014**, *22*, 847. [CrossRef]
37. Yeatman, E.M. Resolution and Sensitivity in Surface Plasmon Microscopy and Sensing. *Biosens. Bioelectron.* **1996**, *11*, 635–649. [CrossRef]
38. RF Module User's Guide, version 6.0, COMSOL, Inc. Available online: https://doc.comsol.com/6.0/doc/com.comsol.help.rf/RFModuleUsersGuide.pdf (accessed on 13 June 2025).
39. Ordal, M.A.; Bell, R.J.; Alexander, R.W.; Long, L.L.; Querry, M.R. Optical Properties of Fourteen Metals in the Infrared and Far Infrared: Al, Co, Cu, Au, Fe, Pb, Mo, Ni, Pd, Pt, Ag, Ti, V, and W. *Appl. Opt.* **1985**, *24*, 4493. [CrossRef]
40. Reznikova, E.; Mohr, J.; Boerner, M.; Nazmov, V.; Jakobs, P.-J. Soft X-ray Lithography of High Aspect Ratio SU8 Submicron Structures. *Microsyst. Technol.* **2008**, *14*, 1683–1688. [CrossRef]
41. Goldenberg, B.G.; Lemzyakov, A.G.; Zelinsky, A.G.; Nazmov, V.P.; Pindyurin, V.F. Multibeam X-ray Lithography to Form Deep Regular Microstructures. *J. Synch. Investig.* **2016**, *10*, 92–95. [CrossRef]
42. Siberian Center for Synchrotron and Terahertz Radiation on the Basis of the Novosibirsk Free Electron Laser at Budker Institute of Nuclear Physics SB RAS. Available online: https://ssrc.biouml.org (accessed on 24 March 2025).
43. Goldenberg, B.G.; Lemzyakov, A.G.; Nazmov, V.P.; Pindyurin, V.F. Multifunctional X-ray Lithography Station at VEPP-3. *Phys. Procedia* **2016**, *84*, 205–212. [CrossRef]
44. Shevchenko, O.A.; Vinokurov, N.A.; Arbuzov, V.S.; Chernov, K.N.; Davidyuk, I.V.; Deichuly, O.I.; Dementyev, E.N.; Dovzhenko, B.A.; Getmanov, Y.V.; Gorbachev, Y.I.; et al. The Novosibirsk Free-Electron Laser Facility. *Bull. Russ. Acad. Sci. Phys.* **2019**, *83*, 228–231. [CrossRef]
45. Tydex. Available online: https://www.tydexoptics.com (accessed on 24 March 2024).
46. Kubarev, V.V.; Kulipanov, G.N.; Kolobanov, E.I.; Matveenko, A.N.; Medvedev, L.E.; Ovchar, V.K.; Salikova, T.V.; Scheglov, M.A.; Serednyakov, S.S.; Vinokurov, N.A. Modulation Instability, Three Mode Regimes and Harmonic Generation at the Novosibirsk Terahertz Free Electron Laser. *Nucl. Instrum. Methods Phys. Res. Sect. A Accel. Spectrometers Detect. Assoc. Equip.* **2009**, *603*, 25–27. [CrossRef]

47. Gerasimov, V.V.; Kameshkov, O.E.; Nikitin, A.K.; Khasanov, I.S.; Lemzyakov, A.G.; Antonova, I.V.; Ivanov, A.I.; Lien, N.T.H.; Nghia, N.T.; Anh, L.T.; et al. First Experimental Demonstration of the Wide-Field Amplitude Surface Plasmon Resonance Microscopy in the Terahertz Range. *Photonics* **2023**, *10*, 723. [CrossRef]
48. Gerasimov, V.V.; Nikitin, A.K.; Vanda, V.S.; Lemzyakov, A.G.; Azarov, I.A. Dispersion of attenuation of surface plasmon polaritons on metal in the terahertz range. *J. Infrared Millim. Terahz. Waves* **2025**, *46*, 32. [CrossRef]
49. Wang, L.; Wang, Y.; Zong, G.; Hu, W.; Lu, Y. Liquid Crystal Based Tunable Terahertz Metadevices. *J. Mater.* **2025**, *11*, 100888. [CrossRef]
50. Li, J.; Chen, J.; Yan, D.; Fan, F.; Chen, K.; Zhong, K.; Wang, Y.; Tian, Z.; Xu, D. A Review: Active Tunable Terahertz Metamaterials. *Adv. Photonics Res.* **2024**, *5*, 2300351. [CrossRef]
51. Tumashov, M.A.; Baena, J.D.; Del Risco, J.P.; Lazorskiy, P.A.; Glybovski, S.B.; Kuznetsov, S.A. Angle-Dependent Metasurface for Nonspectroscopic THz Sensing of Submicrometer Films. *IEEE Sens. J.* **2023**, *23*, 27262–27272. [CrossRef]
52. Xie, Y.; Liu, X.; Zhou, J.; Zhang, H.; Lin, J.-Y.; Chen, W.; Zhu, L.-G.; Meng, K.; Liu, Q.H.; Zhu, J. Enhancing Trace Terahertz Fingerprint Sensing by the Lossy Silicon Metagrating with a Gold Mirror. *IEEE Trans. Microw. Theory Tech.* **2024**, *72*, 2368–2377. [CrossRef]
53. Zhu, W.M.; Liu, A.Q.; Zhang, X.M.; Tsai, D.P.; Bourouina, T.; Teng, J.H.; Zhang, X.H.; Guo, H.C.; Tanoto, H.; Mei, T.; et al. Switchable Magnetic Metamaterials Using Micromachining Processes. *Adv. Mater.* **2011**, *23*, 1792–1796. [CrossRef]
54. Pitchappa, P.; Manjappa, M.; Krishnamoorthy, H.N.S.; Chang, Y.; Lee, C.; Singh, R. Bidirectional Reconfiguration and Thermal Tuning of Microcantilever Metamaterial Device Operating from 77 K to 400 K. *Appl. Phys. Lett.* **2017**, *111*, 261101. [CrossRef]
55. Wang, L.; Ge, S.; Hu, W.; Nakajima, M.; Lu, Y. Graphene-Assisted High-Efficiency Liquid Crystal Tunable Terahertz Metamaterial Absorber. *Opt. Express* **2017**, *25*, 23873. [CrossRef]
56. Lewis, R.A. A Review of Terahertz Sources. *J. Phys. D Appl. Phys.* **2014**, *47*, 374001. [CrossRef]

photonics

MDPI

Communication

Beyond Green's Functions: Inverse Helmholtz and "Om" -Potential Methods for Macroscopic Electromagnetism in Isotropy-Broken Media

Maxim Durach

Center for Advanced Materials Science, Department of Biochemistry, Chemistry and Physics, Georgia Southern University, Statesboro, GA 30460, USA; mdurach@georgiasouthern.edu

Abstract: The applicability ranges of macroscopic and microscopic electromagnetism are contrasting. While microscopic electromagnetism deals with point sources, singular fields, and discrete atomistic materials, macroscopic electromagnetism concerns smooth average distributions of sources, fields, and homogenized effective metamaterials. Green's function method (GFM) involves finding fields of point sources and applying the superposition principle to find fields of distributed sources. When utilized to solve microscopic problems, GFM is well within the applicability range. Extension of GFM to simple macroscopic problems is convenient, but not fully logically sound, since point sources and singular fields are technically not a subject of macroscopic electromagnetism. This explains the difficulty of both finding the Green's functions and applying the superposition principle in complex isotropy-broken media, which are very different from microscopic environments. In this manuscript, we lay out a path to the solution of macroscopic Maxwell's equations for distributed sources, bypassing GFM by introducing an inverse approach and a method based on "Om" ॐ -potential, which we describe here. To the researchers of electromagnetism, this provides access to powerful analytical tools and a broad new space of solutions for Maxwell's equations.

Keywords: electromagnetism; photonics; metamaterials; optics; Green's function

1. Introduction

The main problem of electromagnetism is predicting the interaction between arbitrary charge distributions placed in arbitrary environments [1]. The path to solving this problem is most typically understood as finding fields of point sources. The fields of complex sources can then be obtained via the superposition principle. This approach, known as the Green's function method, is not only mathematically natural for singular microscopic fields but is grounded in the physics of elementary particles, which do not have dimensions according to relativistic considerations [2]. The ubiquitous position of Green's functions in microscopic electromagnetism is best expressed in the essay by Julian Schwinger titled "The Greening of Quantum Field Theory: George and I" [3]. The "Greening" of macroscopic electromagnetism is less obvious since photonics researchers do not deal with elementary particles or singular microscopic fields and do not claim the applicability of macroscopic photonics to elementary particles. Not unexpectedly, the extension of the microscopic Green's function approach to macroscopic electromagnetism faces difficulties due to fundamental differences between the corresponding sets of Maxwell's equations.

A lot of effort is invested in extending Green's function method to various macroscopic electromagnetic media; this cannot be considered very successful, however, at the rugged frontier of isotropy-broken media, due to the complexity and inherent non-locality of these media [4]. Although an integral representation of Green's function can be obtained for the most general case of isotropy-broken media [5], closed-form expressions of dyadic Green's functions are only available for a limited set of relatively simple isotropy-broken media [6,7]. Another complication arises from utilizing the superposition principle in the Green's function method, as finding fields created by non-point sources involves untenable integration, even in the depolarization dyadics approximation [7]. Finding fields of non-point sources in isotropic media, in many cases, relies on symmetries of those sources [1]. This can be extended to isotropy-broken media by means of spectral eigenfunction representations; however, this results in infinite series, requiring truncation [7,8].

In this manuscript, we introduce two approaches to directly obtain fields created by a very broad class of sources immersed in generic isotropy-broken media. First, we apply the inverse approach to the inhomogeneous Helmholtz equation for the vector potential to obtain sources that create desired vector potentials. In the second approach, we draw inspiration from the teachings of Hindu philosophy about the primordial fore-sound of the universe encompassing all creation. We introduce the "Om" ॐ -potential that underlies both sources and fields and provides for the direct method of evaluation of the solutions of macroscopic Maxwell's equations in isotropy-broken media. Please note that the introduction of auxiliary vector fields to aid the solution or analysis of Maxwell equations is not unprecedented in the history of science, as exemplified by the scalar potential, vector potential, Hertz potential [9], and Beltrami fields [10]. The power of our methods is demonstrated by the mappings we uncover between different sources that create identical potential across all materials and between field-source pairs that come from the same "Om" ॐ in materials as they transition between symmetries, topology classes, and so on.

It is important to clarify that the Inverse Helmholtz and Om-potential methods proposed here are fully analytical in nature. Their primary purpose is to reveal structural and symmetry-based relationships between macroscopic sources and fields, not to replace numerical solvers such as FEM or FDTD or fast methods for band field calculations [11,12]. Accordingly, no computational speed or convergence comparisons are made or implied in this work.

2. Helmholtz Equation and Green's Functions in Isotropy-Broken Media

Macroscopic fields satisfy Maxwell's equations

$$\nabla \times \boldsymbol{H} - \frac{1}{c}\frac{\partial \boldsymbol{D}}{\partial t} = \frac{4\pi}{c}\boldsymbol{j}, \nabla \times \boldsymbol{E} + \frac{1}{c}\frac{\partial \boldsymbol{B}}{\partial t} = 0 \quad (1)$$

The constitutive relations are generally expressed as

$$\begin{pmatrix} \boldsymbol{D} \\ \boldsymbol{B} \end{pmatrix} = \hat{M}\begin{pmatrix} \boldsymbol{E} \\ \boldsymbol{H} \end{pmatrix} = \begin{pmatrix} \hat{\epsilon} & \hat{X} \\ \hat{Y} & \hat{\mu} \end{pmatrix}\begin{pmatrix} \boldsymbol{E} \\ \boldsymbol{H} \end{pmatrix} \text{or} \begin{pmatrix} \boldsymbol{D} \\ \boldsymbol{H} \end{pmatrix} = \begin{pmatrix} \hat{C}_{DE} & \hat{C}_{DB} \\ \hat{C}_{HE} & \hat{C}_{HB} \end{pmatrix}\begin{pmatrix} \boldsymbol{E} \\ \boldsymbol{B} \end{pmatrix} = \begin{pmatrix} (\hat{\epsilon} - \hat{X}\hat{\mu}^{-1}\hat{Y}) & \hat{X}\hat{\mu}^{-1} \\ -\hat{\mu}^{-1}\hat{Y} & \hat{\mu}^{-1} \end{pmatrix}\begin{pmatrix} \boldsymbol{E} \\ \boldsymbol{B} \end{pmatrix} \quad (2)$$

Under the Weyl gauge, the relationship between fields and the vector potential reduces to $\boldsymbol{B} = \nabla \times \boldsymbol{A}$ and $\boldsymbol{E} = -\frac{1}{c}\frac{\partial \boldsymbol{A}}{\partial t}$. Combining Maxwell's equations and constitutive relations, we obtain the wave equation for the vector potential in isotropy-broken media

$$\left(\frac{1}{c}\frac{\partial}{\partial t}, -\nabla \times \hat{I}\right)\begin{pmatrix} \hat{C}_{DE} & \hat{C}_{DB} \\ \hat{C}_{HE} & \hat{C}_{HB} \end{pmatrix}\begin{pmatrix} -\frac{1}{c}\frac{\partial}{\partial t} \\ \nabla \times \hat{I} \end{pmatrix}\boldsymbol{A} = \hat{L}\left(\nabla, \frac{1}{c}\frac{\partial}{\partial t}\right)\boldsymbol{A} = -\frac{4\pi}{c}\boldsymbol{j} \quad (3)$$

or

$$-\hat{C}_{DE}\frac{1}{c^2}\frac{\partial^2 A}{\partial t^2}+\hat{C}_{DB}\nabla\times\frac{1}{c}\frac{\partial A}{\partial t}+\nabla\times\hat{C}_{HE}\frac{1}{c}\frac{\partial A}{\partial t}-\nabla\times\hat{C}_{HB}\nabla\times A=-\frac{4\pi}{c}j \quad (4)$$

More conventionally it is expressed as

$$-\left(\nabla\times\hat{I}-\frac{1}{c}\frac{\partial}{\partial t}\hat{X}\right)\hat{\mu}^{-1}\left(\nabla\times\hat{I}+\hat{Y}\frac{1}{c}\frac{\partial}{\partial t}\right)A-\hat{\epsilon}\frac{1}{c^2}\frac{\partial^2 A}{\partial t^2}=-\frac{4\pi}{c}j \quad (5)$$

Transforming into the Fourier domain as $\hat{L}\left(\nabla,\frac{1}{c}\frac{\partial}{\partial t}\right)\to\hat{L}(ik,-ik_0)$ results in the Helmholtz operator for isotropy-broken media

$$\hat{L}(ik,-ik_0)=(k_0,k\times\hat{I})\begin{pmatrix}\hat{C}_{DE} & \hat{C}_{DB}\\ \hat{C}_{HE} & \hat{C}_{HB}\end{pmatrix}\begin{pmatrix}k_0\\ k\times\hat{I}\end{pmatrix}=(k\times\hat{I}+k_0\hat{X})\hat{\mu}^{-1}(k\times\hat{I}-k_0\hat{Y})+k_0^2\hat{\epsilon} \quad (6)$$

The vector potential can be now expressed as $A(k,k_0)=-\frac{4\pi}{c}\frac{\text{adj }\hat{L}}{|\hat{L}|}j(k,k_0)$ and

$$A(r,k_0)=-\frac{4\pi}{c}\int\frac{d^3k}{(2\pi)^3}\frac{\text{adj }\hat{L}}{|\hat{L}|}j(k,k_0)\exp(ikr), \quad (7)$$

where the $|\hat{L}|=\frac{1}{k_0^2}\sum_{i+j+l+m=4}\left[\alpha_{ijlm}k_x^ik_y^jk_z^lk_0^m\right]$ is the determinant of the operator Equation (4), with α_{ijlm} being the Tamm–Rubilar tensor [13,14], and the corresponding adjoint operator is $\text{adj}\hat{L}=\frac{1}{k_0^4}\sum_{i+j+l+m=4}\left[\hat{\beta}_{ijlm}k_x^ik_y^jk_z^lk_0^m\right]$, where $\hat{\beta}_{ijlm}$ are 3×3 matrices.

The usual approach is to use the *superposition principle* to express Equation (5) in terms of dyadic Green's function $\hat{G}(r,r')$

$$A(r)=-\frac{4\pi}{c}\int dr'\,\hat{G}\left(r,r'\right)j\left(r'\right) \quad (8)$$

Correspondingly, from Equations (7) and (8), the dyadic Green's function can be expressed as [5,7]

$$\hat{G}\left(r,r'\right)=\int\frac{d^3k}{(2\pi)^3}\frac{\text{adj }\hat{L}}{|\hat{L}|}\exp\left(ik\left(r-r'\right)\right), \quad (9)$$

The problem of finding fields created by arbitrary sources is thus reduced to finding the Green's function, which is a response to *a delta-functional point source* and always has a singular part [7].

$$\hat{L}(\nabla,-ik_0)\,\hat{G}\left(r,r'\right)=\hat{I}\,\delta\left(r-r'\right) \quad (10)$$

As described in the introduction, in general, both finding the Green's function from Equations (9) and (10) and utilizing the superposition principle, Equation (8), are challenges in macroscopic electromagnetism and are unnatural due to the limited validity of point sources and singular fields in macroscopic environments.

3. The "Om" ॐ – Potential

To bypass the complications related to GFM we introduce differential operators based on Fourier space operators $|\hat{L}|$ and adj $\hat{L}$

$$D(\partial_x,\partial_y,\partial_z)=\frac{1}{k_0^2}\sum_{i+j+l+m=4}(-i)^{i+j+l}\left[\alpha_{ijlm}\partial_x^i\partial_y^j\partial_z^lk_0^m\right] \quad (11)$$

$$\hat{U}(\partial_x,\partial_y,\partial_z)=\frac{1}{k_0^4}\sum_{i+j+l+m=4}(-i)^{i+j+l}\left[\hat{\beta}_{ijlm}\partial_x^i\partial_y^j\partial_z^lk_0^m\right] \quad (12)$$

Note that differential operators D and $\hat{U}$ have constant coefficients in homogeneous media and, therefore, commute. We recast Equation (7) as

$$D\left(\partial_x, \partial_y, \partial_z\right) A(\boldsymbol{r}) = -\frac{4\pi}{c} \hat{U}\left(\partial_x, \partial_y, \partial_z\right) \boldsymbol{j}(\boldsymbol{r}) \tag{13}$$

Instead of representing the source $\boldsymbol{j}(\boldsymbol{r})$ as a superposition of point charges, as is done in the Green's function method, we express the source via the underlying "Om" ॐ – potential vector field

$$\boldsymbol{j}(\boldsymbol{r}) = D\left(\partial_x, \partial_y, \partial_z\right) \text{ॐ}(\boldsymbol{r}), \tag{14}$$

From Equation (13) an expression for the vector potential corresponding to source current Equation (14) can be obtained as

$$\boldsymbol{A}(\boldsymbol{r}) = -\frac{4\pi}{c} \hat{U}\left(\partial_x, \partial_y, \partial_z\right) \text{ॐ}(\boldsymbol{r}) \tag{15}$$

where the Devanagari script "Om" ॐ $(\boldsymbol{r})$ is a vector field, which underlies both the source $\boldsymbol{j}$ in Equation (14) and the vector potential $\boldsymbol{A}$ in Equation (15) in a unified paradigm of Equations (13)–(15).

The Om-potential we introduce in this work can be interpreted as a vector field that underlies both the sources and the fields [see Equations (14)–(15)], which are observables in macroscopic electromagnetism. This distantly mirrors the role of the wavefunction in quantum mechanics, which generates all observables via operator action, or the role of partition functionals in statistical mechanics and quantum field theory, from which correlation functions are derived. Like the Hertz vector potential, the Om-potential is not directly observable but simplifies the analytical framework and reveals invariant structures that may otherwise remain hidden. The Om-potential formalism, therefore, acts as a unifying substrate across varying isotropy-broken electromagnetic media, organizing both the geometric and material complexity of Maxwell's theory into a single functional framework.

Note that for an arbitrary source, the underlying "Om" vector field ॐ $(\boldsymbol{r})$ can be found as

$$\text{ॐ}(\boldsymbol{r}) = \int d\boldsymbol{r}'\, g_{\text{ॐ}}\left(\boldsymbol{r}, \boldsymbol{r}'\right) \boldsymbol{j}\left(\boldsymbol{r}'\right),$$

where the scalar "Om" Green's function is $g_{\text{ॐ}}\left(\boldsymbol{r}, \boldsymbol{r}'\right)$

$$g_{\text{ॐ}}\left(\boldsymbol{r}, \boldsymbol{r}'\right) = \int \frac{d^3k}{(2\pi)^3} \frac{1}{|\hat{L}|} \exp\left(i\boldsymbol{k}\left(\boldsymbol{r} - \boldsymbol{r}'\right)\right)$$

It is important to clarify that using $g_{\text{ॐ}}\left(\boldsymbol{r}, \boldsymbol{r}'\right)$ is challenging for complex media, as described in the introduction, and we do not use it in any of the methods we propose here.

The summary of relationships between the "Om" ॐ -potential, vector potential $\boldsymbol{A}$, and sources $\boldsymbol{j}$ is shown in Figure 1.

To improve readability, we list major quantities discussed in this paper into Table 1.

In vacuum, the operator $\hat{L}$ has the following properties:

$$\hat{L}(i\boldsymbol{k}, -ik_0) = \left(k_0, \boldsymbol{k} \times \hat{I}\right) \begin{pmatrix} k_0 \\ \boldsymbol{k} \times \hat{I} \end{pmatrix} = \left(\boldsymbol{k} \times \hat{I}\right)\left(\boldsymbol{k} \times \hat{I}\right) + k_0^2 \hat{I} = \boldsymbol{k}\boldsymbol{k} + \left(k_0^2 - k^2\right)\hat{I} \tag{16}$$

$$|\hat{L}| = k_0^2 \left(k^2 - k_0^2\right)^2, \ \mathrm{adj}\, \hat{L} = \left(k^2 - k_0^2\right)\left(\boldsymbol{k}\boldsymbol{k} + k_0^2 \hat{I}\right)$$

This means that for a vacuum, Equation (13) can be rewritten as

$$\left(\nabla^2 + k_0^2\right)^2 \boldsymbol{A}(\boldsymbol{r}) = \frac{4\pi}{c}\left(\nabla^2 + k_0^2\right)\left(\hat{I} - \frac{1}{k_0^2}\nabla\nabla\right)\boldsymbol{j}(\boldsymbol{r})$$

$$D_{vac}(\partial_x, \partial_y, \partial_z) = k_0^2\left(\nabla^2 + k_0^2\right)^2,\ \hat{U}_{vac}(\partial_x, \partial_y, \partial_z) = \left(\nabla^2 + k_0^2\right)\left(k_0^2\hat{I} - \nabla\nabla\right)$$

For a point source at $\boldsymbol{r}_0$ polarized in direction $\hat{\boldsymbol{e}}$ in vacuum $\boldsymbol{j} = \hat{\boldsymbol{e}}\delta(\boldsymbol{r} - \boldsymbol{r}_0)$, the "Om" vector field ॐ $(\boldsymbol{r})$ is a spherical wave propagating from the source location

$$\text{ॐ}_{vac-point}(\boldsymbol{r}) = \left(\nabla^2 + k_0^2\right)^{-1} \hat{\boldsymbol{e}}\frac{e^{ik_0|\boldsymbol{r}-\boldsymbol{r}_0|}}{4\pi|\boldsymbol{r} - \boldsymbol{r}_0|} = \hat{\boldsymbol{e}}\, e^{ik_0|\boldsymbol{r}-\boldsymbol{r}_0|}$$

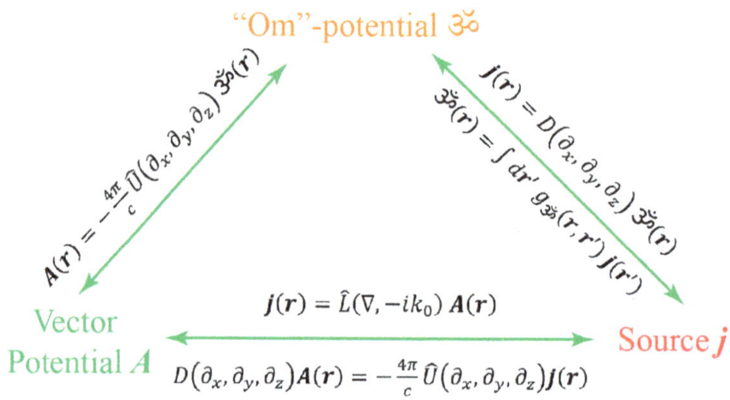

Figure 1. The schematic of the relations between the sources $\boldsymbol{j}(\boldsymbol{r})$, vector potentials $\boldsymbol{A}(\boldsymbol{r})$, and the "Om" ॐ -potential introduced in this manuscript.

Table 1. Symbols and notation used in the manuscript.

Symbol	Description
$\boldsymbol{A}(\boldsymbol{r})$	Vector potential
$\boldsymbol{j}(\boldsymbol{r})$	Source current density
ॐ $(\boldsymbol{r})$	Om-potential (underlying vector field generating both $\boldsymbol{A}(\boldsymbol{r})$ and $\boldsymbol{j}(\boldsymbol{r})$)
$\hat{L}$	Helmholtz operator for isotropy − broken media : $\hat{L}\boldsymbol{A} = -\frac{4\pi}{c}\boldsymbol{j}$
D	Differential operator used to generate $\boldsymbol{j}(\boldsymbol{r})$from ॐ $(\boldsymbol{r})$; its Fourier transform is $\det\hat{L}$
$\hat{U}$	Differential operator used to generate $\boldsymbol{A}(\boldsymbol{r})$from ॐ $(\boldsymbol{r})$; its Fourier transform is $\text{adj}\hat{L}$

4. Inverse Helmholtz Equation Method

The first method to find solutions of Equation (8) relies on inverse approach to the Helmholtz equation

$$\boldsymbol{j}(\boldsymbol{r}) = \hat{L}(\nabla, -ik_0)\, \boldsymbol{A}(\boldsymbol{r}), \tag{17}$$

where instead of looking for vector potential $\boldsymbol{A}(\boldsymbol{r})$ for a given source $\boldsymbol{j}(\boldsymbol{r})$, we set the vector potential $\boldsymbol{A}(\boldsymbol{r})$ and obtain sources $\boldsymbol{j}(\boldsymbol{r})$, which create the desired vector potential.

To proceed, we utilize the Hermite functions $\phi_{n_x n_y n_z}(\boldsymbol{r}) = \psi_{n_x}(x)\psi_{n_y}(y)\psi_{n_z}(z)$, which are eigenfunctions of the quantum harmonic oscillator $\psi_{n_x}(x) = \left(2^{n_x} n_x! \sqrt{\pi} w_x\right)^{-\frac{1}{2}} e^{-\frac{x^2}{2w_x^2}} H_n(x/w_x)$.

The del operator applied to Hermite functions is

$$\nabla\psi_{n_x n_y n_z}(\boldsymbol{r}) = \frac{1}{\sqrt{2}}\left\{\left(\hat{a}_x - \hat{a}_x^+\right)/w_x, \left(\hat{a}_y - \hat{a}_y^+\right)/w_y, \left(\hat{a}_z - \hat{a}_z^+\right)/w_z\right\}\psi_{n_x n_y n_z}(\boldsymbol{r})$$

where the ladder operators $\hat{a}_x\psi_n = \sqrt{n+1}\psi_{n+1}, \hat{a}_x^+\psi_n = \sqrt{n}\psi_{n-1}$.

For a vector potential in vacuum polarized in direction $\hat{\boldsymbol{x}}$ and given by $\boldsymbol{A}(\boldsymbol{r}) = \hat{\boldsymbol{x}}\psi_{000}(\boldsymbol{r})$, the source can be found as

$$\begin{aligned} \boldsymbol{j}(\boldsymbol{r}) &= \hat{L}(\nabla, -ik_0)\boldsymbol{A}(\boldsymbol{r}) = \left\{\left(k_0^2 + \nabla^2\right)\hat{I} - \nabla\nabla\right\}\boldsymbol{A}(\boldsymbol{r}) \\ &= \frac{1}{\sqrt{2}}\frac{1}{w^2}\left\{\sqrt{2}\left(k_0^2 w^2 - 1\right)\psi_{000} + \psi_{020} + \psi_{002}, -\psi_{110}, -\psi_{101}\right\} \end{aligned} \tag{18}$$

If the vector potential $\boldsymbol{A}(\boldsymbol{r})$ is fixed, the only material-dependent factor in the RHS of Equation (17) is the operator $\hat{L}(\nabla, -ik_0)$. This allows us to create a cross-material mapping between sources $\boldsymbol{j}(\boldsymbol{r})$, which create the same vector potential in different media. To demonstrate this, we consider a material with $\hat{M}_\kappa = (1-\kappa)\hat{1} + \kappa\hat{M}_1$, where

$$\hat{M}_1 = \begin{pmatrix} 0.63 & 0.11 & 0.59 & 0.82i & -0.8i & -0.56i \\ 0.11 & -0.078 & -0.37 & 0.81i & -0.65i & 0.87i \\ 0.59 & -0.37 & -0.021 & 0.79i & 0.49i & 0.68i \\ -0.82i & -0.81i & -0.79i & 0.27 & 0.046 & -0.98 \\ 0.8i & 0.65i & -0.49i & 0.046 & 0.22 & -0.73 \\ 0.56i & -0.87i & -0.68i & -0.98 & -0.73 & -0.58 \end{pmatrix}.$$

Note that the matrix $\hat{M}_1$ is also color-coded in Figure 2e. As κ changes from 0 to 1, the material undergoes several topological transitions (see Ref. [15]) from non-hyperbolic to mono-hyperbolic [Figure 2a], to bi-hyperbolic [Figure 2b], to tri-hyperbolic [Figure 2c], to tetra-hyperbolic [Figure 2d].

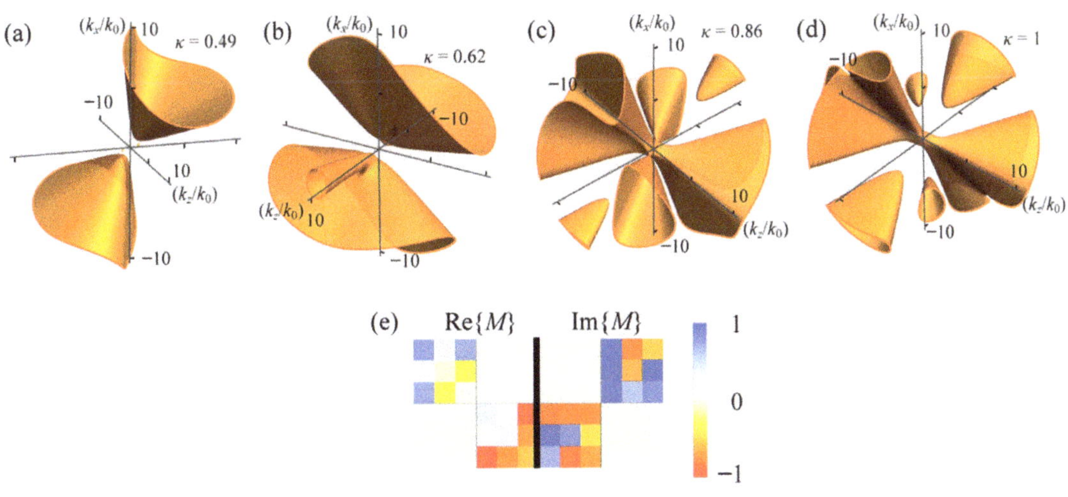

Figure 2. **(a–d)** Topological transitions of material $\hat{M}_\kappa = (1-\kappa)\hat{1} + \kappa\hat{M}$ for different κ as indicated in the panels. **(e)** Color-coded visualization for matrix $\hat{M}_1$.

In Figure 3, we show how the source $\boldsymbol{j}(\boldsymbol{r})$ required to create the vector potential $\boldsymbol{A}(\boldsymbol{r}) = \hat{\boldsymbol{x}}\psi_{000}(\boldsymbol{r})$ changes as κ changes. The leftmost panels of Figure 3 correspond to vacuum $\kappa = 0$ and follow Equation (18). One can see that the source distribution is deformed and rotated as the materials pass through the topological transitions shown in Figure 2.

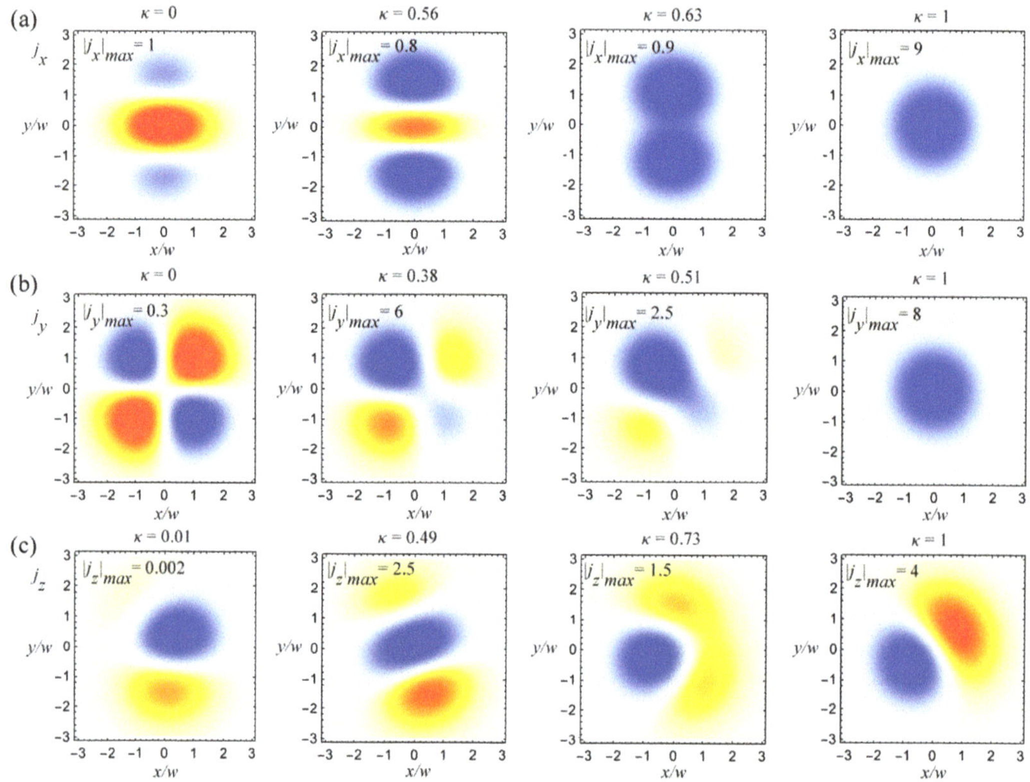

Figure 3. The x–y plane cross-section of the sources $j(r)$ needed to create potential $A(r) = \hat{x}\psi_{000}(r)$ in different materials $\hat{M}_\kappa$. Panel (**a**) shows the x-component j_x; (**b**) j_y; (**c**) j_z. Blue (red) corresponds to negative (positive) values, with maximum magnitude indicated on the panels.

5. The "Om" ॐ – Potential Method

The second method to find solutions to Equation (8) is to use Equations (14) and (15). We select the "Om" ॐ (r)-potential and find the corresponding source $j(r)$ and vector potential $A(r)$. If the "Om" ॐ (r)-potential is fixed, the only material-dependent factors in the RHS of Equations (14) and (15) are operators $D(\partial_x, \partial_y, \partial_z)$ and $\hat{U}(\partial_x, \partial_y, \partial_z)$. This creates cross-material mapping between the source–vector potential pairs $j(r)$ and $A(r)$, which corresponds to the same "Om" ॐ (r)-potential as the material is modified.

In Figure 4, we show the x-y cross-sectional distributions of the sources $j(r)$ and vector potentials $A(r)$ corresponding to ॐ $(r) = \hat{x}\psi_{000}(r)$ for different materials $\hat{M}_\kappa$. We see drastic modifications of both $j(r)$ and $A(r)$, which undergo both deformation and rotation. Interestingly, the rate of change in x-y cross-sections of $j(r)$ and $A(r)$ is not the same as κ is changed. For $\kappa = 0.63 - 0.69$, when the material is in the topological transition into the bi-hyperbolic phase, the source $j(r)$ is strongly modified, as can be seen from the leftmost panels in Figure 4b–d. At the same time, the vector potential $A(r)$ has minimal changes in the same range of κ.

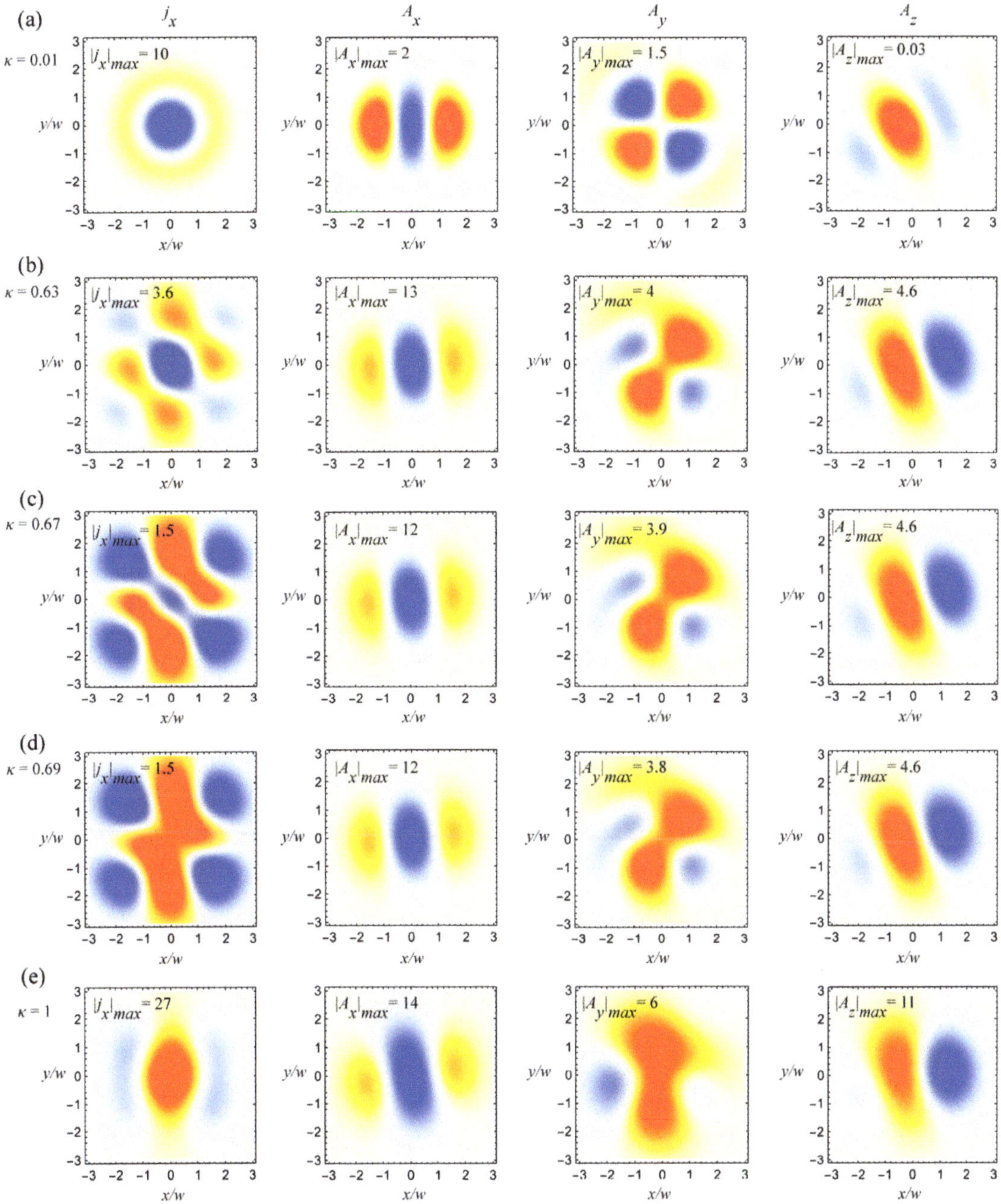

Figure 4. The x-y plane cross-sections of the x-component of the sources j_x (leftmost panels) and the components of the vector potential $\boldsymbol{A}(\boldsymbol{r})$ (three rightmost panels) for the "Om"-potential given by ॐ$(\boldsymbol{r}) = \hat{x}\psi_{000}(\boldsymbol{r})$ for different materials $\hat{M}_\kappa$. In panel (**a**) $\kappa = 0.01$; (**b**) $\kappa = 0.63$; (**c**) $\kappa = 0.67$; (**d**) $\kappa = 0.69$; (**e**) $\kappa = 1$. Blue (red) corresponds to negative (positive) values, with maximum magnitude indicated on the panels.

6. Discussion

We would like to summarize the three methods discussed in this manuscript using Table 2 below.

Table 2. The three approaches to solving macroscopic Maxwell's equations.

Method	Uses Point Sources	Suitable for Isotropy-Broken Media	Structure	Notes
Green's Function Method (GFM)	Yes	Limited	Obtain fields from sources via an integral with singularities/No closed form	Natural in microscopic EM; struggles in complex macroscopic media
Inverse Helmholtz Method	No	Yes	Obtain sources from fields via analytical differentiation/Closed form	Solves for the source from the desired vector potential. Applicable to complex macroscopic media
Om-Potential Method	No	Yes	Obtain source-field pairs from Om-potential via analytical differentiation/Closed form	Introduces unified Om-potential for source and potential. Applicable to complex macroscopic media

While the operators D and $\hat{U}$ introduced in this paper contain high-order derivatives, no numerical discretization is performed in this work. If implemented numerically in future studies, we note that staggered-grid finite-difference schemes and spectral methods have been used successfully for similar operators, with proper stability constraints [16,17].

In conclusion, the Green's function method with point sources and singular fields is inherent to microscopic electromagnetism and fundamentally stems from the properties of dimensionless elementary particles. Extension of GFM to macroscopic electromagnetism faces obvious and fundamental challenges. In this manuscript, we demonstrate that solutions to problems of macroscopic electromagnetism can be found without the use of Green's functions by introducing two new approaches: the Inverse Helmholtz equation method and the "Om" ॐ -potential method. Although this work is restricted to the frequency domain, the analytical methods may be extended to time-dependent and dispersive media (see [18] as an example). These extensions are left for future investigation.

Funding: This research received no external funding. The APC was waived by MDPI. The authors appreciate the support of the Georgia Southern University in all other respects.

Institutional Review Board Statement: Not applicable.

Informed Consent Statement: Not applicable.

Data Availability Statement: All data included in this text.

Conflicts of Interest: The authors declare no conflict of interest.

References

1. Griffiths, D.J. *Introduction to Electrodynamics*; Cambridge University Press: Cambridge, UK, 2023.
2. Landau, L.D.; Lifshitz, E.M. *The Classical Theory of Fields*; Nauka: Moscow, Russia, 1988; Volume 2.
3. Schwinger, J. The greening of quantum field theory: George and I. In *Julian Schwinger: The Physicist, the Teacher, and the Man*; World Scientific: Toh Tuck Link, Singapore, 1996; pp. 13–27.
4. Silveirinha, M.G. Nonlocal homogenization theory of structured materials. In *Theory and Phenomena of Metamaterials*; Capolino, F., Ed.; CRC Press: Boca Raton, FL, USA, 2009.
5. Kong, J.A. Theorems of bianisotropic media. *Proc. IEEE* **1972**, *60*, 1036. [CrossRef]
6. Chen, H.C. *Theory of Electromagnetic Waves: A Coordinate-Free Approach*; McGraw Hill: New York, NY, USA, 1983.
7. Faryad, M.; Lakhtakia, A. *Infinite-Space Dyadic Green Functions in Electromagnetism*; Morgan & Claypool Publishers: San Rafael, CA, USA, 2018.
8. Durach, M. Electromagnetic Scattering by Bianisotropic Spheres. *Appl. Sci.* **2023**, *13*, 5169. [CrossRef]

9. Born, M.; Wolf, E. *Principles of Optics: Electromagnetic Theory of Propagation, Interference and Diffraction of Light*; Elsevier: Amsterdam, The Netherlands, 2013.
10. Lakhtakia, A. *Beltrami Fields in Chiral Media*; World Scientific: Toh Tuck Link, Singapore, 1994.
11. Tsang, L.; Liao, T.H.; Tan, S.; Xu, X.; Bai, X.; Gao, R. Fast Calculations of Vector Electromagnetics in 3D Periodic Structures Based on Multiple Scattering Theory and Broadband Green's Function. *Prog. Electromagn. Res.* **2024**, *179*, 19–36. [CrossRef]
12. Zheng, H.; Bai, X.; Tan, S.; Tsang, L. An Efficient Hybrid Numerical T-Matrix Approach for 3D Multiple Scattering Analysis. *Prog. Electromagn. Res.* **2024**, *181*, 61–72. [CrossRef]
13. Rubilar, G.F. Linear pre-metric electrodynamics and deduction of the light cone. *Ann. Phys.* **2002**, *514*, 717. [CrossRef]
14. Hehl, F.W.; Obukhov, Y.N. *Foundations of Classical Electrodynamics-Charge, Flux, and Metric*; Birkhauser: Boston, MA, USA, 2003.
15. Durach, M.; Williamson, R.F.; Laballe, M.; Mulkey, T. Tri-and tetrahyperbolic isofrequency topologies complete classification of bianisotropic materials. *Appl. Sci.* **2020**, *10*, 763. [CrossRef]
16. Kreiss, H.-O.; Ortiz, O.E. Some Mathematical and Numerical Questions Connected with First and Second Order Time-Dependent Systems of Partial Differential Equations. In *The Conformal Structure of Space-Time*; Lecture Notes in Computational Science and Engineering 31; Springer: Berlin/Heidelberg, Germany, 2002.
17. Boyd, J.P. *Chebyshev and Fourier Spectral Methods*, 2nd ed.; Dover Publications: Mineola, NY, USA, 2001.
18. Besieris, I.M. (3+1)-Dimensional Nonparaxial Spatiotemporally Localized Waves in Transparent Dispersive Media. *Prog. Electromagn. Res.* **2024**, *181*, 43–59. [CrossRef]

photonics

Article

Flexible Color Filter Using Lithium Niobate Metamaterial with Ultrahigh Purity and Brightness Characteristics

Siqiang Zhao , Daoye Zheng , Yunche Zhu, Shuyan Zou and Yu-Sheng Lin *

Pittsburgh Institute, Sichuan University, Chengdu 610207, China; 2023141520257@stu.scu.edu.cn (S.Z.); daoye.zheng@scupi.cn (D.Z.); yunche.zhu@scupi.cn (Y.Z.); 2023141520303@stu.scu.edu.cn (S.Z.)
* Correspondence: yusheng.lin@scupi.cn

Abstract: We propose a simulation-based design for a flexible color filter (FCF) using a lithium niobate metamaterial (LNM) to investigate its color filtering potential. The FCF is composed of three periodically arranged half-ellipse LN arrays on a polydimethylsiloxane (PDMS) substrate, denoted as LNM-1, LNM-2, and LNM-3. The electromagnetic responses of the FCF can be controlled by adjusting the periods of the LNMs. Our simulations predict high-quality (Q) factors in transmission spectra, ranging from 100 to 200 for LNM-1, 290 to 360 for LNM-2, and 140 to 300 for LNM-3. When the FCF is exposed to the surrounding environments with different refractive indexes, it exhibits a theoretical figure of merit (FOM) up to 900 RIU^{-1} and a sensitivity reaching 130 nm/RIU. The electromagnetic field distributions reveal strong confinement within the LNM nanostructures, confirming an efficient light–matter interaction. These results indicate that the proposed LNM-based FCF presents a promising design concept for high-performance color sensing and filtering applications.

Keywords: metamaterial; metasurface; color filter; color sensor; refractive index sensor

1. Introduction

Color filtering technologies have been fundamental to modern optical systems, from consumer electronics displays to advanced sensing applications [1]. The conventional filters leverage the selective absorption properties of organic molecules to achieve color separation, making them particularly attractive for mass production [2–10]. However, despite their commercial success, these traditional filters face several significant limitations that increasingly constrain their applications in emerging technologies [11]. Additionally, these conventional filters may not be tunable and require more arrays to realize all-optical realization. [12]. The integration capabilities also present substantial challenges in modern miniaturized devices, where their relatively thick functional layers often impose constraints on device miniaturization [13–15].

Currently, metamaterial-based color filtering technologies have been proposed as a promising alternative to address these fundamental limitations. These artificially engineered structures, typically consisting of subwavelength periodic arrays, offer control over light–matter interactions through careful geometric design rather than material properties alone [16–22]. Metamaterial color filters have demonstrated remarkable optical characteristics, including extraordinarily narrow spectral bandwidths, high transmission efficiency, and precise wavelength selectivity [23,24]. The ability to precisely control the optical response through nanoscale geometric parameters also offers unprecedented flexibility in design optimization [25].

In metallic metamaterials, surface plasmon polaritons originate from collective oscillation of free electrons driven by external optical fields. The optical field can be strongly confined to the metal–dielectric interface, generating specific reflection and transmission properties. However, metallic structures suffer from intrinsic ohmic losses that limit their Q-factors and efficiency [26]. All-dielectric metamaterials based on Mie scattering theory have been proposed to address these limitations [27]. The interaction between electromagnetic waves and displacement currents can stimulate both electric (E) and magnetic (H) dipole resonances. According to physical mechanisms such as surface plasmon and Mie scattering theories, metamaterials can control the frequency, amplitude, phase, and polarization from microwave to visible light through geometric designs [28–30]. By optimizing the geometric parameters of metamaterials, narrow-band color filtering with high transmission efficiency can be achieved [31].

Among various dielectric materials, we selected lithium niobate (LN) for this study due to a combination of favorable properties. While materials like silicon (Si) and titanium dioxide (TiO_2) offer higher refractive indices, LN possesses excellent transparency with negligible material absorption across the entire visible spectrum, which is critical for achieving high-Q resonances in a transmissive or reflective color filter. Furthermore, LN is chemically stable and benefits from mature, well-established fabrication techniques that allow us to produce high-quality, single-crystal thin films with low defect densities. Although this work does not explicitly utilize LN's strong electro-optic or nonlinear coefficients, its selection provides a clear and promising pathway for future work on actively tunable devices, positioning this passive filter design as a foundational platform for more advanced functionalities.

In this study, we present a flexible color filter (FCF) based on a lithium niobate metamaterial (LNM). An FCF is composed of three periodically half-ellipse LN arrays on a polydimethylsiloxane (PDMS) substrate. They are denoted as LNM-1, LNM-2, and LNM-3, respectively. The optical properties of the FCF can be precisely controlled by adjusting the periods of three LNMs along the y-axis direction. Three LNMs show the blue (B), green (G), and red (R) colors for LNM-1, LNM-2, and LNM-3, respectively. Such LNM-based FCFs reveal high quality (Q) factors, ultrahigh purity, and brightness characteristics. When an LNM-based FCF is exposed to different ambient environments, it exhibits ultrahigh and stable sensitivity in the background with the refractive index range from 1.0 to 1.4 [32–36]. This refractive index range is strategically chosen for its direct relevance to diverse real-world conditions anticipated for the FCF's key applications. Specifically, values from 1.0 (approximating air or vacuum, a common reference for sensing in gaseous media) up to 1.4 cover a broad spectrum of environments. This includes crucial aqueous media, where water has a refractive index of approximately 1.33, which are fundamental to biomedical sensing of biological samples and environmental monitoring of substances in liquids. Investigating the device's performance across this 1.0 to 1.4 span is therefore critical for validating its capability to function effectively and reliably in practical applications [37–40]. This design not only demonstrates excellent optical performance but also offers a straightforward tuning mechanism to adjust the RGB colors. These outstanding optical characteristics make the proposed LNM-based FCF particularly promising for various practical applications. In biomedical sensing, the high Q-factor and sensitivity enable real-time monitoring of subtle changes in biological samples, potentially revolutionizing disease diagnosis and drug development processes. The capability to operate across the entire visible spectrum makes it ideal for advanced spectroscopy applications, where precise wavelength selectivity is crucial for chemical and molecular analyses. In environmental monitoring, the device with a high figure of merit (FOM) allows for accurate detection of pollutants and contaminants in both aqueous and gaseous media. The narrow bandwidth of the spectrum and the tunable mechanism also make it suitable for display technologies, offering improved color purity and contrast ratio compared to conventional filtering methods.

2. Design and Method

The proposed LNM-based FCF is on a PDMS substrate, where the electromagnetic properties can be controlled through an electromechanical mechanism to perform optimization by stretching the PDMS substrate. The fundamental principle underlying the anticipated structural resilience lies in the strategic material distribution and geometric configuration of the composite system. During mechanical stretching operations, it is hypothesized that mechanical deformations would be primarily accommodated by the highly flexible PDMS material in the regions between the stiffer LN nanostructures. This selective deformation mechanism would create a protective environment where the high-modulus LN nanostructures remain largely undisturbed while the surrounding elastomeric matrix absorbs and distributes the applied mechanical stress [19].

The half-ellipse geometry of the LN nanostructures is expected to play a crucial role in this stress distribution paradigm. Unlike continuous thin films that would experience uniform strain distribution, the discrete nature of these nanostructures would create localized stress concentration points that are effectively isolated from one another by the intervening PDMS regions [41–44]. This segmentation could prevent the propagation of mechanical failures across the entire LN array, thereby enhancing the system's overall reliability [31]. While a detailed mechanical simulation of the stress–strain behavior is a subject for future investigation, this qualitative analysis suggests that the design is well-suited for mechanical flexibility, which is essential for maintaining consistent optical performance.

The periodically half-ellipse LN arrays are strategically designed to enhance light–matter interactions and modify the local electromagnetic field distributions within the LNM nanostructures. The optical properties of the LNM-based FCF are simulated using the Ansys Lumerical simulation software (2020 R2, for x64 disk), a commercial-grade finite-different time-domain (FDTD) solver. The simulation environment is configured with a three-dimensional computational domain, where periodic boundary conditions are implemented along the x- and y-axis directions, while perfectly matched layer (PML) boundary conditions are applied in the z-axis direction to eliminate unwanted reflections. The mesh resolution is optimized through convergence testing, with a minimum mesh size of $\lambda/20$ in all directions to ensure computational accuracy. A frequency-domain resolution of 0.1 nm is set to accurately resolve transmission and reflection spectra.

It is important to acknowledge that these simulations assume perfect, idealized geometries and materials. In a practical fabrication scenario, deviations from the designed parameters are inevitable. For instance, variations of $\pm$5–10 nm in the nanostructure dimensions, roughness at the edges of the slots, or non-uniformity in the PDMS substrate thickness could influence the device's performance. Given the high-Q nature of the supported resonances, such imperfections would likely lead to resonance broadening (a decrease in the Q-factor) and a potential reduction in peak efficiency. A comprehensive tolerance analysis would be required in the experimental phase to quantify these effects, but the robustness of the underlying Mie and bound states in the continuum (BIC) resonances suggests that the fundamental filtering characteristics would be preserved.

Figure 1a shows the proposed LNM-based FCF composed of three periodically half-ellipse LN arrays on a PDMS substrate. The periods of LNMs are w_1, w_2, and w_3 for LNM-1, LNM-2, and LNM-3, respectively. The incident wave is a transverse magnetic (TM) polarization state. Figure 1b–d are the detailed geometric parameters of the three LNMs, and each is responsible for one primary color. The horizontal period along the x-direction between LNM units is 80 nm for both LNM-1 and LNM-2 and 100 nm for LNM-3, and the thickness for all three LNMs is 200 nm. The short axis of the three LNMs is 60 nm for LNM-1 and LNM-2 and 80 nm for LNM-3. The long axis of the three LNMs is 240 nm for LNM-1, 300 nm for LNM-2, and 360 nm for LNM-3. Each LNM unit is composed of two half-ellipse LN nanostructures with a slot, whose width is 5 nm for LNM-1 and LNM-3 and 15 nm for LNM-2. By manipulating the w_1, w_2, and w_3 values, the vertical period along the y-axis direction of LNMs, the resonant wavelength generated by plasmonic resonance can cover all the visible light spectra for both transmission and reflection with relatively small full width at half maximum (FWHM) and a high Q-factor.

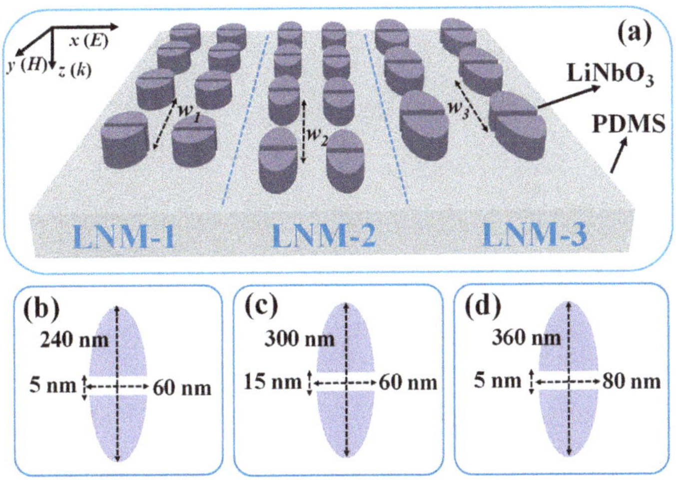

Figure 1. Schematic drawings of (**a**) the proposed LNM-based FCF. Geometric parameters of (**b**) LNM-1, (**c**) LNM-2, and (**d**) LNM-3, respectively. The horizontal period along the x-direction between LNM units is 80 nm for both LNM-1 and LNM-2 and 100 nm for LNM-3, and the thickness of the LNM layers is 200 nm.

The optical responses of the proposed LNM-based FCF can be characterized by its transmission coefficient (T) and reflection coefficient (R), which are expressed by

$$T(\omega) = |t(\omega)|^2 = \left|\frac{E_t(\omega)}{E_i(\omega)}\right|^2 \tag{1}$$

$$R\omega = \mathrm{r}\omega 2R(\omega) = |r(\omega)|^2 = \left|\frac{E_r(\omega)}{E_i(\omega)}\right|^2 \tag{2}$$

where E_t, E_r, and E_i represent the transmitted, reflected, and incident E-fields, respectively [36].

3. Results and Discussions

Figure 2a exhibits the reflection spectra of the LNM-based FCF, demonstrating their ability to cover the entire visible spectrum with high efficiency. LNM-1, LNM-2, and LNM-3 are designed for blue, green, red color filtering characteristics. By changing the vertical period along the y-axis direction of LNMs, i.e., w_1 for LNM-1, w_2 for LNM-2, and w_3 for LNM-3, LNMs show strong reflection peaks from the wavelengths of 440 nm to 540 nm for LNM-1, 550 nm to 590 nm for LNM-2, and 620 nm to 740 nm for LNM-3. These optical characteristics prove the capability of the LNMs to selectively filter specific colors with high precision. The progressive redshift in resonance wavelengths from LNM-1 to LNM-3 is a direct result of the geometric tuning enabled by adjusting the vertical period along the y-axis direction of LNMs, which modifies the Mie scattering resonances and the associated electromagnetic field distributions. This geometric tunability is a key advantage of the LNM-based FCF design, as it allows for precise control over the optical response across the visible spectrum, making it highly adaptable for various applications in color filtering and sensing fields.

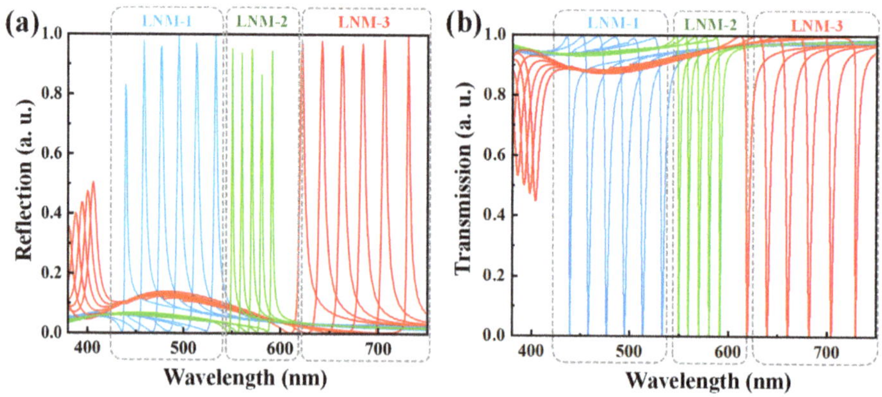

Figure 2. (**a**) Reflection and (**b**) transmission spectra of LNM-1, LNM-2, and LNM-3 to exhibit the corresponding blue (B), green (G), and red (R) colors. For both (**a**,**b**), the w_1 of LNM-1 is changed from 260 nm to 360 nm, the w_2 of LNM-2 is changed from 360 nm to 400 nm, and the w_3 of LNM-3 is changed from 400 nm to 500 nm.

Figure 2b shows the transmission spectra of the LNM-based FCF, enabling high-contrast color filtering with ultrahigh purity and brightness characteristics. The ability to span the entire visible spectrum in both reflection and transmission spectra highlights the robustness of these LNM-based FCFs. The redshift in resonant wavelengths is governed by the interaction between incident light and the periodic LNM nanostructures. In such all-dielectric systems, this interaction is fundamentally rooted in Mie scattering, where incident light excites a series of multipolar resonant modes within the individual nanostructures.

Our design elevates this fundamental interaction by engineering these modes to operate in the regime of quasi-BIC, which is the key to achieving the observed exceptionally sharp spectral features. A true BIC is a theoretical "dark" mode with an infinite quality (Q) factor, as it is perfectly confined and unable to couple with external radiation due to fundamental symmetry incompatibilities with the continuum of free-space modes. To make such a mode accessible, we introduce a controlled perturbation. In our design, the asymmetric slot within each half-ellipse nanostructure serves this purpose explicitly. This geometric asymmetry intentionally breaks the requisite symmetry that protects the BIC, creating a controlled leakage channel. This transformation from a perfect BIC into a quasi-BIC allows the previously confined mode to couple with incident light, manifesting as a high-Q resonance in the optical spectra. The degree of this asymmetry directly dictates the coupling strength and, consequently, the Q-factor of the resonance. The ability to precisely control the resonant wavelength and its Q-factor through geometric tuning is a significant advantage of this FCF design, enabling its application in advanced tunable optoelectronic devices.

Within the LNM nanostructure array, the interactions of Rayleigh–Wood Anomaly (RWA) modes have been observed to produce specific relationships between primary resonances and secondary RWA resonances throughout the visible spectrum. Through the coupling mechanism analysis, the coupling strength of RWAs has been determined to be dependent on the vertical period along the y-axis direction of LNMs between adjacent units, whereby both resonance positions and relative intensities are affected.

The operational parameters of the LNM-1, LNM-2, and LNM-3 designs are systematically optimized through simulations to leverage symmetry-protected BICs, with each configuration precisely targeting distinct spectral regimes. For LNM-1, a geometric scaling parameter w_1 is calibrated to facilitate the evolution from fundamental Mie resonances to high-Q nonlocal BIC hybrid modes in the wavelength range of 400–500 nm. This enhancement is achieved by engineering destructive interference between competing radiative channels. Specifically, the structural asymmetry opens pathways for energy to radiate away from the resonator. By carefully tailoring the geometry, these pathways are forced to interfere destructively in the far-field region, effectively canceling radiation loss and trapping energy within the nanostructure. This process is equivalent to establishing near-orthogonality between the resonant mode and the continuum of free-space radiation channels, which is the mechanism that suppresses radiative losses and enables ultrahigh Q-factors.

For LNM-2 and LNM-3, analogous activation mechanisms are implemented by fine-tuning w_2 and w_3, respectively, where hybrid coupling between symmetry-protected BICs and propagating resonances produces spectrally separated responses in the green (520–580 nm) and red (620–750 nm) regions. The BIC-dominated regime emerges optimally when the ratio of the lattice period to the operational wavelength approaches 0.7–0.9. This parametric framework demonstrates the universal scalability of BIC physics across the visible spectrum, providing a robust platform for tailoring the optical response of all-dielectric metamaterials through geometric optimization.

The slot of the half-ellipse LN nanostructure plays a pivotal role in activating BICs. The slot creates a momentum mismatch that traps light as non-radiative BIC mode. This design suppresses energy leakage, achieving ultrahigh quality factors through destructive interference of radiative channels, enabling wavelength-specific light confinement without relying on BIC coupling. Such symmetry-engineered BICs provide a low-loss platform for nonlinear optics and sensing [40]. Second, by creating specific boundaries within the structure, the slot of the half-ellipse LN nanostructure confines the BIC mode to a designated region. This spatial confinement significantly enhances light–matter interaction strength, improving the optical response efficiency of the proposed device [41]. Third, precise control of the geometric dimensions of the half-ellipse LN nanostructure enables tailored momentum mismatch between the symmetry-protected BIC mode and the radiative channel. The radiative losses are suppressed through destructive interference, while non-radiative BIC confinement enhances near-field intensity and Q-factors exceeding 103. This geometric tuning bypasses RWA coupling limitations, achieving intrinsic Q-factor optimization via the BIC-dominated hybrid mode, as validated by pump–probe spectroscopy in recent low-loss metasurface studies [42].

The relationships between the Q-factor, FWHM, and the vertical period along the y-axis direction of LNMs are systematically analyzed in Figure 3. For reflection spectra (Figure 3a–c), LNM-1 (blue) exhibits a gradual decrease in the Q-factor from 155 to 90 as w_1 varies from 260 nm to 360 nm, accompanied by an increase in FWHM from 2.6 to 5.5 nm (Figure 3a). Similarly, LNM-2 (green) demonstrates the Q-factor ranging from 260 nm to 330 nm as w_2 varies from 360 nm to 420 nm with FWHM spanning from 1.65 to 2.22 nm, reflecting the superior performance in the green spectrum (Figure 3b). LNM-3 (red) shows a moderate Q-factor range from 100 to 200 as w_3 increases from 400 nm to 500 nm, while FWHM shows its rise to 7 nm first, followed by its fall to 3.5 nm (Figure 3c). The above-mentioned characteristics highlight the tunability of the proposed LNM-based FCF for RGB-spectrum applications. These trends confirm that increasing w strengthens the coupling between adjacent resonant units and can cause redshifting. The inverse proportionality between the Q-factor and FWHM across all designs underscores the geometric tunability of LNM-based FCFs, enabling tailored performance for specific spectral bands.

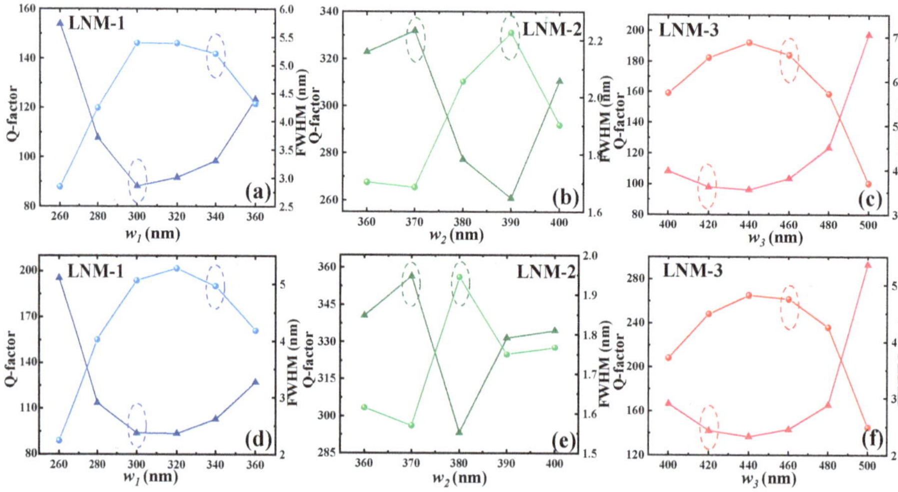

Figure 3. Relationships of Q-factors and FWHM values of (**a**,**d**) LNM-1, (**b**,**e**) LNM-2, and (**c**,**f**) LNM-3 changing w_1 from 260 nm to 360 nm for LNM-1, w_2 from 360 nm to 400 nm for LNM-2, w_3 from 400 nm to 500 nm for LNM-3, summarized by (**a**–**c**) reflection and (**d**–**f**) transmission spectra, respectively.

For transmission spectra (Figure 3d–f), the Q-factor and FWHM exhibit similar inverse relationships to those observed in the reflection spectra (Figure 3a–c), but with distinct numerical ranges that demonstrate enhanced performance in transmission spectra. For LNM-1, the Q-factor rises from 100 to 200 as w_1 increases from 260 nm to 360 nm, with FWHM decreasing from 5.1 nm to 2.5 nm (Figure 3d), while LNM-2 achieves exceptional Q-factors from 290 to 360 (FWHM is between 1.95 nm to 1.55 nm) as w_2 varies from 360 nm to 400 nm (Figure 3e). LNM-3 exhibits a broader tunability range, with Q-factors spanning from 140 to 300 and FWHM narrowing from 4.7 nm to 2.5 nm as w_3 varies from 400 nm to 500 nm (Figure 3f). The enhanced Q-factors of these LNM-based FCFs in transmission spectra compared to reflection spectra suggest stronger light confinement within the nanostructures, driven by optimized w-dependent near-field interactions. The consistent variation in FWHM with increasing w across all designs further validates the role of geometric tuning in achieving high-purity spectral filtering characteristics.

The distinct trends in the Q-factor and FWHM for each LNM design, as shown in Figure 3, are a direct consequence of the tunable coupling between the intrinsic quasi-BIC resonance of the half-ellipse nanostructures and the collective lattice resonances, specifically the Rayleigh-Wood Anomaly (RWA). The vertical period (w) directly controls the spectral position of the RWA, which can introduce a significant radiative loss channel. The Q-factor of the device is critically dependent on the interference between the quasi-BIC mode and this RWA.

For LNM-1 and LNM-3, the Q-factors exhibit a distinct non-monotonic trend, decreasing to a minimum before rising again, while the FWHM shows the inverse behavior. This profile is a classic signature of a strong Fano-type interference. As the period (w_1 or w_3) is tuned, the RWA is swept across the spectral position of the main quasi-BIC resonance. The Q-factor is at its minimum (and FWHM is at its maximum) at the point of spectral degeneracy, where the coupling between the two modes is strongest, maximizing the radiative losses. As the RWA moves away from the resonance on either side, the coupling weakens, and the Q-factor recovers, returning to a state dominated by the intrinsic low-loss nature of the quasi-BIC.

The behavior of LNM-2 is qualitatively different and highlights its specialized design for superior performance. It does not exhibit the pronounced Q-factor dip seen in the other two designs; instead, it maintains an exceptionally high and more stable Q-factor across its entire tuning range. This unique stability is attributed to its distinct geometry, most notably the significantly wider slot of 15 nm compared to the 5 nm slot in LNM-1 and LNM-3. In the physics of BICs, the slot's geometry directly defines the "asymmetry parameter" that transforms a perfect, non-radiating BIC into a quasi-BIC with a finite Q-factor. The wider slot in LNM-2 fundamentally alters this parameter, resulting in a much weaker coupling coefficient between the quasi-BIC mode and the radiative RWA channel. Because of this engineered weak coupling, the resonance of LNM-2 is less perturbed by the RWA, even as it sweeps past. The system remains in a robust, high-Q state, explaining its "strange" yet superior and more stable performance compared to LNM-1 and LNM-3.

To further contextualize the performance of our proposed LNM-FCF, a comparative analysis against several published and relevant works is presented. The key performance metrics, including the Q-factor, sensitivity, and the FOM, alongside other pertinent characteristics such as material systems and device flexibility are summarized in Table 1. This comparison aims to highlight the competitive advantages and unique contributions of our LNM-FCF within the current landscape of advanced optical filters and sensors. The LNM-FCF demonstrates a superior combination of a high experimental Q-factor (up to ~360), excellent sensitivity (130 nm/RIU), and an outstanding FOM (~900 RIU^{-1}).

Table 1. Comparison of the proposed LNM-FCF with literature works. "N/A" indicates that the value was not reported in the cited reference. An asterisk (*) denotes values obtained from simulations, and all other values are experimentally measured.

Ref.	Materials	Max. Q-Factor	Max. S (nm/RIU)	FWHM (nm)	Max. FOM (RIU^{-1})	Flexible
[1]	Dielectric (SiN/Poly-Si) Gratings/Glass	~27.5	N/A	~20	N/A	No
[2]	All-Dielectric (Si) pixel	~18 *	N/A	~30.6 *	N/A	No
[6]	Al/SiO_2	~35 *	~45*	~14.3 *	50 *	No
[18]	LN/PDMS	~160	~65	~3.1	~130	Yes
[42]	TiO_2 MRWG/Glass	93 *	N/A	~5.9 *	N/A	No
[43]	Porous TiO_2 films	~20	N/A	~25	N/A	No
[44]	Si metasurface/Al reflector	~80 *	N/A	~6.9 *	N/A	No
[45]	Al plasmonic pixels	~20	N/A	~25	N/A	No
This work	LN/PDMS	~360 *	130 *	~1.55 *	~900 *	Yes

The transmission spectra of LNM-1, LNM-2, and LNM-3 exhibit significant wavelength shifts in their resonance dips as the refractive index (n) of the surrounding medium increases, as shown in Figure 4a. For LNM-1 (blue), the transmission spectra are shifted from the wavelength of 430 nm at n = 1.0 to 482 nm at n = 1.4, corresponding to a redshift of 52 nm. Similarly, LNM-2 (green) shows a resonant shift of 45 nm from the wavelength of 530 nm (n = 1.0) to 575 nm (n = 1.4), while LNM-3 (red) demonstrates the largest shift of 75 nm, moving from the wavelength of 620 nm to 695 nm. These redshifts arise from the increased effective n value, which modifies the phase-matching conditions for resonant modes. The intensity of each resonance remains stable while changing the n value, indicating robust light confinement despite environmental changes. The linear dependence of the resonant wavelength on the n value highlights greater sensitivity due to stronger field interactions and the tunability of these LNM-based FCFs for environmental sensing applications. The sensitivity and FOM are defined as

$$S = \frac{\Delta\lambda}{\Delta n} \quad (3)$$

$$FOM = \frac{S}{FWHM} \quad (4)$$

where $\Delta\lambda$ is the shift in resonant wavelength, Δn is the change in the refractive index of the surrounding medium, and FWHM is the full width at half maximum of the resonance dip.

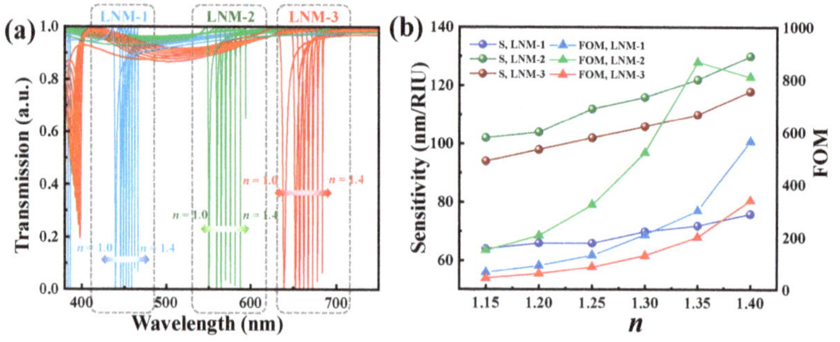

Figure 4. (**a**) Transmission spectra of LNM-1, LNM-2, and LNM-3 exposed to the surrounding environment with different refractive indexes (n). (**b**) Relationships of sensitivities and FOM values of (**a**).

In Figure 4b, the trends of sensitivities show quite distinct and linear characteristics. LNM-2 achieves the highest sensitivity of 130 nm/RIU. In contrast, LNM-1 and LNM-3 exhibit sensitivities of 75 nm/RIU and 115 nm/RIU, respectively. The FOM can quantify the trade-off between sensitivity and resonance linewidth. The FOM peak is 900 RIU^{-1} for LNM-2 at n = 1.35, owing to its narrow FWHM of 3 nm and moderate sensitivity. The trend of the FOM for LNM-2, which peaks and then declines, is noteworthy. This behavior arises because, for refractive indices greater than 1.35, the resonance linewidth (FWHM) begins to broaden more rapidly than the linear increase in wavelength sensitivity (S). Such FWHM broadening at higher ambient refractive indices is a known phenomenon in high-Q resonant systems, where the delicate destructive interference sustaining the quasi-BIC mode becomes perturbed by the changing dielectric environment, leading to increased radiative losses [43].

The observed peak in the FOM for LNM-2 is directly linked to the physics of the quasi-BIC resonance upon which this design is based on. A quasi-BIC achieves its characteristically high Q-factor through the near-perfect destructive interference between different radiation pathways, a condition established by the specific geometry of the nanostructure, such as the half-ellipse slot. For LNM-2, the system is tuned so effectively that this interference condition appears to be maximally fulfilled not in air (n = 1.0) but in a background medium of $n \approx 1.35$. At this optimal point, the peak FOM of 900 RIU^{-1} and a sensitivity of 130 nm/RIU imply an exceptionally narrow FWHM of approximately 0.144 nm (S/FOM). LNM-1 and LNM-3 show FOM values between 50–590 RIU^{-1} and 10–370 RIU^{-1}, respectively. These results show the role of geometric design in balancing sensitivity and spectral resolution. The enhanced performance of LNM-2 in the green spectrum aligns with optimized BIC coupling efficiency at intermediate wavelengths.

The E- and H-field distributions of the LNM-based FCF at resonances in TM mode are shown in Figure 5. Figure 5a–c depict the E-field distributions of LNM-1, LNM-2, and LNM-3, where E-field energies concentrate predominantly on the upper and lower nanostructures of LNMs. The first and second rows of this figure represents the field distributions of LNMs in the lower- and upper-bound geometric parameter conditions (w), respectively. The difference in field distribution in the same LNM structure with different w values demonstrates the tunability of the proposed LNM-based FCFs. A progressive enhancement in field confinement is observed from LNM-1 to LNM-3, with the energy density becoming more localized at the edges of the nanostructures. This increasing confinement corresponds to the higher Q-factors observed in the spectral analysis. Figure 5d–f illustrate the H-field distributions of LNM-1, LNM-2, and LNM-3, demonstrating that H-field energies are concentrated in the slots and become more condensed in the middle slot of LNMs from LNM-1 to LNM-3. This complementary distribution of the E- and H-fields provides direct evidence of the BIC hybridization mechanism proposed earlier, where the slot acts as a coupling channel between propagating waves and bound states. The observed asymmetric field distribution pattern, with E-field density stronger than H-field density, confirms the formation of standing waves within the LNM nanostructure, a characteristic signature of the BIC resonator phenomenon that enables high spectral purity.

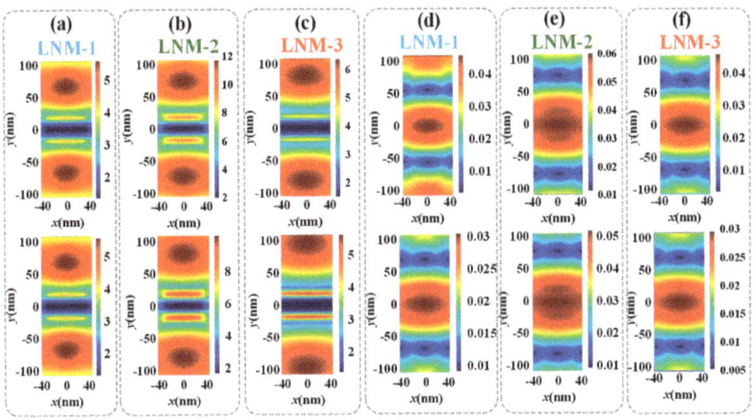

Figure 5. (**a–c**) E-field and (**d–f**) H-field distributions of LNM-1, LNM-2, and LNM-3, respectively. The first and the second rows of this figure represents the field distributions of LNMs in the lower- and upper-bound geometric parameter conditions (w), respectively.

4. Conclusions

In conclusion, we propose an FCF composed of three LNMs to create RGB colors spanning the visible spectrum from 380 nm to 780 nm with ultrahigh purity and brightness characteristics. Through the manipulation of the FCF along the y-axis direction, LNMs show high Q-factors from 90 to 360. The electromagnetic field distributions reveal strong field confinement within the LNMs. When the LNM-based FCF is exposed to different backgrounds with n values from 1.0 to 1.4, it shows high sensitivity up to 130 nm/RIU and the FOM reaching 900 RIU^{-1}. It indicates that the proposed LNM-based FCF possesses the potential for precise control in color sensing applications. The LNM-based FCF design shows significant promise for applications in color filters, particularly in scenarios requiring precise wavelength selectivity. Its conceptual combination of simplicity and high simulated performance makes it a promising candidate for further investigation of the next-generation optical devices.

Author Contributions: S.Z. (Siqiang Zhao), D.Z., Y.Z., S.Z. (Shuyan Zou), and Y.-S.L. carried out related simulations and data analysis. S.Z. (Siqiang Zhao) drafted the manuscript. Y.-S.L. supervised the simulations and the revising of the manuscript. Y.-S.L. also provided suggestions and guidance for the simulations and data analysis. All authors have read and agreed to the published version of the manuscript.

Funding: This study was funded by the Guangdong Science and Technology Program (2024A0505050007) and the Guangdong Basic and Applied Basic Research Foundation (2024A1515013019).

Institutional Review Board Statement: Not applicable.

Informed Consent Statement: Not applicable.

Data Availability Statement: Data will be made available upon request.

Conflicts of Interest: The authors declare no conflicts of interest.

References

1. Horie, Y.; Han, S.; Lee, J.-Y.; Kim, J.; Kim, Y.; Arbabi, A.; Shin, C.; Shi, L.; Arbabi, E.; Kamali, S.M.; et al. Visible Wavelength Color Filters Using Dielectric Subwavelength Gratings for Backside-Illuminated CMOS Image Sensor Technologies. *Nano Lett.* **2017**, *17*, 3159–3164. [CrossRef]
2. Nagasaki, Y.; Suzuki, M.; Takahare, J. All-Dielectric Dual-Color Pixel with Subwavelength Resolution. *Nano Lett.* **2017**, *17*, 7500–7506. [CrossRef] [PubMed]
3. Vashistha, V.; Vaidya, G.; Hegde, R.-S.; Serebryannikov, A.-E.; Bonod, N.; Krawczyk, M. All-Dielectric Metasurfaces Based on Cross-Shaped Resonators for Color Pixels with Extended Gamut. *ACS Photonics* **2017**, *4*, 1076–1082. [CrossRef]
4. Chen, H.-T.; Taylor, A.-J.; Yu, N. A review of metasurfaces: Physics and applications. *Rep. Prog. Phys.* **2016**, *79*, 076401. [CrossRef] [PubMed]
5. Venkatesh, S.; Lu, X.; Saeidi, H.; Sengupta, K. A high-speed programmable and scalable terahertz holographic metasurface based on tiled CMOS chips. *Nat. Electron.* **2020**, *3*, 785–793. [CrossRef]
6. Shrestha, V.R.; Lee, S.-S.; Kim, E.-S.; Choi, D.-Y. Aluminum Plasmonics Based Highly Transmissive Polarization-Independent Subtractive Color Filters Exploiting a Nanopatch Array. *Nano Lett.* **2014**, *14*, 6672–6678. [CrossRef]
7. Song, M.; Wang, D.; Peana, S.; Choudhury, S.; Nyga, P.; Kudyshev, Z.; Yu, H.; Boltasseva, A.; Shalaev, V.; Kildishev, A. Colors with plasmonic nanostructures: A full-spectrum review. *Appl. Phys. Rev.* **2019**, *6*, 041308. [CrossRef]
8. Li, Z.; Clark, A.W.; Cooper, J.M. Dual Color Plasmonic Pixels Create a Polarization Controlled Nano Color Palette. *ACS Nano* **2016**, *10*, 492–498. [CrossRef]
9. Xu, R.; Lin, Y.-S. Characterizations of reconfigurable infrared metamaterial absorbers. *Opt. Lett.* **2018**, *43*, 4783–4786. [CrossRef]
10. Grahn, P.; Shevchenko, A.; Kaivola, M. Electromagnetic multipole theory for optical nanomaterials. *New J. Phys.* **2012**, *14*, 093033. [CrossRef]
11. Yang, W.; Xiao, S.; Song, Q.; Liu, Y.; Wu, Y.; Wang, S.; Yu, J.; Han, J.; Tsai, D.-P. All-dielectric metasurface for high-performance structural color. *Nat. Commun.* **2020**, *11*, 1864. [CrossRef]
12. Haddadin, Z.; Khan, S.; Poulikakos, L.V. Cutting Corners to Suppress High-Order Modes in Mie Resonator Arrays. *ACS Photonics* **2024**, *11*, 187–195. [CrossRef]
13. Yang, B.; Liu, W.; Li, Z.; Cheng, H.; Choi, D.-Y.; Chen, S.; Tian, J. Ultrahighly Saturated Structural Colors Enhanced by Multipolar-Modulated Metasurfaces. *Nano Lett.* **2019**, *19*, 4221–4228. [CrossRef] [PubMed]
14. Yang, J.-H.; Babicheva, V.E.; Yu, M.-W.; Lu, T.-C.; Lin, T.-R.; Chen, K.-P. Structural Colors Enabled by Lattice Resonance on Silicon Nitride Metasurfaces. *ACS Nano* **2020**, *14*, 5678–5685. [CrossRef] [PubMed]
15. Wen, Y.; Lin, J.; Chen, K.; Lin, Y.-S.; Yang, B.-R. Full color metasurface with high-transmission and omnidirectional characteristics. *Opt. Laser Technol.* **2022**, *150*, 108004. [CrossRef]
16. Lee, C.-R.; Lin, S.-H.; Wang, S.-M.; Lin, J.-D.; Chen, Y.-S.; Hsu, M.-C.; Liu, J.-K.; Mo, T.-S.; Huang, C.-Y. Optically controllable photonic crystals and passively tunable terahertz metamaterials using dye-doped liquid crystal cells. *J. Mater. Chem. C* **2018**, *6*, 4959–4966. [CrossRef]
17. Lu, L.; Dong, Z.; Tijiptoharsono, F.; Ng, R.J.H.; Wang, H.; Rezaei, S.D.; Wang, Y.; Leong, H.S.; Lim, P.C.; Yang, J.K.W.; et al. Reversible Tuning of Mie Resonances in the Visible Spectrum. *ACS Nano* **2021**, *15*, 19722–19732. [CrossRef] [PubMed]
18. Zheng, D.; Huang, W.; Chen, K.; Lin, Y.-S. Visible Metamaterial Using a Lithium Niobate Nanoring Structure for Stretchable Color Sensing Application. *ACS Mater. Lett.* **2023**, *5*, 1831–1840. [CrossRef]
19. Zheng, D.; Wen, Y.; Xu, X.; Lin, Y.-S. Metamaterial grating for colorimetric chemical sensing applications. *Mater. Today Phys.* **2023**, *33*, 101056. [CrossRef]

20. Zhang, X.; Lin, Y.-S.; Yang, B.-R. Tunable color switch using split-ring metamaterial. *Opt. Laser Technol.* **2020**, *131*, 106461. [CrossRef]
21. Yang, W.; Lin, Y.-S. Tunable metamaterial filter for optical communication in the terahertz frequency range. *Opt. Express* **2020**, *28*, 17620–17629. [CrossRef]
22. Lin, Y.-S.; Dai, J.; Zeng, Z.; Yang, B.-R. Metasurface Color Filters Using Aluminum and Lithium Niobate Configurations. *Nanoscale Res. Lett.* **2020**, *15*, 77. [CrossRef]
23. Li, B.; Zeng, Q.; Lin, Y.-S. Tunable terahertz meta-resonator with switchable single-and dual-resonance characteristic. *Mater. Today Commun.* **2024**, *41*, 110197. [CrossRef]
24. Wu, L.; Lin, Y.-S. Design and investigation of dual-axis electrostatic driving MEMS scanning micromirror. *Int. J. Optomechatronics* **2024**, *18*, 2350749. [CrossRef]
25. Zheng, D.; Lin, Y.-S. Angular-Dispersive Narrowband Absorption Induced by Quasi-BIC in SiO2-Au Metamaterial. *Small Struct.* **2024**, *5*, 2300529. [CrossRef]
26. Xu, R.; Luo, J.; Sha, J.; Zhong, J.; Xu, Z.; Tong, Y.; Lin, Y.-S. Stretchable IR metamaterial with ultra-narrowband perfect absorption. *Appl. Phys. Lett.* **2018**, *113*, 101907. [CrossRef]
27. Wu, L.; Guan, J.; Lin, Y.-S. Active logic modulator of terahertz metamaterial for future 6G communication system. *Sens. Actuators A Phys.* **2024**, *379*, 115992. [CrossRef]
28. Jang, J.; Badloe, T.; Yang, Y.; Lee, T.; Mun, J.; Rho, J. Spectral Modulation through the Hybridization of Mie-Scatterers and Quasi-Guided Mode Resonances: Realizing Full and Gradients of Structural Color. *ACS Nano* **2020**, *14*, 15317–15326. [CrossRef]
29. Hedayati, M.K.; Elbahri, M. Review of Metasurface Plasmonic Structural Color. *Plasmonics* **2017**, *12*, 1463–1479. [CrossRef]
30. Hu, Y.; Luo, X.; Chen, Y.; Liu, Q.; Li, X.; Wang, Y.; Liu, N.; Duan, H. 3D-Integrated metasurfaces for full-colour holography. *Light Sci. Appl.* **2019**, *8*, 86. [CrossRef]
31. Qiu, C.-W.; Zhang, T.; Hu, G.; Kivshar, Y. Quo Vadis, Metasurfaces? *Nano Lett.* **2021**, *21*, 5461–5474. [CrossRef] [PubMed]
32. Liu, Y.; Lin, Y.-S. Multiplexed terahertz multifocal metalens based on stretchable single-, dual-, and quad-axis synthetic rectangular optical sparse aperture. *Mater. Des.* **2024**, *237*, 112631. [CrossRef]
33. Ebbesen, T.W.; Lezec, H.J.; Ghaemi, H.F.; Thio, T.; Wolff, P.A. Extraordinary optical transmission through sub-wavelength hole arrays. *Nature* **1998**, *391*, 667–669. [CrossRef]
34. Barnes, W.L.; Dereux, A.; Ebbesen, T.W. Surface plasmon subwavelength optics. *Nature* **2003**, *424*, 824–830. [CrossRef]
35. Pitchappa, P.; Manjappa, M.; Krishnamoorthy, H.N.S.; Chang, Y.; Lee, C.; Singh, R. Bidirectional reconfiguration and thermal tuning of microcantilever metamaterial device operating from 77 K to 400 K. *Appl. Phys. Lett.* **2017**, *111*, 261101. [CrossRef]
36. Cong, L.; Pitchappa, P.; Wu, Y.; Ke, L.; Lee, C.; Singh, N.; Yang, H.; Singh, R. Active Multifunctional Microelectromechanical System Metadevices: Applications in Polarization Control, Wavefront Deflection, and Holograms. *Adv. Opt. Mater.* **2017**, *5*, 1600716. [CrossRef]
37. Yi, Z.; Lin, H.; Niu, G.; Chen, X.; Zhou, Z.; Ye, X.; Duan, T.; Yi, Y.; Tang, Y.; Yi, Y. Graphene-based tunable triple-band plasmonic perfect metamaterial absorber with good angle-polarization-tolerance. *Results Phys.* **2019**, *13*, 102149. [CrossRef]
38. Miroshnichenko, A.E.; Flach, S.; Kivshar, Y.S. Fano resonances in nanoscale structures. *Rev. Mod. Phys.* **2010**, *82*, 2257–2298. [CrossRef]
39. Rybin, M.V.; Koshelev, K.L.; Sadrieva, Z.F.; Samusev, K.B.; Bogdanov, A.A.; Limonov, M.F.; Kivshar, Y.S. High-Q Supercavity Modes in Subwavelength Dielectric Resonators. *Phys. Rev. Lett.* **2017**, *119*, 243901. [CrossRef]
40. Hsu, C.W.; Zhen, B.; Stone, A.D.; Joannopoulos, J.D.; Soljačić, M. Bound states in the continuum. *Nat. Rev. Mater.* **2016**, *1*, 16048. [CrossRef]
41. Zelmon, D.E.; Small, D.L.; Jundt, D. Infrared corrected Sellmeier coefficients for congruently grown lithium niobate and 5 mol% magnesium oxide-doped lithium niobate. *J. Opt. Soc. Am. B* **1997**, *14*, 3319–3322. [CrossRef]
42. Park, J.; Li, M.; Wang, Z.; Li, Z.; Fan, J.A. Topology-Optimized Metasurface Resonant Waveguide Gratings. *Nano Lett.* **2024**, *24*, 4803–4810.
43. Yoo, Y.J.; Lim, J.H.; Lee, G.J.; Jang, K.-I.; Song, Y.M. Ultra-thin films with highly absorbent porous media fine-tunable for coloration and enhanced color purity. *Nanoscale* **2017**, *9*, 2986–2991. [CrossRef]
44. Yue, W.; Gao, S.; Lee, S.-S.; Kim, E.-S.; Choi, D.-Y. Highly reflective subtractive color filters capitalizing on a silicon metasurface integrated with nanostructured aluminum mirrors. *Laser Photonics Rev.* **2017**, *11*, 1600285. [CrossRef]
45. Olson, J.; Manjavacas, A.; Basu, T.; Huang, D.; Schlather, A.E.; Zheng, B.; Halas, N.J.; Nordlander, P.; Link, S. High Chromaticity Aluminum Plasmonic Pixels for Active Liquid Crystal Displays. *ACS Nano* **2016**, *10*, 1108–1117. [CrossRef]

photonics MDPI

Article

Reconfigurable High-Efficiency Power Dividers Using Waveguide Epsilon-Near-Zero Media for On-Demand Splitting

Lin Jiang [1,2], Qi Hu [1,*] and Yijun Feng [1,*]

1 School of Electronic Science and Engineering, Nanjing University, Nanjing 210093, China; linjiang@smail.nju.edu.cn
2 School of Electronic Engineering, Yili Normal University, Yining 835000, China
* Correspondence: qihu@nju.edu.cn (Q.H.); yjfeng@nju.edu.cn (Y.F.)

Abstract: Although epsilon-near-zero (ENZ) media have emerged as a promising platform for power dividers, the majority of existing designs are confined to fixed power splitting. In this work, two dynamically tunable power dividers using waveguide ENZ media are proposed by precisely modulating the internal magnetic field and the widths of the output waveguides. The first approach features a mechanically reconfigurable ring-shaped ENZ waveguide. By continuously re-distributing the magnetic field within the ENZ tunneling channels utilizing rotatable copper plates, arbitrary power division among multiple output ports is constructed. The second design integrates a rectangular-loop ENZ cavity into a substrate-integrated waveguide, with four positive–intrinsic–negative diodes embedded to dynamically activate specific output ports. This configuration steers electromagnetic energy toward output ports with varying cross-sectional areas, enabling on-demand control over both the power division and the number of output ports. Both analytical and full-wave simulation results confirm dynamic power division, with transmission efficiencies exceeding 93%. Despite differences in structure and actuation mechanisms, both designs exhibit flexible field control, high reconfigurability, and excellent transmission performance, highlighting their potential in advanced applications such as real-time wireless communications, multi-input–multi-output systems, and reconfigurable antennas.

Keywords: epsilon-near-zero (ENZ) media; reconfigurable power divider (RPD); waveguide; electromagnetic (EM) field control

1. Introduction

Owing to their ability to divide a single input signal into multiple output signals with a balanced or imbalanced power ratio, power dividers have been widely explored in radio-frequency (RF), microwave, and optical circuits [1–3]. For example, in reflectarrays and transmitarrays, power dividers are essential parts of the feeding network and excite every element [4]. Conventional microwave power dividers, typically implemented using microstrip technology such as the Wilkinson type [5], provide equal power division and port isolation through quarter-wavelength transmission lines and matching resistors. Despite their widespread use, high energy loss and structural constraints may limit their further applications in high-quality electromagnetic (EM) wave manipulation and compact system integration [6]. In contrast, metamaterial-based power dividers exploit engineered EM responses to achieve functionalities beyond those of conventional designs. For instance, a 1:4 series power divider based on zero-degree metamaterial structures delivers equal power to all four output ports over a broad frequency range [7]. Furthermore, PT-symmetric non-Hermitian photonic systems enable control of the near-field routing of hyperbolic polaritons, breaking intrinsic symmetry constraints and allowing subwavelength-scale energy guidance [8].

Epsilon-near-zero (ENZ) media [9–13] are elaborately engineered materials with a near-zero dielectric constant and wavenumber, resulting in an extremely small optical index and a nearly uniform phase distribution. Due to their natural temporal and spatial decoupling properties [14–19], ENZ media have emerged as promising platforms for high-efficiency signal transmission [20–25], division [26], and combination [27], exhibiting great potential in massive multi-input–multi-output (MIMO) systems, the Internet of Things (IoT), etc. As one of the most effective construction strategies, waveguides operating at the cut-off frequency can be equivalently considered ENZ media [28–32]. It has been experimentally demonstrated in waveguides that their equivalent dielectric constant near the cut-off frequency can be considered near zero, resulting in a near-infinity wavelength and longitudinal propagation constant. Therefore, featuring merits of low loss and geometry insensitivity, many recent efforts have been devoted to ENZ-based power dividers. For example, an arbitrarily shaped multi-port power divider is proposed by connecting multiple waveguides through an ENZ host medium, which is embedded with a dielectric impurity to facilitate impedance matching [33]. Arbitrary power allocation could be realized with pre-designed output waveguide widths. Additionally, an acoustic power divider and multiplexer have been developed utilizing an index-near-zero medium [34], capable of tunneling sound to an arbitrary number of output ports with phase shifts of 0° or 180°. Precise power and phase control are implemented by adjusting each output channel's cross-sectional area and arranging the output port positions. Despite significant advancements, the above approaches are restricted to fixed power division after fabrication, which may constrain their possibilities in dynamic application scenarios, for example, real-time wireless communications.

Here, we generalize the concept of dynamic high-efficiency on-demand power division with arbitrarily shaped waveguide ENZ media by proposing two distinct types of reconfigurable ENZ waveguide-based power dividers. With the inherent magnetic field homogeneity, tunable power division ratios can be achieved by adjusting the internal magnetic field and/or the widths of the output waveguides. A mechanically reconfigurable power divider is presented by coupling rectangular waveguides via a ring-shaped air-filled narrow ENZ tunnel. Rotating the embedded copper plates continuously tunes the magnetic field in the sub-channels, directly controlling the power delivered to the two output waveguides and enabling arbitrary split ratios. The transmission efficiency remains above 93% across all power divisions. As another approach, an electrically reconfigurable power divider is proposed with output waveguides of different cross-sectional areas. It consists of a substrate-integrated waveguide (SIW) and a rectangular-loop air-filled ENZ cavity with four embedded positive–intrinsic–negative (PIN) diodes. Selective switching of these diodes dynamically reconfigures both the number and cross-sectional areas of the output SIWs, enabling flexible output port selection and on-demand power division. Regardless of the output power split ratios, the transmission efficiencies remain above 93%. Analytical and numerical models are conducted to characterize the EM performances of the proposed reconfigurable ENZ-based power dividers, validating their desired EM regulation with high transmission efficiency, real-time tunability, on-demand power allocation, and flexible geometry parameters. The presented methods may provide new strategies for advanced reconfigurable power dividers (RPDs).

2. Concept and Theoretical Analysis

As conceptually illustrated in Figure 1, we first consider a two-dimensional (2-D) RPD comprising a narrow ENZ channel with an arbitrary geometry, which further interconnects three identical standard waveguides. The waveguides are terminated by ports labeled 1 to 3, with Port 1 designated as the input and Ports 2–3 serving as the output. The external boundaries of the structure, except for at the waveguide ports, are enclosed by a perfect electric conductor (PEC). The incident wave is polarized with the magnetic field along the out-of-plane axis.

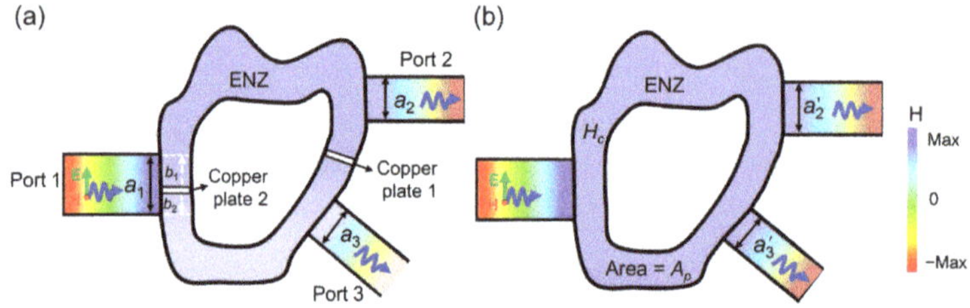

Figure 1. Schematic illustration of the proposed 2-D ENZ-based RPD. (**a**) Magnetic field-based tunability with equal output waveguide widths. (**b**) Geometry-based tunability with a uniform magnetic field.

Transmission efficiency is an essential index of power dividers, which is quantified as $\eta = |S_{21}|^2 + |S_{31}|^2 = 1 - |S_{11}|^2 - \alpha_{loss}$ [35], where α_{loss} denotes the material loss factor. To maximize signal transmission efficiency, reflected waves and energy loss should be suppressed as low as possible. Due to the low-loss characteristic of the ENZ waveguide media, α_{loss} approaches zero. The reflection coefficient R at Port 1 can be expressed as [36–38]

$$R = \frac{(a_1 - a_2 - \ldots - a_n) + ik_0\mu_{r,p}A_p}{(a_1 + a_2 + \ldots + a_n) - ik_0\mu_{r,p}A_p}, \tag{1}$$

which can be further written as

$$R = \frac{1 + ik_0\mu_{r,p}A_p/a_1 - \sum_{n=2}^{N} a_n/a_1}{1 - ik_0\mu_{r,p}A_p/a_1 + \sum_{n=2}^{N} a_n/a_1}, \tag{2}$$

where k_0 is the free-space wavenumber, $\mu_{r,p}$ denotes the relative permeability of the medium inside the ENZ channel, A_p is the cross-sectional area of the channel, a_1 is the width of the input waveguide, and the widths of the output waveguides are $a_2, a_3, \ldots a_n$ (with $n = 3$ in the conceptual illustration). When the output waveguide widths satisfy $\sum_{n=2}^{N} a_n = a_1$ and the ENZ channel area approaches zero ($k_0\mu_{r,p}A_p \to 0$), the reflection coefficient R tends toward zero, allowing EM waves to be efficiently squeezed and transmitted through a narrow, arbitrarily shaped ENZ channel.

Because the magnetic fields are equal at the interfaces between the ENZ region and the input/output waveguides, the power flow densities are also identical at both interfaces under the perfect tunneling condition. Thus, the time-averaged input power can be calculated as [37,38]

$$P_{in} = \eta_0 |H_c|^2 (1 - |R|^2) a_1. \tag{3}$$

Accordingly, the output power of each port can be written as

$$P_n = \eta_0 |H_c(1 + R)|^2 a_n, \tag{4}$$

where η_0 is the free-space impedance, and H_c denotes the magnetic field inside the ENZ region. By reducing R to near zero, the output power can be simplified as

$$P_n = \eta_0 |H_c|^2 a_n. \tag{5}$$

This relation indicates that the output power can be modulated by tuning the magnetic field (H_c) within the sub-channels and/or the widths (a_n) of the output waveguides.

Figure 1a illustrates the method of achieving arbitrary power division by only tuning the internal magnetic field (H_c). To minimize reflection, the input waveguide width a_1 is set equal to the sum of the output widths, i.e., $a_1 = a_2 + a_3$. Equal output widths ($a_2 = a_3$) are chosen to ensure identical output cross-sectional areas. To modulate the magnetic field, two copper plates are inserted into the ENZ channel. Copper plate 1 is located between the output waveguides and acts as an EM barrier to suppress inter-port crosstalk. Copper plate 2 is positioned near the entrance of each sub-channel, where it alters the effective entrance width (b_1 or b_2) and introduces localized field perturbations. Due to the quasi-static field distribution and the high sensitivity of the ENZ region to local changes, such perturbations can significantly redistribute the internal magnetic field, thereby enabling tunable power P_n delivery to each output port. Therefore, the output power ratio between Port 2 and Port 3 can be estimated from the width ratio of b_1/b_2. In this manner, precise and reconfigurable power division can be achieved by tuning the internal magnetic field.

As conceptually sketched in Figure 1b, dynamic power division can also be achieved by tuning the output waveguide widths (a_n). Similarly, to suppress reflected waves, these widths are set to satisfy the relation $a_1 = a_2^{'} + a_3^{'}$. Given the spatially uniform magnetic field profile inside the ENZ channel, the power distribution among the output ports is determined solely by their cross-sectional areas. Accordingly, arbitrary power splitting ratios between Port 2 and Port 3 are attainable with the required output waveguide width ratio of $a_2^{'}/a_3^{'}$.

Guided by the theoretical analysis, two types of highly efficient ENZ-based power dividers were developed with dynamic and on-demand power division by tuning the internal magnetic field (H_c) and the widths of the output waveguides (a_n), respectively.

3. Practical Designs of Reconfigurable ENZ-Based Waveguide Power Dividers

3.1. Mechanically Reconfigurable ENZ-Based Power Divider Based on Magnetic Field Redistribution

A mechanically reconfigurable ENZ-based power divider is proposed to enable continuous and arbitrary tuning of power division ratios via adjustment of the internal magnetic field (H_c). As displayed in Figure 2a,b, the entire structure is enclosed by metallic walls with copper, except for at the interfaces of three rectangular waveguide ports. An air-filled cylindrical waveguide with a height of $h = 50$ mm and a radius of $r_1 = 20$ mm is integrated with a copper post (radius of $r_2 = 18$ mm and conductivity of $\sigma = 5.998 \times 10^7$ S/m) to form a 2 mm wide narrow annular channel. Operating near the TE_{10} cut-off frequency (i.e., $f_c = 0.5c/h = 3$ GHz), the channel exhibits equivalent ENZ behavior, enabling efficient wave tunneling across geometric discontinuities. The ENZ channel is coupled to three rectangular waveguide ports, namely, one input port (Port 1) and two output ports (Ports 2 and 3). As indicated in the previous theoretical analysis, to minimize the reflection coefficient according to Equation (2), the input port width is set to 20 mm, while the output ports are designed with equal widths of 10 mm each. All waveguides are filled with polytetrafluoroethylene (PTFE, $\varepsilon = 2.1$, and $\tan\delta = 0.0004$) and are symmetrically arranged at 120° intervals.

Two 1 mm thick copper plates are symmetrically installed within the ENZ channel and electrically connected to the surrounding metallic enclosure. Copper plate 1 is introduced to suppress coupling between the output ports, and copper plate 2 divides the annular ENZ channel into two sub-channels. By rotating copper plate 2, the magnetic field is redistributed between the two sub-channels, enabling direct modulation of the output power delivered to each output waveguide (as indicated by the dashed arrows). Accordingly, the power ratio between Port 2 and Port 3 is governed by the ratio of their effective input widths (b_1/b_2). Notably, both plates are synchronously rotated via a central copper post for practical feasibility, and the position of copper plate 1 has little influence on the EM performance. To enable manually tunable control, the central copper post is mechanically connected to a disk and base via a rotatable supporting shaft so that disk rotation simultaneously drives both the post and copper plates, allowing precise and repeatable adjustment of the plate rotation angles in real time. Overall, despite minor angular deviations and a tiny shift of the central post during rotation, the ENZ channel maintains highly efficient tunneling with negligible impact on the overall transmission.

Full-wave simulations were performed using the commercial software COMSOL (Version 6.1) to evaluate the performance of the RPD at an operating frequency of 3 GHz. As shown in Figure 2c, the output power varies continuously with the rotation angle θ ranging from $-29°$ to $29°$. When Port 1 is excited with a total power of 1 W, the output power at Port 2 increases from 0 to 0.93 W, while that at Port 3 simultaneously decreases from 0.93 to 0 W. This indicates a continuous and controllable redistribution of EM power between the two output ports, with the reflected power at Port 1 remaining below approximately 0.07 W. Owing to the matched input and output waveguide widths, the channel supports efficient tunneling of EM waves through the low-loss ENZ waveguide. While the finite channel width leads to a slight increase in the reflection coefficient R, the transmission efficiency consistently exceeds 93% across the entire tuning range.

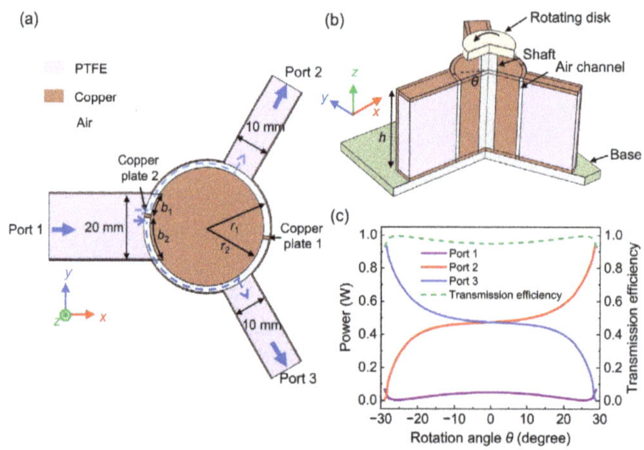

Figure 2. Mechanically reconfigurable ENZ-based power divider for arbitrary power division via magnetic field modulation. (**a**) Top view and (**b**) perspective view (with part of it removed to illustrate its inside) of the proposed power divider. (**c**) Simulated power distribution and transmission efficiency with different copper plate rotation angles (θ). Mechanical rotation is achieved via a copper rotating disk, a shaft, and a fixed base, providing rigid and stable support without affecting the EM performance.

To characterize the frequency-dependent performance, the simulated results of the proposed power divider with exemplary power ratios of 1:1, 1:2, and 0:1 between Ports 2 and 3 are presented in Figure 3a–c, respectively. As the first example of equal power splitting, copper plate 2 is aligned with the input waveguide axis with $\theta = 0°$ and $b_1 = b_2$. As shown in Figure 3a, simulated S_{21} and S_{31} are identical to -3.07 dB at the ENZ frequency, which conforms to the design target and leads to a transmission efficiency of 98%. However, the reflection introduced by copper plate 2 slightly shifts the optimal operating frequency.

Additionally, a balanced power flow density between the two sub-channels is observed in Figure 3d, and the elevated power flow density within the ENZ channel is mainly due to its reduced cross-sectional area. The power split ratio of 1:2 between Port 2 and Port 3 is investigated as the second example of imbalanced power division. By setting the rotation angle θ to $-23.5°$, simulated $S_{21} = -4.86$ dB and $S_{31} = -1.78$ dB are obtained at 3 GHz, as plotted in Figure 3b, and the simulated power flow densities within the sub-channels are approximately 0.65×10^4 W/m^2 at Port 2 and 1.3×10^4 W/m^2 at Port 3, as exhibited in Figure 3e, respectively. The desired power splitting is implemented with a high transmission efficiency of 99%.

A single output port configuration is explored as the third example, in which copper plate 2 is rotated to $-29°$. This further rotation places the plate near the input waveguide edge, effectively suppressing the EM propagation toward Port 2. It is seen in Figure 3c that the simulated S_{21} drops below -170 dB, and Port 3 retains near-unity transmission with $S_{31} = -0.32$ dB. As shown in Figure 3f, the power flow within the sub-channels diminishes at Port 2, and nearly all the transmitted energy is effectively directed to Port 3. The transmission efficiency is calculated as 93% and is slightly lower than that in previous cases. According to Equation (2), the reduced transmission efficiency is primarily attributed to increased reflections, which is caused by a reduced number of output ports and mismatches in waveguide widths. It is noted that positive rotation angles invert the magnetic field within the sub-channels, yielding a power splitting pattern exactly opposite to that with negative rotation angles.

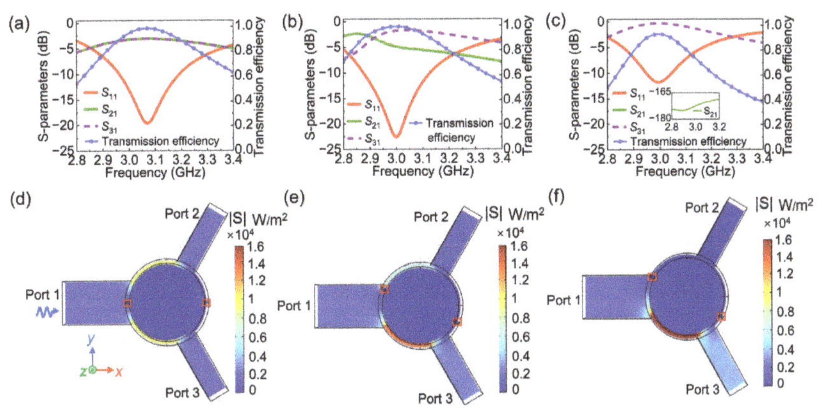

Figure 3. Simulated S-parameters, transmission efficiency, and power flow distributions at the ENZ frequency. S-parameters and transmission efficiency with (**a**) $\theta = 0°$, (**b**) $\theta = -23.5°$, and (**c**) $\theta = -29°$. Power flow distributions with (**d**) $\theta = 0°$ (**e**) $\theta = -23.5°$, and (**f**) $\theta = -29°$.

In this design, efficient arbitrary power division is achieved by redistributing the magnetic field within the ENZ channel in a mechanically reconfigurable manner. All reconfiguration modes exhibit a reasonable bandwidth around the ENZ frequency, achieving transmission efficiencies above 93%, even when considering conductor loss, dielectric loss, and dimensional tolerances.

3.2. *Electrically Reconfigurable ENZ-Based Power Divider Based on Cross-Sectional Area Modulation*

An electrically reconfigurable ENZ-based power divider is further proposed to facilitate on-demand control over both power division modes and the number of output ports by dynamically tuning the cross-sectional areas of the output waveguides.

To increase the number of output ports, we propose a double-layer SIW power divider. The equivalent width a' of the SIW is calculated from the width a of the conventional rectangular waveguide as [39]

$$a' = \frac{2a}{\pi}\mathrm{arcctg}(\frac{\pi W}{4a}\ln\frac{W}{4R}), \quad (6)$$

where R denotes the radius of the metallic vias, and W is the center-to-center spacing between adjacent vias. The formulation is valid under the constraint $W \leqslant 1/20\lambda_g$, where λ_g represents the guided wavelength. When satisfying $R = W/4$, the effective SIW width a' equals a, ensuring propagation characteristics equivalent to a conventional rectangular waveguide. Accordingly, the design adopts R = 1.25 mm, W = 5 mm, and a' = 50 mm, which correspond to a standard rectangular waveguide with a width of 50 mm. Owing to the magnetic field homogeneity at the cut-off frequency, the power delivered to each output port is primarily determined by the cross-sectional areas of the output waveguides. By precisely tailoring these dimensions, various power division ratios can be effectively realized, including both equal and unequal splitting among the output ports.

As illustrated in Figure 4, the proposed electrically reconfigurable SIW power divider is fully enclosed by metallic walls with a height of h = 6 mm, except for at the SIW interfaces. A rectangular copper block embedded in the center of an air-filled waveguide forms a rectangular-loop-shaped transmission channel with a width of w_1 = 0.2 mm and a length of l = 50 mm. This configuration yields a cut-off frequency of 3 GHz and facilitates equivalent ENZ tunneling behavior. The central ENZ channel interconnects four SIWs filled with F4B ($\varepsilon = 2.2$ and tanδ = 0.001), with Port 1 designated as the input and Ports 2 to 4 designated as the outputs. To suppress inter-channel coupling and ensure EM compatibility, metallic isolation layers with heights of h_5 = 2 mm and h_6 = 3 mm are incorporated between adjacent SIWs. The cross-sectional areas of the output waveguides are precisely engineered with h_1 = 3 mm, $h_2 = h_3$ = 1 mm, and h_4 = 2 mm.

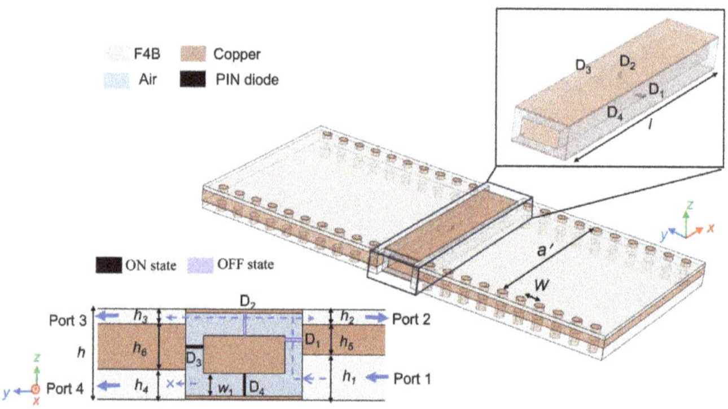

Figure 4. Schematic diagram of the electrically reconfigurable power divider with on-demand power division via tunable output waveguide cross-sectional areas. The enlarged inset highlights the rectangular-loop channel, with four PIN diodes (D_1–D_4) embedded on its four faces. The dashed arrows depict the propagation direction of EM waves within the rectangular-loop channel under the condition that D_1 and D_2 are OFF while D_3 and D_4 are ON.

To enable effective dynamic modulation of the EM responses, four PIN diodes (D_1–D_4) are strategically positioned near the region with a concentrated electric field intensity. Specifically, D_2 and D_4 are mounted at the midpoints of the upper and lower channel walls, whereas D_1 and D_3 are mounted at the midpoints between the central isolating layer and the rectangular copper block. Each diode (SMP 1320-079LF, Skyworks) is modeled as a series resistor–inductor–capacitor (RLC) circuit, with parameters R_d = 0.5 Ω and L = 0.7 nH in the ON state and L = 0.5 nH and C = 0.24 pF in the OFF state. The cathodes of these diodes are connected to the inner rectangular copper block, while the anodes are individually connected to the four outer walls of the ENZ channel. When the PIN diodes are switched on, they effectively behave as conductive "copper plates", blocking or redirecting EM wave propagation within the rectangular-loop channel. Conversely, when the diodes are switched off, nearly unobstructed wave transmission through the rectangular-loop ENZ channel is enabled. By appropriately switching the diodes, both the output ports and internal EM field distributions can be dynamically reconfigured, enacting on-demand power division with varying cross-sectional areas.

As an example of equal power splitting between two output ports, State 1 is realized by switching on D_3 and D_4 while turning off D_1 and D_2. In this configuration, Port 4 is effectively isolated and results in a two-port output of Ports 2 and 3. The power ratio between them is determined by their corresponding cross-sectional areas, which are designed to be equal and confirmed by the simulation results shown in Figure 5a. In this state, the transmission efficiency reaches 93%. In contrast, State 2 is configured by turning on D_1 and D_2 while switching off D_3 and D_4, thereby blocking Port 2 and redirecting the energy to Ports 3 and 4. The area ratio between Ports 4 and 3 is approximately 2:1, leading to an unequal power split of 2:1, as illustrated in Figure 5b. Notably, the total output width in this case exactly matches that of the input, minimizing the reflection coefficient and achieving a high transmission efficiency of 97%. When all diodes are turned off (State 3), Ports 2, 3, and 4 all serve as output ports, resulting in a three-port power distribution. The simulated results exhibit an asymmetric power ratio close to 1:1:2, as shown in Figure 5c. This state approximately satisfies the ideal matching condition, resulting in a slightly lower transmission efficiency of 95%. Due to the ENZ tunneling effect, energy can freely propagate to the activated ports, leading to unavoidable inter-port coupling and reduced isolation. Despite the differences in output port numbers and power splitting ratios, it is observed in Figure 5d–f that the simulated magnetic fields in the output waveguides are in phase and of equal amplitude.

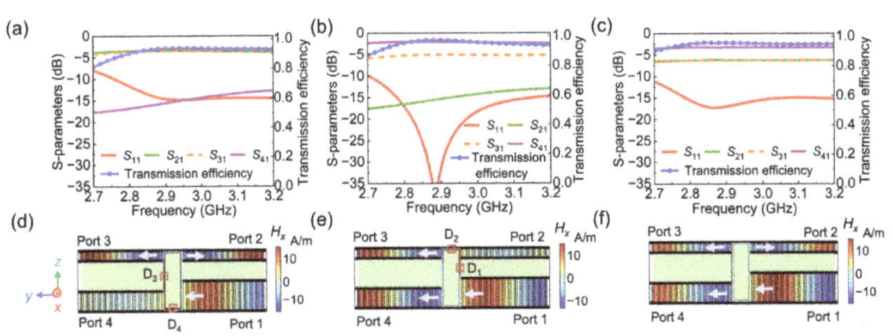

Figure 5. Simulated S-parameters, transmission efficiency, and magnetic field distributions of the SIW-based RPD. S-parameters and transmission efficiency of the power divider with (**a**) D_3 and D_4 turned on, (**b**) D_1 and D_2 turned on, and (**c**) all diodes turned off. Magnetic field distributions with (**d**) D_3 and D_4 turned on, (**e**) D_1 and D_2 turned on, and (**f**) all diodes turned off. Arrows positioned at the input and output ports represent the direction of power flow.

These simulated results collectively confirm that the proposed ENZ-based RPD exhibits stable and efficient transmission performance in multiple operation modes. Notably, delicate variation in the cross-sectional areas of the output SIWs allows for controllable and on-demand power division among the output ports. Furthermore, the number of output ports can be dynamically reconfigured, allowing flexible transitions between two-port and three-port power division configurations. The achievable power split ratios are not limited to those analyzed herein, and, in general, more possibilities could be realized by varying the polarity configurations of the PIN diodes. In addition to high reconfigurability and mode selectivity, the proposed power divider features a simple structure and practical feasibility.

4. Conclusions

In summary, this work presents versatile, high-efficiency, reconfigurable power division based on the tunneling characteristics of waveguide ENZ media. Two reconfigurable control mechanisms are introduced, supporting a flexible power distribution by tuning the magnetic field within the ENZ channel or adjusting the widths of the output waveguides. To validate these theoretical concepts, two distinct experimentally feasible implementations are presented. Compared with previous works reporting insertion losses of 0.88 dB [7] and 0.60 dB [33], the proposed design achieves a transmission efficiency of 93%, corresponding to an insertion loss of approximately 0.31 dB. It also enables dynamic tunability of power division ratios and flexible reconfigurability of output ports while maintaining a reasonable bandwidth around the ENZ frequency and a simple profile. These results confirm the feasibility of efficient ENZ-based RPDs, which may pave the way for RF front ends, real-time wireless communications, and integrated ENZ-based microwave systems.

Author Contributions: Conceptualization, Y.F. and L.J.; methodology, Y.F. and L.J.; software, L.J.; validation, L.J. and Q.H.; data curation, L.J.; writing—original draft preparation, L.J. and Q.H.; writing—review and editing, Q.H. and Y.F.; supervision, Q.H. and Y.F. All authors have read and agreed to the published version of the manuscript.

Funding: This work was supported in part by the Jiangsu Funding Program for Excellent Postdoctoral Talent under Grant 2024ZB619, the Postdoctoral Fellowship Program of China Postdoctoral Science Foundation under Grant GZC20240676, the China Postdoctoral Science Foundation under Grant 2024M761395, the National Natural Science Foundation of China (NSFC) under Grant 62271243 and 62471216, and the Jiangsu Provincial Key Laboratory of Advanced Manipulating Technique of Electromagnetic Wave.

Institutional Review Board Statement: Not applicable.

Informed Consent Statement: Not applicable.

Data Availability Statement: The original contributions presented in this study are included in the article. Further inquiries can be directed to the corresponding author.

Conflicts of Interest: The authors declare no conflicts of interest.

Abbreviations

The following abbreviations are used in this manuscript:

ENZ	epsilon-near-zero
RPD	reconfigurable power divider
SIW	substrate-integrated waveguide
PIN	positive–intrinsic–negative
EM	electromagnetic
RF	radio-frequency
MIMO	multi-input–multi-output
IoT	Internet of Things
PTFE	polytetrafluoroethylene
PEC	perfect electric conductor
2-D	two-dimensional

References

1. Han, L.; Kuo, B.P.P.; Alic, N.; Radic, S. Ultra-broadband multimode 3dB optical power splitter using an adiabatic coupler and a Y-branch. *Opt. Express* **2018**, *26*, 14800. [CrossRef]
2. Pallavi, M.; Kumar, P.; Ali, T.; Shenoy, S.B.; Shivakumar, B. Design and validation of a miniaturized reconfigurable power divider with arbitrary power split ratio and flexible output phase difference. *Int. J. Electron. Commun.* **2024**, *183*, 155373. [CrossRef]
3. Wu, Y.; Jiao, L.; Zhuang, Z.; Liu, Y. The art of power dividing: A review for state-of-the-art planar power dividers. *China Commun.* **2017**, *14*, 1–16. [CrossRef]
4. Lee, S.; Park, J.; Hong, S. Millimeter-Wave Multi-Band Reconfigurable Differential Power Divider for 5G Communication. *IEEE Trans. Microw. Theory Tech.* **2022**, *70*, 886–894. [CrossRef]
5. Jamshidi, M.B.; Roshani, S.; Talla, J.; Roshani, S.; Peroutka, Z. Size reduction and performance improvement of a microstrip Wilkinson power divider using a hybrid design technique. *Sci. Rep.* **2021**, *11*, 7773. [CrossRef]
6. Wang, S.; Zhang, Y.; Liu, Z.; Zhang, J.; Zhou, Z.; Chen, W.; Zhang, Z.; Chen, X.; Wei, K.; Li, Y. Antenna Thousandfold Miniaturization with Ohmic-Biased Transistor Circuit. *Electromagn. Sci.* **2025**, *3*, 0100583. [CrossRef]
7. Antoniades, M.; Eleftheriades, G. A broadband series power divider using zero-degree metamaterial phase-shifting lines. *IEEE Microw. Wireless Compon.* **2005**, *15*, 808–810. [CrossRef]
8. Guo, Z.; Jiang, J.; Wang, Y.; Alvarez-Cuervo, J.; Martin-Luengo, A.T.; Hu, S.; Jiang, J.; Gonzalez, P.A.; Duan, J.; Chen, H. Exceptional point empowered near-field routing of hyperbolic polaritons. *Sci. Bull.* **2024**, *69*, 3491–3495. [CrossRef]
9. Kinsey, N.; DeVault, C.; Boltasseva, A.; Shalaev, V.M. Near-zero-index materials for photonics. *Nat. Rev. Mater.* **2019**, *4*, 742–760. [CrossRef]
10. Liberal, I.; Engheta, N. Near-zero refractive index photonics. *Nat. Photonics* **2017**, *11*, 149–158. [CrossRef]
11. Liberal, I.; Mahmoud, A.M.; Li, Y.; Edwards, B.; Engheta, N. Photonic doping of epsilon-near-zero media. *Science* **2017**, *355*, 1058–1062. [CrossRef]
12. Maas, R.; Parsons, J.; Engheta, N.; Polman, A. Experimental realization of an epsilon-near-zero metamaterial at visible wavelengths. *Nat. Photonics* **2013**, *7*, 907–912. [CrossRef]
13. Ziolkowski, R.W. Propagation in and scattering from a matched metamaterial having a zero index of refraction. *Phys. Rev. E* **2004**, *70*, 046608. [CrossRef]
14. Li, H.; Zhou, Z.; He, Y.; Sun, W.; Li, Y.; Liberal, I.; Engheta, N. Geometry-independent antenna based on Epsilon-near-zero medium. *Nat. Commun.* **2022**, *13*, 3568. [CrossRef]
15. Li, H.; Zhou, Z.; Li, Y. Length-Irrelevant Dual-Polarized Antenna Based on Antiphase Epsilon-Near-Zero Mode. *IEEE Trans. Antennas Propag.* **2022**, *70*, 720–725. [CrossRef]
16. Liberal, I.; Li, Y.; Engheta, N. Magnetic field concentration assisted by epsilon-near-zero media. *Philos. Trans. R. Soc. Math. Phys. Eng. Sci.* **2017**, *375*, 20160059. [CrossRef] [PubMed]
17. Zhao, L.; Feng, Y.J. Optical field enhancements and applications by epsilon-near-zero medium with dielectric dopant. *Acta Phys. Sin.* **2020**, *69*, 154101. [CrossRef]
18. Zhao, L.; Feng, Y.; Zhu, B.; Zhao, J. Electromagnetic properties of magnetic epsilon-near-zero medium with dielectric dopants. *Opt. Express* **2019**, *27*, 20073. [CrossRef]
19. Zhou, Z.; Li, Y. A Photonic-Doping-Inspired SIW Antenna with Length-Invariant Operating Frequency. *IEEE Trans. Antennas Propag.* **2020**, *68*, 5151–5158. [CrossRef]
20. Edwards, B.; Alù, A.; Silveirinha, M.G.; Engheta, N. Reflectionless sharp bends and corners in waveguides using epsilon-near-zero effects. *J. Appl. Phys.* **2009**, *105*, 044905. [CrossRef]
21. Mitrovic, M.; Jokanovic, B.; Vojnovic, N. Wideband Tuning of the Tunneling Frequency in a Narrowed Epsilon-Near-Zero Channel. *IEEE Antennas Wirel. Propag. Lett.* **2013**, *12*, 631–634. [CrossRef]
22. Yan, W.; Zhou, Z.; Li, H.; Li, Y. Transmission-type photonic doping for high-efficiency epsilon-near-zero supercoupling. *Nat. Commun.* **2023**, *14*, 6154. [CrossRef] [PubMed]
23. Zhou, Z.; Li, H.; Sun, W.; He, Y.; Liberal, I.; Engheta, N.; Feng, Z.; Li, Y. Dispersion coding of ENZ media via multiple photonic dopants. *Light Sci. Appl.* **2022**, *11*, 207. [CrossRef]
24. Zhou, Z.; Li, Y.; Li, H.; Sun, W.; Liberal, I.; Engheta, N. Substrate-integrated photonic doping for near-zero-index devices. *Nat. Commun.* **2019**, *10*, 4132. [CrossRef]
25. Zhou, Z.; Li, Y.; Nahvi, E.; Li, H.; He, Y.; Liberal, I.; Engheta, N. General Impedance Matching via Doped Epsilon-Near-Zero Media. *Phys. Rev. Appl.* **2020**, *13*, 034005. [CrossRef]
26. Wang, Y.; Lin, J.; Xu, P. Transmission-reflection decoupling of non-Hermitian photonic doping epsilon-near-zero media. *Front. Phys.* **2024**, *19*, 33206. [CrossRef]
27. Hong, B.; Sun, L.; Wang, W.; Qiu, Y.; Feng, N.; Su, D.; Somjit, N.; Robertson, I.; Wang, G.P. Five-channel frequency-division multiplexing using low-loss epsilon-near-zero metamaterial waveguide. *Sci. China Phys. Mech. Astron.* **2022**, *65*, 274211. [CrossRef]

28. Edwards, B.; Alù, A.; Young, M.E.; Silveirinha, M.; Engheta, N. Experimental Verification of Epsilon-Near-Zero Metamaterial Coupling and Energy Squeezing Using a Microwave Waveguide. *Phys. Rev. Lett.* **2008**, *100*, 033903. [CrossRef]
29. Li, P.; Yan, W.; Wang, S.; Fu, P.; Zhang, Y.; Li, Y. Engineering Epsilon-Near-Zero Media with Waveguides. *Adv. Phys. Res.* **2024**, *3*, 2400070. [CrossRef]
30. Li, Y.; Liberal, I.; Della Giovampaola, C.; Engheta, N. Waveguide metatronics: Lumped circuitry based on structural dispersion. *Sci. Adv.* **2016**, *2*, e1501790. [CrossRef] [PubMed]
31. Liu, R.; Cheng, Q.; Hand, T.; Mock, J.J.; Cui, T.J.; Cummer, S.A.; Smith, D.R. Experimental Demonstration of Electromagnetic Tunneling Through an Epsilon-Near-Zero Metamaterial at Microwave Frequencies. *Phys. Rev. Lett.* **2008**, *100*, 023903. [CrossRef] [PubMed]
32. Rotman, W. Plasma simulation by artificial dielectrics and parallel-plate media. *IRE Trans. Antennas Propag.* **1962**, *10*, 82–95. [CrossRef]
33. Zhou, Z.; Li, Y. N -Port Equal/Unequal-Split Power Dividers Using Epsilon-Near-Zero Metamaterials. *IEEE Trans. Microw. Theory Tech.* **2021**, *69*, 1529–1537. [CrossRef]
34. Esfahlani, H.; Byrne, M.S.; Alù, A. Acoustic Power Divider Based on Compressibility-Near-Zero Propagation. *Phys. Rev. Appl.* **2020**, *14*, 024057. [CrossRef]
35. Collin, R.E. *Foundations for Microwave Engineering*, 2nd ed.; John Wiley & Sons: Hoboken, NJ, USA, 2007.
36. Ourir, A.; Maurel, A.; Pagneux, V. Tunneling of electromagnetic energy in multiple connected leads using ε-near-zero materials. *Opt. Lett.* **2013**, *38*, 2092. [CrossRef] [PubMed]
37. Silveirinha, M.; Engheta, N. Tunneling of electromagnetic energy through subwavelength channels and bends using ε-near-zero materials. *Phys. Rev. Lett.* **2006**, *97*, 157403. [CrossRef]
38. Silveirinha, M.G.; Engheta, N. Theory of supercoupling, squeezing wave energy, and field confinement in narrow channels and tight bends using ε near-zero metamaterials. *Phys. Rev. B* **2007**, *76*, 245109. [CrossRef]
39. Che, W.; Xu, L.; Wang, D.; Deng, K.; Chow, Y. Short-circuit equivalence between rectangular waveguides of regular sidewalls (rectangular waveguide) and sidewalls of cylinders (substrate-integrated rectangular waveguides), plus its extension to cavity. *IET Microw. Antennas Propag.* **2007**, *1*, 639–644. [CrossRef]

photonics

MDPI

Article

Tunable Ultra-Wideband VO_2–Graphene Hybrid Metasurface Terahertz Absorption Devices Based on Dual Regulation

Kele Chen [1,2], Zhengning Wang [1,*], Meizhang Guan [2], Shubo Cheng [3], Hongyu Ma [4], Zao Yi [4,5] and Boxun Li [6]

1 School of Information and Communication Engineering, University of Electronic Science and Technology of China, Chengdu 611731, China; chenkele1988ioe@sina.com
2 Sichuan Intronics Electronic Technology Co., Ltd., Suining 629000, China; gm794998199@163.com
3 School of Physics and Optoelectronic Engineering, Yangtze University, Jingzhou 434023, China; shubocheng@yangtzeu.edu.cn
4 Joint Laboratory for Extreme Conditions Matter Properties, Southwest University of Science and Technology, Mianyang 621010, China; 15082734560@163.com (H.M.); yizaomy@swust.edu.cn (Z.Y.)
5 School of Chemistry and Chemical Engineering, Jishou University, Jishou 416000, China
6 School of Physics and Optoelectronics, Xiangtan University, Xiangtan 411105, China; lbxcsu@xtu.edu.cn
* Correspondence: zhengning.wang@uestc.edu.cn; Tel./Fax: +86-0816-2480830

Abstract: In this study, a dynamically tunable terahertz device based on a VO_2–graphene hybrid metasurface is proposed, which realizes the dual functions of ultra-wideband absorption and efficient transmission through VO_2 phase transformation. At 345 K (metallic state), the device attains an absorption efficiency exceeding 90% (average 97.06%) in the range of 2.25–6.07 THz (bandwidth 3.82 THz), showing excellent absorption performance. At 318 K (insulated state), the device achieves 67.66–69.51% transmittance in the 0.1–2.14 THz and 7.51–10 THz bands while maintaining a broadband absorption of 3.6–5.08 THz (an average of 81.99%). Compared with traditional devices, the design breaks through the performance limitations by integrating phase change material control with 2D materials. The patterned graphene design simplifies the fabrication process. System analysis reveals that the device is polarization-insensitive and tunable via graphene Fermi energy and relaxation time. The device's excellent temperature response and wide angular stability provide a novel solution for terahertz switching, stealth technology, and sensing applications.

Keywords: phase change materials VO_2; graphene; terahertz absorber; tunable broadband

1. Introduction

In recent years, electromagnetic wave absorption technology has emerged as a cutting-edge research focus, demonstrating significant potential for applications in diverse domains, including energy conversion, environmental monitoring, and so on [1–3]. Among them, devices with broadband electromagnetic wave absorption characteristics are widely used in communications, military, medical diagnostics, and imaging fields [4–6]. In view of the depth of basic research and the potential for industrialization of this technology, the global scientific research team is continuously improving its performance indicators through various strategies, such as structural design innovation and material system optimization. At present, related research has given rise to a variety of new types of devices. According to bandwidth, these can be divided into wide/narrowband absorbers [7–9]. According to the response frequency band, they can be divided into single-frequency/multi-frequency absorbers [10,11]. According to configuration differences, metasurface absorbers and dynamically tunable absorbers are covered, and these breakthroughs have provided a solid basis for the advancement of electromagnetic wave absorption technology [12,13].

Among many functional materials, vanadium dioxide (VO_2), a phase change material, offers a novel approach to designing broadband absorbers due to its unique dynamic control characteristics [14,15]. As a typical strongly correlated phase change material, VO_2 has garnered significant interest in smart materials owing to its unique metal–insulator transition (MIT) characteristics [16,17]. When the temperature reaches a critical point (about 68 °C), VO_2 undergoes a reversible structural transition from monoclinic to tetragonal, accompanied by a sudden change in conductivity of up to five orders of magnitude (from 10^2 S/m to 10^5 S/m). This phase change process has significant thermal hysteresis properties—the transition starts at 340 K when heated and 334 K when cooled, with a temperature difference window of 6 K, ensuring stable material properties [18]. Even more strikingly, this phase transition can be completed in picoseconds, making it an ideal material for the development of highly sensitive terahertz devices [19,20]. Through multi-physics manipulation of optical, electrical, and thermal fields, VO_2 can achieve precise switching between the metallic state and the insulating state, and this dynamic tunability provides a novel approach to designing new intelligent devices, such as reconfigurable metamaterials and adaptive thermal control coatings [21,22].

Moreover, graphene materials also play a significant role in electromagnetic absorption. As a special form of carbon, graphene is unique in materials research with its unique single-atomic-layer two-dimensional structure [23–25]. This sp^2-bonded 2D material with a hexagonal honeycomb structure typically has a sheet size between 10 and 100 nanometers, and this microstructure characteristic allows it to exhibit excellent electromagnetic wave absorption characteristics in the mid-infrared to terahertz frequency band [26,27]. Compared with conventional functional materials, the material not only has excellent environmental stability and carrier migration characteristics but also exhibits excellent mechanical strength and electrical and thermal conductivity. It is particularly noteworthy that the unique properties of the Fermi level are controlled by external bias voltage, so electromagnetic wave absorption modulation devices based on this material can achieve dynamically adjustable absorption characteristics [28,29]. Notably, the combination of graphene and phase change VO_2 represents a relatively novel design paradigm in current terahertz absorber research. For instance, Shen et al. proposed a VO_2–graphene hybrid metamaterial absorber with excellent overall absorption performance, yet its seven-layered structure leads to high fabrication complexity, leaving room for structural optimization [30]. Complementing the phase transition control of VO_2, graphene provides another option for absorber design. Although there are still technical bottlenecks in the large-scale preparation and cost control of graphene, with the continuous optimization of the preparation process, this revolutionary material is moving from the laboratory to industrial application [31,32].

In this study, we combine graphene with the phase change material VO_2 to develop a new type of ultra-wideband response phase change device. The device consists of a classic three-layer thin film structure, with specially patterned graphene on the top, which is composed of several simple rectangular shapes. SiO_2 is employed as the dielectric layer and VO_2 as the substrate of the device. At the relatively high temperature of T = 345 K, VO_2 displays a metallic state, and this device shows perfect ultra-wideband absorption of terahertz bands, with an absorption efficiency of more than 90% in the frequency range of 2.25 THz–6.07 THz and a total of 3.82 THz, demonstrating an averaged absorption of 97.06%. At the relatively low temperature of T = 318 K, VO_2 exhibits a dielectric state, and this device demonstrates broadband absorption across the frequency band of 3.6 THz–5.08 THz, with an average absorption rate of 81.99%. At the same time, the device also exhibits good transmission ability at relatively low temperatures, exhibiting a mean transmittance of 67.66% and 69.51% across the frequency bands of 0.1–2.14 THz and 7.5–10 THz, respectively. First, we describe the fabrication process of the device, which is made easier to fabricate by using simple patterned graphene. Second, the phase transition mechanism of VO_2 is explained in detail, and the performance transition caused by this change is explored. Subsequently, we investigate the relative impedance matching and the interfacial E-field in the device at relatively high temperatures, analyze the absorption mechanism, and confirm the polarization-independent behavior of the device. On this basis, the effects of the chemical parameters of graphene, such as the Fermi energy level, relaxation time, and important physical parameters of the device's impact on absorption performance, are analyzed. At the end of the article, we analyze the effects of different polarizations and angles of incidence on the absorption performance of the device. Compared with previous articles, this study demonstrates that the device has excellent characteristics, such as high absorption and transmission efficiency, bias manufacturing, and phase change control, and it is believed that the device can shine in the fields of terahertz communication, terahertz switching, terahertz stealth, and medical detection.

2. Numerical Model and Structural Design

The proposed ultra-wideband phase change device is presented in Figure 1. Figure 1a shows a cell-level layout diagram of the ultra-wideband phase change device, which consists of a fundamental three-component thin film structure. As shown in Figure 1b, the topmost layer is a patterned 1 nm graphene layer composed of three simple rectangular graphene layers. The middle dielectric layer of the device is SiO_2, and the relevant dielectric parameter data comes from Kitamura [33]. The substrate layer of the device is VO_2, and the relevant dielectric constant is from reference [14]. The device exhibits excellent absorption and transmission characteristics at 0.1–10 THz. See Table 1 for the specific device parameters.

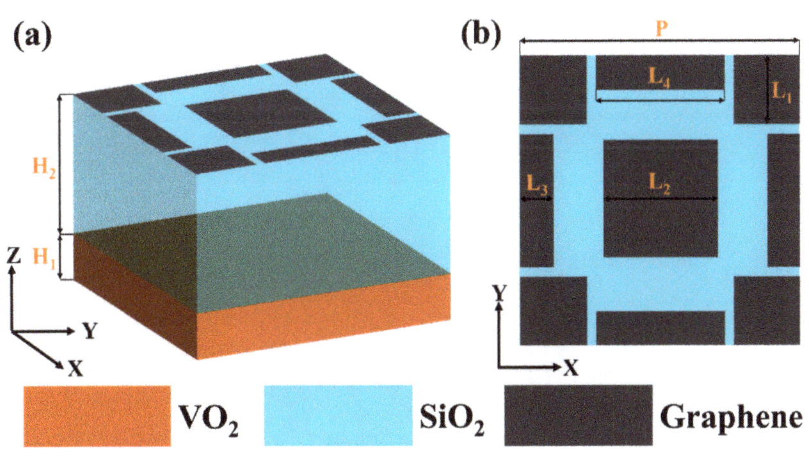

Figure 1. (**a**) A cell-level layout diagram of the ultra-wideband phase change device. (**b**) XY plane top view of the device.

Table 1. Parameters of an ultra-wideband phase change device.

Parameter	L1	L2	L3	L4	P	H1	H2
Value (μm)	5	8.5	2.5	9.6	21	5	12

As shown in Figure 2, we demonstrate a simple and feasible process for fabricating this ultra-wideband phase change device. Firstly, the VO_X film was deposited on the substrate by magnetron sputtering, and the VO_2 layer was formed by annealing in an O_2 atmosphere. A 12-micron-thick SiO_2 dielectric layer was then deposited by magnetron sputtering. Graphene sheets were prepared independently using CVD technology, the desired pattern was formed by electron beam lithography, and finally, the patterned graphene was transferred to the SiO_2 surface to complete the device fabrication [34,35].

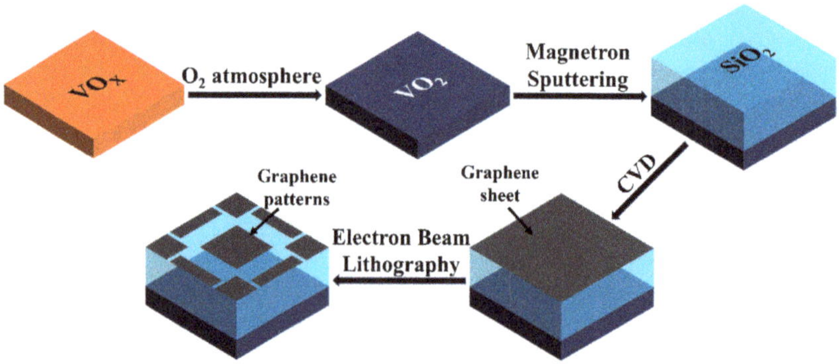

Figure 2. Simple layout representation of the fabrication process of the ultra-wideband phase change device.

After constructing the basic model structure diagram, we used CST simulation software to simulate the electromagnetic behavior of the model and used the FDID algorithm to simulate the model [36,37]. First, periodic boundary conditions were applied along the X and Y axes, while the Z-axis remained open, along which the electromagnetic wave travels in the negative Z-axis direction [38]. Based on this design, we were able to simulate and calculate the entire periodic array using only a single device unit, as shown in Figure 1a, which greatly reduces the time required for calculation and speeds up the research while maintaining the computational precision. In addition, we used the S-parameter inversion method to represent the reflectance (R) and transmittance (T) of the device, respectively; thus, the absorption rate (A) can be easily obtained, demonstrating the device's capability, and the specific formula is as follows [39,40]:

$$A(\omega) = 1 - R(\omega) - T(\omega) = 1 - |S_{11}|^2 - |S_{21}|^2 \tag{1}$$

In the above formula, $T(\omega)$ and $R(\omega)$ are the transmittance and reflectivity of the impinging optical wave when irradiating the structure of the absorbing unit, respectively, ω is the angular frequency, and $|S_{11}|^2$ and $|S_{21}|^2$ indicate the reflection and transmission coefficients, respectively.

In addition, we used the Drude model to represent the surface conductivity of the graphene layer used in the ultra-wideband phase change device during the simulation, as follows [14,41]:

$$\sigma(\omega) = \frac{e^2 E_F}{\pi h^2} \frac{i}{\omega + i\tau^{-1}} \tag{2}$$

In Equation (2), where ω and h denote the incident wave frequency and reduced Planck constant, respectively, defining E_F as the Fermi level and τ the relaxation time in graphene, these two parameters are as follows [42]:

$$E_F = V_F \sqrt{\frac{\pi \varepsilon_0 \varepsilon_\gamma V_0}{e_0 n_s}} \tag{3}$$

$$\tau = \frac{\mu E_F}{e_0 V_i^2} \tag{4}$$

In Equation (3), V_0 represents the magnitude of the external voltage, e_0 refers to the number of electron charges, V_F represents the Fermi velocity, and n_s is used to represent the number of layers of SiO_2 in the structure. In Equation (4), $\mu = 10^4$ cm^2/(V.S) represents the carrier mobility, $V_i = 10^6$ m/s is the aforementioned Fermi velocity, and E_F is the Fermi level in Equation (3). Based on this, we can easily come up with the corresponding dielectric constant formula [43]:

$$\varepsilon = 1 + \frac{i\sigma(\omega)}{\varepsilon_0 t\omega} \tag{5}$$

In Equation (5), ε_0 represents the conductivity in a vacuum environment and t denotes the graphene layer's thickness. The graphene used in this paper has a thickness of 1 nm, a relaxation time of 0.1 ps, and a Fermi level of 1.0 ev.

After the construction of graphene materials was completed, we also employed the Drude model to establish VO_2 materials in the terahertz band, and the relevant formulas are as follows [14,44]:

$$\varepsilon(\omega)_{VO_2} = \varepsilon_\infty - \frac{\omega_p^2}{\omega(\omega + i\gamma)} = \varepsilon_\infty - \frac{\sigma\omega_{p0}^2}{\sigma_0\omega(\omega + i\gamma)} \tag{6}$$

where ε_∞ is used to express the dielectric constant at high frequencies with a value of 12. γ is the collision frequency; the value is 5.75×10^{13} rad/s. ω_p is the plasma frequency, ω_{p0}^2 is 1.4×10^{15} rad/s, and the σ_0 value is 3×10^5 s/m. Based on the above data, we obtained the relationship between the conductivity of VO_2 and the permittivity using Equation (6) above. In the simulation software, we can represent VO_2 at different temperatures by changing the conductivity of VO_2, and the related temperature and state relationship is shown in Figure 3. As the temperature increases from 318 K to 345 K, the dielectric-to-metal transition occurs in VO_2 due to its phase transformation properties, with the corresponding thermal and dielectric property curves during heating/cooling shown in Figure 3a.

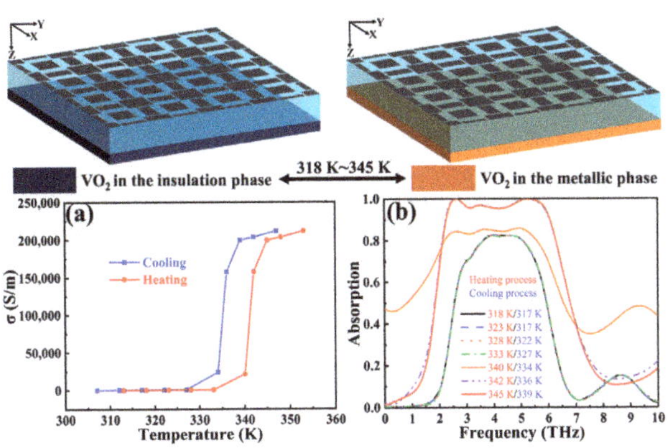

Figure 3. Vanadium dioxide undergoes phase transformation during a temperature change from 318 K to 345 K. (**a**) The conductivity variations with temperature are plotted as curves for VO_2 film under heating and cooling. (**b**) Variation of the absorption spectrum (unitless) of the ultra-wideband phase change device at different temperatures under heating and cooling.

3. Parameter Calculation and Result Analysis

After the above analysis and simulation process was completed, we calculated the absorption efficiency of the ultra-wideband phase change device in the cooling (317–339 K) and heating (318–345 K) states, respectively, and plotted the absorption spectrum of the device under the corresponding conditions, as illustrated in Figure 3b. The data in the figure illustrates that when heated to 342–345 K, VO_2 transitions to the metallic phase, and the device exhibits ultra-wideband perfect absorption in the terahertz band, with absorption efficiency exceeding 99% at a specific frequency point. In the relatively low temperature range of T = 318–333 K, VO_2 shows an insulating phase, and the device still maintains excellent broadband absorption, exhibiting an average absorption rate of 81.99%. It is worth noting that the cooling and heating processes exhibit a thermal hysteresis effect of 6 K, but the overall variation law is basically the same. These results fully show that the device has temperature control characteristics, and the absorption of the device in the terahertz band exhibits temperature-dependent characteristics, and the absorption efficiency can be changed from 0 to 99% over a wide range.

The thermal modulation of absorption was quantified through temperature-dependent spectra at 345 K (Figure 4a), highlighting the device's optothermal response. At 345 K, the device produces an ultra-wideband absorption of more than 90% spanning 2.25–6.07 THz and a total of 3.82 THz, averaging 97.06% absorption in this band. The absorption spectrum fluctuates three times in this range, and the device exhibits perfect absorption of 99.93% at P1 = 2.62 THz, 96.88% at P2 = 3.49 THz, and 99.85% at P3 = 5.24 THz. In addition, we used TE and TM waves to act on the device, respectively, and the absorption efficiency curve remained almost unchanged, which shows the polarization insensitivity of the device [45,46].

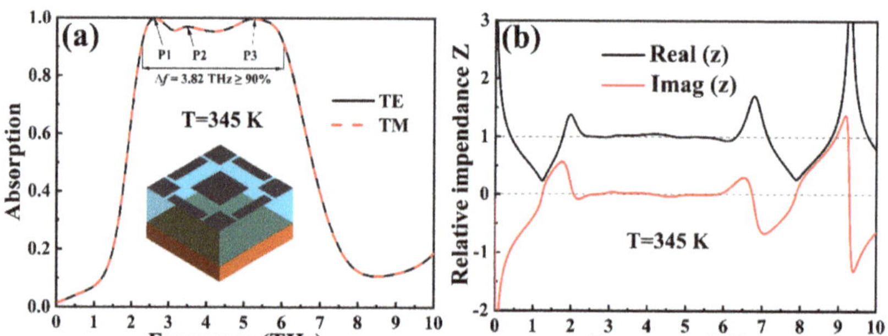

Figure 4. (**a**) TE/TM-polarized light absorption (unitless) in an ultra-wideband phase change device across 0.1–10 THz and 345 K. (**b**) Relative impedance diagram of the device in the same case.

To investigate the absorption mechanism in depth and validate the reliability of our computational findings, we introduce the effective impedance matching theory to invert the obtained absorption data, and the relevant formulas of the impedance matching theory are as follows [47,48]:

$$Z = \pm\sqrt{\frac{(1+S_{11})^2 - S_{21}^2}{(1-S_{11})^2 - S_{21}^2}} \tag{7}$$

In Equation (7), S_{11} and S_{21} indicate the reflection and transmission coefficients, respectively. For the impedance matching analysis, when the imaginary part of the impedance Imag (z) is close to 0 and the real part of the impedance Real (z) is close to 1, it is considered that the impedance of the ultra-wideband phase change device is perfectly matched with the spatial impedance, which also corresponds to the generation of perfect absorption [49,50]. In Figure 4b, we have plotted the relative impedance of the device, and it is obvious that the impedance matching plot in the band of 2.25–6.07 THz is the same as the theoretical effect, which not only explains why the device can produce an excellent effect, demonstrating >97% average absorption in this frequency band, but also verifies that our calculation results accurately match the relative impedance matching theory.

Subsequently, we numerically simulated and visualized the device's surface E-field distribution, as illustrated in Figure 5. First, we regionally number the top-layer graphene of the ultra-wideband phase change device, which provides a clear reference framework for the subsequent analysis of the E-field distribution characteristics of the surface, as illustrated in Figure 5a. The constructive superposition mechanism of different resonant modes originates from the spatial distribution decoupling and frequency-domain response complementarity. In the low-frequency band of P1 = 2.62 THz, a symmetric dipole resonance is formed at the edge of graphene (such as the periphery of block 3), with the electric field showing a symmetric distribution. In the middle-frequency band of P2 = 3.49 THz, the nanogap between block 1 and block 3 excites localized surface plasmon resonance (LSPR), capturing electromagnetic energy at the subwavelength scale through near-field coupling [51,52]. In the high-frequency band of P3 = 5.24 THz, it is characterized by cross-unit multipolar resonance coupling, with reverse electric field oscillations formed at the center of block 2 and the edges of block 1/3. As the frequency increases from P1 to P3, the electric field distribution gradually shifts from the single-block edge to the double-block gap and cross-unit region, avoiding phase cancellation caused by spatial overlap. In terms of the frequency domain, the half-widths of the three are 1.2 THz, 1.5 THz, and 1.1 THz, respectively, the frequency band overlap rate exceeds 60%, the real part of the impedance in each frequency band is close to 1, and the imaginary part is approximately 0 (as shown in Figure 4b), meeting the wide-frequency impedance matching condition and finally achieving continuous and efficient absorption at 3.82 THz. The E-field distributions corresponding to the three characteristic absorption peaks P1 = 2.62 THz, P2 = 3.49 THz, and P3 = 5.24 THz are shown in Figure 5b–d, respectively. At the low-frequency absorption peak P1 = 2.62 THz (Figure 5b), the E-field is mainly localized at the peripheral edge of block 3 (the maximum field strength is about 4×10^6 V/m), showing a symmetrically distributed dipole resonance pattern. When the frequency is increased to P2 = 3.49 THz (Figure 5c), the concentrated region of the E-field shifts to the nanogap between blocks 1 and 3, and the field strength is significantly enhanced to 5×10^6 V/m, which is due to the local surface plasmon resonance (LSPR) excited at the gap, and the electromagnetic energy is efficiently captured at the subwavelength scale through near-field coupling [53,54]. When the frequency is further increased to P3 = 5.24 THz (Figure 5d), the E-field distribution shows the characteristics of multi-region synergistic resonance. The longitudinal edge of the central region of block 2 has a strong field area, and the edges of block 1 and block 3 form an inverted oscillation, which is also concentrated in the longitudinal gap between block 1 and block 3 and maintains the maximum field strength. This multipole resonance coupling across cells effectively extends the absorption bandwidth of the device [55]. It is worth noting that the geometric arrangement of blocks 1 and 3 in the graphene layer and the central region 2 forms a cascading resonance structure. As the incident wave frequency shifts from P1 to P3, the local position of the E-field changes from single block edge (P1) to double block gap (P2) and cross-cell multi-region (P3), and this controlled tuning of the spatial distribution enables the device to achieve ultra-wideband absorption spanning 2.25–6.07 THz. The gradient of E-field intensity (0–5×10^6 V/m) shown in the chroma bar on the right further confirms the local enhancement effect of nanostructures on electromagnetic fields.

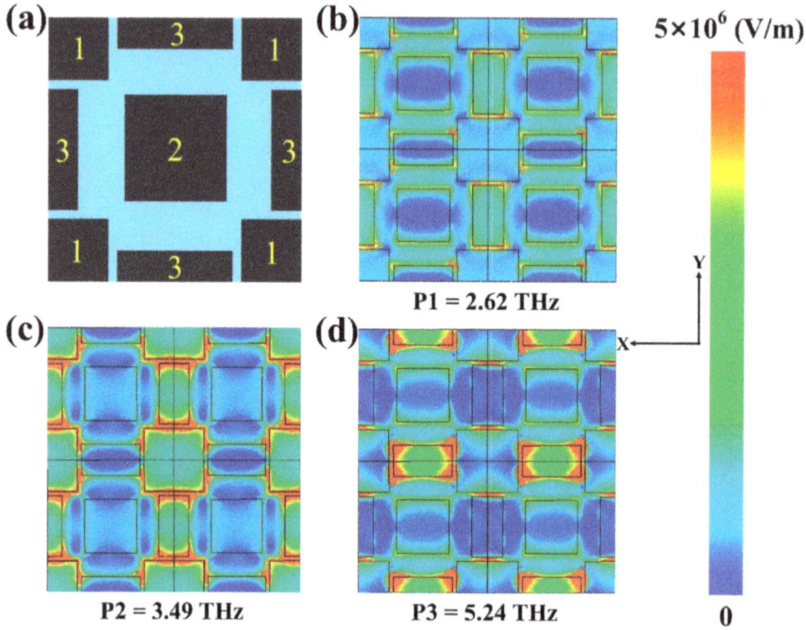

Figure 5. (**a**) Convenient numbering of graphene in the top layer of an ultra-wideband phase change device. (**b**–**d**) XY surface E-field distribution of the three absorption peaks at P1 = 2.62 THz, P2 = 3.49 THz, and P3 = 5.24 THz at T = 345 K.

As shown in Figure 6, to investigate the influence of the top-layer graphene structure on the terahertz band absorption efficiency of the device, we designed comparative experiments with different structures and tested their absorption performance. The red curve in the figure corresponds to the absorption characteristics of the "three-rectangle patterned graphene" structure proposed in this paper (Figure 1b). The absorption mechanism and performance advantages of this structure have been detailed in the above text (e.g., Section 3, "Parameter Calculation and Result Analysis", and the E-field distribution analysis in Figure 5) and will not be repeated here. It can be seen from the curve that the absorption effect of this structure in the target frequency band is significantly better than that of other comparative structures. Regarding the other three comparative curves, the blue curve corresponds to the structure with "the central block 2 of the top-layer graphene removed". In the 2.25–6.07 THz band, its average absorption rate is 88.91%. Although it maintains good absorption performance, it is 8.15 percentage points lower than that of the red curve. The green curve corresponds to the structure with "the surrounding block 3s of the top-layer graphene removed". In the 2.25–6.07 THz band, its average absorption rate is 85.57%, and the absorption rate in the 2.91–5.18 THz band can exceed 90%, still exhibiting a certain broadband absorption capability. However, its overall absorption efficiency is 11.49 percentage points lower than that of the red curve. The black curve corresponds to the device with "the top-layer graphene replaced by a full-coverage unpatterned structure". For this structure, the absorption efficiency decreases significantly in the 2.25–6.07 THz band, with an average absorption rate of less than 80% and no obvious high-absorption frequency band. Its overall performance is poor and has no discussion value. In conclusion, by comparing the absorption performance of different top-layer graphene structures, it can be concluded that the "three-rectangle patterned graphene" structure proposed in this paper achieves the optimal absorption effect in the terahertz band. The resonant modes (edge dipole resonance, gap LSPR, and cross-unit multipolar resonance) synergistically excited by its multi-rectangle units are key to realizing ultra-wideband and high-efficiency absorption, which further verifies the rationality and necessity of this structural design.

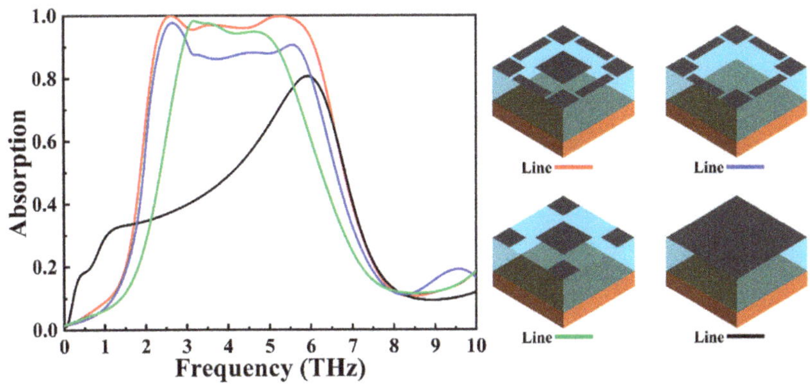

Figure 6. Comparison of absorption spectra of different graphene structures (T = 345 K).

After the study of the device at a relatively high temperature, the absorption spectra of the device at a relatively low temperature of T = 318 K were plotted and analyzed separately, as shown in Figure 7. Figure 7a also shows the absorption spectrum of the device in the frequency band of 0.1–10 THz under this temperature state, and the experimental data indicate that the device still maintains the broadband absorption characteristics after changing the temperature, and the absorption efficiency exceeds 80% over 1.48 THz between 3.60 and 5.08 THz, and the average absorption rate in this range is 81.99%. TE waves and TM waves were also used to act on the device, and nearly consistent absorption spectra were obtained, which once again proves that the polarization insensitivity of the device is differentiated [56]. Combined with Equation (7) above, we analyzed the impedance matching of the device at 318 K, as illustrated in Figure 7b. Obviously, the device still conforms to the relative impedance matching theory at this time, which proves that our device is consistent with classical physics theories regardless of whether it is in a high or low temperature state, and it also shows that the results of our simulation are accurate [57]. In addition, we calculated and plotted the transmission spectrum of the device at low temperatures, as shown in Figure 7c. It can be seen from the figure that the device has dual-frequency transmission performance at low temperatures, with an average transmittance of 67.66% at 0.1–2.14 THz and 69.51% at 7.51–10 THz. In summary, the device still conforms to the impedance matching theory at low temperatures, is not polarization sensitive, and has good broadband absorption and ultra-wideband transmission characteristics, which further proves that the device can be controlled by temperature, indicating the advantages.

Based on the above research, we investigated the impact of fundamental device parameters on overall absorption performance to explore control methods beyond temperature tuning. As shown in Figure 8, we first analyzed the effect of graphene's top-layer Fermi level on absorption at 345 K. Notably, we can modulate the Fermi level of graphene by applying an external bias voltage, as described in Equation (3). Figure 8 presents a parametric sweep clearly reflecting absorption efficiency changes with EF. At T = 345 K (Figure 8a), increasing the Fermi level from 0.6 eV to 1.0 eV progressively enhances absorption. In the 2.25–6.07 THz band, the average absorption rises from 85.10% (0.6 eV) to 97.06% (1.0 eV), stabilizing at ~97% for higher E_F. This trend aligns with the linear dispersion relation in graphene's electronic band structure, where a higher E_F increases conduction electron density near the Dirac point, strengthening Drude-like free-carrier absorption of terahertz waves [58–60]. Additionally, elevated electron density modifies the graphene metasurface's plasmonic resonance, optimizing impedance matching with VO_2 and extending the absorption bandwidth.

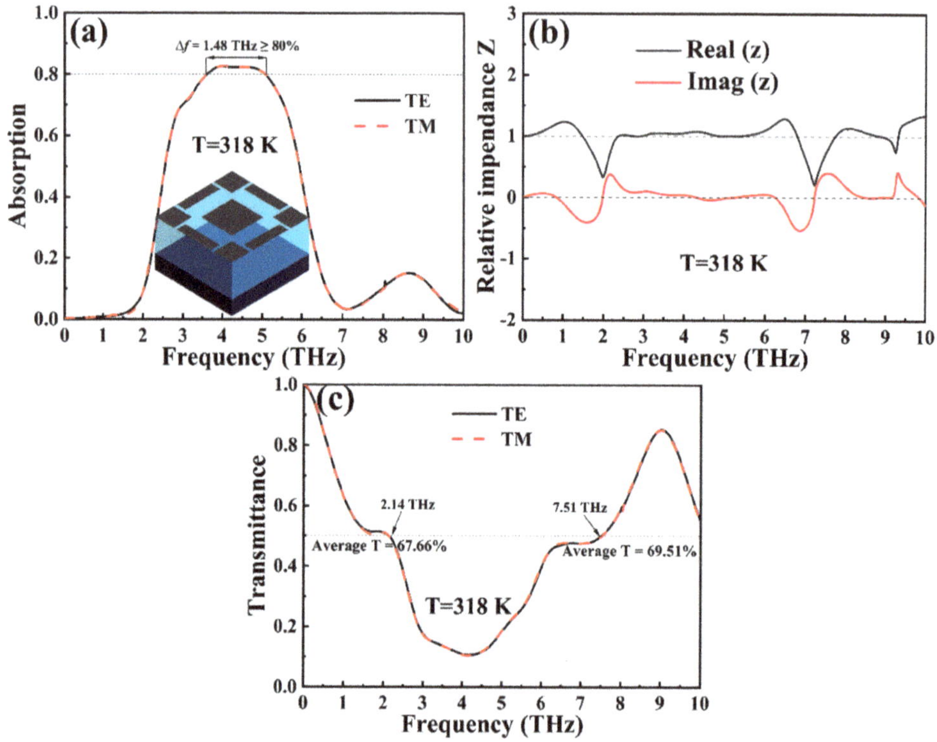

Figure 7. (**a**) TE and TM absorption spectra (unitless) of an ultra-wideband phase change device at 0.1–10 THz and T = 318 K. (**b**) Relative impedance diagram of the device in the same case. (**c**) TE/TM transmission spectra of ultra-wideband phase change devices.

After exploring the graphene Fermi level, we immediately began to explore the time of relaxation for graphene. As shown in Figure 8b, we plotted the spectrum of the absorption efficiency of a device with an increase in the time of relaxation for graphene from 0.1 ps to 0.9 ps, and the related formula for the relaxation time of graphene is shown in Equation (4) above. It is evident from Figure 8b that when the time of relaxation for graphene is 0.1 ps, the overall broadband absorption effect is the best. As the relaxation time increases, the broadband absorption effect gradually deteriorates, and the overall narrowband absorption trend is presented. This could be attributed to the extension of the relaxation time, which speeds up the energy dissipation rate within the material, resulting in the deterioration of the broadband absorption effect, which results in a narrow-band absorption peak with good absorption [61]. The alteration of relaxation time has a tendency to narrow the absorption bandwidth, which may provide some inspiration and reference for future research and will not be further studied in this paper. It has basically the same effect at low temperatures and will not be explored here.

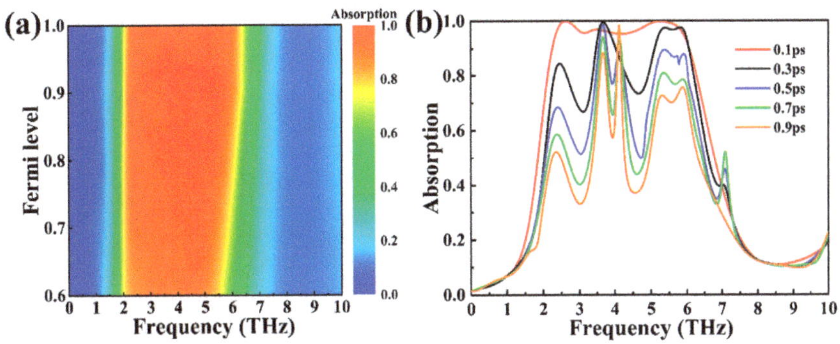

Figure 8. (**a**) Effect of different graphene Fermi levels on top of the ultra-wideband phase change device at T = 345 K. (**b**) Effect of different relaxation times of graphene on the absorption efficiency (unitless) of the device at a temperature of 345 K.

Continuing to investigate the effects of changes in the physical parameters of the device, as shown in Figures 9 and 10, we explored the influence of changes in the physical parameters of the device on the absorption efficiency, as presented in Table 1 above. Figure 9 plots the spectral curves corresponding to the parameter changes of the four graphene sheets L1–L4. As illustrated in Figure 9a, we change the L1 parameter from 4.6 to 5.4 μm, and the absorption of the device shows a certain degree of redshift. In the same way, the average absorption rates of the device in the 2.25–6.07 THz band proposed above for L1 of 4.6, 4.8, 5.0, 5.2, and 5.4 μm were 96.21%, 96.94%, 97.06%, 96.68%, and 94.98%, respectively, and the average absorption rate reached the maximum value at L1 = 5.0 μm. At the same time, the change of L1 mainly affected the absorption effect of the characteristic absorption peak P1, and there was little change between P2 and P3. Figure 9b shows the change in absorption efficiency of L2 from 7.5 μm to 9.5 μm. Since the change in L2 has a slight influence on the change in the overall absorption efficiency of the piece, we drew the contour plot after the scanning parameters to demonstrate the consistency of the absorption rate more intuitively. We selected a median of 8.5 μm as the specific parameter of L2, and the effect of L2 on the absorption efficiency demonstrates that the device can tolerate a certain manufacturing error on the L2 parameter in the manufacturing process, and an error in the range of 2 μm will not have much impact on the overall absorption effect of the device [62]. The L3 parameters are shown in Figure 9c. For the same frequency band, when L3 is 1.5 μm, 2.0 μm, 2.5 μm, 3.0 μm, and 3.5 μm, the corresponding average absorption rates are 92.52%, 95.50%, 97.06%, 95.84%, and 92.56%, respectively. The average absorption first rises and then decreases, and when the L3 value is 2.5 μm, the average absorption effect is the best, and the absorption influence of the characteristic absorption peaks P2 and P3 is also the best. Figure 9d shows the change in parameter L4, which is similar to that of (b). Changing this parameter within this range has little effect on the absorption efficiency of the device, and it is believed that similar conclusions can be drawn. When the L4 parameters are 9.2, 9.4, 9.6, 9.8, and 10.0 μm, the absorption rates corresponding to P1 are 99.26%, 99.91%, 99.93%, 99.63%, and 97.87%, respectively, and the rate of absorption attains the maximum value when the L4 parameter is 9.6 μm, so we believe that 9.6 μm should be the optimal parameter for L4. Summarizing the analysis of the impact of alterations in the four parameters of the graphene layers, we prove that the four graphene layer size parameters given in Table 1 of this article are the optimal values after our calculation and discussion, and changes in these four parameters within the range we discuss have no great impact on the overall device of the device, which can basically maintain the mean absorption rate of the device in the explored terahertz band above 95%, and it can also be said that Figure 9 provides some references for the acceptable error range in the device manufacturing process.

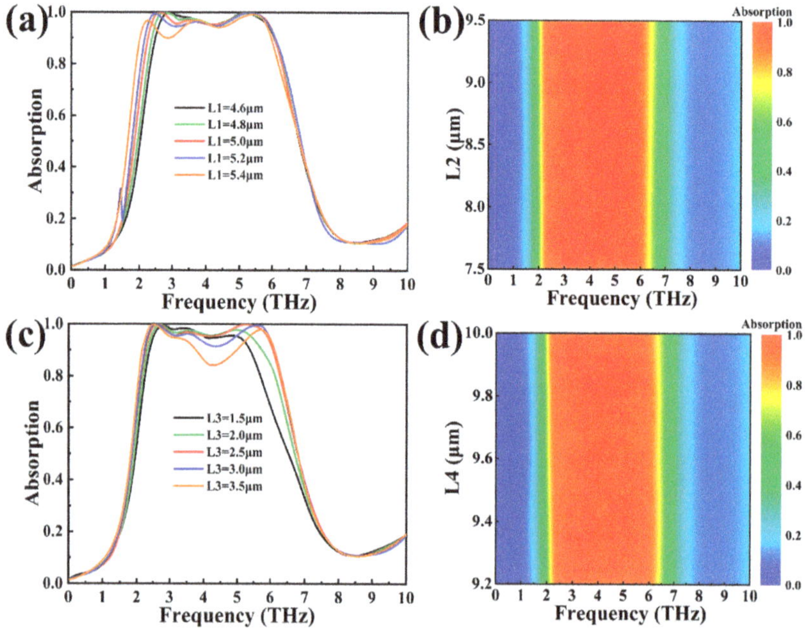

Figure 9. Influence of structural parameters of top-layer graphene on the absorption efficiency (unitless) of ultra-wideband phase change devices at T = 345 K. Each subfigure shows the influence of different parameter variations: (**a**) L1; (**b**) L2; (**c**) L3; (**d**) L4.

To further improve the device parameters, as shown in Figure 10, after exploring L1–L4, we studied the effect of changing the H1, H2, and P parameters on the absorption rate of the device. Figure 10a shows the corresponding results for the variation of H1. It can be observed that the device's absorption performance remains almost unchanged with the variation of H1. This is primarily because when T = 345 K, VO_2 is in the metallic phase, serving as a metallic substrate for the device. The thickness of VO_2 (H1 = 5 μm) is significantly greater than the skin depth in this frequency band. Specifically, at 3 THz, the skin depth of VO_2 in the metallic phase is approximately 20 nm, while the thickness of H1 is 5 μm (5000 nm), which is 250 times larger than the skin depth. This allows VO_2 to fully reflect the incident terahertz waves, enabling the absorption layer to reabsorb the reflected waves for perfect absorption, with VO_2 itself having almost no impact on the absorption performance. Figure 10b shows the correlation of changes in H2. As can be seen in the figure, when the H2 layer gradually increases, the high-frequency part of the absorption curve has a more obvious redshift, which may be due to the fact that the high-frequency terahertz wave reflected back to the absorption layer by the substrate is partially consumed in the dielectric layer when its thickness has increased, so the relative overall bandwidth becomes smaller [63,64]. When H2 is 10, 11, 12, 13, and 14 μm, the average absorption rates in the band range of 2.25–6.07 THz are 91.98%, 95.30%, 97.06%, 95.63%, and 91.53%, respectively, and the mean absorption rate reaches a peak when the H2 value is 12 μm. Finally, we explored the corresponding effect of the P size of the device cycle, as shown in Figure 10c. When P is 21, 22, 23, 24, and 25 μm, the mean absorption efficiency of the device in the discussed frequency band is 97.06%, 93.78%, 92.70%, 91.32%, and 88.86%, respectively, showing a downward trend, and when P is 21 μm, that is, the size of the

SiO_2 dielectric layer and the VO_2 substrate is just the size of the top layer of graphene, the device's absorption effectiveness is the best.

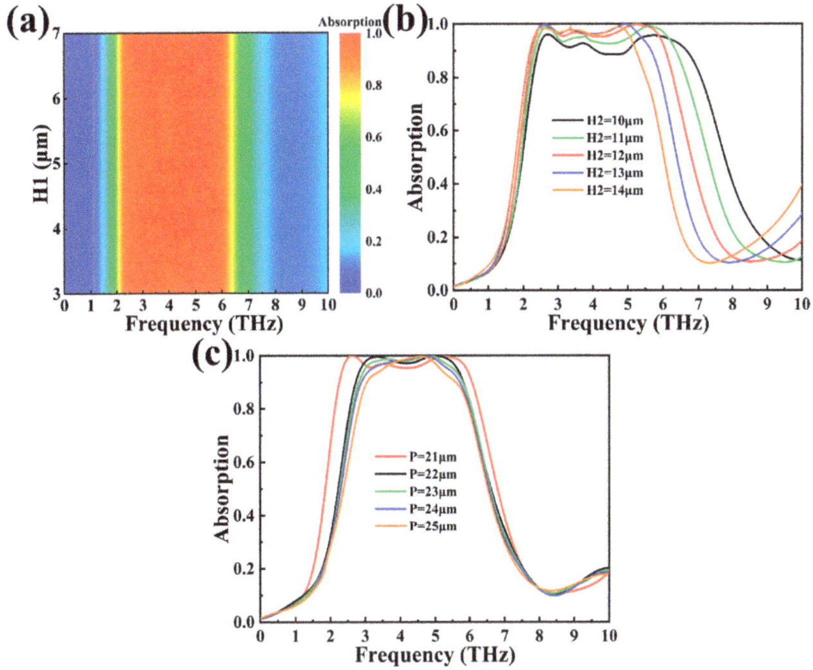

Figure 10. Influence of structural parameters of an ultra-wideband phase change device on absorption efficiency (unitless) at a temperature of 345 K. (**a**) Parameter scan of the effect of H1 on the efficiency of absorption. (**b**) Spectral pattern of the effect of H2 on absorption efficiency. (**c**) Spectral diagram of the effect of P on the efficiency of absorption.

In addition to temperature and device parameters, changes in the polarization angle and incidence angle of light are also important indicators for evaluating devices. In actual applications, the incidence angle of electromagnetic waves may not be perpendicular to the incidence angle of the device, and a device that can adapt to multiple incidence angles is suitable for more application scenarios [65–67]. As shown in Figure 11, we plotted the contour plot of the impact of angle of polarization and angle of incidence changes on the absorption efficiency of the apparatus when the temperature is 345 K. Figure 11a shows the spectrum of the absorption efficiency of the device for polarization angles of 0 degrees to 90 degrees, and it is clearly apparent that the absorption of the device is not influenced by the polarization angle, because the device we designed is a center-symmetrical structure, and the change of polarization angle will not change the plane wave received by the absorbing layer, so the device is polarization insensitive [68–72]. Figure 11b is a scan of the absorption spectrum of the device with a change of incidence angle of 0–80 degrees, and it is obvious that when the incidence angle increases from 0 to 50 degrees, the absorption remains basically unchanged, and the average absorption rate is maintained at 90% or more. When the angle is greater than 50, more than 90% of the absorption frequency band is split from about 4.5 THz, and the angle gradually increases from 50 degrees to 70 degrees, and the low frequency band and high frequency band can still maintain a relatively good absorption trend, but the absorption of the high frequency band continues to move to the higher frequency band with the change of angle, and the high absorption bandwidth of the low frequency band gradually decreases. When the incidence angle increases to 80 degrees, the high absorption frequency band of the device is mainly displayed in the high frequency band, and the overall absorption effect becomes worse, mainly because the incidence angle is too large, so only a small part of the electromagnetic wave can irradiate the surface of the device, and we believe that an incidence angle greater than or equal to 80 degrees is too extreme to warrant continuing the discussion. In general, the ultra-wideband phase change device is impervious to the alteration of the polarization angle and can adapt to a wide range of changes in the incidence angle from 0 to 60 degrees (the average absorption efficiency is maintained at 80% and above), which is suitable for a variety of application scenarios [73–75].

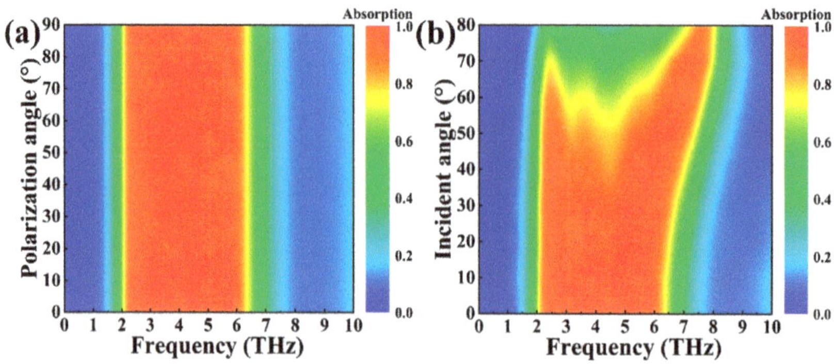

Figure 11. (**a**) Scans of absorption efficiency (unitless) at different polarization angles at a temperature of 345 K. (**b**) Scans of absorption at different angles of incidence at T = 345 K.

To conclude the article, we compare the performance of this ultra-wideband phase change device with the other studies, as shown in Table 2 [30,76–79]. The combination of phase change material VO_2 and graphene enables the absorption device to have better performance and more flexible control methods.

Table 2. Comparison of the performance of this ultra-wideband phase change device with other devices.

Ref.	Materials Used	Number of Device Layers	Absorption Range (>90%)	Bandwidth (>90%)	Average Absorption	Angle of Incidence
[76]	Graphene	5	0.82–1.87 THz	1.05 THz	Over 90%	0–60°
[77]	Graphene	3	3.29–5.25 THz	1.96 THz	Over 90%	0–60°
[78]	VO_2	5	2.37–4.56 THz	2.19 THz	Over 90%	0–55°
[30]	VO_2–Graphene	7	2.42–4.83 THz	2.41 THz	Over 90%	0–60°
[79]	VO_2	3	3.09–4.61 THz and 5.36–5.79 THz	1.95 THz	Over 90%	0–50°
This work	VO_2–Graphene	3	2.25–6.07 THz	3.82 THz	97.06%	0–60°

4. Conclusions

In this study, a dynamically tunable terahertz ultra-wideband phase change absorber based on a VO_2–graphene hybrid metasurface was successfully designed, and the dual functions of ultra-wideband absorption and efficient transmission were realized through VO_2 phase change. At 345 K (metallic state), the device exhibits an average excellent absorption performance of 97.06% in the range of 2.25–6.07 THz (bandwidth 3.82 THz). At 318 K (insulated state), it can not only maintain 3.6–5.08 THz broadband absorption, with an average absorption rate of 81.99%, but also achieves high transmittance of 67.66–69.51% in the 0.1–2.14 THz and 7.51–1 Hz frequency bands. Through the synergistic control of phase change materials and two-dimensional materials, the design uses patterned graphene to simplify the preparation process, which has both polarization insensitivity and wide incidence angle stability. The physical mechanism of ultra-wideband absorption is revealed by impedance matching theory and surface E-field distribution analysis, and the tunability of the device is verified by a systematic study of Fermi level, relaxation time, and structural parameters. It is believed that this research will lay a foundation for the optimal design of related devices in the future and will provide new technical ideas for the development of terahertz switching, stealth technology, and sensing applications.

Author Contributions: Conceptualization, K.C., Z.W., M.G. and Z.Y.; data curation, K.C., Z.W., M.G. and H.M.; formal analysis, K.C., Z.W. and M.G.; methodology, K.C., Z.W., M.G. and Z.Y.; resources, K.C., Z.W. and M.G.; software, K.C., Z.W., M.G., Z.Y. and H.M.; data curation, K.C., Z.W., M.G., S.C., Z.Y. and B.L.; writing—original draft preparation, K.C., Z.W. and M.G.; writing—review and editing, K.C., Z.W., M.G., H.M. and Z.Y. All authors have read and agreed to the published version of the manuscript.

Funding: The authors are grateful for the support of the National Natural Science Foundation of China (No. 21506257).

Institutional Review Board Statement: Not applicable.

Informed Consent Statement: Not applicable.

Data Availability Statement: Publicly available datasets were analyzed in this study. The data can be found here: [https://www.lumerical.com/]. (accessed on 1 January 2020).

Conflicts of Interest: Author Kele Chen and Meizhang Guan are employed by the company Sichuan Intronics Electronic Technology Co., Ltd. The remaining authors declare that the research was conducted in the absence of any commercial or financial relationships that could be construed as a potential conflict of interest.

References

1. Yang, Q.; Yu, M.; Chen, Z.; Ai, S.; Kentsch, U.; Zhou, S.; Jia, Y.; Chen, F.; Liu, H. A novel approach towards robust construction of physical colors on lithium niobate crystal. *Opto-Electron. Adv.* **2025**, *8*, 240193. [CrossRef]
2. Balashov, I.S.; Chezhegov, A.A.; Chizhov, A.S.; Grunin, A.A.; Anokhin, K.V.; Fedyanin, A.A. Light-stimulated adaptive artificial synapse based on nanocrystalline metal-oxide film. *Opto-Electron. Sci.* **2023**, *2*, 230016. [CrossRef]
3. Zhao, J.; Lai, H.R.; Li, M. Anchoring 1T-MoS2 petals on N-doped reduced graphene oxide for exceptional electromagnetic wave absorption. *Int. J. Miner. Metall. Mater.* **2025**, *32*, 619–630. [CrossRef]
4. Wang, X.Y.; Lin, Q.; Wang, L.L.; Liu, G.D. Dynamic control of polarization conversion based on borophene nanostructures in optical communication bands. *Phys. Scr.* **2024**, *99*, 085531. [CrossRef]
5. Li, Y.C.; Chen, F. Tunable plasmon-induced transparency for advanced sensing in MoS2 based metamaterial. *Phys. Lett. A* **2025**, *553*, 130689. [CrossRef]
6. Xiao, T.X.; Tu, S.; Liang, S.Z.; Guo, R.J.; Tian, T.; Müller-Buschbaum, P. Solar cell-based hybrid energy harvesters towards sustainability. *Opto-Electron. Sci.* **2023**, *2*, 230011. [CrossRef]
7. Yang, C.; Luo, M.H.; Ju, X.W.; Hu, J.Y. Ultra-narrow dual-band perfect absorber based on double-slotted silicon nanodisk arrays. *J. Phys. D Appl. Phys.* **2024**, *57*, 345104. [CrossRef]
8. Zhang, J.; Chen, Y.; Chen, S.; Hou, J.; Liu, J. A simple ultra-broadband metamaterial solar perfect absorber with highly efficient photothermal conversion performance. *Phys. B Condens. Matter* **2025**, *699*, 416845. [CrossRef]
9. Yu, Z.; Li, M.; Xing, Z.; Gao, H.; Liu, Z.; Pu, S.; Mao, H.; Cai, H.; Ma, Q.; Ren, W.; et al. Genetic algorithm assisted meta-atom design for high-performance metasurface optics. *Opto-Electron. Sci.* **2024**, *3*, 240016. [CrossRef]
10. Wang, X.Y.; Huang, S.L.; Chen, Y.; Chen, S.J.; Dai, W.; Hou, J.; Wang, J.Y.; Yang, W.M.; Song, S.Y. Perfect absorber based on toroidal dipole in metamaterial of silicon and gallium phosphide. *Photonics Nanostruct. Fundam. Appl.* **2025**, *66*, 101410. [CrossRef]
11. Li, Y.M.; Tan, C.X.; Hu, J.Y.; Bai, W.D.; Zhang, R.L.; Lin, Q.; Zhang, Y.; Wang, L.L. Ultra-narrow band perfect absorbance induced by magnetic lattice resonances in dielectric dimer metamaterials. *Results Phys.* **2022**, *39*, 105730. [CrossRef]
12. Wang, Z.; Pan, W.K.; He, Y.; Zhu, Z.; Jin, X.; Liu, M.; Ma, S.; He, Q.; Sun, S.; Zhou, L.; et al. Efficient generation of vectorial terahertz beams using surface-wave excited metasurfaces. *Opto-Electron. Sci.* **2025**, *4*, 240024. [CrossRef]

13. Liu, Y.; Ma, X.; Chao, K.; Sun, F.; Chen, Z.; Shan, J.; Chen, H.; Zhao, G.; Chen, S. Simultaneously Realizing Thermal and Electromagnetic Cloaking by Multi-Physical Null Medium. *Opto-Electron. Sci* **2024**, *3*, 230027. [CrossRef]
14. Huffman, T.J.; Xu, P.; Qazilbash, M.M.; Walter, E.J.; Krakauer, H.; Wei, J.; Cobden, D.H.; Bechtel, H.A.; Martin, M.C.; Carr, G.L.; et al. Anisotropic Infrared Response of Vanadium Dioxide Microcrystals. *Phys. Rev. B* **2013**, *87*, 115121. [CrossRef]
15. Qazilbash, M.M.; Brehm, M.; Chae, B.-G.; Ho, P.-C.; Andreev, G.O.; Kim, B.-J.; Yun, S.J.; Balatsky, A.V.; Maple, M.B.; Keilmann, F.; et al. Mott Transition in VO_2 Revealed by Infrared Spectroscopy and Nano-Imaging. *Science* **2007**, *318*, 1750–1753. [CrossRef]
16. Yang, C.; Lin, Q.; Du, W.J.; Wang, L.L.; Liu, G.D. Bi-tunable absorber based on borophene and VO_2 in the optical telecommunication band. *J. Opt. Soc. Am. B* **2022**, *39*, 2969–2974. [CrossRef]
17. Li, W.X.; Cheng, S.B.; Yi, Z.; Zhang, H.F.; Song, Q.J.; Hao, Z.Q.; Sun, T.Y.; Wu, P.H.; Zeng, Q.D.; Raza, R. Advanced optical reinforcement materials based on three-dimensional four-way weaving structure and metasurface technology. *Appl. Phys. Lett.* **2025**, *126*, 033503. [CrossRef]
18. Ai, Z.; Yang, H.; Liu, M.S.; Cheng, S.B.; Wang, J.Q.; Tang, C.J.; Gao, F.; Li, B.X. Phase-Transition-Enabled Dual-Band Camouflage in VO2/Ag Multilayered Nanostructures. *Phys. E Low-Dimens. Syst. Nanostruct.* **2025**, *173*, 116327. [CrossRef]
19. Zhang, X.; Wang, G.; Liu, J.; Zuo, S.; Li, M.; Yang, S.; Jia, Y.; Gao, Y. Switchable and Tunable Terahertz Metamaterial Based on Vanadium Dioxide and Photosensitive Silicon. *Nanomaterials* **2023**, *13*, 2144. [CrossRef] [PubMed]
20. Kats, M.A.; Sharma, D.; Lin, J.; Genevet, P.; Blanchard, R.; Yang, Z.; Qazilbash, M.M.; Basov, D.N.; Ramanathan, S.; Capasso, F. Ultra-thin perfect absorber employing a tunable phase change material. *Appl. Phys. Lett.* **2012**, *101*, 221101. [CrossRef]
21. Driscoll, T.; Palit, S.; Qazilbash, M.M.; Brehm, M.; Keilmann, F.; Chae, B.G.; Yun, S.J.; Kim, H.T.; Cho, S.Y.; Jokerst, N.M.; et al. Dynamic tuning of an infrared hybrid-metamaterial resonance using vanadium dioxide. *Appl. Phys. Lett.* **2008**, *93*, 024101. [CrossRef]
22. Hu, J.Y.; Bai, W.D.; Tan, C.X.; Li, Y.M.; Lin, Q.; Wang, L.L. Highly electric field enhancement induced by anapole modes coupling in the hybrid dielectric-metal nanoantenna. *Opt. Commun.* **2022**, *511*, 127987. [CrossRef]
23. Jiang, B.; Hou, Y.; Wu, J.; Ma, Y.; Gan, X.; Zhao, J. In-fiber photoelectric device based on graphene-coated tilted fiber grating. *Opto-Electron. Sci.* **2023**, *2*, 230012. [CrossRef]
24. Wang, H.Y.; Ma, R.; Liu, G.D.; Wang, L.L.; Lin, Q. Optical force conversion and conveyor belt effect with coupled graphene plasmon waveguide modes. *Opt. Express* **2023**, *31*, 32422. [CrossRef]
25. Zeng, T.Y.; Liu, G.D.; Wang, L.L.; Lin, Q. Light-matter interactions enhanced by quasi-bound states in the continuum in a graphene-dielectric metasurface. *Opt. Express* **2021**, *29*, 40177–40186. [CrossRef]
26. Shao, M.R.; Ji, C.; Tan, J.B.; Du, B.Q.; Zhao, X.F.; Yu, J.; Man, B.; Xu, K.; Zhang, C.; Li, Z. Ferroelectrically modulate the Fermi level of graphene ox-ide to enhance SERS response. *Opto-Electron. Adv.* **2023**, *6*, 230094. [CrossRef]
27. Wang, P.; Yang, M.; Tang, S.; Li, Y.; Lin, X.; Zhang, H.; Zhu, Z.; Chen, F. Z-scheme heterojunctions composed of 3D graphene aerogel/g-C3N4 nanosheets/porous ZnO nanospheres for the efficient photocatalytic reduction of CO_2 with H_2O under visible light irradiation. *J. Alloys Compd.* **2022**, *918*, 165607. [CrossRef]
28. Li, W.; Yi, Y.; Yang, H.; Cheng, S.; Yang, W.X.; Zhang, H.; Yi, Z.; Yi, Y.; Li, H. Active Tunable Terahertz Band-width Absorber Based on single layer Graphene. *Commun. Theor. Phys.* **2023**, *75*, 045503. [CrossRef]
29. Zeng, Y.; Ling, Z.X.; Liu, G.D.; Wang, L.L.; Lin, Q. Tunable plasmonically induced transparency with giant group delay in gain-assisted graphene metamaterials. *Opt. Express* **2022**, *30*, 455954. [CrossRef] [PubMed]
30. Shen, J.; Li, H.; Qiu, X.; Lu, J.; Wang, Y.; Jin, C. VO2-Graphene based four-state ultra-wideband terahertz metamaterial with switchable absorption, reflection, and transmission. *Micro Nanostruct.* **2025**, *203*, 208143. [CrossRef]
31. Li, Z.T.; Cheng, S.B.; Zhang, H.F.; Yang, W.X.; Yi, Z.; Yi, Y.G.; Wang, J.Q.; Ahmad, S.; Raza, R. Ultrathin broadband terahertz metamaterial based on single-layer nested patterned graphene. *Phys. Lett. A* **2025**, *534*, 130262. [CrossRef]
32. Chen, Z.Y.; Cheng, S.B.; Zhang, H.F.; Yi, Z.; Tang, B.; Chen, J.; Zhang, J.G.; Tang, C.J. Ultra wideband absorption absorber based on Dirac semimetallic and graphene metamaterials. *Phys. Lett. A* **2024**, *517*, 129675. [CrossRef]
33. Kitamura, R.; Pilon, L.; Jonasz, M. Optical constants of silica glass from extreme ultraviolet to far infrared at near room tem-perature. *Appl. Opt.* **2007**, *46*, 8118–8133. [CrossRef] [PubMed]
34. Chen, S.; Wu, X.H.; Fu, C.J. Active tuning of anisotropic phonon polaritons in natural van der Waals crystals with negative permittivity substrates and its application in energy transport. *Opto-Electron. Sci.* **2024**, *3*, 240002. [CrossRef]
35. Ren, J.L.; Ma, Q.F.; Sun, X.F.; Ma, J.Y.; Liu, G.R.; Yang, H. In^{3+}-doping and oxygen vacancies co-engineering active sites of Bi_2WO_6 hollow nanospheres to achieve efficient photoreduction of CO_2 to CO with nearly 100% selectivity. *Fuel* **2025**, *397*, 135454. [CrossRef]
36. Zeng, Z.L.; Liu, H.F.; Zhang, H.F.; Cheng, S.B.; Yi, Y.G.; Yi, Z.; Wang, J.Q.; Zhang, J.G. Tunable ultra-sensitive four-band terahertz sensors based on Dirac semimetals. *Photonics Nanostruct. Fundam. Appl.* **2025**, *63*, 101347. [CrossRef]
37. Dai, Y.Y.; Liu, G.R.; Sun, X.F.; Ma, J.Y.; Xian, T.; Yang, H. Mn doping and ZnS nanoparticles modification on Bi_2MoO_6 to achieve an highly-efficient photocatalyst for TC degradation. *Appl. Surf. Sci.* **2025**, *681*, 161611. [CrossRef]

38. Yang, J.Q.; Lin, Q.; Wang, L.L.; Liu, G.D. Research of bound states in the continuum and their polarization control in dielectric metamaterial. *Phys. Scr.* **2025**, *100*, 065551. [CrossRef]
39. Luo, M.H.; Hu, J.Y.; Li, Y.M.; Bai, W.D.; Zhang, R.L.; Lin, Q.; Wang, L.L. Anapole-assisted ultra-narrow-band lattice resonance in slotted silicon nanodisk arrays. *J. Phys. D Appl. Phys.* **2023**, *56*, 375102. [CrossRef]
40. Liu, G.R.; Dai, Y.Y.; Sun, X.F.; Ma, J.Y.; Xian, T.; Yang, H. Synergistically regulating energy band structure and forming quantum wells to enhance the photocatalytic activity of Bi_2MoO_6 for tetracycline removal. *Sep. Purif. Technol.* **2025**, *361*, 131622. [CrossRef]
41. Hu, H.; Chen, F.; Li, Y.; Li, J.; Cui, L.; Jiang, D.; Lin, X.; Gao, J. Construction of graphene supported TiO_2 nanosheet array/CdS/Ni_2P composite with dual heterojunctions for boosting photocatalytic hydrogen evolution. *J. Alloys Compd.* **2025**, *1024*, 180216. [CrossRef]
42. Liu, S.H.; Chen, F. Dynamically tunable Fano resonance effect based on monolayer graphene with disk defect robustness. *Phys. B Condens. Matter* **2025**, *715*, 417638. [CrossRef]
43. Hu, J.Y.; Tan, C.X.; Bai, W.D.; Li, Y.M.; Lin, Q.; Wang, L.L. Dielectric nanocavity-coupled surface lattice resonances for high-efficiency plasmonic sensing. *J. Phys. D Appl. Phys.* **2022**, *55*, 075105. [CrossRef]
44. Qazilbash, M.M.; Li, Z.; Podzorov, V.; Brehm, M.; Keilmann, F.; Chae, B.; Kim, H.-T.; Basov, D. Electrostatic modification of infrared response in gated structures based on VO_2. Appl. *Phys. Lett.* **2008**, *92*, 24. [CrossRef]
45. Xiang, T.; Sun, Z.; Wang, L.L.; Lin, Q.; Liu, G.D. Polarization independent perfect absorption of borophene metamaterials operating in the communication band. *Phys. Scr.* **2024**, *99*, 085519. [CrossRef]
46. Huang, Z.; Liu, G.D.; Wang, L. Active modulation of quasi-bound state in the continuum based on bulk Dirac semimetals metamaterial. *Appl. Phys. Express* **2022**, *15*, 032006. [CrossRef]
47. Zhang, B.W.; Luo, Y.N. Dynamic optical tuning and sensing in L-shaped dirac semimetal-based terahertz metasurfaces. *Phys. Lett. A* **2025**, *541*, 130419. [CrossRef]
48. Liu, Y.J.; Liu, M.S.; Yang, H.; Yi, Z.; Zhang, H.; Tang, C.J.; Deng, J.; Wang, J.Q.; Li, B.X. Photoelectric simulation of perovskite solar cells based on two inverted pyramid structures. *Phys. Lett. A* **2025**, *552*, 130653. [CrossRef]
49. Gu, X.; Liu, X.; Yan, X.F.; Du, W.J.; Lin, Q.; Wang, L.L.; Liu, G.D. Polaritonic coherent perfect absorption based on self-hybridization of a quasi-bound state in the continuum and exciton. *Opt. Express* **2023**, *31*, 4691–4700. [CrossRef]
50. Liu, B.W.; Chen, F. Adjustable slow light and optical switch in a black phosphorus metamaterial based on double plasmon-induced transparency. *Phys. B Condens. Matter* **2025**, *714*, 417423. [CrossRef]
51. Ma, R.; Zhang, L.; Liu, G.; Wang, L.; Lin, Q. The total optical force exerted on black phosphorus coated dielectric cylinder pairs enhanced by localized surface plasmon. *J. Appl. Phys.* **2021**, *130*, 113103. [CrossRef]
52. Li, W.; Cheng, S.; Zhang, H.; Yi, Z.; Tang, B.; Ma, C.; Wu, P.; Zeng, Q.; Raza, R. Multi-functional metasurface: Ultra-wideband/multi-band absorption switching by adjusting guided mode resonance and local surface plasmon resonance effects. *Commun. Theor. Phys.* **2024**, *76*, 065701. [CrossRef]
53. Long, T.; Zhang, L.; Wang, L.L.; Lin, Q. Tunable narrow transparency windows induced by the coupled quasi-guided modes in borophene plasmonic nanostructure. *J. Phys. D Appl. Phys.* **2022**, *55*, 315101. [CrossRef]
54. Yan, X.F.; Lin, Q.; Wang, L.L.; Liu, G.D. Tunable strong plasmon–exciton coupling based modulator employing borophene and deep subwavelength perovskite grating. *J. Phys. D Appl. Phys.* **2023**, *56*, 435106. [CrossRef]
55. Cheng, S.B.; Li, W.X.; Zhang, H.F.; Akhtar, M.N.; Yi, Z.; Zeng, Q.D.; Ma, C.; Sun, T.Y.; Wu, P.H.; Ahmad, S. High sensitivity five band tunable metamaterial absorption device based on block like Dirac semimetals. *Opt. Commun.* **2024**, *569*, 130816. [CrossRef]
56. Yuan, Y.; Liu, H.F.; Liu, M.S.; Zhang, W.B.; Li, X.H.; Cheng, S.B. Design of wide-angle broadband titanium-nitride solar absorber based on column-cavity structure. *Phys. Lett. A* **2025**, *556*, 130832. [CrossRef]
57. Gu, X.; Liu, G.D.; Wang, L.L.; Lin, Q. Robust Fano resonance induced by topologically protected interface modes interference at gigahertz. *Appl. Phys. Express* **2022**, *15*, 082004. [CrossRef]
58. Li, L.Y.; Chen, F. Tunable four-band metamaterial absorber and sensor based on a stacking double-ring Dirac semimetal structure design. *Phys. Lett. A* **2025**, *544*, 130489. [CrossRef]
59. Li, Z.T.; Li, X.; Liu, G.D.; Wang, L.L.; Lin, Q. Analytical investigation of unidirectional reflectionless phenomenon near the exceptional points in graphene plasmonic system. *Opt. Express* **2023**, *31*, 30458. [CrossRef] [PubMed]
60. Ai, Z.; Liu, H.F.; Cheng, S.B.; Zhang, H.F.; Yi, Z.; Zeng, Q.D.; Wu, P.H.; Zhang, J.G.; Tang, C.J.; Hao, Z.Q. Four peak and high angle tilted insensitive surface plasmon resonance graphene absorber based on circular etching square window. *J. Phys. D Appl. Phys.* **2025**, *58*, 185305. [CrossRef]
61. Li, X.; Li, Z.T.; Wang, L.L.; Lin, Q. Unidirectional reflectionless propagation in borophene plasmonic metamaterials. *J. Phys. D Appl. Phys.* **2023**, *57*, 015103. [CrossRef]
62. Ren, J.L.; Ma, Q.F.; Sun, X.F.; Wang, S.F.; Liu, G.R.; Yang, H. Interface-engineering enhanced photocatalytic conversion of CO_2 into solar fuels over S-type Co-Bi_2WO_6@Ce-MOF heterostructured photocatalysts. *J. Colloid Interf. Sci.* **2025**, *691*, 137452. [CrossRef]
63. Ma, Q.; Ren, J.; Sun, X.; Wang, S.; Chen, X.; Liu, G.; Yang, H. Enhanced CO_2 photoreduction over S-scheme Sdoped-BiVO4/AgCl heterostructures and interface interaction mediated selective generation of CH_4. *Chem. Eng. J.* **2025**, *509*, 161444. [CrossRef]

64. Zhao, J.; Lu, Y.X.; Zhou, Z. Correlation filters based on temporal regularization and background awareness. *Comput. Electr. Eng.* **2020**, *86*, 106757. [CrossRef]
65. Sun, X.; Zhang, J.; Luo, M.; Ma, J.; Xian, T.; Liu, G.; Yang, H. Elevating photocatalytic H2 evolution over ZnIn2S4@Au@Cd0.7Zn0.3S multilayer nanotubes via Au-mediating H–S antibonding-orbital occupancy. *Chem. Eng. J.* **2024**, *499*, 156455. [CrossRef]
66. Chen, Y.; Song, S.Y.; Zhou, Z.K.; Chen, S.J.; Hou, J.; Yang, W.X. Multidielectric-metal laminated solar perfect absorber with ultra-high manufacturing error tolerance. *Phys. Scr.* **2025**, *100*, 065538. [CrossRef]
67. Ning, X.; Sun, T.Y.; Song, Q.J.; Yi, Z.; Cheng, S.B.; Wang, J.Q.; Zeng, Q.D.; Yi, Y.G. A modulated broadband polarimetric insensitive metamaterial absorber based on a monolayer of graphene. *Commun. Theor. Phys.* **2025**, *77*, 095702. [CrossRef]
68. Zhou, Z.; Xue, B.; Wang, H.; Zhao, J. Bidirectional multi-scale deformable attention for video super-resolution. *Multimed. Tools Appl.* **2024**, *83*, 27809–27830. [CrossRef]
69. Li, Z.; Song, Q.J.; Jia, L.B.; Yi, Z.; Cheng, S.B.; Wang, J.Q.; Li, B.X. Actively tunable multi-frequency narrowband terahertz absorber using graphene metamaterials. *Opt. Commun.* **2025**, *583*, 131768. [CrossRef]
70. Han, H.; Zhao, J.; Zhai, W.; Xiong, Z.H.; Niyato, D.; Renzo, M.D.; Pham, Q.V.; Lu, W.; Lam, K.Y. Reconfigurable Intelligent Surface Aided Power Control for Physical-Layer Broadcasting. *IEEE Trans. Commun.* **2021**, *69*, 7821–7836. [CrossRef]
71. Jiang, X.W.; Yan, T.H.; Zhu, J.J.; He, B.; Li, W.H.; Du, H.P.; Sun, S.S. Densely Connected Deep Extreme Learning Machine Algorithm. *Cogn. Comput.* **2020**, *12*, 979–990. [CrossRef]
72. Ma, Q.; Ren, J.; Sun, X.; Chen, X.; Liu, G.; Wang, S.; Yang, H. Strong evidence for interface-field-induced photocarrier sepa-ration in new $AgFeO_2$-$BiVO_4$ heterostructures and their efficient photo-Fenton degradation of ciprofoxacin. *Appl. Surf. Sci.* **2024**, *679*, 161275. [CrossRef]
73. Huang, S.L.; Chen, Y.; Yu, C.C.; Chen, S.J.; Zhou, Z.K.; Liang, J.; Dai, W. Optimized metamaterial solar absorber with ultra-wideband, polarization-independent and large incident angle-insensitive. *Chin. J. Phys.* **2024**, *89*, 740–747. [CrossRef]
74. Sun, X.F.; Xian, T.; Sun, C.Y.; Zhang, J.Q.; Liu, G.R.; Yang, H. Enhancing CO_2 photoreduction on Au@CdZnS@MnO_2 hollow nanospheres via electron configuration modulation. *J. Mater. Sci. Technol.* **2025**, *228*, 256–268. [CrossRef]
75. Li, Y.C.; Chen, F.; Yang, W.X. A multifunctional bidirectional metamaterial perfect absorber for efficient narrowband and broadband absorption. *Opt. Commun.* **2025**, *586*, 131878. [CrossRef]
76. Cao, F.L.; Chen, B.J. New architecture of deep recursive convolution networks for super-resolution. *Knowl. Based Syst.* **2019**, *178*, 98–110. [CrossRef]
77. Zhang, K.; Dong, S.; Wu, X.; Yu, K.; Liu, Y. Graphene-based tunable broadband metamaterial absorber for terahertz waves. *Opt. Laser Technol.* **2025**, *180*, 111490. [CrossRef]
78. Dai, S.; Liao, S.Y.; Pan, J.R.; Zhang, H. Theoretical study on ultra-wideband absorption and tripling octave frequency linear to circular polarization conversion tunable metastructure based on vanadium dioxide. *Nanoscale* **2025**, *17*, 12673–12683. [CrossRef]
79. Valathuru, M.; Pardhasaradhi, P.; Prasad, N.; Madhav, B.T.P.; Das, S.; Soliman, N.F.; Ghzaoui, M.E. Design and Analysis of Dual-band Hexagon-shaped Polarization-insensitive Metamaterial Absorber using Vanadium dioxide (VO_2) for Terahertz Applications. *Plasmonics* **2025**, *20*, 4221–4240. [CrossRef]

photonics

MDPI

Article

Analytical Design of Optically Transparent, Wideband, and Tunable Microwave Absorber Based on Graphene Spiral Resonator Metasurface

Ioannis S. Fosteris and George S. Kliros *

Department of Aeronautical Sciences, Hellenic Air Force Academy, Dekeleia Air Force Base, Dekeleia, GR-13671 Athens, Greece
* Correspondence: gskliros@ieee.org or georgios.kliros@hafa.haf.gr

Abstract: We present the design of an optically transparent, flexible, and tunable microwave absorber covering the X and Ku frequency bands. The absorber is based on a metasurface composed of a periodic array of graphene spiral resonators (GSRs) attached to an ultrathin PET film placed over an ITO-backed dielectric spacer. An equivalent circuit model (ECM), described by closed-form equations, is proposed to optimize the structure for maximum absorption within the target frequency range. The optimized absorber achieves a peak absorbance of 99.7% for normally incident waves while maintaining over 90% absorption at various incident angles in the frequency range from 8.5 GHz to 17.4 GHz. In addition, a double-layer graphene spiral resonator (DGSR) metasurface is proposed to extend the absorber's operational bandwidth, demonstrating a bandwidth enhancement of approximately 3 GHz and a relative bandwidth of 90% without compromising miniaturization or incident angle stability. Given their remarkable attributes, both GSR and DGSR configurations show great potential for applications in radar stealth technology and transparent electromagnetic compatibility.

Keywords: microwave absorber; graphene; spiral resonator; metasurface; wideband; transparent absorber; tunable absorber; equivalent circuit model

1. Introduction

Metamaterial absorbers (MAs) have attracted significant attention in recent years due to their promising applications in stealth technology, sensing, and multispectral thermal detection [1,2]. Microwave absorbers with wideband tunable absorption are of great importance for a wide range of practical applications in both military and daily life. For instance, MAs can be used to reduce radar cross-section (RCS) and electromagnetic interference (EMI) [3–6]. Nowadays, several absorbers in the microwave and optical regimes have been developed utilizing the special characteristics of metamaterials [7,8]. In certain applications, such as aircraft cockpit windows, optical detectors, and aerospace equipment, an optically transparent and flexible absorber is highly desirable.

Metasurfaces [9], regarded as the two-dimensional equivalents of metamaterials, are composed of multiple subwavelength components periodically placed on a planar surface. In comparison to three-dimensional metamaterials, metasurfaces are ultrathin and lightweight and have a relatively simple structure. They have been extensively used in various electromagnetic (EM) applications such as polarization conversion, wideband EM wave absorption, and RCS reduction [10–12].

Graphene, a one-atom-thick sheet of carbon atoms arranged in a honeycomb lattice, has unique properties. It offers exceptional mechanical strength, robust thermal stability, and excellent electrical conductivity. Due to the low density of states near the Dirac point, its Fermi level can be tuned with a bias voltage, allowing control of electromagnetic waves across a broad spectral range. These distinctive features, along with its adjustable electromagnetic response, have made graphene a strong candidate for tunable absorbers with high optical transparency [13]. Graphene's tunable conductivity enables dynamic control of the absorption bands, while metallic and dielectric metamaterials can achieve only non-dynamic control of their absorption bands through fine-tuning their geometrical parameters. Moreover, multi-band perfect absorption can be achieved by using single-layer graphene gratings due to the excitation of standing-wave graphene surface plasmon polaritons [14,15].

Graphene-based metasurfaces are being suggested for designing microwave absorbers with tunable absorption and RCS reduction, while also allowing for transparency and flexibility [16,17]. For example, integrating a graphene sandwich structure can facilitate adaptive microwave absorption through electrostatic doping of conductivity; Balci [18] proposed a reconfigurable graphene-based absorber with this approach. In [19], a transparent tunable absorber is created by patterning graphene into discrete patches, using two techniques: the 'surface resistance approach' and the 'stacking graphene metasurface' approach. Lu et al. [20] combined a graphene metasurface with an oxide–metal-oxide ground plane to produce a wideband microwave absorber with high transparency and sufficient flexibility. In [21], a transparent, tunable microwave absorber was created using a combination of patterned graphene and indium tin oxide (ITO). Wang et al. [22] developed a dual-tunable microwave absorber using a graphene metasurface, in which applying separate bias voltages to either continuous or patterned graphene enables independent control of the absorption frequency and amplitude. Geng et al. [23] created an effective microwave absorber based on graphene that can change its center frequency within the Ku band. However, this design is not optically transparent because the patterned graphene is directly deposited onto a layer of periodic metal patches.

In this work, we present the design of an optically transparent and flexible microwave tunable absorber, covering both the X and Ku frequency bands. The absorber is based on a metasurface composed of a periodic array of graphene spiral resonators (GSRs) attached to an ultrathin polyethylene terephthalate (PET) film, which is placed over an ITO-backed dielectric spacer. An equivalent circuit model (ECM) described by closed-form equations is proposed to help optimize the structure to achieve maximum absorption within the targeted frequency range. The incident angle insensitivity of the absorber for both TE and TM wave polarizations is also investigated. In addition, to broaden the absorber's bandwidth, a double-layer graphene spiral resonator (DGSR) metasurface is designed, and its absorption performance is evaluated.

2. Absorber Design and Equivalent Circuit Model

A perspective view of the proposed MA's schematic is illustrated in Figure 1, while its unit cell, which consists of three stacking functional layers, is depicted in Figure 2. The upper layer consists of a square graphene spiral resonator (GSR) attached to a 0.125 mm thick PET film. Square spiral resonators have been extensively used as metamaterial unit cells to offer a high level of miniaturization [24,25]. PET was chosen because of its optical transparency and mechanical flexibility. Moreover, monolayer graphene attached to PET film is nowadays commercially available [26]. The middle layer (spacer) is a transparent, flexible dielectric layer (PVC) with a thickness equal to one-quarter wavelength in the medium. The bottom surface of the dielectric spacer is backed by an indium tin oxide (ITO) ultrathin layer, which has a low sheet resistance (~6 Ω/sq) and acts as a reflective ground plane (PEC), providing zero transmission. The use of a low-sheet-resistance ITO layer instead of a metal plane has the advantages of high optical transparency and flexibility. The relative dielectric constants of PET and PVC are taken to be $\varepsilon_r = 2.8$ and $\varepsilon_{rs} = 3.2$, respectively. In the microwave regime, the loss tangents of PET and PVC films are $tan\delta_{pet}$ = 0.02 and $tan\delta_{PVC}$ = 0.00033, respectively [27]. The geometric parameters of the square GSR are as follows: the number of turns N of the spiral, the side length of the external turn l, the width of the strips w, and the separation between two adjacent turns s. The unit cell size of the MA and the PVC-substrate thickness are denoted by $a = l + 2s$ and h_1, respectively. It should be noticed that the maximum number of spiral turns should satisfy the structure's filling factor restriction: $N_{\max} = IntegerPart\{[l - (w + s)]/2(w + s)\}$.

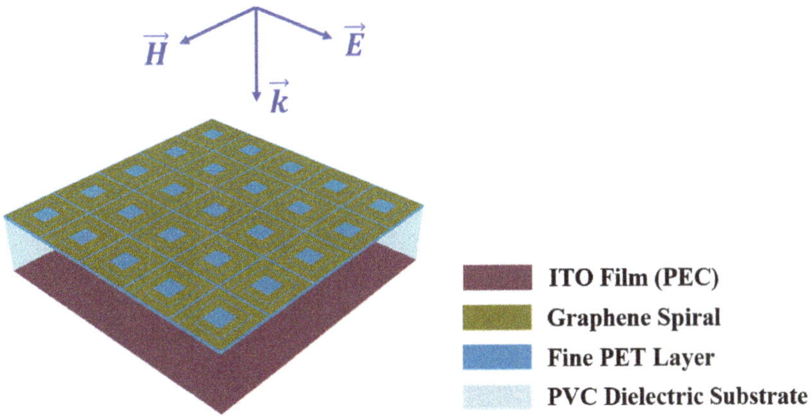

Figure 1. Perspective view of the proposed absorber (periodic layout).

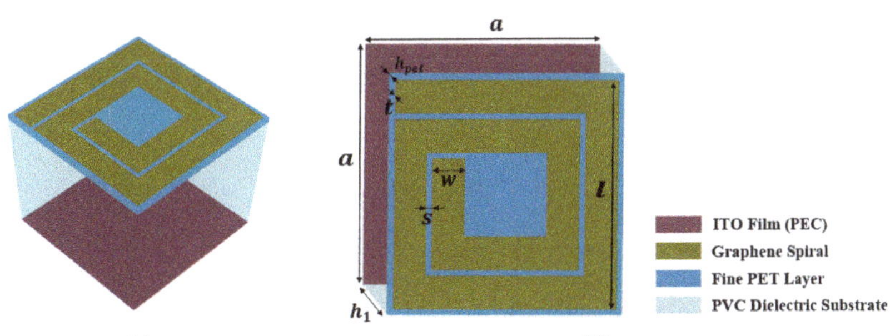

Figure 2. (**a**) Perspective view of the unit cell of the proposed absorber; (**b**) vertical view of the unit cell structure specifying its geometric parameters. The geometrical dimensions of the absorber are as follows: $l = 2.573$ mm, $s = 76.6$ µm, $w = 0.438$ mm, $t = 0.35$ nm, $h_1 = 3.2$ mm, $h_{pet} = 0.125$ mm $a = l + 2s = 2.726$ mm, $N = 2$.

It is well known that the dimensions of the structural components of the unit cell significantly affect the maximum absorption and the corresponding operating frequency range. To optimize the structure for maximum absorption across a broad frequency range, we developed an ECM using analytical expressions for resistive, capacitive, and inductive elements. Equivalent circuit models have been successfully utilized to design MAs [28,29]. They are useful tools that provide accurate results and yield valuable physical insights into the absorption characteristics of the structure and the impact that different parameters have.

In our ECM, the GSR is represented by the series combination of a resistor R_g, an inductor L_g, and a capacitor C_g with a shunt resistance R_d that takes into account the dissipation in the lossy PET thin substrate. Following Ref. [30], the equivalent impedance of the GSR is given by

$$Z_{g,eq} = R_g + \frac{R_d}{1+(\omega C_g R_d)^2} + j\omega\left[L_g - \frac{{R_d}^2 C_g}{1+(\omega C_g R_d)^2}\right] \tag{1}$$

where

$$C_g = \frac{l}{4(w+s)}\frac{N^2}{N^2+1}\left[l(N-1) - \frac{N^2-1}{2}(w+s)\right]C_0 \tag{2}$$

with

$$C_0 = \varepsilon_0\left\{1+\frac{2}{\pi}arctg\left[\frac{h_{pet}}{2\pi(w+s)}\right](\varepsilon_r-1)\right\}\frac{K\left(\sqrt{1-k^2}\right)}{K(k)} \tag{3}$$

where $K(k)$ is the complete elliptic integral of the first kind and $k = s/(s+2w)$.

$$L_g = \frac{\mu_0}{2\pi}l_{avg}\left[\frac{1}{2}+\ln\left(\frac{l_{avg}}{2w}\right)\right],\ l_{avg} = \frac{4lN - [2N(1+N)-3](s+w)}{N} \tag{4}$$

$$R_d = \frac{1}{\sigma_d}\frac{s}{4h_{pet}[l-(2w+s)]}\frac{l_{avg}}{4l},\ \sigma_d = 2\pi f\varepsilon_0\varepsilon_r tan\delta_{pet} \tag{5}$$

$$R_g = \frac{l_{avg}}{2\pi w\sigma_g}\left[\frac{1}{2}+\ln\left(\frac{l_{avg}}{2w}\right)\right] \tag{6}$$

The above closed-form equations have been validated in previous studies of metallic metamaterials based on the spiral geometry [25,31].

Based on Kubo's formula [32], there are two primary factors contributing to the surface conductivity of graphene: intraband and interband electronic transitions. In the microwave regime, interband transitions are extremely weak. In contrast, the contribution due to intraband transitions is the dominant one. Thus, considering $\exp(j\omega t)$ as the time harmonic dependency, the surface conductivity of the monolayer graphene can be expressed as [32]

$$\sigma_g(\omega, E_F, \Gamma, T) = \frac{-je^2 k_B T}{\pi\hbar^2(\omega - j2\Gamma)}\left[\frac{E_F}{k_B T} + 2\ln\left(e^{-\frac{E_F}{k_B T}}+1\right)\right] \tag{7}$$

where ω is the angular frequency; E_F is the Fermi energy; Γ is the electron scattering rate that is assumed to be energy independent; T is the temperature; e is the electron charge; $\hbar = h/(2\pi)$ is the reduced Planck constant; and k_B is the Boltzmann constant. $\Gamma = 1/(2\tau)$, where τ is the electron–phonon relaxation time. In this work, $\tau = 0.2\ ps$ is used, which is consistent with measurements for monolayer graphene produced by chemical vapor deposition (CVD) [19], and the ambient temperature ($T = 300$ K) is assumed. Although the relaxation time depends on both the Fermi energy and the frequency [33], such dependencies are typically weak; thus, we do not expect them to influence our main conclusions.

In the ECM, the PVC grounded substrate (spacer) can be modeled as a short-circuited transmission line because the ITO ground plane acts as a PEC. The impedance of the PVC spacer is given by

$$Z_{PVC} = jZ_{TE,TM}\tan(\beta h_1) \tag{8}$$

where $Z_{TE,TM}$ is the characteristic impedance of the PVC spacer for transverse electric (TE) or transverse magnetic (TM) polarized waves; β is the propagation constant, $\beta = \sqrt{k_s^2 - k_0^2\sin^2\theta_i}$ with $k_s = \omega\sqrt{\mu_0\mu_{rs}\varepsilon_0\varepsilon_{rs}} \approx k_0\sqrt{\varepsilon_{rs}}$; h_1 is the thickness of the PVC spacer; and θ_i denotes the incidence angle of the obliquely incoming waves into the structure. Moreover, the following equations give the characteristic impedances for TM- and TE-polarized waves:

$$Z_{TE} = \frac{\omega\mu_0\mu_{rs}}{\beta} \qquad Z_{TM} = \frac{\beta}{\omega\varepsilon_0\varepsilon_{rs}} \tag{9}$$

It is obvious from Equations (8) and (9) that the characteristic impedance of the PVC spacer is a function of the incident angle for both polarizations. It should be noticed that the equivalent capacitance of the GSR metasurface should be modified for the case of oblique incidence as follows [34,35]:

$$C_g^{TE}(\theta_i) = C_g\left(1 - \frac{1}{\varepsilon_r + 1} sin^2\theta_i\right), \quad C_g^{TM}(\theta_i) = C_g \tag{10}$$

Furthermore, the sheet impedance of graphene remains almost constant with the incident angle and polarization state due to the negligible spatial dispersion of graphene.

We utilize the ECM depicted in Figure 3 to optimize the proposed absorber for high absorption across a broad frequency range, covering both X and Ku bands. The absorbance of an MA can be expressed in terms of S-scattering parameters as

$$A(f) = 1 - |S_{11}(f)|^2 - |S_{21}(f)|^2 \tag{11}$$

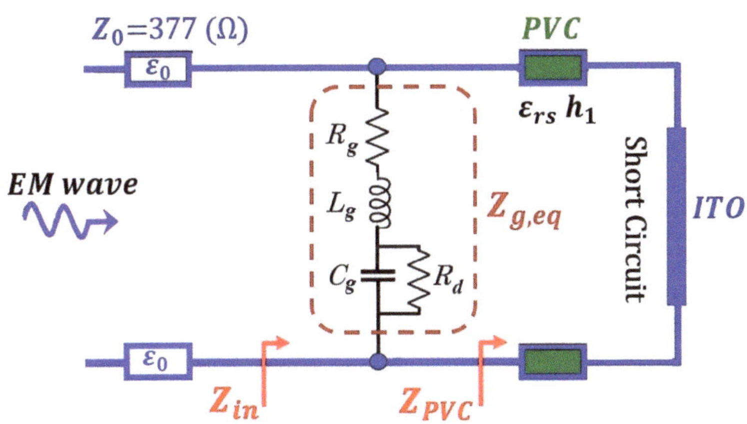

Figure 3. Equivalent circuit model (ECM) representation of the proposed GSR absorber.

Since the ITO ground plane results in almost vanishing transmission $|S_{21}(f)|^2 = 0$, Equation (11) for normal incidence can be written as

$$A(f) = 1 - |S_{11}(f)|^2 = 1 - \left|\frac{Z_{in}(f) - Z_0}{Z_{in}(f) + Z_0}\right|^2 \tag{12}$$

The input impedance of the absorber $Z_{in}(f)$ can be calculated by

$$Z_{in}(f) = Z_{g,eq} \parallel Z_{PVC} = \frac{Z_{g,eq}(f) Z_{PVC}(f)}{Z_{g,eq}(f) + Z_{PVC}(f)} \tag{13}$$

where $Z_{PVC}(f)$ is given by Equation (8), and $Z_0 = \sqrt{\mu_0/\varepsilon_0} = 377\ \Omega$ is the free space wave impedance. In the case of oblique incidence, the wave impedances in free space should be modified for both TE and TM wave polarizations as follows [34,35]:

$$Z_0^{TE}(\theta_i) = \frac{Z_0}{\cos\theta_i},\ Z_0^{TM}(\theta_i) = Z_0\cos\theta_i \tag{14}$$

and the scattering parameter $S_{11}(f)$ in Equation (12) should be replaced by

$$S_{11}^{TE}(\theta_i) = \frac{Z_{in}^{TE}(\theta_i) - \frac{Z_0}{\cos\theta_i}}{Z_{in}^{TE}(\theta_i) + \frac{Z_0}{\cos\theta_i}},\ S_{11}^{TM}(\theta_i) = \frac{Z_{in}^{TM}(\theta_i) - Z_0\cos\theta_i}{Z_{in}^{TM}(\theta_i) + Z_0\cos\theta_i} \tag{15}$$

for TE and TM wave polarizations, respectively.

In the following, using the analytical expressions of our ECM, we can identify optimized geometrical and electrical parameters of the structure for achieving wideband high absorption (>90%) covering both the X (8–12 GHz) and Ku (12–18 GHz) bands. At first, the thickness of the PVC dielectric spacer is selected to be equal to one-quarter wavelength according to interference cancellation theory [36]:

$$h_1 = \frac{c}{4f_c\sqrt{\varepsilon_{rs}}} \tag{16}$$

where $f_c = 13$ GHz is taken to be the central frequency of the desired operating frequency band. Hence, from Equation (16), we obtain $h_1 = 3.2$ mm. After investigating the dependence of the resonant frequency $f_r = 1/\left(2\pi\sqrt{L_g\, C_g}\right)$ on the GSR's number of turns N and its geometric parameters, a set of initial parameters can be selected: $N = 2$, $l = 2.6$ mm, $s = 35$ μm, and $w = 0.25$ mm. The thickness of the graphene spirals is taken to be $t = 0.35$ nm. According to Equation (12), perfect absorption could be achieved when the impedance matching condition $Z_{in}(f) = Z_0$ is satisfied; i.e., the real and imaginary parts of the input impedance should be $Re\{Z_{in}(f)\} = Z_0 = 377\ \Omega$ and $Im\{Z_{in}(f)\} = 0$, respectively, for a wideband frequency range. The optimized values of the geometric parameters can be easily found with the aid of a simple optimization algorithm with the cost function defined by the above impedance matching condition. Finally, the optimum values of the structural parameters of the proposed absorber are found to be $N = 2$, $l = 2.573$ mm, $s = 76.6$ μm, $w = 0.438$ mm, $t = 0.35$ nm, $h_1 = 3.2$ mm, $h_{pet} = 0.125$ mm.

3. Absorber Performance and Parameter Retrieval Method

In this section, we demonstrate the performance of the designed MA by considering a plane EM wave in the microwave regime impinging on the GSR metasurface. Initially, we assume normal incidence of an EM wave with transverse electric (TE) or transverse magnetic (TM) polarization. Figure 4a depicts the tunability of the absorption spectra by adjusting the Fermi energy of graphene from 0.4 eV to 1.2 eV. Note that, based on experimental evidence, the Fermi energy of graphene can be effectively modulated from 0.4 to 1.2 eV by adjusting a DC voltage [37]. The DC bias voltage can be applied directly between the patterned graphene metasurface and the ITO ground plane. However, a more efficient application scheme of the DC bias voltage is the utilization of an ionic liquid electrolyte (or gel). In this scheme, an ultrathin layer of ionic gel can be coated on the graphene as a high-capacitance spacer layer between a top metal electrode and the patterned graphene, as described in Refs. [19,21]. The ionic gel generates high carrier densities in graphene, causing an increase in the Fermi energy. It is worth noting that, with the aid of ionic liquids, a significant range of Fermi energies can be obtained by applying a moderate DC voltage [18].

As shown in Figure 4a, both the maximum absorbance and the bandwidth increase as the Fermi energy E_F is increased, and for $E_F = 1.2$ eV, the MA demonstrates more than 90% absorbance with a bandwidth of 8.9 GHz, covering both the X and Ku bands. To characterize the absorption performance for the MA, we can calculate the relative absorption bandwidth (RBW) defined as $RBW = 2(f_U - f_L)/(f_U + f_L)$, where f_U and f_L are the upper and lower frequencies corresponding to the absorption frequency range with absorbance more than 90%. The proposed MA achieves an RBW of 68%, and thus, ultrawideband is achieved, which is a demanding feature for modern microwave applications. Moreover, the MA shows a peak absorbance of 99.7% at $E_F = 1.2$ eV.

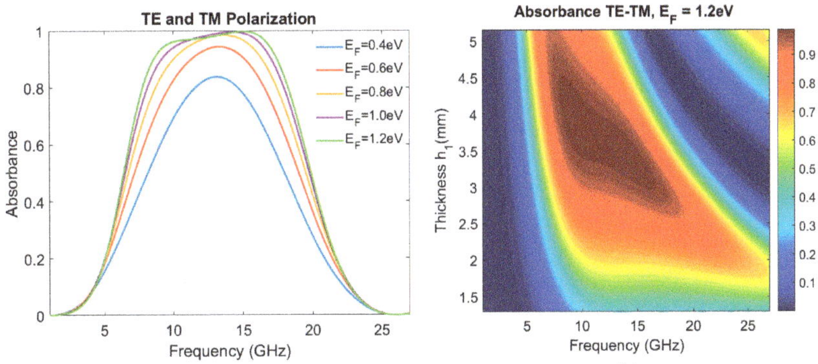

(a) (b)

Figure 4. (**a**) Absorbance spectrum at normal incidence at different Fermi energies; (**b**) contour plot of absorbance spectrum as a function of PVC spacer thickness.

The thickness of the PVC spacer is a crucial geometrical parameter determining the absorption spectrum due to impedance mismatch. In our design, the thickness is chosen to be one-quarter wavelength in the medium, i.e., $h_1 = 3.2$ mm. By varying the thickness h_1, while keeping other parameters unchanged, the absorption spectrum is significantly affected, as shown in Figure 4b. When the PVC thickness is $h_1 = 3.2$ mm, the absorber performs effectively in the frequency band between 8.5 GHz and 17.35 GHz. If the PVC thickness is smaller than 3.2 mm, the absorption band is shifted to a higher frequency range, whereas if the PVC thickness is larger than 3.2 mm, the absorption band is shifted to a lower frequency range.

To gain a deeper insight into the tuning mechanism of absorbance amplitude, the real and imaginary parts of the normalized input impedance, $Z_{in}(f)/Z_0$, at different Fermi energies, are illustrated in Figure 5a,b, respectively. It can be seen that, as the graphene Fermi energy increases, the real part of impedance fluctuates around 1 while the imaginary part fluctuates around 0 across the desired frequency range. This suppresses the reflection of the incident waves by the absorber, resulting in enhanced absorption. Thus, by changing the graphene conductivity σ_g, the absorption amplitude can be tuned in accordance with the input impedance matching, enabling efficient wideband absorption.

It is well known that the response of EM metamaterials can be characterized using equivalent EM parameters, such as effective permittivity ε_{eff} and effective permeability μ_{eff}. These effective constitutive parameters can also serve as a significant tool for validating the absorbing mechanism [38]. The Nicolson–Ross–Weir (NRW) retrieval method [39,40] provides a feasible method for extracting these parameters. According to the NRW technique, the effective permittivity and effective permeability of the proposed MA can be evaluated through the use of the S-parameters as follows [40]:

$$\varepsilon_{r,eff} = \frac{2}{jk_0 d} \cdot \frac{1 - S_{11} - S_{21}}{1 + S_{11} + S_{21}}, \quad \mu_{r,eff} = \frac{2}{jk_0 d} \cdot \frac{1 - S_{21} + S_{11}}{1 + S_{21} - S_{11}} \tag{17}$$

where d is the thickness of the absorber, and $k_0 = \omega / c_0$ denotes the propagation constant in free space. Since $S_{21} = 0$ due to the ITO ground plane (PEC), Equation (17) can be simplified as

$$\varepsilon_{r,eff} = \frac{2}{jk_0 d} \cdot \frac{1 - S_{11}}{1 + S_{11}}, \quad \mu_{r,eff} = \frac{2}{jk_0 d} \cdot \frac{1 + S_{11}}{1 - S_{11}} \tag{18}$$

where S_{11} can be calculated analytically from the input impedance of the structure under normal incidence.

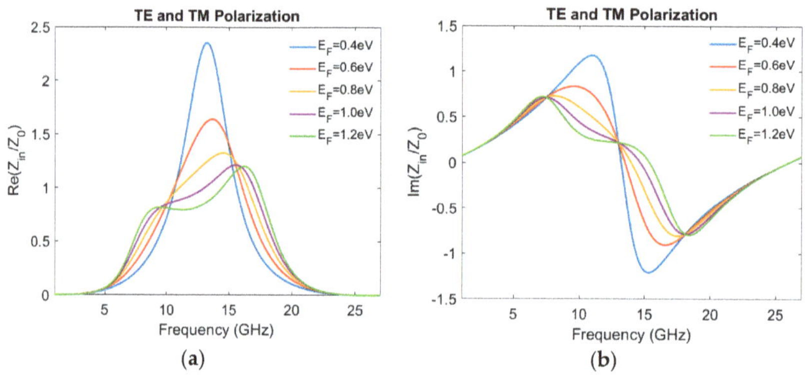

Figure 5. Real part (**a**) and imaginary part (**b**) of the normalized input impedance Z_{in} / Z_0 as a function of frequency at different Fermi energies.

Figure 6a,b show the effective permittivity $\varepsilon_{r,eff}(f)$ and the effective permeability $\mu_{r,eff}(f)$ of the proposed absorber as a function of frequency. It can be seen that, in the frequency range of interest (X and Ku bands), the structure exhibits a single-negative (SNG) metamaterial property characterized by a negative value of either the real permittivity or the real permeability [41]. The imaginary parts of the permittivity and permeability account for the losses and represent the material's impact on absorbing the incident electromagnetic waves. Using the extracted electromagnetic parameters $\varepsilon_{r,eff}(f)$ and $\mu_{r,eff}(f)$, the normalized effective impedance as a function of frequency can be calculated as [39]

$$\frac{Z_{eff}(f)}{Z_0} = \sqrt{\frac{\mu_{r,eff}(f)}{\varepsilon_{r,eff}(f)}} \tag{19}$$

Figure 6c presents the frequency dependence of both the real and imaginary parts of the normalized effective impedance Z_{eff} / Z_0, which is derived from the obtained effective parameters. These curves coincide with the corresponding ones for the real and imaginary parts of the input impedance of the absorber obtained by the ECM at the Fermi energy $E_F = 1.2$ eV. Consequently, the results of the effective parameter retrieval method support the reliability of our ECM.

Next, the absorbance of the designed MA under oblique incidence is considered. To investigate the angular stability of the absorber, the absorbance is calculated as a function of frequency and incident angle for both TE and TM wave polarizations. Figure 7a illustrates the absorption spectrum at different incident angles from 0° to 70° for an EM wave with

TE polarization, while Figure 7b presents the contour map of the absorbance spectrum as a function of the incident angle. The corresponding plots for an EM wave with TM polarization are presented in Figure 8a,b, respectively.

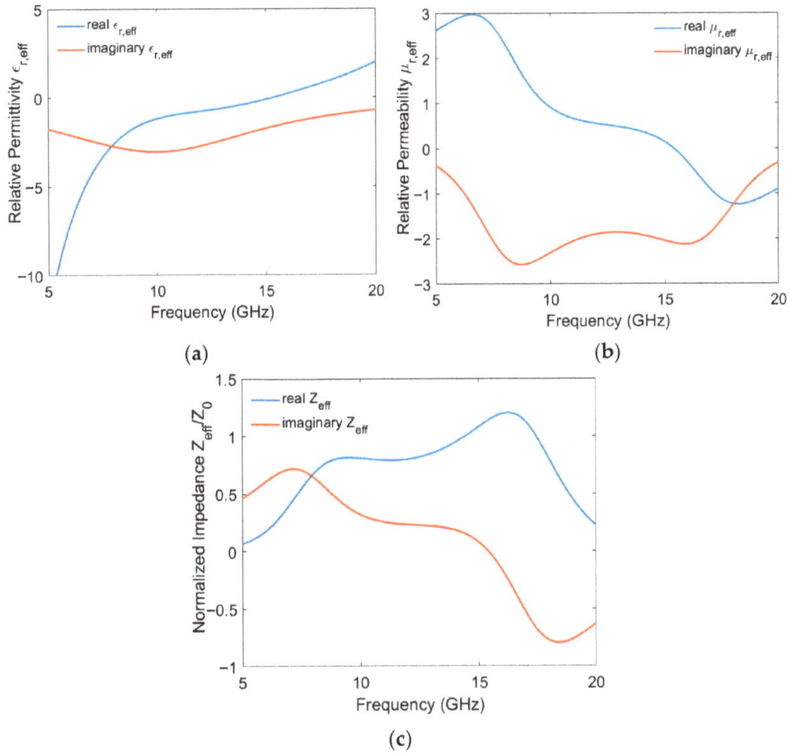

Figure 6. Retrieved electromagnetic parameters of the absorber. (**a**) Real and imaginary parts of effective permittivity, (**b**) real and imaginary parts of effective permeability, and (**c**) normalized effective impedance Z_{eff} / Z_0 as functions of frequency. The Fermi energy is set to be E_F = 1.2 eV.

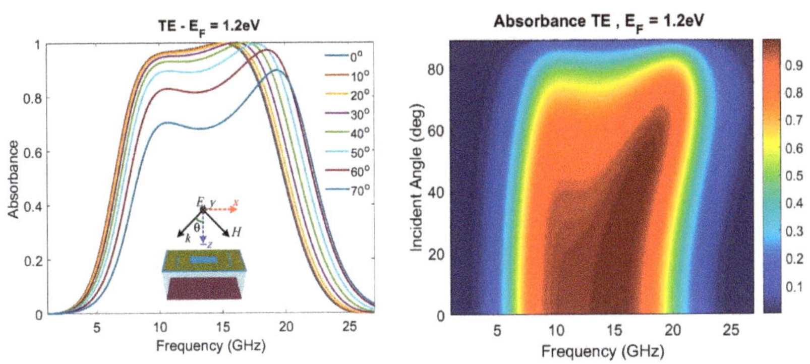

Figure 7. (**a**) The absorption spectrum at different incident angles of the EM wave with TE polarization. (**b**) Contour map of the absorbance spectrum as a function of angle of incidence. The Fermi energy is set to be $E_F = 1.2$ eV.

With the increase in the incident angle, the absorbance gradually decreases. This reduction observed at greater incident angles is anticipated due to increased reflection and less interaction between the incoming waves and the metasurface at steeper angles. The absorption performance under TM incidence is inferior to that under TE incidence for the same incident angle. Interestingly, in TE polarization, the absorption bandwidth is increased, and the absorbance peak is blue-shifted as the incident angle increases. On the contrary, the absorption bandwidth of TM polarization is slightly narrowed as the angle of incidence increases. The proposed absorber maintains a high absorbance of above 90% at an incident angle up to 50° for both TE and TM polarizations. Moreover, the structure guarantees an absorbance of more than 80% for both TE and TM polarizations, when the incident angle is increased to 60°. Notably, this wide-angle stability covers the target frequency bands (X-Ku) with an RBW of about 68% for TE waves and of about 60% for TM waves.

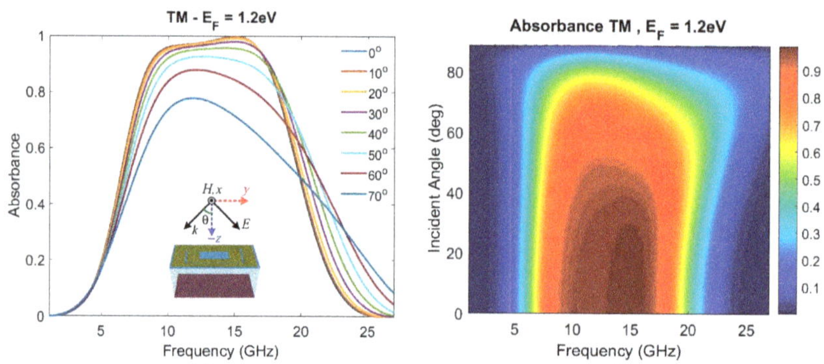

Figure 8. (**a**) The absorption spectrum at different incident angles of the EM wave with TM polarization. (**b**) Contour map of the absorbance spectrum as a function of angle of incidence. The Fermi energy is set to be $E_F = 1.2$ eV.

4. Design of a Double-Layer GSR Metasurface Absorber

In this section, the 'stacking graphene metasurface approach' [19] is used to broaden the absorption bandwidth. Above the designed GSR metasurface absorber, we place another similar GSR metasurface with a dielectric spacer (PVC) of thickness h_2. The configuration of this double-layer GSR metasurface absorber (DGSR) is illustrated in Figure 9. Incorporating two layers of GSR, the controllability of the input impedance of the structure can be enhanced by independently adjusting the Fermi energy of each graphene layer.

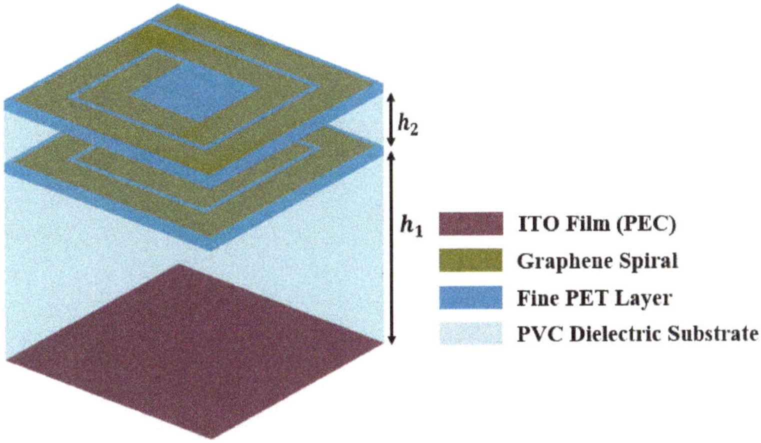

Figure 9. Perspective view of the unit cell of the double-layer GSR metasurface absorber (DGSR).

The ECM model depicted in Figure 10 can be used to optimize the DGSR absorber's performance. The absorbance $A(f)$ of the structure is given by Equation (12), where the input impedance $Z_{in}\ (f)$ can be calculated as

$$Z_{in} = Z_{gb} \parallel Z_b = \frac{Z_{gb}Z_b}{Z_{gb} + Z_b} \tag{20}$$

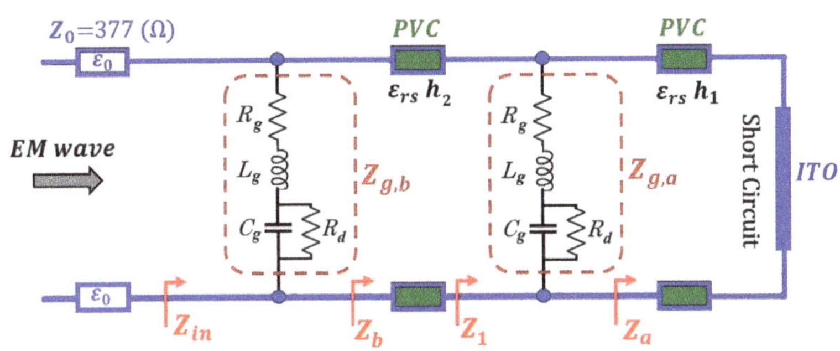

Figure 10. Equivalent circuit model of the double-layer GSR metasurface absorber (DGSR).

The impedance Z_b can be expressed by the transmission line correlation formula as

$$Z_b = Z_{TE,TM}\frac{Z_1 + jZ_{TE,TM}\tan(\beta h_2)}{Z_{TE,TM} + jZ_1 tan(\beta h_2)} \tag{21}$$

where

$$Z_1 = Z_{ga} \parallel Z_a = \frac{Z_{ga}Z_a}{Z_{ga} + Z_a},\ Z_a = jZ_{TE,TM}\tan(\beta h_1) \tag{22}$$

and the impedances $Z_{TE,TM}$ are given by Equation (9). Moreover, to cover the oblique incidence case, the modifications described by Equations (14) and (15) should be made.

The optimized parameters can be determined through an iterative optimization technique using our ECM. Keeping the thickness h_1 equal to one-quarter wavelength in the medium, the thickness $h_2 < h_1 = 3.2$ mm is swept for different pairs of Fermi energies E_{Fa}, E_{Fb} in the range (0.2 eV–1.2 eV) to find the optimum values that achieve the maximum absorption bandwidth with absorbance $A \geq 90\%$. It is found that, for $E_{Fa} = E_{Fb} = 0.9$ eV and $h_2 = 21.5\mu m$, a maximum absorption bandwidth of 11.7 GHz is achieved, covering the X and Ku bands. Figure 11a shows the absorbance spectrum of the DGSR absorber for the case of normal incidence at different Fermi energies E_{Fb} with $E_{Fa} = 0.9$ eV. Figure 11b provides a comparison between the absorbance spectra of the DGSR and the GSR absorbers. As can be seen, taking an absorbance of 90% as a reference, a significant bandwidth enhancement of approximately 3 GHz is achieved. Moreover, the absorber possesses an RBW of about 90%. However, it should be noted that the absorbance of the DGSR absorber is slightly decreased compared with that of the GSR absorber.

The absorption spectra at different incident angles of the DGSR MA are plotted in Figure 12a,b for TE and TM incident EM waves, respectively. It is demonstrated that, in the case of the TE wave, the average absorbance remains above 90% for incident angles less than 30°.

The absorbance starts to decrease significantly as the incident angle is increased above 30° however, it remains higher than 80% for incident angles up to 50°. On the other hand, the absorber presents better angle stability for the case of the TM wave, since the absorbance remains above 90% for incident angles up to 50° across the desired frequency range. Overall, the DGSR MA is relatively stable under different incident angles.

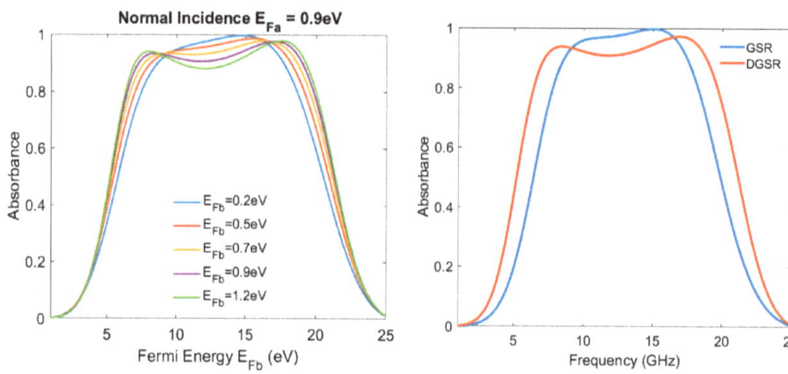

(a) (b)

Figure 11. (**a**) The absorption spectrum at different Fermi energies E_{Fb}. The Fermi energy of the first GSR metasurface is set to be $E_{Fa} = 0.9$ eV. (**b**) Comparison of the absorbance spectra of the DGSR and the GSR absorbers.

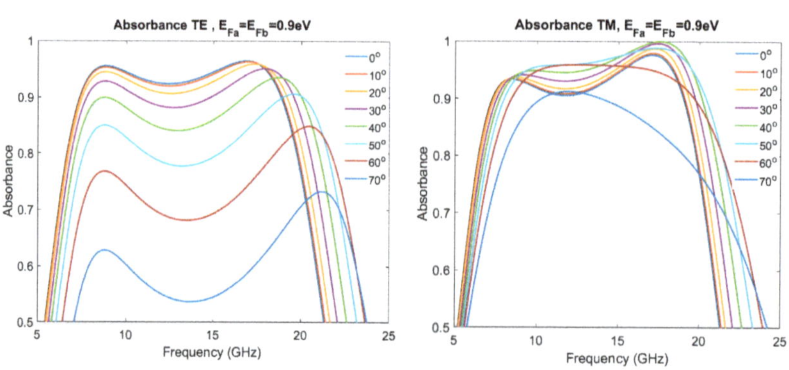

(a) (b)

Figure 12. The absorption spectrum of the DGSR MA at different incident angles (**a**) for an EM wave with TE polarization and (**b**) for an EM wave with TM polarization. The Fermi energies are set to be $E_{Fa} = E_{Fb} = 0.9$ eV.

5. Discussion and Conclusions

We have designed and analytically studied a tunable microwave MA based on a metasurface made of graphene spiral resonators (GSRs). An ECM described by closed-form equations has been utilized to optimize the structure and demonstrate its absorption performance. The amplitude of the absorption spectra can be modified by applying a bias voltage between the graphene metasurface and the grounded ITO plane, allowing for control over the Fermi energy of the graphene. The whole structure possesses optical transparency and mechanical flexibility, achieved by incorporating transparent and flexible materials. In our design, the lossy GSR metasurface is responsible for the high absorption, since most of the impinging EM energy is dissipated through the induced currents flowing on the resistive graphene spiral resonators. In contrast, the PVC dielectric spacer exhibits negligible consumption of EM waves. Additionally, to broaden the absorption bandwidth, a double-layer GSR (DGSR) is designed using two similar GSR metasurfaces. The optimized DGSR absorber achieves a bandwidth enhancement of roughly 3 GHz, resulting in an RBW of about 90% while remaining ultrathin. Such a BW enhancement has been previously achieved by utilizing thicker structures that share comparable fabrication complexity [42]. It is worth noting that both absorber designs have a subwavelength thickness (approximately one-tenth of a wavelength), which meets the Rozanov limit [43], and thus, the validity of our ECM is guaranteed [44].

Concerning fabrication feasibility, the proposed absorber can be realized using well-established techniques. The monolayer graphene can be grown on copper foil using the conventional chemical vapor deposition (CVD) method. Afterward, the copper foil is etched away. The graphene film is then transferred onto the PET film. However, graphene/PET layers are now commercially available [26]. The infrared laser beam cutting method can be used to pattern the graphene into a square spiral shape. Next, the patterned graphene can be deposited on the top of the PVC dielectric spacer via a lamination process. Lastly, an ITO film with low square resistance should be attached to the bottom of the structure to act as a ground plane. Some deviations may be found between our simulated results and the experimental ones. There could be discrepancies in the absorbance amplitude and maximum absorbance frequencies. These may be caused by (a) inhomogeneity of the patterned graphene/PET layer sample due to imperfect fabrication; (b) differences in the dielectric constants of the materials from those used in our calculations; (c) the finite size of the fabricated sample, resulting in edge diffraction; and (d) degradation of graphene's surface resistance due to the presence of impurities in air.

The performance of the proposed GSR and DGSR metasurface absorbers was compared with those of some previously reported graphene-based microwave absorbers, as shown in Table 1. Evaluated alongside other relevant research, the proposed GSR absorber achieves ultrawideband absorption covering both the X and Ku bands with a relative bandwidth of 68% and high absorbance up to 99.7% at the peaks. Moreover, the designed GSR absorber is characterized by wide-angle stability, optical transparency, mechanical flexibility, and low thickness of 0.09λ, where λ is the wavelength corresponding to the lowest frequency of the absorption band. Furthermore, the designed DGSR absorber has an enhanced absorption bandwidth with a relative bandwidth of 90% and high absorbance up to 97.6% at the peaks. It also remains miniaturized, optically transparent, and flexible. Both GSR and DGSR configurations, due to their attractive features, may have prospects for applications in stealth technology and EM compatibility. Additionally, due to their optical transparency and flexibility, they can contribute to practical applications, including the cockpit windows of low-detectable aircraft and aerospace devices.

Table 1. Comparison of the proposed absorber with previously reported graphene-based metasurface microwave absorbers.

Reference	Bandwidth (GHz)($A \geq 90\%$)	Central Frequency (GHz)	Angular Stability (θ_i)	Thickness (mm)	Transparent	Flexibility
[19]	3.0–13.0	8.0	-	3.4	Yes	No
[20]	5.0–16.0	10.5	<45°	4.82	Yes	Yes
[21]	10.9–13.1	12.0	<60°	2.76	No	Yes
[22]	9.8–11.3	10.5	<60°	3.2	Yes	Yes
[23]	13.9–16.9	15.4	-	1.2	No	No
This work GSR	8.5–17.4	12.9	<60°	3.325	Yes	Yes
This work DGSR	7.2–19.0	13.1	<50°	3.45	Yes	Yes

Author Contributions: Conceptualization, G.S.K.; methodology, G.S.K. and I.S.F.; formal analysis, I.S.F.; investigation, I.S.F. and G.S.K.; writing—original draft preparation, G.S.K.; writing—review and editing, I.S.F. and G.S.K.; supervision, G.S.K. All authors have read and agreed to the published version of the manuscript.

Funding: This research received no external funding.

Institutional Review Board Statement: Not applicable.

Informed Consent Statement: Not applicable.

Data Availability Statement: All data are included in this text.

Conflicts of Interest: The authors declare no conflicts of interest.

Abbreviations

The following abbreviations are used in this manuscript:

MA	Metamaterial Absorber
EM	Electromagnetic
GSR	Graphene Spiral Resonator
DGSR	Double-layer Graphene Spiral Resonator
ECM	Equivalent Circuit Model
PET	Polyethylene Terephthalate
ITO	Indium Tin Oxide
PEC	Perfect Electric Conductor
RBW	Relative Bandwidth
NRW	Nicolson–Ross–Weir

References

1. Lee, Y.P.; Rhee, J.Y.; Yoo, Y.J.; Kim, K.W. Metamaterials for Perfect Absorption. In *Springer Series in Materials Science 236*; Springer: Cham, Switzerland, 2016.
2. Wang, B.-X.; Xu, C.; Duan, G.; Xu, W.; Pi, F. Review of Broadband Metamaterial Absorbers: From Principles, Design Strategies, and Tunable Properties to Functional Applications. *Adv. Funct. Mater.* **2023**, *33*, 2213818. [CrossRef]
3. Fan, Y.; Wang, J.; Fu, X.; Li, Y.; Pang, Y.; Zheng, L.; Yan, M.; Zhang, J.; Qu, S. Recent developments of metamaterials/metasurfaces for RCS reduction. *EPJ Appl. Metamat.* **2019**, *6*, 15. [CrossRef]
4. Shittu, I.; Abou-Khousa, M.; Viegas, J.; Hernandez-Figueroa, H.E.; Elfadel, I.M. Radar Cross Section Reduction Metamaterials: A Review of Principles, Design Methods, and Applications Beyond. *IEEE Aerosp. Electron. Syst. Mag.* **2025**, *1*, 1–21. [CrossRef]

5. Kliros, G.S.; Maniotis, K.I. A Triple-Band Ultra-Thin Metamaterial Absorber for Radar Cross-Section Reduction. In Proceedings of the 2018 26th Telecommunications Forum (TELFOR), Belgrade, Serbia, 20–21 November 2018; pp. 509–512. [CrossRef]
6. Saadeldin, A.S.; Sayed, A.M.; Amr, A.M.; Sayed, M.O.; Hameed, M.F.O.; Obayya, S.S.A. Wideband Ultrathin and Polarization Insensitive Metamaterial Absorber for Ku-Band Applications. *J. Mater. Sci. Mater. Electron.* **2023**, *34*, 1797. [CrossRef]
7. Pham, T.S.; Khuyen, B.X.; Lam, V.D.; Chen, L.; Lee, Y. Wide-Angle, Polarization-Independent Broadband Metamaterial Absorber by Using Plasmonic Metasurface-Based Split-Circular Structure. *Photonics* **2025**, *12*, 334. [CrossRef]
8. Li, Z.; Wang, J.; Zheng, W. Deformable Transparent Metasurface with Multipolarization Absorptive Characteristics for RCS Reduction. *IEEE Trans. Antenn. Propag.* **2025**, *73*, 1768–1776. [CrossRef]
9. Chen, H.-T.; Taylor, A.J.; Yu, N. A review of metasurfaces: Physics and applications. *Rep. Prog. Phys.* **2016**, *79*, 076401. [CrossRef]
10. Zhao, J.; Li, N.; Cheng, Y. Ultrabroadband chiral metasurface for linear polarization conversion and asymmetric transmission based on enhanced interference theory. *Chin. Opt. Lett.* **2023**, *21*, 113602. [CrossRef]
11. Ma, Y.; Wang, J.; Shi, L.; Xue, S.; Ran, Y.; Li, J.; Liu, Y. Ultra-wideband, optically transparent, and flexible microwave metasurface absorber. *Opt. Mater. Exp.* **2021**, *11*, 2206–2218. [CrossRef]
12. Xue, H.; Ge, J.; Zhang, Y.; Jiang, X.; Zhang, C.; Dong, H.; Zhang, L. An Optically Transparent Metasurface for Microwave Amplitude-Phase Manipulation. *Photonics* **2025**, *12*, 384. [CrossRef]
13. Shi, K.; Su, J.; Liang, H.; Hu, K.; Xu, J. Highly optically transparent graphene mesh for electromagnetic interference shielding. *Diam. Relat. Mater.* **2022**, *123*, 108849. [CrossRef]
14. Xia, S.-X.; Zhai, X.; Huang, Y.; Liu, J.-Q.; Wang, L.-L.; Wen, S.-C. Multi-band perfect plasmonic absorptions using rectangular graphene gratings. *Opt. Lett.* **2017**, *42*, 3052–3055. [CrossRef]
15. Xia, S.-X.; Zhang, D.; Zhai, X.; Wang, L.-L.; Wen, S.-C. Phase-controlled topological plasmons in 1D graphene nanoribbon array. *Appl. Phys. Lett.* **2023**, *123*, 101102. [CrossRef]
16. Chen, H.; Ma, W.; Huang, Z.; Zhang, Y.; Huang, Y.; Chen, Y. Graphene-based materials toward microwave and terahertz absorbing stealth technologies. *Adv. Opt. Mater.* **2019**, *7*, 1801318. [CrossRef]
17. Shao, L.; Zhu, W. Graphene-derived microwave metamaterials and meta-devices: Emerging applications and properties. *Electron* **2025**, *3*, e60. [CrossRef]
18. Balci, O.; Polat, E.O.; Kakenov, N.; Kocabas, C. Graphene-enabled electrically switchable radar-absorbing surfaces. *Nat. Commun.* **2015**, *6*, 6628. [PubMed]
19. Yi, D.; Wei, X.C.; Xu, Y.L. Tunable microwave absorber based on patterned graphene. *IEEE Trans. Microw. Theory Tech.* **2017**, *65*, 2819–2826. [CrossRef]
20. Lu, W.B.; Wang, J.W.; Zhang, J.; Liu, Z.G.; Chen, H.; Song, W.J.; Jiang, Z.H. Flexible and optically transparent microwave absorber with wide bandwidth based on graphene. *Carbon* **2019**, *152*, 70–76. [CrossRef]
21. Zhong, S.; Zhang, Y.; Ma, Y. Optically Transparent Frequency-Tunable Microwave Absorber Based on Patterned Graphene-ITO Structure. *IEEE Trans. Antenn. Propag.* **2022**, *70*, 9959–9964. [CrossRef]
22. Wang, P.; Tao, H.; Han, W.; Zhang, C.; Xu, Y.; Wang, Q. Polarization-insensitive, frequency- and amplitude-tunable transparent microwave absorber based on graphene metasurface. *J. Appl. Phys.* **2024**, *135*, 043103. [CrossRef]
23. Geng, M.Y.; Liu, Z.G.; Wu, W.J.; Chen, H.; Wu, B.; Lu, W.B. A dynamically tunable microwave absorber based on graphene. *IEEE Trans. Antenn. Propag.* **2020**, *68*, 4706–4713. [CrossRef]
24. Baena, J.D.; Marqués, R.; Medina, F.; Martel, J. Artificial magnetic metamaterial design by using spiral resonators. *Phys. Rev. B* **2004**, *69*, 014402. [CrossRef]
25. Bilotti, F.; Toscano, A.; Vegni, L. Design of spiral and multiple split-ring resonators for the realization of miniaturized metamaterial samples. *IEEE Trans. Antenn. Propag.* **2007**, *55*, 2258–2267. [CrossRef]
26. Banerjee, I.; Faris, T.; Stoeva, Z.; Harris, P.G.; Chen, J.; Sharma, A.K.; Ray, A.K. Graphene films printable on flexible substrates for sensor applications. *2D Mater.* **2017**, *4*, 015036. [CrossRef]
27. Yang, F.; Xu, H.; Zhu, X.; Wang, D.; Zhang, Y.; Feng, Z. Graphene-based flexible optically transparent dual-tunable metasurface. *Adv. Mater. Technol.* **2024**, *9*, 2301911. [CrossRef]
28. Costa, F.; Monorchio, A.; Manara, G. Analysis and design of ultra-thin electromagnetic absorbers comprising resistively loaded high impedance surfaces. *IEEE Trans. Antenn. Propag.* **2010**, *58*, 1551–1558. [CrossRef]
29. Costa, F.; Genovesi, S.; Monorchio, A.; Manara, G. A Circuit-based model for the interpretation of perfect metamaterial absorbers. *IEEE Trans. Antenn. Propag.* **2013**, *61*, 1201–1209. [CrossRef]
30. Bilotti, F.; Toscano, A.; Vegni, L.; Alici, K.B.; Aydin, K.; Ozbay, E. Equivalent circuit models for the design of metamaterials based on artificial magnetic inclusions. *IEEE Trans. Microw. Theory Tech.* **2007**, *MTT-55*, 2865–2873. [CrossRef]
31. Bilotti, F.; Toscano, A.; Alici, K.B.; Ozbay, E.; Vegni, L. Design of miniaturized narrowband absorbers based on resonant magnetic inclusions. *IEEE Trans. Antenn. Propag.* **2011**, *53*, 63–71. [CrossRef]
32. Hanson, G.W. Dyadic Green's functions and guided surface waves for a surface conductivity model of graphene. *J. Appl. Phys.* **2008**, *103*, 064302. [CrossRef]

33. Tan, Y.W.; Zhang, Y.; Bolotin, K.; Zhao, Y.; Adam, S.; Hwang, E.H.; Das Sarma, S.; Stormer, H.L.; Kim, P. Measurement of scattering rate and minimum conductivity in graphene. *Phys. Rev. Lett.* **2007**, *99*, 246803. [CrossRef] [PubMed]
34. Luukkonen, O.; Simovski, C.; Granet, G.; Goussetis, G.; Lioubtchenko, D.; Raisanen, A.V.; Tretyakov, S.A. Simple and accurate analytical model of planar grids and high-impedance surfaces comprising metal strips or patches. *IEEE Trans. Antenn. Propag.* **2008**, *56*, 1624–1632. [CrossRef]
35. Wang, X.; Tretyakov, S.A. Toward Ultimate Control of Terahertz Wave Absorption in Graphene. *IEEE Tran. Antenn. Propag.* **2019**, *67*, 2452–2461. [CrossRef]
36. Chen, H.-T. Interference theory of metamaterial perfect absorbers. *Opt. Express* **2012**, *20*, 7165–7172. [CrossRef] [PubMed]
37. Balci, S.; Balci, O.; Kakenov, N.; Atar, F.B.; Kocabas, C. Dynamic tuning of plasmon resonance in the visible using graphene. *Opt. Lett.* **2016**, *41*, 1241–1244. [CrossRef]
38. Smith, D.; Vier, D.; Koschny, T.; Soukoulis, C. Electromagnetic Parameter Retrieval from Inhomogeneous Metamaterials. *Physical Review, E.* **2005**, *71*, 036617. [CrossRef]
39. Nicolson, A.M.; Ross, G.F. Measurement of the Intrinsic Properties of Materials by Time Domain Techniques. *IEEE Trans. Instrum. Meas.* **1970**, *19*, 377–382. [CrossRef]
40. Ziolkowski, R.W. Design, fabrication, and testing of double negative metamaterials. *IEEE Trans. Antenn. Propag.* **2003**, *51*, 1516–1529. [CrossRef]
41. Hannan, S.; Islam, M.T.; Hoque, A.; Singh, M.J.; Almutairi, A.F. Design of a Novel Double Negative Metamaterial Absorber Atom for Ku and K Band Applications. *Electronics* **2019**, *8*, 853. [CrossRef]
42. Yi, D.; Wei, X.-C.; Xu, Y.-L. Transparent Microwave Absorber Based on Patterned Graphene: Design, Measurement, and Enhancement. *IEEE Trans. Nanotechnol.* **2017**, *16*, 484–490. [CrossRef]
43. Rozanov, K.N. Ultimate thickness to bandwidth ratio of radar absorbers. *IEEE Trans. Antenn. Propag.* **2000**, *48*, 1230–1234. [CrossRef]
44. Ye, D.; Wang, Z.; Xu, K.; Li, H.; Huangfu, J.; Wang, Z.; Ran, L. Ultrawideband Dispersion Control of a Metamaterial Surface for Perfectly-Matched-Layer-Like Absorption. *Phys. Rev. Lett.* **2013**, *111*, 187402. [CrossRef] [PubMed]

photonics

MDPI

Article

Design Methodology of a VIS Hybrid Refractive–Metalens System with a Wide FOV

Xingyi Li [1,2], Peixuan Wu [1], Yuanyuan Xing [2], Peng Shi [2], Xinjian Yao [2] and Yaoguang Ma [1,*]

1 State Key Laboratory for Extreme Photonics and Instrumentation, College of Optical Science and Engineering, ZJU-Hangzhou Global Scientific and Technological Innovation Center, Zhejiang University, Hangzhou 310027, China; 0623524@zju.edu.cn (X.L.); 22330087@zju.edu.cn (P.W.)
2 Najing Technology Co., Ltd., Hangzhou 310027, China; xyy@najing.cn (Y.X.); ship@najing.cn (P.S.); yaoxj@najing.cn (X.Y.)
* Correspondence: mayaoguang@zju.edu.cn

Abstract: The emergence of metalenses has opened new possibilities for miniaturizing optical systems. However, the limited group delay provided by meta-atoms restricts their aperture size under broadband operation. This challenge has stimulated the development of hybrid refractive–metalens systems, which overcome the performance limitations of individual metalenses while achieving a more compact form factor than conventional refractive lens assemblies. Here, we propose a design methodology for hybrid lenses that combines ray tracing with full-wave simulation. We analyze key aspects of the metalens within the hybrid system for a wide wavelength band—specifically, dispersion and transmission efficiency. Based on this approach, we designed a high-resolution hybrid lens operating in the 435–656 nm visible band with a 35° field of view. The results demonstrate that the proposed lens achieves imaging performance equivalent to that of conventional refractive systems while reducing the total track length by 29%. This validates the effectiveness of our design method, indicating its strong potential for application in compact and lightweight optical systems.

Keywords: hybrid refractive–metalens; aberration-corrected; visible imaging

1. Introduction

The rapid advancement of imaging sensor technology has intensified the need for optical systems that deliver exceptionally high resolution while conforming to stringent size, weight, and power constraints, particularly in applications such as unmanned aerial vehicles and next-generation consumer electronics. Conventional refractive lenses form images by gradually bending light through curved surfaces, relying on cumulative phase accumulation as waves propagate through media of different refractive indices. To correct various optical aberrations, an intricate combination of multiple, precisely crafted lens elements is required to precisely manipulate the optical path differences and phase accumulation across both the aperture and the spectrum. This systematic correction, while effective, fundamentally compromises the core goals of miniaturization and cost efficiency, leading to increased volume, weight, and manufacturing complexity.

The emergence of metasurfaces, planar metamaterials engineered with subwavelength nanostructures, has heralded a paradigm shift, promising unprecedented control over the phase, amplitude, and polarization of light [1]. These ultra-thin platforms offer exceptional design freedom, enabling high-performance metalenses [2–4], polarization-selective devices [5–8], and holography [9–11] within dramatically reduced form factors. Recent advancements, such as nonlocal metasurfaces utilizing multilayer coupling, further open pathways toward ideal wavefront manipulation and system integration [12–14]. However, the promise of metasurfaces is tempered by formidable challenges. The design and simulation complexity escalate prohibitively for large-aperture metalenses. While dispersion can be tailored using nanostructures with specific group delays, this approach imposes stringent fabrication constraints, requiring extreme alignment precision and deeply subwavelength features, often at the expense of efficiency and practical manufacturability [15–18].

In this landscape, a synergistic approach emerges: the hybrid refractive–metalens system (HRMS). By marrying the robust light-bending capabilities of conventional refractive lenses with the nanoscale wavefront manipulation of metasurfaces, HRMS offers a compelling path to compact, aberration-corrected optics [19]. This synergy reduces both the number of refractive elements required and the fabrication complexity demanded of the metasurface [20–22]. More recently, several innovative approaches to hybrid lens design have been proposed. A metalens integrated onto substrates with variable height has been designed to achieve broadband imaging capabilities in the near-infrared regime [23]. For visible-light imaging, hybrid lenses incorporating nonlinear materials have demonstrated the ability to convert near-infrared wavelengths into the visible range for detection and imaging purposes [24]. Yet, current explorations of HRMSs remain limited, predominantly focused on infrared wavelengths and on-axis performance at normal incidence, with wide-field imaging scenarios and critical factors like metasurface-induced transmission loss often overlooked.

Here, we present a comprehensive design methodology for a visible-spectrum HRMS that addresses these critical gaps. Our work emphasizes the optimization of optical transmission and the effective mitigation of chromatic aberration across the entire visible band. We outline fundamental principles for intelligently distributing optical power between refractive and metasurface components. This strategy not only simplifies the lens system, minimizing the number of elements, particularly challenging aspheric lenses, but also relaxes the fabrication tolerances for the metasurface without compromising its diffraction efficiency. The metasurface is specifically engineered to provide chromatic compensation, leveraging the complementary advantages of both technologies. To validate our framework, we designed and tested an exemplary HRMS. The results demonstrate a significant reduction in the total track length (TTL) compared to conventional refractive designs, alongside relaxed manufacturing requirements and a simplified metasurface architecture. This work provides a practical and efficient strategy for realizing broadband, achromatic imaging in next-generation compact optical systems, bridging the gap between high performance and stringent size, weight, and power (SWaP) constraints.

2. Design Principles

The design of an HRMS leverages the exceptional phase manipulation freedom offered by metasurfaces, which surpasses the limitations of conventional freeform optics in both achievable phase profiles and practical manufacturability. On the one hand, for chromatic aberration correction, the metalens and the refractive lens possess opposite dispersion characteristics, which effectively counteracts the axial chromatic aberration introduced by the refractive lens. On the other hand, to address off-axis aberrations (e.g., coma, astigmatism) induced by wide-angle incidence, the metasurface offers unparalleled phase control freedom. Its phase profile can be precisely designed as a function of position, wavelength, and angle of incidence, enabling the simultaneous synthesis of a phase front that compensates for large-angle aberrations. This enhanced freedom is key to simultaneously reducing the TTL and improving the modulation transfer function (MTF) across the field.

The design workflow of the HRMS is illustrated in Figure 1. A key challenge in such systems lies in bridging the macroscopic optical layout with the microscopic behavior of meta-atoms. Macroscopic optical simulations typically rely on ray tracing, whereas the response of individual meta-atoms requires full-wave vectorial simulation methods, which are not directly applicable to large-scale systems. To address this multiscale problem, we adopt a combined approach integrating ray tracing and full-wave electromagnetic simulations. Although the light propagation through a metalens is governed by vectorial Maxwell's equations, the deflection of light at the metalens interface follows generalized Snell's law. This makes it feasible to derive the phase gradient point by point and reconstruct a continuous phase profile. By treating the metalens as a phase surface in ray-tracing software, it becomes possible to co-optimize the metalens with conventional refractive elements, thereby bridging macro- and microscale optical design. Note, however, that this phase-surface abstraction ignores effects such as diffraction efficiency and higher-order diffraction. It is therefore essential that the actual phase response of the fabricated metalens closely match the ideal phase profile.

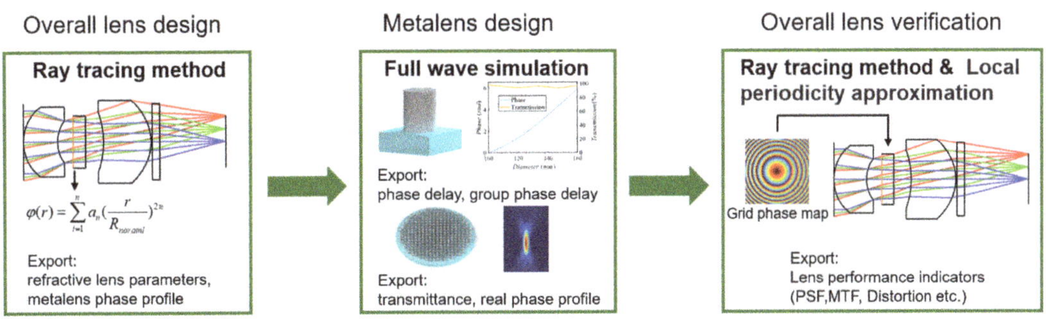

Figure 1. Design flowchart of the HRMS.

The first step involves using ray-tracing software to optimize the entire optical system by representing the metalens as a phase surface. Here, the commercial software ZEMAX (Zemax OpticStudio 19.4) is used for optical design. The optimization procedure mirrors that of conventional refractive lenses, involving the selection of appropriate variables and operands to minimize aberrations. To achieve high imaging quality across a wide field of view (FOV), different from the parabolic profile used in axial focusing, an even-order polynomial phase profile is adopted in the design [25]. Consistent with the methodology used in single-layer metalens designs for large FOV applications [26], this approach enables effective aberration correction through the optimization of the even-order coefficients. Special attention is also paid to the angles of incidence and dispersion at the phase surface to ensure the manufacturability of the metalens.

In the second step, the metalens is designed at the meta-atom level. Full-wave simulations are performed to characterize the optical response of individual meta-atoms; here, the finite-difference time-domain method (FDTD) is used for simulating the optical performance of the metalens. By scanning their geometric parameters, a phase and transmission library is obtained. The target phase profile obtained from ray tracing is then used to optimize the arrangement of meta-atoms across the surface, resulting in a physically realizable metasurface layout.

The third step is the validation of the HRMS. The actual phase profile of the manufactured metalens is either discretized into a grid phase matrix based on the local periodicity approximation or fitted with a new set of polynomial coefficients. These representations are re-imported into the ray-tracing software for approximate simulation, enabling performance verification before physical fabrication.

Throughout the optimization of the metalens, two key performance metrics are emphasized: focusing efficiency and polychromatic MTF across the entire FOV. High focusing efficiency directly contributes to an improved signal-to-noise ratio in the captured images, while the full-field MTF indicates the imaging resolution of the system under real-world conditions. Many existing single-layer metasurface designs for achromatic operation achieve color correction at the cost of reduced focusing efficiency and often only optimize the MTF for on-axis incident light. As a result, although compact, such systems tend to exhibit lower imaging signal-to-noise ratios and degraded off-axis performance. To overcome these limitations and achieve higher overall image quality, our design simultaneously prioritizes both focusing efficiency and full-field MTF, ensuring that the metalens performs robustly with high resolution across varied practical imaging scenarios.

Building upon the design workflow and figure of merit outlined previously, we now delve into the specific characteristics of the metalens. The operation of metasurfaces is governed by diffraction; they exhibit dispersion behavior analogous to that of conventional diffractive lenses. This similarity allows well-established design methodologies for diffractive optics to be informatively applied to metalenses. In the ray tracing design, a broadband light source is employed, while the metalens is set in Zemax as a wavelength-independent binary phase profile. This phase profile is defined by an even-order polynomial independent of wavelength, which can be written as

$$\varphi(r) = \sum_{n=1}^{m} a_n (\frac{r}{r_{normal}})^{2n} \tag{1}$$

where a_n denotes phase coefficients of the even-order polynomial, r is the radial coordinate of the metalens, and rnormal is the normalized radius, respectively. Note that the phase profile here is wavelength-independent; when different incident wavelengths are taken into account, the phase profile can be depicted as follows:

$$\varphi(r,\lambda) = \sum_{n=1}^{m} \frac{2\pi b_n}{\lambda} (\frac{r}{r_{normal}})^{2n} \tag{2}$$

In Equation (2), the phase coefficient a_n is transformed into a quantity that is related to the wavelength. It can be observed that the phase profile exhibits a linear relationship with the reciprocal of wavelength, which is consistent with the phase response of the meta-atom across different wavelengths. Therefore, the meta-atom dimension at a given position of the metalens can be determined by matching both the phase value at the central wavelength and the group delay to the target phase profile. It is also essential to maintain high transmission efficiency over the operating bandwidth to ensure the overall performance of the metalens. A wavelength-dependent constant c(λ) can be into Equation (2), and intelligent algorithms such as particle swarm optimization can be used to achieve better matching between the meta-atom and the phase profile [27]. If the required group delay is excessively large, a quasi-achromatic approach based on folding the group delay in the time domain can be employed [28]. By folding the group delay with respect to the frequency difference, the requirement on its range is relaxed, thereby enabling achromatic performance across a broader bandwidth. On the other hand, based on Equation (2), the metalens could maintain chromatic aberration in the HRMS. To better illustrate the chromatic behavior, we consider a representative single-layer metalens designed at a central wavelength of 550 nm to compare the focal length variation between FDTD simulation and ZEMAX. Figure 2a illustrates the schematic of the meta-atoms composed of niobium oxide (Nb_2O_5) nanopillars with cylindrical, square, and hexagonal cross-sections on a 350 nm period silica substrate with a height of 700 nm. In order to ensure stable and reproducible fabrication, the minimum lateral size of all three nanostructure geometries was set to 100 nm, while the maximum lateral dimension was constrained to 250 nm. As shown in Figure 2b, each geometry exhibits distinct transmittance and phase responses. Achieving high overall transmission requires not only accurate phase modulation but also high transmittance at the meta-atom level, necessitating a diverse set of structural options. Figure 2c presents the focal length shift across wavelengths obtained through full-wave simulation. The metalens, designed for 550 nm, exhibits progressively shorter focal lengths as wavelength increases. Figure 2d compares full-wave simulation results with ZEMAX simulations using a binary 2 surface, revealing a consistent trend. This agreement confirms that the metalens preserves certain chromatic properties that can be utilized by sharing the same phase profile across multiple wavelengths.

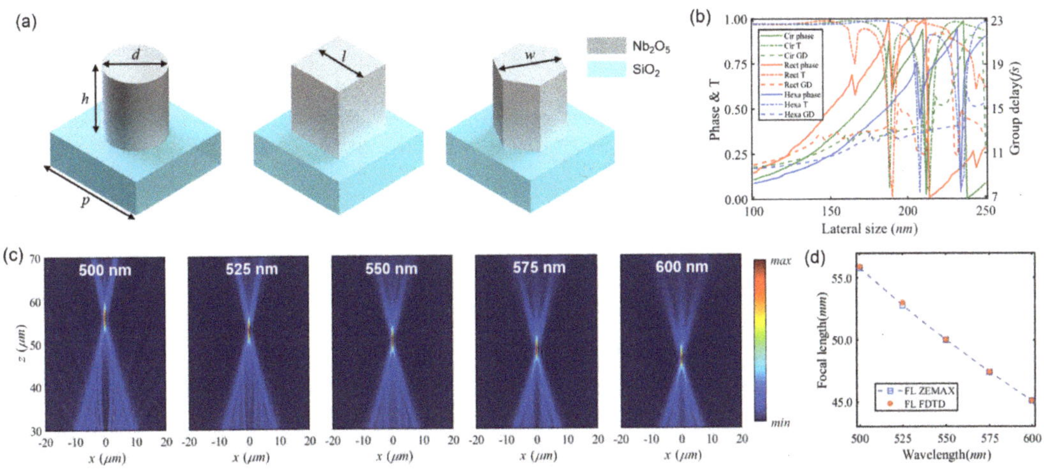

Figure 2. Verification of the chromatic aberration of the metalens. (**a**) Schematic diagram of the meta-atom. (**b**) Transmittance, phase retardation, and group delay with meta-atoms of different shapes and lateral sizes. (**c**) Intensity at axial plane at different wavelengths using the FDTD method. (**d**) Focal length for FDTD simulation results with Zemax simulations.

Unlike single metalens imaging systems [29], the hybrid design allows the use of a broadband source during ray-tracing optimization. The opposite signs of chromatic aberration between refractive and meta-diffractive elements enable effective chromatic compensation within the HRMS. Consequently, a satisfactory polychromatic MTF can be achieved through co-optimization, demonstrating the potential of the HRMS in correcting chromatic aberrations.

In addition to chromatic aberration, transmittance is another critical performance metric for metasurfaces. High transmittance helps reduce stray light and ensures a high signal-to-noise ratio in the resulting images. During the design process, the period of the meta-atoms must be carefully chosen to satisfy the Nyquist sampling criterion under oblique illumination in the whole wavelength range. As the angle of incidence increases, the effective normal wavevector decreases; thus, the period should be small enough to suppress higher-order diffraction and stray light under large angle incidence. As illustrated in Figure 3a, which analyzes cylindrical meta-atoms at 550 nm under tilted incidence, a notable decrease in transmittance is observed beyond 30° incidence [30]. Furthermore, the optical power distribution of the metalens must be properly allocated. As the numerical aperture (NA) increases, the focusing efficiency tends to decrease [31], which means a higher optical focal degree will reduce the diffraction efficiency of the metalens. Figure 3b shows the focusing efficiency, defined as the ratio of energy within three times the full width at half maximum (FWHM) to the total incident energy, for metalenses with a diameter of 50 μm at 550 nm. It is evident that the focusing efficiency declines with increasing NA. Therefore, a balanced co-design of refractive and meta optics is essential to achieve both high image quality and device miniaturization.

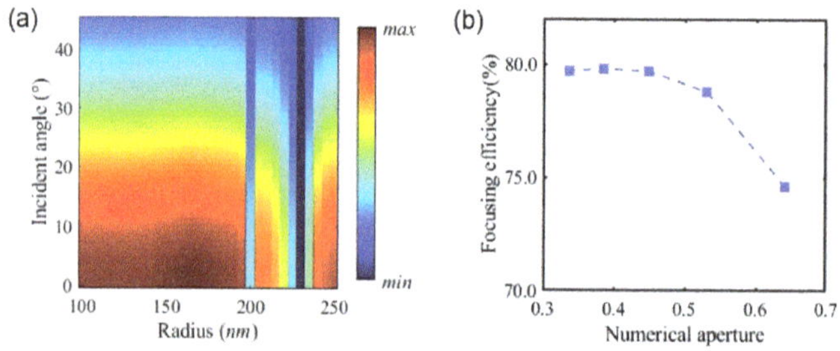

Figure 3. Incident angle analysis of the metalens. (**a**) Transmittance of the circular meta-atom with different incident angles at a wavelength of 550 nm. (**b**) Focusing efficiency of the metalens with different numerical apertures.

3. Specific Design, Fabrication, and Characterization

To validate the proposed design methodology, an HRMS operating across the visible spectrum from 430 nm to 650 nm was designed for an 8-megapixel sensor. The specifications are summarized in Table 1. The lens provides a 35° full FOV, a focal length of 15 mm, and an F-number of 1.67. Key performance metrics include a polychromatic MTF exceeding 0.4 at 120 lp/mm over the entire field and distortion controlled below 5%. To realize a compact system, the TTL was minimized and the number of elements reduced without employing any aspheric surfaces.

Table 1. Specifications for the HRMS.

Specification Parameter	HRMS
1. Spectral range (nm)	435–656
a. Design wavelength (nm)	435, 486, 546, 587, 656
b. Weighting	1:1:1:1:1
2. F-number	1.67 ± 0.1
3. Field of view (°)	± 17.5
4. EFL (mm)	15.0 ± 0.03
5. MTF	120 lp/mm
a. 0 Field	≥ 0.4
b. 0.7 Field	≥ 0.4
c. 1 Field	≥ 0.4
6. F-θ distortion, nominal maximum (%)	<5%

As illustrated in Figure 4a, the system incorporates one metalens and five spherical lenses, achieving a TTL of 22.4 mm. As shown in the layout diagram, light rays passing through the metalens do not exhibit pronounced bending, implying that the optical power introduced by the metalens is relatively small, and its primary role lies in aberration correction. In comparison, a conventional refractive design with equivalent performance requires four spherical and two aspherical glass lenses, resulting in a TTL of 31.5 mm, as shown in Figure 4c. The reduction in the total track length (TTL) is primarily achieved through the unique advantages of metalenses in off-axis aberration correction. In conventional systems, correcting off-axis aberrations often requires multiple aspheric elements to maintain manufacturability. Metalenses, however, overcome this limitation by enabling arbitrary phase manipulation with subwavelength structures, thereby integrating the functionality of multiple aspheric surfaces into a single flat optical element. Consequently, the HRMS significantly shortens the TTL without compromising optical performance. It should be noted that if all elements in a system already employ aspheric surfaces, the potential for further TTL reduction using a metalens becomes more limited. Figure 4b presents the MTF curves of the lens at 0°, 12.25°, and 17.5° field angles obtained through ray-tracing simulation. The MTF values at 120 lp/mm for the three fields are 0.67, 0.52, and 0.44, respectively, indicating sufficient image resolution across the FOV. It is worth noting that the system operates effectively under broadband visible light (430–650 nm) without significant degradation in imaging performance, demonstrating its robustness over a wide spectral range. For comparison, the MTF of the refractive lens is presented in Figure 4d, showing performance nearly identical to that of the HRMS lens. However, to correct aberrations in the peripheral field of view, the refractive design requires a longer optical path, resulting in a greater total lens length. Figure 4e illustrates the variation in focal length with wavelength across the visible spectrum. The focal shift is less than 2 μm, and the depth of focus for the central wavelength calculated based on the Rayleigh criterion is 6 μm, which indicates the achromatic performance of the system. This effect stems from the complementary chromatic aberration contributions between the refractive elements and the metalens. Furthermore, the point spread function (PSF) and back focal distance under on-axis illumination were simulated at different wavelengths, as summarized in Figure 4d. It is noted that the RMS diameters of the PSF at five wavelengths are 2.71 μm, 2.43 μm, 2.74 μm, 2.68 μm and 2.46 μm, which remain small across all wavelengths, ensuring high imaging resolution.

Subsequently, the designed HRMS was fabricated. Figure 5a shows the overall appearance of the metalens. The refractive lens part, consisting entirely of spherical surfaces, was manufactured using conventional precision grinding and polishing techniques. The metasurface was fabricated via electron-beam lithography (EBL). Specifically, a 700 nm thick Nb_2O_5 layer was deposited on a glass substrate using plasma-enhanced chemical vapor deposition (PECVD). The refractive index of the film, measured by spectroscopic ellipsometry (HORIBA Scientific UVISEL, Paris, France), was 2.32 at the central wavelength of 550 nm. The sample was spin-coated with ma-N2405 electron-beam resist (Micro Resist Technology, Berlin, Germany). The metalens pattern was exposed using EBL (Raith Voyager, Dortmund, Germany) and developed in ma-D525 (Micro Resist Technology, Berlin, Germany). Inductively coupled plasma reactive ion etching (Samco RIE-230IP, Kyoto, Japan) was employed to transfer the pattern into the niobium oxide layer. Finally, the residual resist was removed via plasma etching. Figure 5b show the optical image of the fabricated metalens whilescanning electron microscopy (SEM) images of the fabricated metasurface are shown in Figure 5c. Owing to the reasonable distribution of optical power in the design, the required group delay for chromatic compensation is relatively modest, which allows most nanostructures to take the form of cylindrical pillars. For system assembly, dedicated mechanical mounts are designed to secure each lens element, minimizing misalignment-induced performance degradation. Since the metalens is fabricated on a square substrate, it is integrated into the optical assembly using UV-curing adhesive. The fabricated HRMS is experimentally characterized using the setup illustrated in Figure 5d.

The FOV and MTF were evaluated by imaging a standard checkerboard target. As shown in Figure 5e, the captured image exhibits clearly resolved grid lines across the entire field, confirming a wide and effective FOV. Quantitative analysis of the image further indicates a diagonal FOV exceeding 35°. And it also can be seen from the picture that the distortion of the HRMS is low. The MTF was measured using the slanted-edge method (SFR) at three representative field positions: on-axis, 0.7 field, and full field. The results are summarized in Figure 5f. At 120 lp/mm, the measured MTF values were 0.46, 0.41, and 0.40, respectively. Although these values are slightly lower than the design predictions, they still meet the specified performance requirements. The minor degradation can be attributed to fabrication imperfections in both the refractive lenses and the metalens, as well as alignment tolerances during assembly. And the focusing efficiency of the HRMS was also characterized by illuminating the system with collimated white light and measuring the energy of the focused spot at the focal plane. The focusing efficiency was determined by comparing this focused energy with the incident light intensity. Based on this method, the fabricated HRMS was found to exhibit an efficiency of 82.3% across the visible spectrum. This can be attributed to the well-designed configuration and high-precision fabrication process employed in the device.

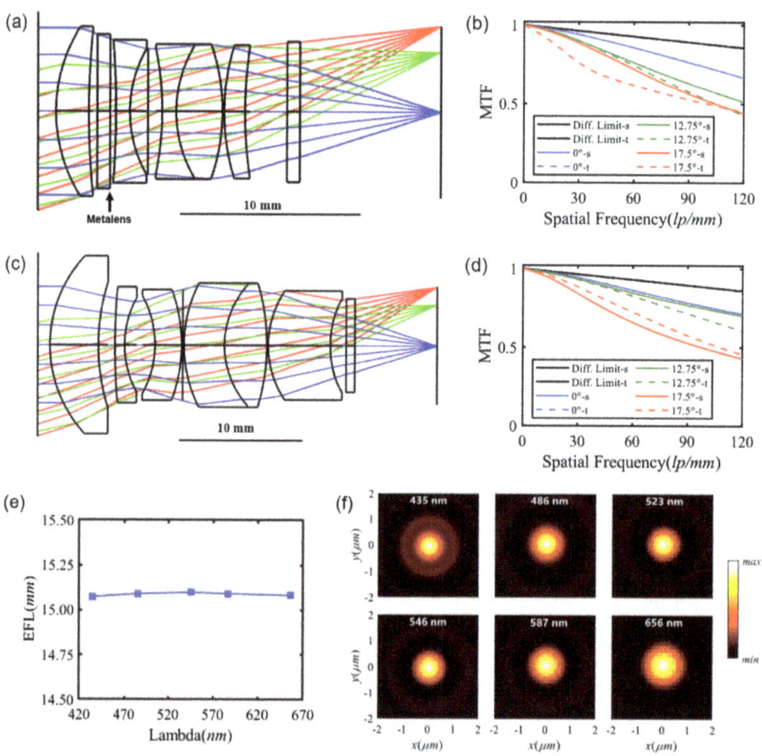

Figure 4. Design results of the HRMS. (**a**) Optical layout of the HRMS of the visible spectrum. (**b**) Simulated MTF for both tangential and sagittal at 0 field, 0.7 field, and 1 field of the HRMS. (**c**) Optical layout of the refractive lens. (**d**) Simulated MTF for tangential and sagittal at 0 field, 0.7 field, and 1 field of the refractive lens. (**e**) Effective focal length of the HRMS with different wavelengths. (**f**) Simulated PSF at different wavelengths of the HRMS.

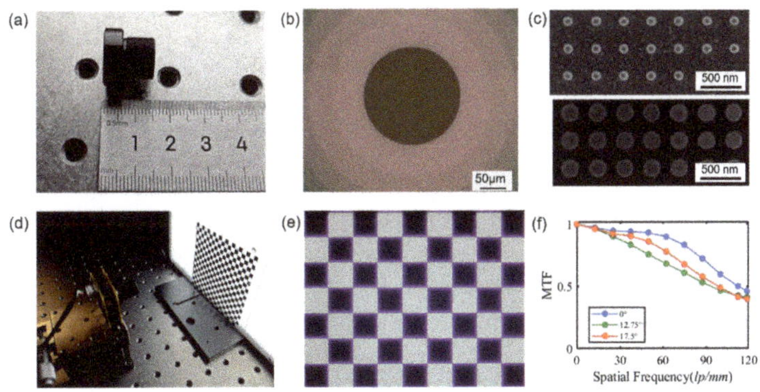

Figure 5. Characterization and testing of the HRMS. (**a**) Schematic of a prototype HRMS. (**b**) Optical microscope image and (**c**) scanning electron microscope image of the metalens in the designed HRMS. (**d**) Experimental platform for resolution test characterization. (**e**) Image of a checkerboard captured by the designed HRMS using halogen lamps for lighting. (**f**) Measured MTF curve at 0 field, 0.7 field, and 1 field.

To evaluate the imaging performance of the fabricated HRMS for color applications, we captured images of pictures projected on a screen, as shown in Figure 6a–d. Compared to a monolayer achromatic metalens design, the HRMS demonstrates higher resolution and reduced chromatic aberration. However, a slight color shift is observed in the captured images, which mainly stems from the wavelength-dependent transmission efficiency of the metalens. This issue can be effectively addressed through neural network-based image enhancement algorithms. The integration of computational optics presents a significant opportunity for performance enhancement. For instance, the PSF of the HRMS can be characterized across the field of view. A neural network can then be employed to perform spatially variant deconvolution, leveraging the estimated PSF to restore image clarity and compensate for residual aberrations across the entire image, even without a co-designed optical encoder. For instance, generative adversarial networks can be employed to further reduce chromatic aberrations and improve image sharpness. Notably, AI-based image processing offers a promising solution to compensate for such optical imperfections and enhance overall image quality [32]. Alternatively, in projection systems, a neural network can be used to pre-correct the input image, thereby compensating for optical imperfections and achieving superior imaging quality [33]. Such a combination of meta-optical design and computational imaging offers a powerful pathway toward high-fidelity, compact imaging systems, making the HRMS particularly promising for applications in wearable devices and other portable technologies.

Figure 6. (**a**–**d**) Color images displayed on a screen captured by the HRMS.

4. Conclusions

In summary, we developed a versatile design methodology for broadband, high-resolution hybrid optical systems with wide field-of-view capabilities. As shown in Table 2, the HRMS designed in this work features a large aperture while maintaining high resolution and efficiency across the visible spectrum. The effectiveness of this approach is validated through the design and characterization of a high-resolution HRMS operating in the visible spectrum. Within a design framework combining ray-tracing and full-wave simulation, we analyzed both macroscopic requirements and microscopic characteristics of the metalens. It is shown that the metalens should assume low optical power and operate under small angles of incidence within the HRMS, while its chromatic behavior can be treated similarly to that of diffractive optical elements. Based on this theoretical foundation, we designed and fabricated a hybrid refractive–metasurface lens working in the visible band, which achieves a full-field MTF above 0.4 at 120 lp/mm. Compared to conventional refractive lenses, the proposed system significantly simplifies the architecture and reduces the TTL. The presented theory and design framework are expected to accelerate the adoption of HRMSs in practical applications. We envision that such systems may progressively replace traditional multi-element refractive lenses, thereby advancing the development of compact and lightweight optical systems.

Table 2. Performance of this work compared with other HRMSs.

Refs.	Wavelength	F/# or NA	FOV	MTF	Efficiency
[19]	460–700 nm	NA = 0.075	/	/	Peak: 35.0%
[20]	8–12 μm	F/# = 1.5	20°	≥0.2@20 lp/mm	Peak: 58.6%
[21]	8–12 μm	F/# = 1.2	24°	≥0.2@20 lp/mm	48.7%
[22]	8–12 μm	F/# = 1.8	0°	≥0.4@22 lp/mm	85%
[23] [25]	1000–1800 nm	NA = 0.22	/	/	peak: 83%
This work	435–656 nm	F/# = 1.65	35°	≥0.4@120 lp/mm	82.3%

Author Contributions: Conceptualization, Y.M. and X.L.; modeling and simulation, X.L. and P.S.; fabrication, Y.X. and X.L.; experimental characterization, P.W. and X.Y.; data analysis, X.L. and Y.M.; writing—original draft preparation, X.L. and P.W.; writing—review and editing, X.L. and Y.M. All authors have read and agreed to the published version of the manuscript.

Funding: This work was supported by a National Key Research and Development Program of China grant (2023YFF0613000); National Natural Science Foundation of China (NSFC) grant (62222511); Natural Science Foundation of Zhejiang Province China grant (LR22F050006); and STI 2030–Major Projects grant (2021ZD0200401).

Institutional Review Board Statement: Not applicable.

Data Availability Statement: All data that support the findings of this study are included within the article.

Acknowledgments: The author would like to acknowledge Weige Lv, Liying Chen, and Wei Wang from the State Key Laboratory for Extreme Photonics and Instrumentation and the College of Optical Science and Engineering, Zhejiang University, for their assistance in the experiments.

Conflicts of Interest: The authors X.L., Y.X., P.S. and X.Y. are from Hangzhou Najing Technology Co., Ltd. The remaining authors declare that the research was conducted in the absence of any commercial or financial relationships that could be construed as a potential conflict of interest.

Abbreviations

The following abbreviations are used in this manuscript:

HRMS	Hybrid refractive–metalens system
SWaP	Size, weight, and power
MTF	Modulation transfer function
FOV	Field of view
NA	Numerical aperture
PSF	Point spread function
EBL	Electron-beam lithography
PECVD	Plasma-enhanced chemical vapor deposition
SEM	Scanning electron microscopy

References

1. Yu, N.; Genevet, P.; Kats, M.A.; Aieta, F.; Tetienne, J.-P.; Capasso, F.; Gaburro, Z. Light propagation with phase discontinuities: Generalized laws of reflection and refraction. *Science* **2011**, *334*, 333–337. [CrossRef]
2. Khorasaninejad, M.; Chen, W.T.; Devlin, R.C.; Oh, J.; Zhu, A.Y.; Capasso, F. Metalenses at visible wavelengths: Diffraction-limited focusing and subwavelength resolution imaging. *Science* **2016**, *352*, 1190–1194. [CrossRef]
3. Li, L.; Liu, Z.; Ren, X.; Wang, S.; Su, V.-C.; Chen, M.-K.; Chu, C.H.; Kuo, H.Y.; Liu, B.; Zang, W. Metalens-array–based high-dimensional and multiphoton quantum source. *Science* **2020**, *368*, 1487–1490. [CrossRef]
4. Ossiander, M.; Meretska, M.L.; Hampel, H.K.; Lim, S.W.D.; Knefz, N.; Jauk, T.; Capasso, F.; Schultze, M. Extreme ultraviolet metalens by vacuum guiding. *Science* **2023**, *380*, 59–63. [CrossRef] [PubMed]
5. Balthasar Mueller, J.; Rubin, N.A.; Devlin, R.C.; Groever, B.; Capasso, F. Metasurface polarization optics: Independent phase control of arbitrary orthogonal states of polarization. *Phys. Rev. Lett.* **2017**, *118*, 113901. [CrossRef]
6. Rubin, N.A.; D'Aversa, G.; Chevalier, P.; Shi, Z.; Chen, W.T.; Capasso, F. Matrix Fourier optics enables a compact full-Stokes polarization camera. *Science* **2019**, *365*, eaax1839. [CrossRef]
7. Xiong, B.; Liu, Y.; Xu, Y.; Deng, L.; Chen, C.-W.; Wang, J.-N.; Peng, R.; Lai, Y.; Liu, Y.; Wang, M. Breaking the limitation of polarization multiplexing in optical metasurfaces with engineered noise. *Science* **2023**, *379*, 294–299. [CrossRef]
8. Fu, B.; Zhou, X.; Li, T.; Zhu, H.; Liu, Z.; Zheng, S.; Zhou, Y.; Yu, Y.; Cao, X.; Wang, S.; et al. Miniaturized high-efficiency snapshot polarimetric stereoscopic imaging. *Optica* **2025**, *12*, 391–398. [CrossRef]
9. Huang, L.; Chen, X.; Mühlenbernd, H.; Zhang, H.; Chen, S.; Bai, B.; Tan, Q.; Jin, G.; Cheah, K.-W.; Qiu, C.-W. Three-dimensional optical holography using a plasmonic metasurface. *Nat. Commun.* **2013**, *4*, 2808. [CrossRef]
10. Zheng, G.; Mühlenbernd, H.; Kenney, M.; Li, G.; Zentgraf, T.; Zhang, S. Metasurface holograms reaching 80% efficiency. *Nat. Nanotechnol.* **2015**, *10*, 308–312. [CrossRef] [PubMed]
11. Zhao, Z.; Shi, Y.; Wang, Z.; Wan, S.; Li, Z.; Li, Z. Azimuth-Resolved On-Chip Meta-Hologram. *Laser Photonics Rev.* **2025**, e01642. [CrossRef]
12. Overvig, A.C.; Malek, S.C.; Yu, N. Multifunctional nonlocal metasurfaces. *Phys. Rev. Lett.* **2020**, *125*, 017402. [CrossRef] [PubMed]
13. Malek, S.C.; Overvig, A.C.; Alù, A.; Yu, N. Multifunctional resonant wavefront-shaping meta-optics based on multilayer and multi-perturbation nonlocal metasurfaces. *Light Sci. Appl.* **2022**, *11*, 246. [CrossRef]
14. Li, T.; Liu, M.; Hou, J.; Yang, X.; Wang, S.; Wang, S.; Zhu, S.; Tsai, D.; Wang, Z. Chip-scale metaphotonic singularities: Topological, dynamical, and practical aspects. *Chip* **2025**, *3*, 100109. [CrossRef]
15. Wang, S.; Wu, P.C.; Su, V.-C.; Lai, Y.-C.; Chen, M.-K.; Kuo, H.Y.; Chen, B.H.; Chen, Y.H.; Huang, T.-T.; Wang, J.-H. A broadband achromatic metalens in the visible. *Nat. Nanotechnol.* **2018**, *13*, 227–232. [CrossRef]
16. Chen, W.T.; Zhu, A.Y.; Sanjeev, V.; Khorasaninejad, M.; Shi, Z.; Lee, E.; Capasso, F. A broadband achromatic metalens for focusing and imaging in the visible. *Nat. Nanotechnol.* **2018**, *13*, 220–226. [CrossRef]
17. Shrestha, S.; Overvig, A.C.; Lu, M.; Stein, A.; Yu, N. Broadband achromatic dielectric metalenses. *Light Sci. Appl.* **2018**, *7*, 85. [CrossRef]
18. Wang, Y.; Chen, Q.; Yang, W.; Ji, Z.; Jin, L.; Ma, X.; Song, Q.; Boltasseva, A.; Han, J.; Shalaev, V.M. High-efficiency broadband achromatic metalens for near-IR biological imaging window. *Nat. Commun.* **2021**, *12*, 5560. [CrossRef]
19. Chen, W.T.; Zhu, A.Y.; Sisler, J.; Huang, Y.-W.; Yousef, K.M.; Lee, E.; Qiu, C.-W.; Capasso, F. Broadband achromatic metasurface-refractive optics. *Nano Lett.* **2018**, *18*, 7801–7808. [CrossRef] [PubMed]
20. Liu, M.; Zhao, W.; Wang, Y.; Huo, P.; Zhang, H.; Lu, Y.Q.; Xu, T. Achromatic and coma-corrected hybrid meta-optics for high-performance thermal imaging. *Nano Lett.* **2024**, *24*, 7609–7615. [CrossRef] [PubMed]
21. Hu, T.; Wen, L.; Li, H.; Wang, S.; Xia, R.; Mei, Z.; Yang, Z.; Zhao, M. Aberration-corrected hybrid metalens for longwave infrared thermal imaging. *Nanophotonics* **2024**, *13*, 3059–3066. [CrossRef]

22. Li, A.; Chen, J.; Liu, M.; Wei, W.; Duan, H.; Jia, H.; Hu, Y. Fundamental design framework of hybrid refractive-metalens system for axial aberrations correction and its validation in LWIR band. *Opt. Express* **2025**, *33*, 27535–27547. [CrossRef]
23. Balli, F.; Sultan, M.; Lami, S.K.; Hastings, J.T. A hybrid achromatic metalens. *Nat. Commun.* **2020**, *11*, 3892. [CrossRef]
24. Karepov, S.; De Angelis, C.; Ellenbogen, T. Hybrid Nonlinear Metasurface Refractive Lens. *Nano Lett.* **2025**, *25*, 8103–8109. [CrossRef] [PubMed]
25. Wu, P.; Li, X.; Xing, Y.; Wang, J.; Zheng, W.; Wang, Z.; Ma, Y. Broadband Achromatic Hybrid Metalens Module with 100° Field of View for Visible Imaging. *Sensors* **2025**, *25*, 3202. [CrossRef]
26. Luo, X.; Zhang, F.; Pu, M.; Guo, Y.; Li, X.; Ma, X. Recent advances of wide-angle metalenses: Principle, design, and applications. *Nanophotonics* **2021**, *11*, 1–20. [CrossRef]
27. Ou, K.; Yu, F.; Li, G.; Wang, W.; Miroshnichenko, A.E.; Huang, L.; Wang, P.; Li, T.; Li, Z.; Chen, X. Mid-infrared polarization-controlled broadband achromatic metadevice. *Sci. Adv.* **2020**, *6*, eabc0711. [CrossRef]
28. Chen, Q.; Gao, Y.; Pian, S.; Ma, Y. Theory and fundamental limit of quasiachromatic metalens by phase delay extension. *Phys. Rev. Lett.* **2023**, *131*, 193801. [CrossRef]
29. Luo, S.; Zhang, F.; Lu, X.; Xie, T.; Pu, M.; Guo, Y.; Wang, Y.; Luo, X. Single-layer metalens for achromatic focusing with wide field of view in the visible range. *J. Phys. D Appl. Phys.* **2022**, *55*, 235106. [CrossRef]
30. Yun, J.-G.; Kang, H.; Lee, K.; Jeong, Y.; Lee, E.; Kim, J.; Choi, M.; Koo, B.; Kim, D.; Choi, J. Compact eye camera with two-third wavelength phase-delay metalens. *Nat. Commun.* **2025**, *16*, 7299. [CrossRef] [PubMed]
31. Egede Johansen, V.; Gür, U.M.; Martínez-Llinás, J.; Fly Hansen, J.; Samadi, A.; Skak Vestergaard Larsen, M.; Nielsen, T.; Mattinson, F.; Schmidlin, M.; Mortensen, N.A. Nanoscale precision brings experimental metalens efficiencies on par with theoretical promises. *Commun. Phys.* **2024**, *7*, 123. [CrossRef]
32. Wei, J.; Huang, H.; Zhang, X.; Ye, D.; Li, Y.; Wang, L.; Ma, Y.; Li, Y. Neural-network-enhanced metalens camera for high-definition, dynamic imaging in the long-wave infrared spectrum. *ACS Photonics* **2025**, *12*, 140–151. [CrossRef]
33. Chen, Q.; Zhou, J.; Pian, S.; Xu, J.; Li, X.; Li, B.; Lu, C.; Wang, Z.; Jiang, Q.; Qin, S. Hybrid Meta-Optics Enabled Compact Augmented Reality Display with Computational Image Reinforcement. *ACS Photonics* **2024**, *11*, 3794–3803. [CrossRef]

photonics MDPI

Article

Mach–Zehnder Interferometer Electro-Optic Modulator Based on Thin-Film Lithium Niobate Valley Photonic Crystal

Ying Yao [1,2,†], Hongming Fei [1,2,3,*,†], Xin Liu [1,2], Mingda Zhang [1,2], Pengqi Dong [4], Junjun Ren [5] and Han Lin [6,*]

[1] College of Physics and Optoelectronics, Taiyuan University of Technology, Taiyuan 030024, China; 2023511427@link.tyut.edu.cn (Y.Y.); liuxin@tyut.edu.cn (X.L.); zhangmingda@tyut.edu.cn (M.Z.)
[2] Shanxi Key Laboratory of Precision Measurement Physics, Taiyuan University of Technology, Taiyuan 030024, China
[3] State Key Laboratory of Quantum Optics and Quantum Optics Devices, Shanxi University, Taiyuan 030006, China
[4] The Institute of Architectural Design and Research of Taiyuan University of Technology Co, Ltd., Taiyuan 030006, China; dongpengqi2025@163.com
[5] Shanxi Chenghaoxin Intelligent Technology Co., Ltd., Taiyuan 030006, China; renjunjun2025@126.com
[6] Centre for Atomaterials and Nanomanufacturing, School of Science, Royal Melbourne Institute of Technology University, Melbourne, VIC 3000, Australia
* Correspondence: feihongming@tyut.edu.cn (H.F.); han.lin2@rmit.edu.au (H.L.)
† These authors contributed equally to this work.

Abstract

Thin-film lithium niobate (TFLN) electro-optic modulators (EOMs) offer distinct advantages, including high speed, broad bandwidth, and low power consumption. However, their large size hinders the density of integration, which trades off with the half-wave voltage. Photonic crystal (PC) structures can effectively reduce the device footprint via the slow-light effect; however, they experience significant losses due to fabrication defects and sharp corners. Here, we theoretically demonstrate an ultracompact Mach–Zehnder interferometer (MZI) EOM based on a TFLN valley photonic crystal (VPC) structure. The design can achieve a high forward transmittance (>0.8) due to defect-immune unidirectional propagation in the VPC, enabled by the unique spin-valley locking effect. The EOM, with a small footprint of 21 μm × 17 μm, achieves an extinction ratio of 16.13 dB and a modulation depth of 80%. The design can be experimentally fabricated using current nanofabrication techniques, making it suitable for broad applications in optical communications.

Keywords: thin-film lithium niobate; topological photonic crystal; valley photonic crystal; Mach–Zehnder interferometer; electro-optic modulators

1. Introduction

Electro-optic modulators (EOMs) are crucial devices in optical communication [1–4] and quantum computing systems [5–7], which utilize electrical signals to modulate light, serving as an interface between electronics and photonics. Due to its strong electro-optic effect, lithium niobate (LN) is considered the most promising material for EOM applications, offering a broad operating wavelength range and a large working bandwidth [8]. The electro-optical coefficient of LN is much higher than that of other semiconductor materials, such as silicon [9–11], silicon nitride [12,13], and indium phosphide [14], enabling the design of high-performance, highly efficient EOMs that meet the needs of high-speed optical communication systems.

Currently, significant progress has been made in modulator technology, and the recent development of thin-film lithium niobate (TFLN) enables a higher refractive-index contrast, achieving low-loss rigid waveguides and a lower half-wave voltage (HWV, the voltage required to achieve a π phase shift) [15–17]. Various TFLN EOMs based on different device architectures have been demonstrated, including the Michelson interferometer (MI) [18,19], ring resonators [20], and the Mach–Zehnder interferometer (MZI) [15]. The basic principle of the MI modulator is to divide the incident light into two coherent beams via amplitude-splitting interference, so that at least one of the beams traverses the modulation path twice, thereby doubling the optical path difference, and finally achieving light intensity modulation through interference. Its advantage is that 'double-pass modulation' can halve the device length, thereby improving modulation efficiency [21]. However, this approach suffers from high intrinsic loss due to superimposed waveguide propagation loss, interface reflection loss, and other factors, resulting in low output optical power. TFLN ring resonator modulators apply an electric field to the ring resonator, which alters the refractive index, changing the phase of light [22], which can achieve a high modulation efficiency and low power consumption. More importantly, it is necessary to use a low Q-factor to improve bandwidth, but at the expense of modulation efficiency, thus failing to meet the requirements of high bandwidth and high efficiency simultaneously.

In comparison, the MZI EOMs exhibit substantial advantages in optical communications, distinguished by their high-speed modulation capability, enabling the transmission of signals with bandwidths ranging from tens to hundreds of GHz and ensuring stable, high-rate transmission. However, conventional MZI EOMs are based on the normal dispersion of the waveguide modes, because of which the required HWV is inversely proportional to the length of the waveguide. As a result, it generally requires a large size to achieve a low HWV [17,23,24]. Consequently, efforts have been made to develop novel structures, such as photonic crystals (PCs) [25–27], metasurfaces [28], and plasmonic devices [29]. The slow-light effect in PC structures can significantly increase the effective refractive index of PC waveguides, thereby reducing the required length to achieve miniaturized designs. However, the transmittance of PC waveguides is highly sensitive to fabrication-related structural defects, compromising their performance [30–33].

Recent developments in topological photonic crystals (TPCs) enable the achievement of defect-immune unidirectional transmission. Among different types of TPC structures, valley photonic crystals (VPCs) offer a simple design process and a large working bandwidth [34,35]. More importantly, VPCs can use dielectric materials without requiring an external magnetic field. As a result, silicon VPC structures have been demonstrated [36,37]. Our previous work has shown that the slow-light effects of topological edge states can effectively enhance the electro-optic effect in lithium niobate VPC waveguides, achieving a low HWV with an ultracompact footprint. However, MZI EOMs based on lithium niobate VPC structures have not been demonstrated.

Here, we theoretically demonstrate an ultracompact MZI electro-optic amplitude modulator based on a thin-film lithium niobate (TFLN) VPC structure. The design can achieve a high forward transmittance (>0.8) due to defect-immune unidirectional propagation in the VPC, enabled by the unique spin-valley locking effect. The EOM with a small footprint of 21 μm × 17 μm achieves an extinction ratio of 16.13 dB and a modulation depth of 80%. The design can be experimentally fabricated using current complementary metal-oxide-semiconductor (CMOS) nanofabrication techniques, making it suitable for a wide range of optical communication applications.

https://doi.org/10.3390/photonics13010033

2. Design of Thin-Film Lithium Niobate MZI EOM Based on VPC Structures

The conceptual demonstration of the MZI EOM based on the VPC structure is shown in Figure 1a. The MZI structure is constructed by two Ω-shaped topological waveguides formed by combining two mirror-inverted VPC structures, which compose triangular airholes in the free-standing TFLN substrate. The gold electrodes are deposited on both sides of one waveguide (3 μm apart) to apply an electric field. Compared with Al and TiN, gold exhibits higher electrical conductivity and superior chemical stability. Additionally, the fabrication process for gold electrodes is well-established and has been widely adopted in lithium niobate modulator research. The strong electro-optic (EO) effect of LN can introduce refractive modulation in response to an external voltage, thereby producing a phase difference between the two waveguides. The simulated electric field distribution (using COMSOL Multiphysics 6.2 software) is shown in Figure 1b. The thin-film design enables the strongest electric field to cross the waveguide, thereby minimizing the required voltage. The proposed device employs a single-arm modulation scheme. The working scenarios are schematically shown in Figure 1c,d, in which the incident beam is split at the first Y-junction and combined to interfere at the second Y-junction. The line and arrows indicate the optical paths. The interference results are decided by the phase difference φ_d between the two waveguides (arms). The EOM is designed to achieve destructive interference (φ_d = nπ, n is an odd number) when the external voltage is 0 V (Figure 1c), and constructive interference is achieved when a HWV is applied, resulting in a phase difference of φ_d = 2 nπ, where n is an integer. This design employs a 220 nm-thick x-cut LN thin film to exploit the largest EO coefficient (r_{33}) [25]. Due to the Pockels effect, LN exhibits an EO response with a direct current (DC) electric field [38], which can be formulated as follows:

$$\begin{cases} n_x = n_o \pm \frac{1}{2} r_{13} n_o^3 E_z \\ n_y = n_o \pm \frac{1}{2} r_{13} n_o^3 E_z \\ n_z = n_e \pm \frac{1}{2} r_{33} n_e^3 E_z \end{cases} \tag{1}$$

Among these, n_x, n_y, and n_z are the refractive indices of LN with an external voltage in the x, y, and z directions; no and ne are the ordinary and extraordinary refractive indices, respectively; r_{13} and r_{33} are the two Pockels coefficients of LN. In our design, since an x-cut LN thin film is used (Figure 1a), the EO coefficient r_{33} is employed for modulation. To quantify the Pockels effect, we calculate the local-field factor within the device using a previously published method [39,40]. For the LN PC structure, we define the effective EO effect as expressed in Equation (2):

$$r_{33}^{PC} = f_1 f_2 f_3 r_{33} \tag{2}$$

$$f = \sqrt{\frac{v_g^{bulk}}{v_g^{PC}}} \tag{3}$$

where f_i (i = 1, 2, 3) are the local field factors of the three waves involved in the second-order process. v_g^{bulk} denotes the group velocity of bulk LN material without nanostructures, which is $v_g^{bulk} = c/n_e$. Meanwhile, v_g^{PC} is the group velocity of the modes within the PC structure, which is $v_g^{PC} = d\omega/dk$. From Equations (2) and (3), it can be seen that maximizing the EO coefficient r_{33}^{PC} is an effective means to achieve maximal effective refractive index modulation. This means that the group velocity of the PC structure must be minimized, since v_g^{bulk} is a constant only depending on the material.

https://doi.org/10.3390/photonics13010033

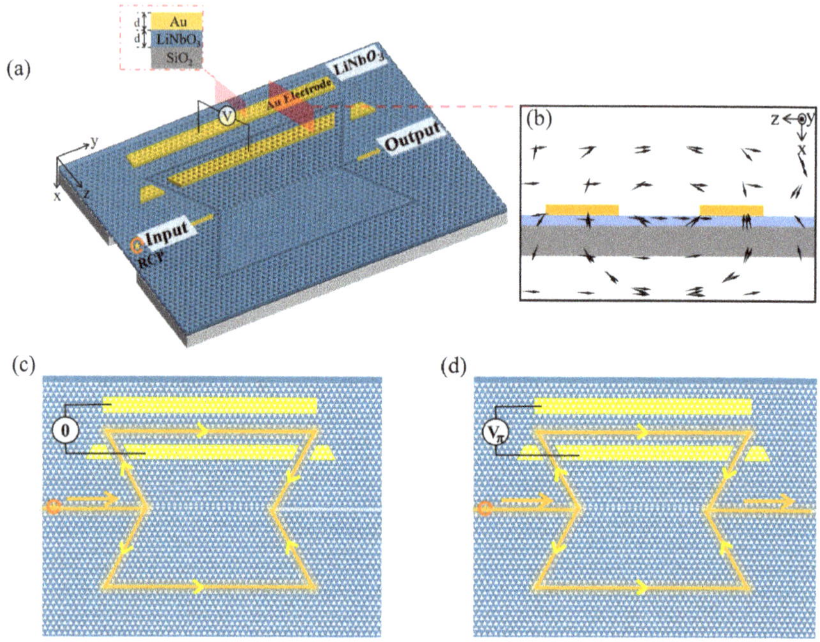

Figure 1. Conceptual design of MZI EOM based on lithium niobate VPC structure. (**a**) 3D schematic of MZI EOM based on a lithium niobate VPC structure and cross-sectional view of the device. The thickness of the TFLN and the gold electrode are both d = 220 nm (the star represents right-handed circularly polarized light). (**b**) Simulated electric field distribution inside the EOM by applying an external voltage, where the arrow indicates the vectorial direction of the electric field. (**c**) Schematic of destructive interference when the external voltage is 0 V. (**d**) Schematic of constructive interference when the external voltage is HWV. The arrows indicate the direction of light propagation.

In order to design the VPC structures, we first create a honey PC structure with C_{6v} symmetry (the unicell comprises two sets of triangular air holes, $r_A = r_B$ = 90 nm, which is defined as the distance between the vertex and geometric center of the triangle), which shows a Dirac cone structure in the photonic band diagram of transverse electric (TE) polarized mode. The lattice constant of the structure is a = 440 nm. By changing the size of the triangular holes (r_A = 180 nm, r_B = 0 nm), valley photonic crystals VPC1 and VPC2 are obtained, which are reduced to C3 symmetry. The K and K′ points degenerate and form a photonic bandgap. Since VPC1 and VPC2 are mirror symmetric, the photonic band diagrams of VPC1 and VPC2 are the same (the details of the design process are shown in Supplementary Materials Section S1) [41–43]. By combining VPC1 and VPC2, two configurations form waveguides that support topological edge states, namely, zigzag- and beard-type boundaries. Figure 2a,b shows the dispersion relations of beard-type and zigzag-type boundaries. Compared to the beard-type boundary, the zigzag-type boundary features high transmittance, low loss, and broad working bandwidth (the details of the design process are shown in Supplementary Materials Section S2). Therefore, the zigzag-type waveguide is selected for constructing the electrically tunable MZI in this study. In this work, commercial software (Lumerical FDTD2020) based on the finite-difference time-domain (FDTD) method is used for simulation.

https://doi.org/10.3390/photonics13010033

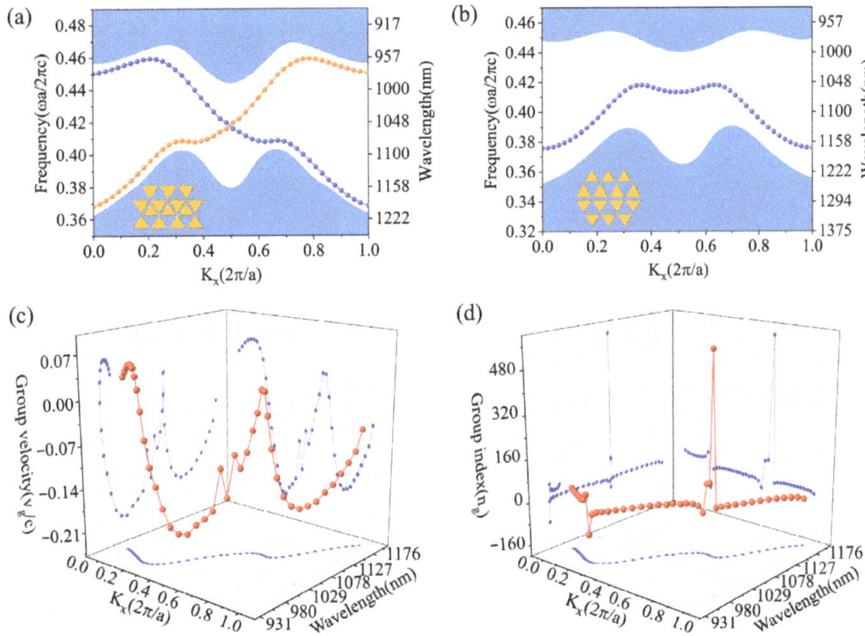

Figure 2. (**a**) Edge state plots (dotted lines) of the zigzag (**a**) and beard (**b**) type boundary structures (The shaded blue regions represent the bulk bands. The blue and orange dotted lines in (**a**) represent the two pseudospin states). Plots (red dotted lines) of normalized group velocity (v_g/c) (**c**) and group refractive index (n_g) (**d**) versus K_x (blue dotted lines) and wavelength (blue dotted lines) of the zigzag-type boundary.

The v_g and n_g of the edge states can be calculated as $v_g = d\omega/dk$ and $n_g = c/v_g$ [44]. Therefore, v_g represents the slope of the edge state and can be zero at the maximum and minimum points on the edge state plot. The v_g and n_g plots of the zigzag-type boundary are shown in Figure 2c,d. The calculations indicate that the edge states of the zigzag-type boundary can achieve a v_g of zero at the point $K_x = 0$, which corresponds to the minimum of the standing wave mode. However, in real applications, this point cannot be utilized since the mode cannot propagate through waveguides. Therefore, we select a point near $v_g = 0$ at 1077 nm to achieve a relatively low v_g and high n_g. At this point, the group velocity v_g/c is 0.00180036, and the group refractive index n_g is 555. According to r_{33} of the x-cut TFLN (32 pm/V [45], the r_{33}^{PC} is calculated as 131,072 pm/V.

To construct an MZI, we first design an Ω-shaped waveguide with a zigzag-type boundary because the light path along each arm of the MZI is along an Ω-shaped trajectory, as shown in Figure 3a, where the orange blocks indicate the electrodes. The forward transmittance spectrum and the electric field intensity distribution under 0 V are shown in Figure 3c (black curve) and Figure 3e. The zigzag-type boundary features high transmittance (0.92), low loss, and a relatively broad bandwidth of 68 nm. (details are provided in Supplementary Materials Section S3). It should be noted that group velocity dispersion (GVD), inherent to slow-light systems, narrows the effective operating bandwidth in high-speed applications. This further demonstrates the VPC's excellent anti-scattering performance, enabling light to pass through sharp corners without strong reflection or scattering. Due to the EO effect, the refractive index of LN changes with the applied voltage, and the corresponding refractive index modulation formula is as follows [39]:

https://doi.org/10.3390/photonics13010033

$$\Delta n = \frac{1}{2} r_{33}^{PC} n_e^3 E_z \tag{4}$$

$$E_z = V/s \tag{5}$$

where s is the gap between electrodes, and there is a linear relationship between the applied voltage and the change in refractive index. As shown in Figure 3d, when a voltage (V_π) is applied in the orange region, a π-phase modulation is generated. We set the voltage change to 0.0538 V (corresponding to the refractive index change of 0.0125) (the details of the design process are shown in Supplementary Materials Section S4). The edge-state curves at different voltages are shown in Figure 3b, and it can be seen that the working wavelength red-shifts as the voltage on the Ω-shaped waveguide is sequentially increased, because the effective refractive index increases with increasing applied voltage. Meanwhile, the forward transmittance spectrum and the electric field intensity distributions at V_π remain similar to those at 0 V, as shown in Figure 3c,f confirm that phase-only modulation can be achieved. In addition, the phase distributions confirm the achievement of π phase modulation as shown in Figure 3d.

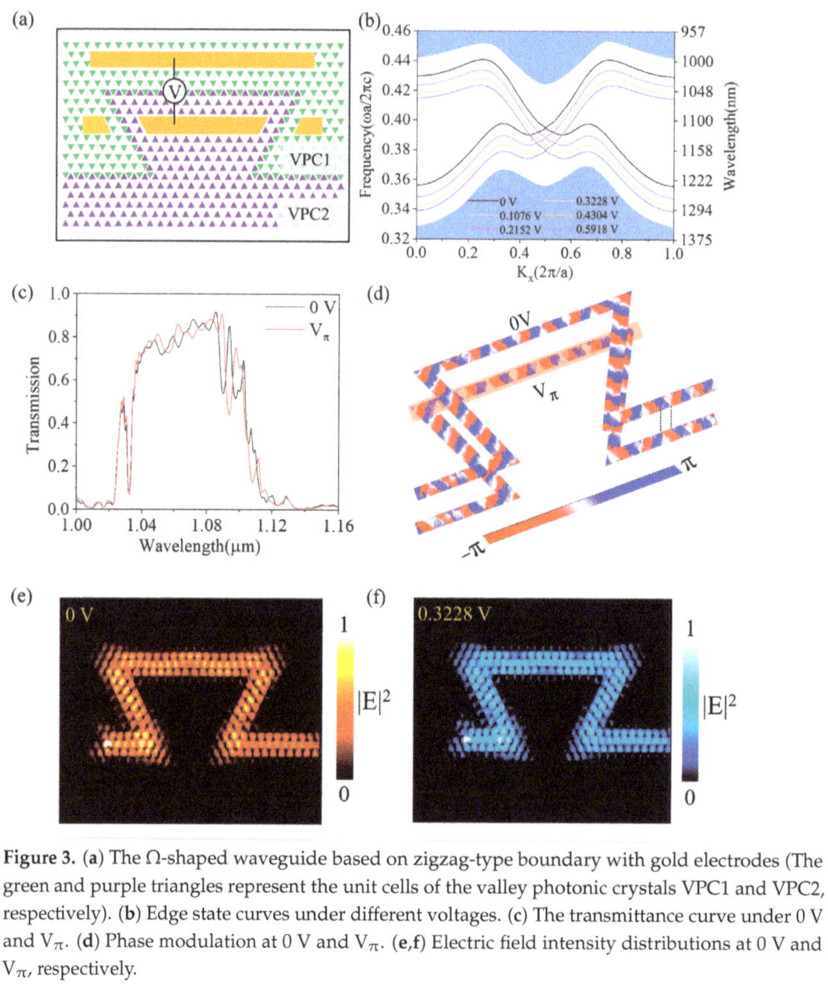

Figure 3. (**a**) The Ω-shaped waveguide based on zigzag-type boundary with gold electrodes (The green and purple triangles represent the unit cells of the valley photonic crystals VPC1 and VPC2, respectively). (**b**) Edge state curves under different voltages. (**c**) The transmittance curve under 0 V and V_π. (**d**) Phase modulation at 0 V and V_π. (**e**,**f**) Electric field intensity distributions at 0 V and V_π, respectively.

https://doi.org/10.3390/photonics13010033

3. Performance Analysis of the MZI EOM

We can see from Figure 4a that the MZI structure consists of two topological straight waveguides and two Ω-shaped interference arms, L1 and L2, represented by black and red lines, respectively. The MZI's principle is based on the interference effect of two optical waveguides. In the MZI, the incident light first passes through a straight waveguide of length 13*a* and is split into two identical beams, which then pass through the interference arms L1 (L1 = 66*a*) and L2 (L2 = 63*a*). The difference in the length of the two arms is 5*a* (ΔL = 5*a*). When light is transmitted, a phase difference is established between the two interference arms. Consequently, following the combination of these two light beams and the subsequent occurrence of interference, the presence of this phase difference gives rise to a change in the intensity of the light power at the final output end. The phase difference between the two interference arms can be expressed as follows:

$$\varphi_d = \frac{2\pi}{\lambda}(n_1 \cdot l_1 - n_2 \cdot l_2) \tag{6}$$

where n_1 and n_2 are the effective refractive indices of the interference arms L1 and L2, $n_1 = n_2$ (V = 0 V). l_1 and l_2 are the lengths of the interference arms L1 and L2, respectively. A π phase difference is usually required between the arms to achieve destructive interference and to obtain a near-zero valley at the designed wavelength.

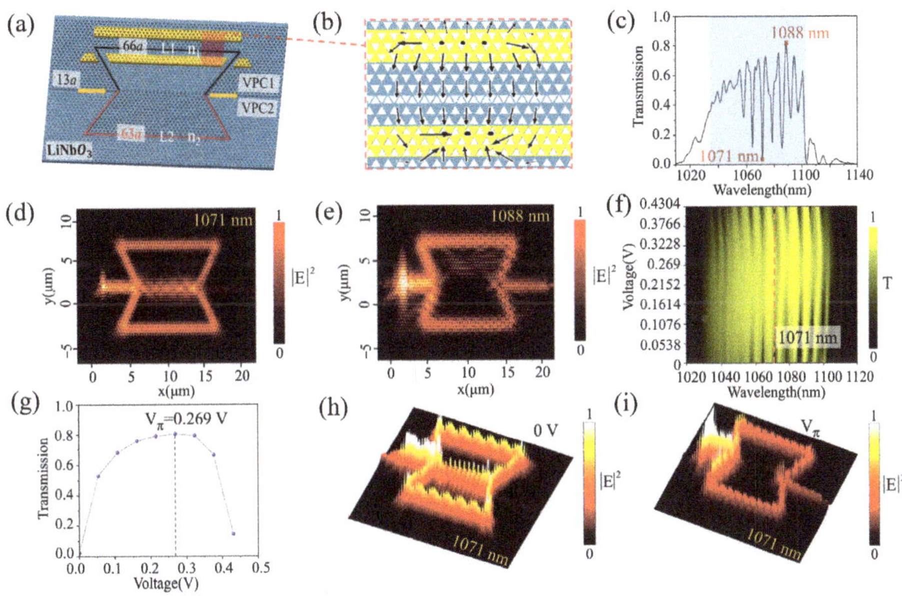

Figure 4. (**a**) 3D schematic of the designed VPC MZI structure. The red and blue colors represent different arms of the MZI structure. (**b**) Simulated electric field distribution inside the EOM by applying an external voltage, where the arrows indicate the vectorial direction of the electric field. (**c**) Transmission spectrum of MZI. Electric field intensity distributions of incident light at the wavelengths of 1071 nm (**d**) and 1088 nm (**e**). (**f**) The transmission surface plot of an MZI EOM at different external voltages and wavelengths. (**g**) Transmission plot at 1071 nm at different external voltages. Electric field intensity distribution at 1071 nm at the voltage of 0 V (**h**) and V_π (0.269 V) (**i**).

The light intensity at the output can be expressed as follows:

$$I_0 = A_1^2 + A_2^2 + 2 \cdot A_1 \cdot A_2 \cdot cos\varphi_d \tag{7}$$

https://doi.org/10.3390/photonics13010033

where A_1 and A_2 denote the amplitude of the light waves in the two interference arms, respectively, and φ_d is the phase difference shown in Equation (6). The forward transmittance spectrum of the VPC MZI structure (V = 0 V) is shown in Figure 4c. Due to the high forward transmittance within the wavelength range of 1043–1099 nm, one can see distinct interference peaks (close to 1) and valleys (near zero) in the spectrum. The electric field distributions of MZI (V = 0 V) at different wavelengths are plotted in Figure 4d,e, which correspond to the constructive interference at 1088 nm and destructive interference at 1071 nm.

The extinction ratio, insertion loss, modulation depth, and sensitivity of MZIs are key performance indicators. When constructive interference (φ_d = an integer multiple of 2π) is achieved, the output power is P_{on}, and when destructive interference ($\varphi_d = \pi$) occurs, the output power is P_{off}. The extinction ratio of the MZI can be expressed as Equation (8):

$$EXT = -10lg\left(P_{off}/P_{on}\right) \tag{8}$$

The insertion loss of the MZI can be expressed as Equation (9):

$$IL = -10lg(T) \tag{9}$$

Assuming that the maximum output power of MZI is P_{max} and the minimum output power is P_{min} during the modulation process, the modulation depth of MZI can be expressed as Equation (9):

$$m = P_{max} - P_{min} \tag{10}$$

Subsequently, a voltage is applied to the arm L1. Due to the strong EO effect of LN, the effective refractive index n_1 of L1 changes ($n_1 \neq n_2$), which affects the phase of L1. Thus, the position of the peaks (constructive interference) and valleys (destructive interference) in the transmission spectrum can be well controlled by applying a voltage to one arm. This approach successfully modulates light intensity at a selected wavelength. The simulated electric field distribution (COMSOL Multiphysics) is shown in Figure 4b, revealing a strong electric field across the waveguide, providing the strongest electro-optical modulation. The applied voltage of the TFLN substrate ranges from 0 V to 0.4304 V with a gradient of 0.0538 V corresponding to the index change of 0.0125 of L1. The forward transmission spectra at different voltages are shown in Figure 4f (details are provided in Supplementary Materials Section S5). Here, 1071 nm in the working bandwidth is selected as an example for specific analysis. As the applied voltage changes, the device's transmission changes accordingly, as shown in Figure 4g.

The figure shows that at 0 V and an input light wavelength of 1071 nm, the device produces destructive interference, resulting in the lowest observed transmission of 0.02. At an applied voltage of 0.269 V, the device exhibits constructive interference, yielding a maximum transmission of 0.82. The distributions of electric field intensity are depicted in Figure 4h,i. The high transmittance in the constructive interference peak is attributed to the efficient light splitting and combining at the Y-junctions, and the back-reflection remains consistently low (<0.1) throughout the working bandwidth (details are provided in Supplementary Materials Section S6). Thus, we realize intensity modulation at the selected wavelength, as calculated using Equations (8)–(10); the device's extinction ratio is 16.13 dB, the insertion loss is 0.86 dB, and the modulation depth is 80%. The V_π is 0.269 V.

It is worth noting that as the refractive index of the LN waveguide increases with applied voltage, the bandgap redshifts. We conducted simulation analyses of edge states under different externally applied voltages, and the resulting dispersion curves are shown in Figure 5a. As shown in Figure 5b, the central wavelength of the bandgap redshifts

https://doi.org/10.3390/photonics13010033

along with the voltage increase. When the externally applied voltage is 0.6994 V, the bandgap exhibits a passband in the range of 0.368 a/λ–0.411 a/λ (1071 nm–1196 nm), which deviates slightly from the target wavelength of 1071 nm. Therefore, the maximum applied voltage should not exceed 0.6994 V. In principle, this method can be applied to design MZI EOM working at any wavelength by changing the geometry of the MZI. Furthermore, by adjusting the VPC parameters, the spectrum can be extended towards 1550 nm. (details are provided in Supplementary Materials Section S7).

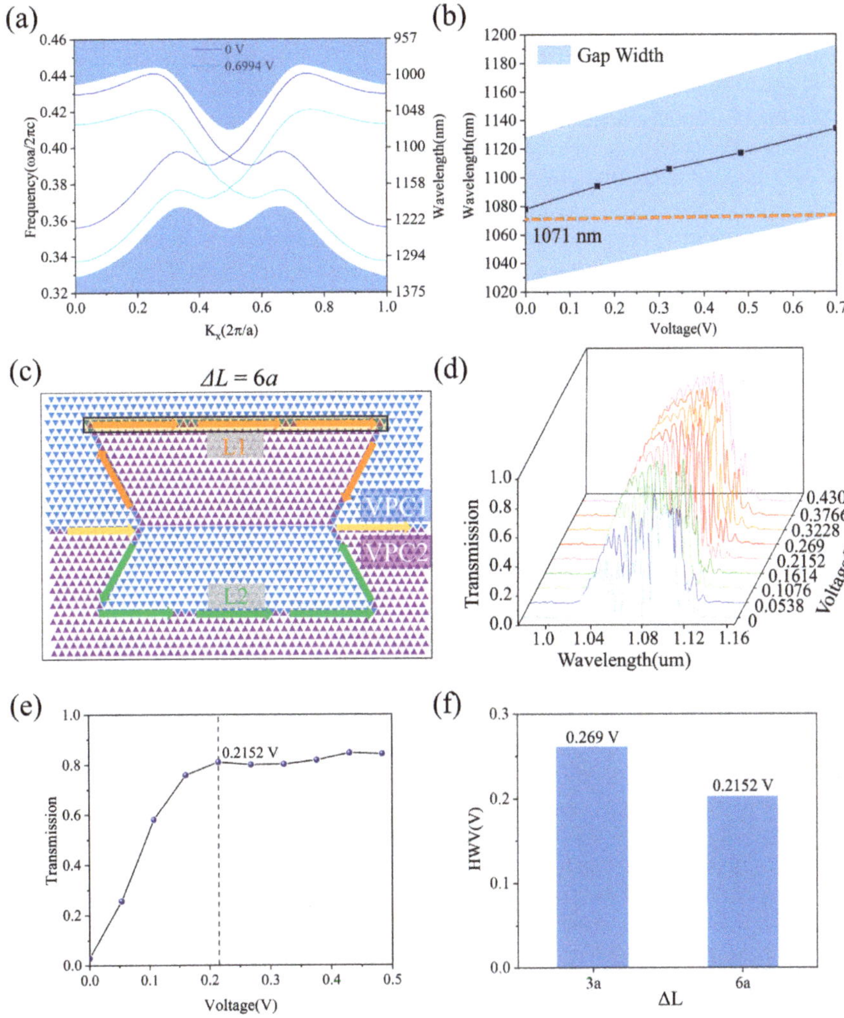

Figure 5. (**a**) Edge state curves under externally applied voltages of 0 V and 0.6994 V; (**b**) The tuning curve of the photonic bandgap range with voltage. (**c**) Schematic diagrams of the structure when the arm length difference ΔL = 6*a*. The blue and purple colors represent VPC1 and VPC2, respectively. The yellow arrows represent the entry and exit waveguides of the MZI structure. The orange and green colors represent the optical waveguides of the arms L1 and L2, respectively. (**d**) The transmission plots of the MZI EOM (ΔL = 6*a*) with different voltages at different wavelengths. (**e**) Localized magnification of the transmittance at 1075 nm. (**f**) The comparison bar chart of the HWV of ΔL = 3*a* and ΔL = 6*a*.

https://doi.org/10.3390/photonics13010033

Based on the optical intensity modulation, we further studied the influence of different arm length differences, ΔL, on the half-wave voltage. The relationship between arm length difference and half-wave voltage can be expressed as follows:

$$V = \frac{\lambda \cdot s \cdot \varphi_d}{\pi \cdot r_{33}^{PC} \cdot n_e^3 \cdot \Delta L} \tag{11}$$

where s is the gap between electrodes and φ_d is the phase difference shown in Equation (6). It can be seen from Equation (11) that the HWV is inversely proportional to ΔL, and the larger the ΔL, the smaller the HWV. Here, we selected the length difference between the two arms of ΔL = 6*a*, whose geometry is shown in Figure 5c. The transmission spectra of the MZI EOM with different voltages at an arm-length difference of ΔL = 6*a* are shown in Figure 5d, which exhibit distinct changes with voltage (details are shown in Supplementary Materials Section S8). Due to the high forward transmittance in the wavelength range of 990–1140 nm, the MZI shows distinct interference peaks (close to 1) and valleys (close to 0). Due to the differences in interference spectra, we chose different wavelengths for this arm length. Through analysis and verification, we selected 1075 nm (for ΔL = 6*a*), as shown in Figure 5e. As shown in Figure 5f, with the increase in the arm length difference, the HWV decreases, which is as expected from Equation (11). Furthermore, we present a discussion on the manufacturing tolerances of the corner rounding in VPC design in the Supplementary Materials Section S9, allowing us to define the required fabrication accuracy.

4. Conclusions

In conclusion, we have designed a thin-film lithium niobate MZI EOM based on a VPC structure. The EO effect of LN was used to design the EOM, and the phase change at the output was achieved by applying a voltage, thereby changing the LN's refractive index. The modulator not only achieves a compact size (21 μm × 17 μm) but also exhibits excellent optical performance, including a peak forward transmittance of 0.82, an extinction ratio of 16.13 dB, and a modulation depth of 80%. The thin-film lithium niobate VPC MZI electro-optical modulators are expected to contribute to next-generation optical networks with their high-speed modulation, low loss, and ease of integration, helping realize higher speeds and broader bandwidth for optical signals.

Supplementary Materials: The following supporting information can be downloaded at: https://www.mdpi.com/article/10.3390/photonics13010033/s1, Figure S1: A schematic of the photonic crystal and photonic band diagram. Figure S2: (a) Schematic diagram of the zigzag-type boundary. (b) Schematic diagram of the beard-type boundary. (c) The transmittance spectrum of the straight waveguide based on the zigzag-type boundary. (d) The transmittance spectrum of the straight waveguide based on the beard-type boundary. Figure S3: (a) Ω-shaped waveguide assembled by zigzag-type boundary, corresponding transmittance (c) and electric field distribution at 1071 nm (e); (b) Ω-shaped waveguide assembled by beard-type boundary, corresponding transmittance (d) and electric field distribution at 1064 nm (f). Figure S4: The refractive index of LN changes with the applied voltage. Figure S5: Transmittance diagram of MZI at different voltages. Figure S6: Transmission analysis of the Y-junction of the MZI structure. Figure S7: Photonic band diagrams of the original honeycomb PC and VPC structures. Figure S8: Transmittance plots of MZI with ΔL=6a at different voltages. Figure S9: A schematic of the honeycomb PC and photonic band diagram.

Author Contributions: Conceptualization, Y.Y., H.F., M.Z. and H.L.; methodology, Y.Y. and H.F.; software, Y.Y.; validation, H.F., X.L., M.Z. and H.L.; formal analysis, Y.Y. and H.F.; investigation, Y.Y. and H.F.; resources, H.F.; data curation, Y.Y.; writing—original draft preparation, Y.Y. and H.F.; writing—review and editing, Y.Y., H.F. and H.L.; visualization, X.L.; supervision, H.F.; project administration, H.F.; funding acquisition, H.F., P.D. and J.R. All authors have read and agreed to the published version of the manuscript.

 https://doi.org/10.3390/photonics13010033

Funding: This research was funded by the National Key Research and Development Program of China (Grant No. 2022YFA1404201), the Australia Research Council (Grant Nos. DP220100603 and FT220100559, LP210200345, and LP210100467), the National Natural Science Foundation of China (Grant No. U23A20375), the Natural Science Foundation of Shanxi Province (Grant No. 202403021211011), the Research Project Supported by Shanxi Scholarship Council of China (Grant No. 2024-032), The Fund Program for the Scientific Activities of Selected Returned Overseas Professionals in Shanxi Province (Grant No. 20240006), the Program of State Key Laboratory of Quantum Optics and Quantum Optics Devices (Grant No. KF202402), the Basic scientific research conditions and major scientific instrument and equipment development of Anhui Science and Technology Department (Grant No. 2023YFF0715700), and the Key Research project of Shanxi Province (Grant No. 2023021501010001).

Institutional Review Board Statement: Not applicable.

Informed Consent Statement: Not applicable.

Data Availability Statement: The data that support the findings of this study are available from the corresponding author upon reasonable request.

Acknowledgments: The authors acknowledge the useful discussions and suggestions from Yuan Tian.

Conflicts of Interest: Author Pengqi Dong was employed by the institute of architectural design and research of Taiyuan university of technology Co, Ltd. Author Junjun Ren was employed by Shanxi Chenghaoxin Intelligent Technology Co., Ltd. The remaining authors declare that the research was conducted in the absence of any commercial or financial relationships that could be construed as a potential conflict of interest.

References

1. Zhang, Y.; Shen, J.; Li, J.; Wang, H.; Feng, C.; Zhang, L.; Sun, L.; Xu, J.; Liu, M.; Wang, Y.; et al. High-speed electro-optic modulation in topological interface states of a one-dimensional lattice. *Light Sci. Appl.* **2023**, *12*, 206. [CrossRef]
2. Sinatkas, G.; Christopoulos, T.; Tsilipakos, O.; Kriezis, E.E. Electro-optic modulation in integrated photonics. *J. Appl. Phys.* **2021**, *130*, 010901. [CrossRef]
3. Miller, D.A.B. Attojoule Optoelectronics for Low-Energy Information Processing and Communications. *J. Light. Technol.* **2017**, *35*, 346–396. [CrossRef]
4. Koeber, S.; Palmer, R.; Lauermann, M.; Heni, W.; Elder, D.L.; Korn, D.; Woessner, M.; Alloatti, L.; Koenig, S.; Schindler, P.C.; et al. Femtojoule electro-optic modulation using a silicon–organic hybrid device. *Light Sci. Appl.* **2015**, *4*, e255. [CrossRef]
5. Xu, W.; Guo, T.; Zhang, K.; Li, Z.; Zhou, T.; Zuo, Q.; Sheng, Y.; Jing, L.; Ma, H.; Yu, M.; et al. Manipulations of a transmon qubit with a null-biased electro-optic fiber link. *Nat. Commun.* **2025**, *16*, 2629. [CrossRef] [PubMed]
6. Pittaluga, M.; Lo, Y.S.; Brzosko, A.; Woodward, R.I.; Scalcon, D.; Winnel, M.S.; Roger, T.; Dynes, J.F.; Owen, K.A.; Juárez, S.; et al. Long-distance coherent quantum communications in deployed telecom networks. *Nature* **2025**, *640*, 911–917. [CrossRef]
7. Kues, M.; Reimer, C.; Roztocki, P.; Cortés, L.R.; Sciara, S.; Wetzel, B.; Zhang, Y.; Cino, A.; Chu, S.T.; Little, B.E.; et al. On-chip generation of high-dimensional entangled quantum states and their coherent control. *Nature* **2017**, *546*, 622–626. [CrossRef]
8. Qi, Y.; Li, Y. Integrated lithium niobate photonics. *Nanophotonics* **2020**, *9*, 1287–1320. [CrossRef]
9. Teng, M.; Fathpour, S.; Safian, R.; Zhuang, L.; Honardoost, A.; Alahmadi, Y.; Polkoo, S.S.; Kojima, K.; Wen, H.; Renshaw, C.K.; et al. Miniaturized Silicon Photonics Devices for Integrated Optical Signal Processors. *J. Light. Technol.* **2020**, *38*, 6–17. [CrossRef]
10. Sun, C.; Wade, M.T.; Lee, Y.; Orcutt, J.S.; Alloatti, L.; Georgas, M.S.; Waterman, A.S.; Shainline, J.M.; Avizienis, R.R.; Lin, S.; et al. Single-chip microprocessor that communicates directly using light. *Nature* **2015**, *528*, 534–538. [CrossRef]
11. Han, C.; Zheng, Z.; Shu, H.; Jin, M.; Qin, J.; Chen, R.; Tao, Y.; Shen, B.; Bai, B.; Yang, F.; et al. Slow-light silicon modulator with 110-GHz bandwidth. *Sci. Adv.* **2023**, *9*, eadi5339. [CrossRef]
12. Skandalos, I.; Bucio, T.D.; Mastronardi, L.; Yu, G.; Zilkie, A.; Gardes, F.Y. A 100 Gb s−1 quantum-confined Stark effect modulator monolithically integrated with silicon nitride on Si. *Commun. Eng.* **2025**, *4*, 82. [CrossRef] [PubMed]
13. Li, B.; Yuan, Z.; Williams, J.; Jin, W.; Beckert, A.; Xie, T.; Guo, J.; Feshali, A.; Paniccia, M.; Faraon, A.; et al. Down-converted photon pairs in a high-Q silicon nitride microresonator. *Nature* **2025**, *639*, 922–927. [CrossRef]
14. Ogiso, Y.; Ozaki, J.; Ueda, Y.; Kashio, N.; Kikuchi, N.; Yamada, E.; Tanobe, H.; Kanazawa, S.; Yamazaki, H.; Ohiso, Y.; et al. Over 67 GHz Bandwidth and 1.5 V Vπ InP-Based Optical IQ Modulator With n-i-p-n Heterostructure. *J. Light. Technol.* **2017**, *35*, 1450–1455. [CrossRef]

https://doi.org/10.3390/photonics13010033

15. Wang, C.; Zhang, M.; Chen, X.; Bertrand, M.; Shams-Ansari, A.; Chandrasekhar, S.; Winzer, P.; Lončar, M. Integrated lithium niobate electro-optic modulators operating at CMOS-compatible voltages. *Nature* **2018**, *562*, 101–104. [CrossRef]
16. Valdez, F.; Mere, V.; Wang, X.; Boynton, N.; Friedmann, T.A.; Arterburn, S.; Dallo, C.; Pomerene, A.T.; Starbuck, A.L.; Trotter, D.C.; et al. 110 GHz, 110 mW hybrid silicon-lithium niobate Mach-Zehnder modulator. *Sci. Rep.* **2022**, *12*, 18611. [CrossRef]
17. Stokowski, H.S.; McKenna, T.P.; Park, T.; Hwang, A.Y.; Dean, D.J.; Celik, O.T.; Ansari, V.; Fejer, M.M.; Safavi-Naeini, A.H. Integrated quantum optical phase sensor in thin film lithium niobate. *Nat. Commun.* **2023**, *14*, 3355. [CrossRef]
18. Huang, X.; Liu, Y.; Li, Z.; Guan, H.; Wei, Q.; Tan, M.; Li, Z. 40 GHz high-efficiency Michelson interferometer modulator on a silicon-rich nitride and thin-film lithium niobate hybrid platform. *Opt. Lett.* **2021**, *46*, 2811–2814. [CrossRef] [PubMed]
19. Ghoname, A.O.; Hassanien, A.E.; Goddard, L.L.; Gong, S. Compact Lithium Niobate Michelson Interferometer Modulators Based on Spiral Waveguides. *IEEE J. Sel. Top. Quantum Electron.* **2024**, *30*, 3400108. [CrossRef]
20. Wang, P.-Y.; Wan, S.; Zhang, M.; Ma, R.; Bo, F.; Shen, Z.; Wan, W.; Sun, F.-W.; Guo, G.-C.; Dong, C.-H. Dynamic Control of Non-Hermitian On-Site Potential in the Lithium Niobate Microresonator. *Laser Photon. Rev.* **2025**, *19*, e00552. [CrossRef]
21. Jian, J.; Xu, M.; Liu, L.; Luo, Y.; Zhang, J.; Liu, L.; Zhou, L.; Chen, H.; Yu, S.; Cai, X. High modulation efficiency lithium niobate Michelson interferometer modulator. *Opt. Express* **2019**, *27*, 18731–18739. [CrossRef]
22. Wang, C.; Zhang, M.; Stern, B.; Lipson, M.; Lončar, M. Nanophotonic lithium niobate electro-optic modulators. *Opt. Express* **2018**, *26*, 1547–1555. [CrossRef]
23. Wei, C.; Feng, H.; Ye, K.; Eijkel, M.; Klaver, Y.; Chen, Z.; Keloth, A.; Wang, C.; Marpaung, D. Programmable multifunctional integrated microwave photonic circuit on thin-film lithium niobate. *Nat. Commun.* **2025**, *16*, 2281. [CrossRef]
24. Tao, Z.; Wang, H.; Feng, H.; Guo, Y.; Shen, B.; Sun, D.; Tao, Y.; Han, C.; He, Y.; Bowers, J.E.; et al. Ultrabroadband on-chip photonics for full-spectrum wireless communications. *Nature* **2025**, *645*, 80–87. [CrossRef]
25. Li, M.; Ling, J.; He, Y.; Javid, U.A.; Xue, S.; Lin, Q. Lithium niobate photonic-crystal electro-optic modulator. *Nat. Commun.* **2020**, *11*, 4123. [CrossRef]
26. Fei, H.; Wu, M.; Lin, H.; Liu, X.; Yang, Y.; Zhang, M.; Cao, B. An on-chip nanophotonic reciprocal optical diode for asymmetric transmission of the circularly polarized light. *Superlattices Microstruct.* **2019**, *132*, 106155. [CrossRef]
27. Du, Z.; Liao, K.; Dai, T.; Wang, Y.; Gao, J.; Huang, H.; Qi, H.; Li, Y.; Wang, X.; Su, X.; et al. Ultracompact and multifunctional integrated photonic platform. *Sci. Adv.* **2024**, *10*, eadm7569. [CrossRef]
28. Weigand, H.; Vogler-Neuling, V.V.; Escalé, M.R.; Pohl, D.; Richter, F.U.; Karvounis, A.; Timpu, F.; Grange, R. Enhanced Electro-Optic Modulation in Resonant Metasurfaces of Lithium Niobate. *ACS Photonics* **2021**, *8*, 3004–3009. [CrossRef]
29. Thomaschewski, M.; Zenin, V.A.; Fiedler, S.; Wolff, C.; Bozhevolnyi, S.I. Plasmonic Lithium Niobate Mach–Zehnder Modulators. *Nano Lett.* **2022**, *22*, 6471–6475. [CrossRef]
30. Butt, M.A.; Khonina, S.N. Recent Advances in Photonic Crystal and Optical Devices. *Crystals* **2024**, *14*, 543. [CrossRef]
31. Deng, R.; Liu, W.; Shi, L. Inverse design in photonic crystals. *Nanophotonics* **2024**, *13*, 1219–1237. [CrossRef]
32. Saghaei, H.; Soroosh, M.; Maleki, M.J.; Mondal, H.; Nurmohammadi, T.; Jafari, B.; Shahi, M.; Sepahvandi, V.; Adibnia, E. High-performance and compact photonic crystal channel drop filter using P-shaped ring resonator. *Results Opt.* **2025**, *21*, 100817. [CrossRef]
33. Soroosh, M.; Al-Shammri, F.K.; Maleki, M.J.; Balaji, V.R.; Adibnia, E. A Compact and Fast Resonant Cavity-Based Encoder in Photonic Crystal Platform. *Crystals* **2024**, *15*, 24. [CrossRef]
34. Wu, M.; Yang, Y.B.; Fei, H.M.; Lin, H.; Zhao, X.D.; Kang, L.J.; Xiao, L.T. On-Chip Ultra-Compact Hexagonal Boron Nitride Topological Ring-Resonator in Visible Region. *J. Light. Technol.* **2022**, *40*, 7610–7618. [CrossRef]
35. Dong, J.-W.; Chen, X.-D.; Zhu, H.; Wang, Y.; Zhang, X. Valley photonic crystals for control of spin and topology. *Nat. Mater.* **2017**, *16*, 298–302. [CrossRef] [PubMed]
36. Yan, B.; Liao, B.; Shi, F.; Xi, X.; Cao, Y.; Xiang, K.; Meng, Y.; Yang, L.; Zhu, Z.; Chen, J.; et al. Realization of Topology-Controlled Photonic Cavities in a Valley Photonic Crystal. *Phys. Rev. Lett.* **2025**, *134*, 033803. [CrossRef] [PubMed]
37. He, X.-T.; Liang, E.-T.; Yuan, J.-J.; Qiu, H.-Y.; Chen, X.-D.; Zhao, F.-L.; Dong, J.-W. A silicon-on-insulator slab for topological valley transport. *Nat. Commun.* **2019**, *10*, 872. [CrossRef] [PubMed]
38. Wang, Y.; Fei, H.; Lin, H.; Bai, J.; Zhang, M.; Liu, X.; Cao, B.; Tian, Y.; Xiao, L. Ultra-compact electro-optic phase modulator based on a lithium niobate topological slow light waveguide. *Opt. Express* **2024**, *32*, 3980–3988. [CrossRef]
39. Roussey, M.; Bernal, M.-P.; Courjal, N.; Van Labeke, D.; Baida, F.I.; Salut, R. Electro-optic effect exaltation on lithium niobate photonic crystals due to slow photons. *Appl. Phys. Lett.* **2006**, *89*, 241110. [CrossRef]
40. Razzari, L.; Träger, D.; Astic, M.; Delaye, P.; Frey, R.; Roosen, G.; André, R. Kerr and four-wave mixing spectroscopy at the band edge of one-dimensional photonic crystals. *Appl. Phys. Lett.* **2005**, *86*, 231106. [CrossRef]
41. Lu, L.; Joannopoulos, J.D.; Soljačić, M. Topological photonics. *Nat. Photonics* **2014**, *8*, 821–829. [CrossRef]
42. Han, Y.; Fei, H.; Lin, H.; Zhang, Y.; Zhang, M.; Yang, Y. Design of broadband all-dielectric valley photonic crystals at telecommunication wavelength. *Opt. Commun.* **2021**, *488*, 126847. [CrossRef]

 https://doi.org/10.3390/photonics13010033

43. Ezawa, M. Topological Kirchhoff law and bulk-edge correspondence for valley Chern and spin-valley Chern numbers. *Phys. Rev. B* **2013**, *88*, 161406. [CrossRef]
44. Nussbaum, E.; Sauer, E.; Hughes, S. Inverse design of broadband and lossless topological photonic crystal waveguide modes. *Opt. Lett.* **2021**, *46*, 1732–1735. [CrossRef] [PubMed]
45. Weis, R.S.; Gaylord, T.K. Lithium niobate: Summary of physical properties and crystal structure. *Appl. Phys. A* **1985**, *37*, 191–203. [CrossRef]

 https://doi.org/10.3390/photonics13010033

hv photonics

MDPI

Article

Effect of Mo Layer Thickness on Bandwidth Tunability and Absorption Properties of Planar Ultra-Wideband Optical Absorbers

Kao-Peng Min [1], Yu-Ting Gao [2], Cheng-Fu Yang [3,4,*], Walter Water [2] and Chi-Ting Ho [5]

1 College of Engineering, National Formosa University, Huwei, Yunlin 632, Taiwan; d1481103@nfu.edu.tw
2 Department of Electronic Engineering, National Formosa University, Yunlin 632, Taiwan
3 Department of Chemical and Materials Engineering, National University of Kaohsiung, Kaohsiung 811, Taiwan
4 Department of Aeronautical Engineering, Chaoyang University of Technology, Taichung 413, Taiwan
5 Department of Mechanical Design Engineering, National Formosa University, Yunlin 632, Taiwan
* Correspondence: cfyang@nuk.edu.tw

Abstract

This study utilizes COMSOL Multiphysics (version 6.0) to design a planar ultra-broadband optical absorber with a multilayer configuration. The proposed structure consists of seven stacked layers arranged from bottom to top: W (*h*1, acting as a reflective substrate and transmission blocker), WSe_2 (*h*2), SiO_2 (*h*3), Ni (*h*4), SiO_2 (*h*5), Mo (*h*6), and SiO_2 (*h*7). One key finding of this study is that, when all other layer thicknesses are fixed, variations in the Mo layer thickness systematically induce a redshift in both the short- and long-wavelength cutoff edges. Notably, the long-wavelength cutoff exhibits a larger shift than the short-wavelength edge, resulting in an increased absorption bandwidth where absorptivity remains above 0.900. The second contribution is the demonstration that this planar structure can be readily engineered to achieve ultra-broadband absorption, spanning from the near-ultraviolet and visible region (360 nm) to the mid-infrared (6300 nm). An important characteristic of the proposed design is that the thickness of the *h*7 SiO_2 layer influences the cutoff wavelength at the short-wavelength edge, while the thickness of the *h*6 Mo layer governs the cutoff position at the long-wavelength edge. This dual modulation capability allows the proposed optical absorber to flexibly tune both the spectral range and the bandwidth in which absorptivity exceeds 0.900, thereby enabling the realization of a wavelength- and bandwidth-tunable optical absorber.

Keywords: COMSOL Multiphysics; Mo layer; absorptivity; bandwidth; planar ultra-wideband optical absorber

1. Introduction

The design of broadband optical absorbers capable of covering a wide spectral range, from the near-ultraviolet (UV) to the middle-infrared (mid-IR), has become an important research focus due to their broad applications in photovoltaics, thermal emitters, photodetectors, and optical sensing. Such absorbers can efficiently capture and utilize incident electromagnetic energy across different wavelengths, enabling enhanced energy conversion, improved stealth technologies, and advanced photonic devices. To achieve this functionality, a wide variety of structures have been developed for optical absorbers. However, most approaches require highly complex designs or geometric structures to reach the desired performance [1–4]. In contrast, fully planar multilayer thin-film structures are often favored because of their relatively simple geometry and compatibility with large-area fabrication techniques [5,6]. However, designing a planar multilayer absorber that maintains high absorptivity over such an exceptionally broad spectrum presents significant challenges [7]. The primary difficulty arises from the need to simultaneously manage multiple optical phenomena, including interference effects, impedance matching, and resonance conditions across widely separated wavelength regimes [8–11].

Furthermore, the refractive index contrast between different materials, along with the thickness sensitivity of each layer, must be carefully optimized to minimize strong reflectance peaks and prevent transmission leakage at specific wavelengths. The wavelength-dependent nature of material properties adds another layer of complexity, as optimal parameters for UV wavelengths may conflict with those required for mid-IR performance. These multifaceted challenges make it non-trivial to balance broadband absorption performance while maintaining both structural simplicity and fabrication feasibility. Therefore, designing a fully planar optical absorber capable of efficiently operating across the broad spectral range from the below 400 nm in the near-UV to beyond 6000 nm in the mid-IR is far from straightforward. The development of efficient, multilayer planar absorbers spanning from the near-UV to the mid-IR remains a critical yet demanding objective in optical materials research, requiring innovative approaches to overcome the inherent trade-offs between spectral breadth, absorption efficiency, and practical implementation [12–14].

Tungsten diselenide (WSe_2) has emerged as a highly promising material for optical absorption owing to its unique electronic and structural properties [15]. One of its most distinctive features is the pronounced spin–orbit coupling, which enables spin-selective light absorption and charge transfer. This property not only broadens the scope of its use in conventional optoelectronic devices but also positions WSe_2 as a candidate for designing an optical absorber. In its monolayer form, WSe_2 possesses a direct bandgap of approximately 1.65 eV, which lies within the visible spectrum and makes it particularly effective for harvesting visible light. Such a direct bandgap allows for efficient photoelectric conversion, thereby enhancing the performance of photovoltaic and photodetection systems. Furthermore, as a two-dimensional layered material, WSe_2 offers remarkable tunability in its optical properties; its bandgap and absorption spectrum can be systematically engineered by controlling the number of atomic layers, providing a versatile platform for customized device design. In addition, the nearly dangling-bond-free surface of WSe_2 ensures superior interface quality when integrated with other semiconductors or metallic layers, significantly reducing interfacial defects and minimizing scattering losses [16–18]. These combined advantages make WSe_2 an attractive and versatile absorber material for advanced optoelectronic and photonic applications.

Molybdenum (Mo) thin films offer several advantages when employed as optical absorbers. Their high absorption coefficient in the visible and near-IR regions enables efficient suppression of reflection and enhancement of overall absorption performance [19,20]. Owing to its high melting point (~2623 °C), Mo exhibits excellent thermal stability, allowing it to withstand high-temperature fabrication processes and resist degradation under intense light or thermal loads. In addition, as a metal with good electrical conductivity, Mo thin films can serve not only as absorptive layers but also as electrodes, which facilitates integration in optoelectronic devices. Mo is also highly compatible with a wide range of materials, including semiconductors such as Si, dielectric layers such as SiO_2 and Al_2O_3, and various metals, making it suitable for multilayer optical structures and metamaterial designs [21,22]. From a fabrication perspective, Mo thin films can be reliably deposited using established methods such as electron-beam evaporation and sputtering, offering stable processes at costs significantly lower than those associated with noble metals like Au

https://doi.org/10.3390/photonics13010086

and Ag. Moreover, Mo exhibits superior corrosion resistance and mechanical robustness compared with many other metals, ensuring long-term durability in practical applications.

This study employs COMSOL Multiphysics® (version 6.0) as the simulation platform to design an ultra-broadband optical absorber [23,24]. The proposed absorber is composed of seven stacked layers, arranged from bottom to top as follows: W (*h*1, serving as the reflective substrate and transmission blocker), WSe_2 (*h*2), SiO_2 (*h*3), Ni (*h*4), SiO_2 (*h*5), Mo (*h*6), and SiO_2 (*h*7). A key aspect of this design is the tunability introduced by the Mo layer. When the thicknesses of all other layers are held constant, increasing the Mo thickness leads to a redshift in both the short- and long-wavelength cutoffs of the absorption spectrum. Importantly, the shift at the long-wavelength edge is more pronounced than at the short-wavelength edge, thereby broadening the overall absorption bandwidth where absorptivity exceeds 0.900. This phenomenon can be attributed to the strong plasmonic response of Mo in the near- and mid-IR regions, which enhances light-matter interaction and extends resonance effects.

The second significant feature is the ability to realize a fully planar, ultra-broadband absorber based on a straightforward multilayer configuration that does not rely on complex nanostructuring. The designed structure achieves a high absorptivity spectrum that extends continuously from the UV region ($\approx$360 nm) through the visible and near-IR, and well into the mid-IR region ($\approx$6300 nm). Such a broad absorption range is particularly advantageous for applications in solar energy harvesting, thermal emitters, photodetection, and stealth technologies, where wide spectral coverage and high efficiency are critical. In previous studies, very few fully planar optical absorbers have demonstrated the capability to simultaneously tune both the absorption band and bandwidth through a comparably direct structural modulation approach. An important characteristic of the proposed design is that jointly adjusting the thicknesses of the top SiO_2 layer (*h*7) and the Mo layer (*h*6) allows effective tuning of the cutoff wavelengths at both spectral edges, thereby enabling flexible control over the bandwidth in which the absorptivity exceeds 0.900. Specifically, the *h*7-SiO_2 layer governs the cutoff wavelength in the short-wavelength edge, while the *h*6-Mo layer dictates the cutoff wavelength in the long-wavelength edge. This dual control mechanism not only simplifies the design compared to more complex multilayered structures but also provides significant versatility, allowing the proposed absorber to adaptively adjust its operational spectral range and bandwidth. As a result, the structure functions as a wavelength- and bandwidth-tunable optical absorber, making it highly promising for practical applications that demand adaptive and broadband absorption performance across different regions of the electromagnetic spectrum.

2. Structure of the Investigated Ultra-Wideband Optical Absorber with the Bandwidth from UV-C to Mid-IR

This work present the design of an ultrabroadband solar absorber based on a planar metamaterial stack. The structure integrated a SiO_2 insulating layer, a WSe_2 layer, and multiple metallic components-specifically W, Ni, and Mo-arranged to achieve superior absorption performance. Figure 1 illustrates the intricate geometric design of the unit cell with p = 300 nm in the proposed multi-layered metamaterial absorber. The structure consisted of a carefully engineered seven-layer configuration, tailored to maximize electromagnetic wave absorption. In designing the optical absorber, the properties of all used materials were obtained directly from COMSOL's built-in material database. These included their absorption coefficients, complex refractive indices, dielectric constants, wavelength-dependent optical dispersion curves, as well as their $n(\lambda)$ and $k(\lambda)$. First, the structure features precisely engineered layer thicknesses: a 150 nm W (*h*1) substrate, followed by alternating functional and dielectric layers. The WSe_2 (*h*2), Ni (*h*4), and Mo (*h*6) layers are set at thicknesses

https://doi.org/10.3390/photonics13010086

of 28 nm, 5 nm, and 1100 nm, respectively, while the three SiO_2 layers (*h*3, *h*5, and *h*7) are designed with thicknesses of 50 nm, 180 nm, and 50 nm. To determine the optimal geometric configuration of the proposed optical absorber, the incident light was polarized normal to the top layer and directed along the negative z-axis.

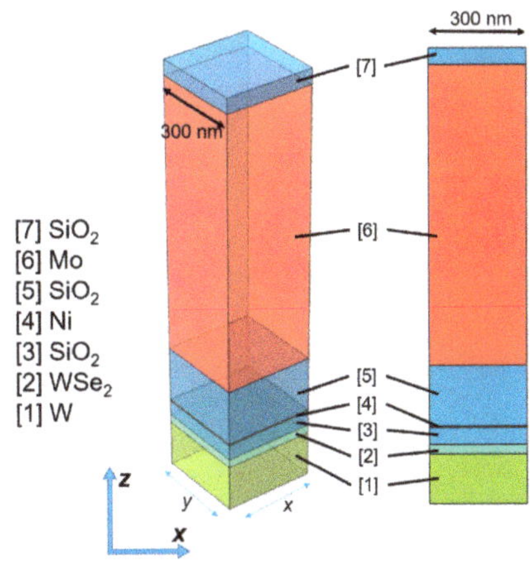

Figure 1. The structure of the investigated optical absorber.

The absorber was then modeled and optimized using COMSOL Multiphysics® (version 6.0), a finite element analysis platform capable of accurately capturing light-matter interactions. Simulation results demonstrated that the removal of the topmost SiO_2 layer led to a pronounced reduction in absorptivity across the broadband region (where absorptivity exceeds 0.900), highlighting its essential role in enhancing absorption. To refine the absorber's performance, a series of systematic parametric sweep analyses were carried out. In these analyses, each structural parameter was varied independently while the others were held constant, enabling precise evaluation of the influence of geometric variations on absorption efficiency and the corresponding electromagnetic response. The structural model of the proposed seven-layer metamaterial absorber was discretized using the finite element method to ensure accurate numerical simulation. The mesh consisted of 26,388 nodes, with element sizes ranging from 0.80 nm to 1.60 nm, and included 147,710 tetrahedral elements, 24,648 triangular elements, 1264 edge elements, and 56 endpoint elements.

For comparison, a preliminary two-dimensional model generated 1425 nodes, 2757 triangular elements, 347 edge elements, and 16 endpoint elements, achieving uniform coverage of the entire structure. Mesh quality was evaluated using skewness as the metric, yielding a minimum of 0.5553 and an average of 0.8168 in the 2D model, and a minimum of 0.1873 with an average of 0.6292 in the full 3D configuration. These results indicate that most elements possessed good geometric quality, with histograms showing that the majority of elements were concentrated in the high-quality range, effectively reducing the risk of distortion or irregularities. The minimum element area was 0.001622 nm^2, while the total mesh area and volume reached 1.887×10^5 nm^2 and 7.965×10^7 nm^3, respectively. Overall, the mesh provided an appropriate resolution and element distribution to ensure stability, convergence, and accuracy in subsequent simulations. All numerical simulations were performed using the finite element method implemented in COMSOL Multiphysics (version 6.0). The optical response of the proposed multilayer planar absorber was analyzed using the Wave Optics Module under frequency-domain formulation. A linearly polarized plane wave was normally incident onto the structure from the air side. Periodic (Floquet) boundary conditions were applied along the lateral directions to emulate an infinite planar array, while perfectly matched layers (PMLs) were employed along the

https://doi.org/10.3390/photonics13010086

propagation direction to eliminate artificial reflections and ensure accurate absorption evaluation. The governing Maxwell's equations were solved using the default direct solver with adaptive frequency stepping. Convergence was achieved when the relative residual satisfied the default tolerance criteria defined by COMSOL. Mesh independence was verified by comparing two-dimensional and full three-dimensional models, confirming that further mesh refinement produced negligible changes in the calculated absorptivity spectra. These settings ensure numerical stability and reliable prediction of the ultra-broadband absorption characteristics of the proposed structure. Similarly, the transmittance $T(\lambda)$, reflectance $R(\lambda)$, and optical impedance were calculated within the same framework.

3. Processes to Investigate the Optimal Parameter of Each Layer

3.1. Effects of the h2–h5 Layers Thicknesses on the Absorption Properties of the Investigated Absorbers

In the proposed absorber, W is employed as the bottom layer to act simultaneously as a reflective substrate and a transmission-blocking layer. Owing to its high optical loss and strong reflectivity over a wide spectral range, the W layer suppresses transmission and promotes energy dissipation within the upper functional layers, thereby establishing the boundary condition required for broadband absorption. The role of the *h*2 WSe_2 layer is primarily associated with interfacial loss enhancement and near-field coupling rather than direct spectral tuning. To verify this role, the thickness of the WSe_2 layer was varied from 5 to 55 nm, as shown in Figure 2a. The results indicate that, under fixed structural conditions, the absorber maintains absorptivity above 0.900 over most of the spectral range from approximately 420 to 7300 nm, with only minor sensitivity to the WSe_2 thickness. This weak dependence suggests that the WSe_2 layer does not dominate the determination of the cutoff wavelengths, but instead serves as an auxiliary lossy layer that stabilizes broadband absorption. A slight reduction in absorptivity near 4000 nm, as highlighted in the inset of Figure 2a, indicates that absorption in this region is governed by other layers with stronger wavelength selectivity, which will be discussed in relation to the Mo and top SiO_2 layers in subsequent sections.

As the thickness of the WSe_2 layer increases, the absorption spectrum exhibits a gradual redshift of the resonance dip accompanied by a slight reduction in absorptivity at the minimum point. These behaviors arise from increased optical path length and altered interference conditions within the multilayer stack, which modify the phase accumulation of the reflected and transmitted waves. However, the overall absorption response shows only weak sensitivity to the WSe_2 thickness, indicating that this layer does not play a dominant role in determining the cutoff wavelengths. From a physical perspective, the primary function of the WSe_2 layer is to introduce additional optical loss and enhance interfacial field coupling rather than to control spectral boundaries. Accordingly, a thickness of 5 nm was selected as a practical compromise that preserves high absorptivity at the resonance minimum while maintaining fabrication feasibility. The residual absorptivity reduction observed near 4000 nm is therefore attributed to wavelength-selective effects governed by other functional layers, and can be effectively compensated through coordinated tuning of the Mo and SiO_2 layers, which directly regulate the long- and short-wavelength cutoff positions.

https://doi.org/10.3390/photonics13010086

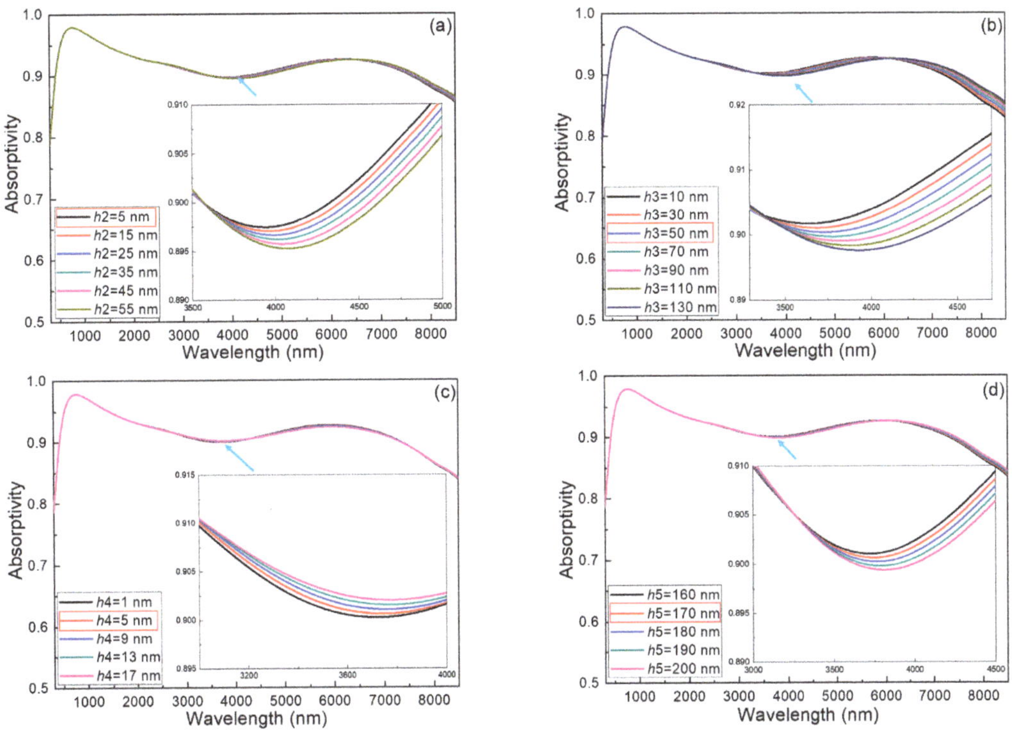

Figure 2. Effects of the thicknesses of (**a**) *h*2 WSe_2, (**b**) *h*3 SiO_2, (**c**) *h*4 Ni, and (**d**) *h*5 SiO_2 layers on the absorption spectra of the investigated absorbers.

The influence of the *h*3 SiO_2 layer thickness was examined by varying its thickness from 10 to 130 nm, as shown in Figure 2b. Similarly to the WSe_2 layer, the absorber maintains absorptivity above 0.900 over most of the spectral range (≈420–7500 nm), indicating that the *h*3 SiO_2 layer does not primarily determine the absorption bandwidth. As the SiO_2 thickness increases, the absorption dip exhibits a gradual redshift accompanied by a reduction in absorptivity at the minimum point. This behavior arises from the increased optical path length in the dielectric spacer, which modifies the interference conditions and weakens field confinement at resonance. In addition, a moderate blueshift of the long-wavelength cutoff is observed, reflecting altered phase-matching conditions at extended wavelengths as the dielectric thickness increases. These results indicate that the *h*3 SiO_2 layer mainly acts as an interference-tuning spacer rather than a cutoff-controlling layer. Accordingly, a thickness of 50 nm was selected to maintain absorptivity above 0.900 at the resonance dip while remaining compatible with practical fabrication considerations. Fine adjustment of the absorption spectrum is subsequently achieved through coordinated tuning of the Mo and top SiO_2 layers, which directly govern the long- and short-wavelength cutoff positions.

The effect of the *h*4 Ni layer thickness was examined by varying it from 1 to 5 nm, as shown in Figure 2c. The results indicate that changing the Ni thickness has a negligible influence on both the absorption valley position and the long-wavelength cutoff, confirming that the Ni layer does not govern the spectral boundaries. Instead, a slight increase in absorptivity at the valley is observed with increasing Ni thickness, which can be attributed to enhanced optical loss and improved impedance matching between the metallic Ni layer and the adjacent dielectric layers. Since all tested thicknesses maintain absorptivity above 0.900 across the ultra-broadband range, a thickness of 5 nm was se-

https://doi.org/10.3390/photonics13010086

lected as a practical compromise that ensures stable absorption while remaining compatible with reliable thin-film fabrication [25]. The influence of the *h*5 SiO_2 layer thickness was investigated over the range of 160–200 nm, as shown in Figure 2d. Variations in this layer primarily affect the absorptivity at the valley without inducing significant shifts in either the valley wavelength or the long-wavelength cutoff. This behavior indicates that the *h*5 SiO_2 layer mainly functions as an interference spacer that modulates local field distribution rather than controlling the absorption bandwidth. A thickness of 170 nm was therefore chosen to maintain absorptivity above 0.900 at the valley while supporting stable broadband absorption.

3.2. Effects of the h6-Mo and h7-SiO_2 Thicknesses on the Absorption Properties of the Investigated Absorbers

Further analysis reveals that the *h*6 Mo and *h*7 SiO_2 layers form a strongly coupled pair that jointly governs the absorption bandwidth. With the Mo thickness fixed at 700 nm, varying the top SiO_2 thickness primarily shifts the short-wavelength cutoff, while only moderately influencing the long-wavelength edge, as shown in Figure 3a. This behavior indicates that the *h*7 SiO_2 layer plays a dominant role in regulating short-wavelength absorption through interference-controlled phase accumulation, which is highly sensitive to dielectric thickness. From a physical perspective, increasing the *h*7 SiO_2 thickness enlarges the effective optical path length of the upper cavity, thereby modifying the interference condition and redshifting the short-wavelength cutoff. In contrast, the long-wavelength cutoff exhibits weaker sensitivity because it is mainly governed by loss and field penetration in the Mo layer. The improvement in absorptivity around the mid-infrared region further reflects enhanced field confinement within the cavity formed by the Mo and dielectric layers. However, excessively increasing the *h*7 SiO_2 thickness shifts the short-wavelength cutoff beyond the near-ultraviolet region, limiting broadband energy harvesting. Consequently, an *h*7 SiO_2 thickness of 40 nm was selected for a Mo thickness of 700 nm, as this configuration provides the widest high-absorptivity (>0.900) bandwidth while maintaining coverage from the near-UV to the mid-infrared.

Increasing the *h*6 Mo layer thickness from 700 to 900 nm significantly alters the absorption characteristics, as shown in Figure 3b, primarily by extending the long-wavelength cutoff and shifting the absorption dip toward longer wavelengths. These changes originate from the increased effective cavity length and enhanced phase accumulation within the lossy Mo layer, which modify the resonance conditions and redistribute the electromagnetic field. As a result, the long-wavelength absorption edge is pushed further into the mid-infrared, while field confinement at the dip is weakened, leading to a reduction in absorptivity at that wavelength. With the Mo thickness fixed at 900 nm, variation in the top *h*7 SiO_2 layer mainly affects the short-wavelength cutoff, confirming its dominant role in regulating short-wavelength absorption through interference-controlled phase accumulation. Excessively thin or thick SiO_2 layers either reduce absorptivity at the resonance dip or shift the short-wavelength cutoff beyond the near-ultraviolet region, thereby limiting broadband coverage. These results further demonstrate that the absorption bandwidth is governed by the coupled interaction between the Mo and top SiO_2 layers: the Mo layer primarily controls the long-wavelength response through loss and field penetration, while the top SiO_2 layer adjusts the short-wavelength cutoff via interference effects. Based on this coupled mechanism, an *h*6 Mo thickness of 900 nm combined with an *h*7 SiO_2 thickness of 50 nm was selected, providing a broad high-absorptivity (>0.900) bandwidth spanning approximately 360–5550 nm.

https://doi.org/10.3390/photonics13010086

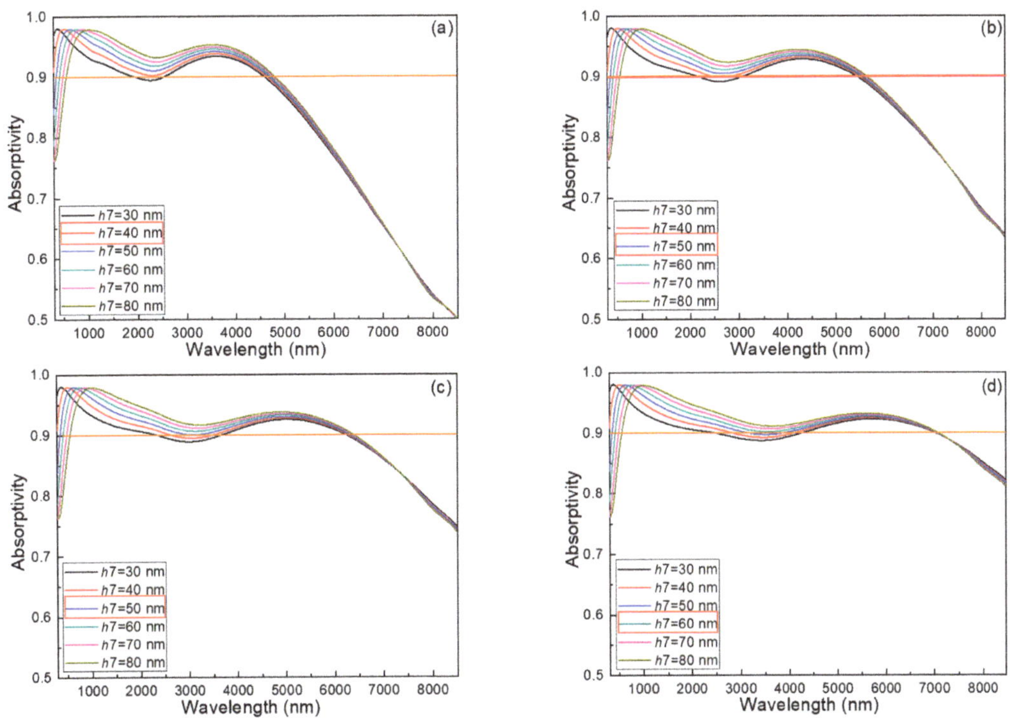

Figure 3. Effects of the different thicknesses of *h*6-Mo and *h*7-SiO_2 layers on the absorption spectra of the investigated absorbers. (**a**) *h*6 = 700 nm, (**b**) *h*6 = 900 nm, (**c**) *h*6 = 1100 nm, and (**d**) *h*6 = 1300 nm.

Finally, we further increased the thickness of the *h*6 Mo layer to 1100 nm and 1300 nm, with the corresponding results shown in Figure 3c,d. Under these two conditions, the spectral response exhibits even more pronounced redshifts. Specifically, the long-wavelength cutoff shifts further, extending from ~6200–6400 nm to ~6900–7000 nm, while the absorption dip also redshifts more significantly, moving to ~3000 nm and ~3950 nm. At the same time, the absorptivity at these dips decreases more noticeably. As shown in Figure 3c, when the *h*7 SiO_2 layer thickness is 30 nm or 40 nm, the absorptivity at the ~3000 nm dip falls below 0.900. Similarly, Figure 3d reveals that with *h*7 SiO_2 thicknesses of 30–50 nm, the absorptivity at the ~3950 nm dip also drops below 0.900. Therefore, when the *h*6 Mo layer thickness is set to 1100 nm, the optimal condition is obtained with *h*7 SiO_2 at 50 nm, yielding a broad high-efficiency absorption bandwidth spanning from ~360 nm to ~6300 nm while maintaining absorptivity above 0.900. When the Mo thickness is increased to 1300 nm, the best configuration occurs at *h*7 = 60 nm, where the bandwidth extends from ~430 nm to ~7040 nm.

To further confirm that the h6 Mo layer plays the most critical role in determining the bandwidth and absorption characteristics of the designed optical absorber, we fixed the other parameters at *h*1 = 150 nm, *h*2 = 5 nm, *h*3 = 50 nm, *h*4 = 5 nm, *h*5 = 170 nm, and *h*7 = 50 nm, while varying only the *h*6 Mo thickness from 300 nm to 1500 nm. The results, shown in Figure 4a, reveal that the short-wavelength cutoff remains nearly unchanged across this range. This observation, when combined with the results of Figure 3, clearly demonstrates that the redshift of the short-wavelength cutoff originates primarily from variations in the *h*7 SiO_2 layer rather than the Mo layer. In contrast, the long-wavelength cutoff increases rapidly with Mo thickness, extending from ~3030 nm to ~7560 nm as

https://doi.org/10.3390/photonics13010086

$h6$ increases from 300 nm to 1500 nm. Similarly, the absorption dips shift progressively toward longer wavelengths as the Mo thickness increases. However, when $h6$ reaches 1300 nm or 1500 nm, the absorptivity at the dip falls below 0.900, which makes these conditions unsuitable for our final design. Figure 4b provides a comparative overview of the relatively optimal absorption spectra selected from the analyses in Figure 3. These include four representative cases: (1) $h6$ = 700 nm and $h7$ = 40 nm, (2) $h6$ = 900 nm and $h7$ = 50 nm, (3) $h6$ = 1100 nm and $h7$ = 50 nm, and (4) $h6$ = 1300 nm and $h7$ = 60 nm.

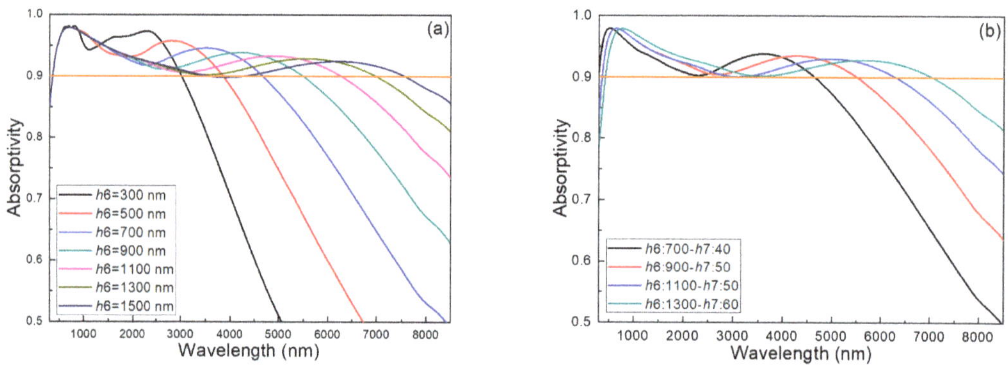

Figure 4. Effects of (**a**) the thickness of $h6$-Mo layer and (**b**) different thickness of $h6$-Mo and $h7$-SiO_2 layers on the absorption spectra of the investigated absorbers.

It is evident that when the SiO_2 spacer thickness $h7$ is kept constant, the short-wavelength cutoff remains nearly identical, further confirming that $h7$ governs the short-wavelength response, whereas $h6$ primarily dictates the long-wavelength cutoff and the position of absorption dips. Since our design objective is to achieve broadband high-efficiency absorption extending into the near-UV region, we ultimately selected the configuration with $h6$ = 1100 nm and $h7$ = 50 nm as the optimal condition. Taken together with the results in Figure 3, these findings highlight a fundamental design principle: the Mo back-reflector thickness ($h6$) predominantly determines the long-wavelength cutoff and absorption dip positions through its influence on optical path length and cavity resonances, while the SiO_2 dielectric layer ($h7$) governs the short-wavelength cutoff. Excessive Mo thickness enhances phase accumulation and redshifts resonances but also reduces absorptivity at the dips due to weakened field confinement. By jointly tuning $h6$ and $h7$, the absorber's operational bandwidth can be flexibly engineered, as shown in Figures 3 and 4, thereby offering a robust strategy to balance spectral coverage and absorption efficiency across the near-UV to mid-IR range.

To further demonstrate the crucial role of the topmost SiO_2 layer ($h7$) in determining the absorption characteristics of the proposed absorber, we performed a comparative study without including this seventh layer. Specifically, the structure was designed with $h1$ = 150 nm (W), $h2$ = 5 nm (WSe_2), $h3$ = 50 nm (SiO_2), $h4$ = 5 nm (Ni), and $h5$ = 170 nm (SiO_2), while the Mo layer ($h6$) thickness was varied from 100 nm to 2100 nm to analyze its effect on absorptivity. As shown in Figure 5, only for $h6$ = 100 nm and 300 nm were narrow absorption windows with absorptivity exceeding 0.900 observed, specifically within 860–1455 nm and 1410–2205 nm, respectively. For other Mo thicknesses, the absorptivity either remained slightly above 0.900 or dropped below this value across the entire spectrum. Moreover, it is evident that increasing the Mo thickness results in a pronounced redshift of the cutoff wavelength at the longer-wavelength end. These results highlight the essential function of the topmost SiO_2 layer. Without this seventh layer, impedance mismatch between the absorber and free space becomes more severe, causing stronger reflection and

https://doi.org/10.3390/photonics13010086

significantly reducing the broadband absorptivity. In contrast, the inclusion of the top SiO_2 layer enables more relative effective impedance matching, thereby suppressing reflection and ensuring high absorptivity across the wide wavelength range. This explains why the complete seven-layer structure exhibits enhanced absorption performance compared with the truncated six-layer configuration, which shows a noticeable reduction in absorptivity.

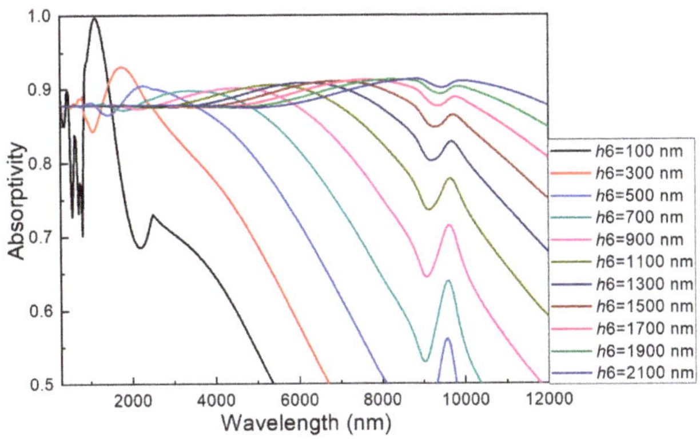

Figure 5. Effect of the thickness of Mo layer on the absorption spectra of the six-layer (without topmost SiO_2 layer) absorber.

These results clearly indicate that once the optimal parameters for the *h*2–*h*6 layers are determined, the absorption band and bandwidth of the proposed optical absorber can be flexibly tuned by adjusting the thicknesses of the *h*6-Mo layer and the *h*7-SiO_2 layer. As shown in Figure 5, when the Mo layer thickness is increased to 2100 nm, the absorption spectrum extends dramatically toward longer wavelengths, reaching into the far-infrared region up to 10,000 nm (10 μm). However, despite this broadband extension, the absorptivity across most of the spectrum remains below 0.900. This observation is consistent with the results in Figures 3 and 4, which revealed that thicker Mo layers reduce absorptivity at intermediate absorption dips. Restoring absorptivity above 0.900 would require increasing the thickness of the SiO_2 spacer, but such an adjustment would simultaneously redshift the short-wavelength cutoff beyond 400 nm, thereby excluding the visible region. This trade-off makes the configuration unsuitable for solar absorber applications, where efficient absorption across the visible spectrum is essential. Nevertheless, the findings highlight the tunability of the design: by leveraging the same principle, the absorber could be further developed in the future to achieve ultra-broadband absorption across the entire infrared spectrum, from near-IR to far-IR.

3.3. Other Relative Properties of the Investigated Absorbers

To further investigate the impedance matching and absorption characteristics of the proposed design, we analyzed the condition where *h*6 = 1100 nm and *h*7 = 50 nm, as identified in Figure 4. Figure 6a illustrates the wavelength-dependent impedance of the absorber. As shown, when the wavelength falls below ~360 nm, the real part of the impedance deviates from unity while the imaginary part departs from zero, indicating a mismatch with free-space impedance. A similar mismatch occurs for wavelengths longer than ~6300 nm. These deviations correspond directly to reduced absorptivity at the spectral edges, since imperfect impedance matching increases reflection losses and prevents efficient coupling of incident light into the absorber. The absorption and impedance characteristics were obtained through numerical simulations using COMSOL Multiphysics® (version 6.0) with built-in material parameters. The electromagnetic wave frequency-domain interface

https://doi.org/10.3390/photonics13010086

was employed to solve Maxwell's equations and model the interaction of light with the multilayer absorber. Absorptivity was calculated according to $A(\lambda) = 1 - R(\lambda) - T(\lambda)$, where $R(\lambda)$ and $T(\lambda)$ denote reflectance and transmittance, respectively. As shown in Figure 6b, the simulated results confirm that $T(\lambda)$ is nearly zero across the analyzed spectrum, while absorptivity remains consistently above 0.900 and with an average absorptivity of 0.925 in the 360–6300 nm range. The reduction in absorptivity outside this range is primarily attributed to increased reflectance (>0.100), which originates from impedance mismatch at the short- and long-wavelength cutoffs.

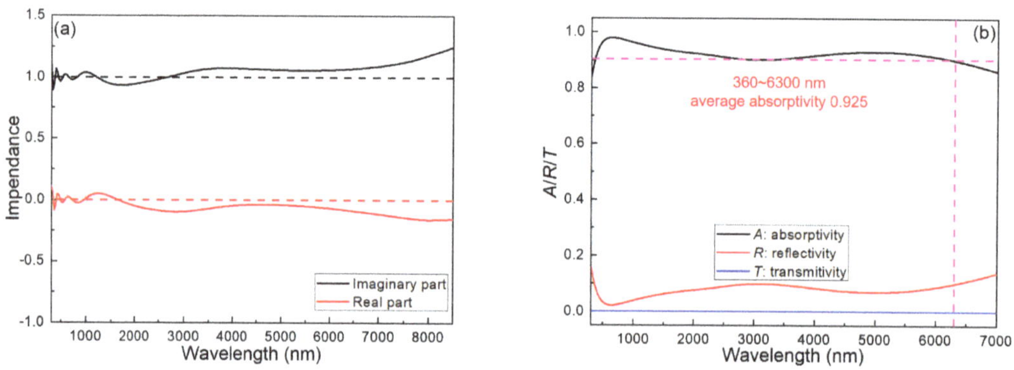

Figure 6. (**a**) The optical impedance and (**b**) the variations in the absorption, reflection, and transmission spectra of the investigated absorber with the optimal structure parameters.

Furthermore, the average absorptivity across the 360–6300 nm range was calculated to be 0.925, which is remarkably high for a broadband absorber composed of only seven planar thin-film layers. Achieving broadband absorption over an ultra-wide spectral range with a seven-layer planar architecture highlights the effectiveness of the proposed design. The underlying absorption mechanism can be explained by impedance matching theory. The relative effective impedance of the absorber is expressed as:

$$Z = \sqrt{\frac{(1 + S_{11})^2 - S_{21}^2}{(1 + S_{11})^2 + S_{21}^2}} \tag{1}$$

where S_{11} and S_{21} are the reflection and transmission coefficients, respectively [26,27]. The absorptivity is then given by:

$$A = 1 - |S_{11}|^2 - |S_{21}|^2 \tag{2}$$

When the structure is engineered so that $\mathrm{Re}(Z) \approx 1$ and $\mathrm{Im}(Z) \approx 0$, the relative effective impedance of the absorber matches that of free space. This eliminates impedance mismatch, minimizes reflection, and allows near-perfect absorption of the incident electromagnetic waves. For example, Figure 6a demonstrates that the absorber achieves $\mathrm{Re}(Z) \approx 1$ and $\mathrm{Im}(Z) \approx 0$ across the 360–6300 nm range, resulting in absorptivity above 0.900 throughout this band. This performance underscores the critical role of impedance matching in achieving ultrabroadband absorption. Within the matched region, the absorber minimizes reflections and effectively traps incident light through multiple internal reflections and resonance effects within the Mo-SiO_2-WSe_2 cavity. Outside this region, the impedance mismatch weakens field confinement, allowing a portion of the energy to be reflected rather than absorbed. The combination of broadband absorption, high average absorptivity, and a relatively seven-layer planar architecture demonstrates the effectiveness of

https://doi.org/10.3390/photonics13010086

the proposed design and highlights its potential for practical applications requiring wide spectral coverage from the near-UV to the mid-IR.

We further investigated the angular and polarization-dependent absorption behavior of the proposed seven-layer planar absorber over a broad spectral range extending from the near-UV (300 nm) to the mid-infrared (9000 nm). The analysis considered incident angles between 0° and 90° under both transverse electric (TE) polarization, as displayed in Figure 7a, and transverse magnetic (TM) polarization, as displayed in Figure 7b. As shown in Figure 6, the absorber exhibits distinct responses for TE and TM modes, reflecting the polarization-dependent interaction of incident light with the layered structure. The results indicate that absorptivity gradually decreases as the incidence angle increases, which can be attributed to the growing impedance mismatch and enhanced reflection at oblique angles. For TE polarization, the reduction in absorptivity is most pronounced in the 300–1800 nm range, whereas for TM polarization it extends up to approximately 3000 nm. Nevertheless, in the critical working band of 300–6300 nm, the structure maintains absorptivity above 0.900 for a wide range of incidence angles, particularly up to ~40° for TE and ~30° for TM polarization. Beyond these angles, reflection becomes dominant, resulting in reduced absorption.

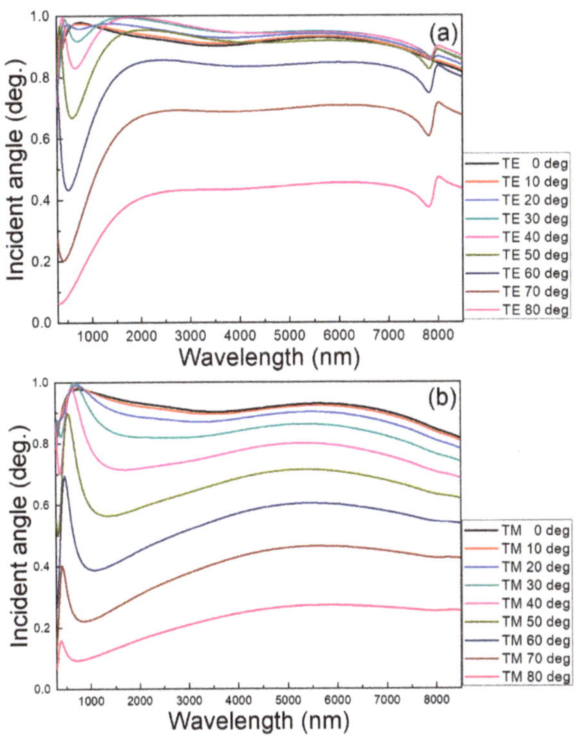

Figure 7. Angle-dependent absorption spectra of the proposed ultra-broadband absorber for (**a**) TE and (**b**) TM polarizations.

This behavior can be understood through impedance matching principles: when the relative effective impedance of the absorber approaches that of free space ($Re(Z) \approx 1$ and $Im(Z) \approx 0$), reflection is minimized and strong absorption occurs. At larger incident angles, however, the deviation of relative effective impedance from free-space values increases, leading to reduced absorptivity. Importantly, the absorber demonstrates excellent angular tolerance within practical operating conditions, as most solar radiation is received at

https://doi.org/10.3390/photonics13010086

moderate incidence angles. Overall, these findings highlight the robustness of the proposed ultra-broadband absorber. Despite its seven-layer planar configuration, the proposed absorber achieves high absorptivity over a broad spectral range and maintains stable absorption performance over a moderate range of incident angles. While the angular tolerance is not exceptional, this level of angular robustness remains beneficial for practical solar energy harvesting scenarios, where variations in the angle of incoming radiation are inevitable.

The average absorptivity was calculated using the equation of $A = \int_{\lambda 2}^{\lambda 1} A(\lambda)d\lambda/(\lambda 1 - \lambda 2)$, where the used $\lambda 1$ value was 300 nm and $\lambda 2$ values were 3000 and 4000 nm. Based on the simulation results of the proposed absorber, the average absorptivity reaches 0.927 over the wavelength range of 300–4000 nm and increases to 0.935 when the range is limited to 300–3000 nm. The absorption spectrum exhibits a prominent peak near 630 nm, where the absorptivity attains a maximum value of approximately 0.979. To evaluate the solar-energy harvesting potential of the proposed ultra-wideband absorber, its spectral absorption under the AM1.5G solar spectrum (300–4000 nm) was analyzed, as shown in Figure 8. The absorber maintains high absorptivity across the entire spectral range, with particularly stable performance in the near-infrared region, where the absorptivity remains close to or above 0.900. When the spectrum is divided into representative wavelength intervals-300–400 nm, 400–700 nm, 700–1400 nm, and 1400–4000 nm-the corresponding absorptivities are 0.918, 0.941, 0.948, and 0.909, respectively. These results indicate that the absorber exhibits consistently strong coupling to solar radiation across the ultraviolet, visible, and infrared regions. The slightly reduced absorptivity at longer wavelengths can be attributed to increased penetration depth and loss redistribution in the multilayer structure, while the high absorption in the visible and near-infrared ranges aligns well with the spectral intensity distribution of the AM1.5G spectrum. Overall, the absorber achieves an average solar absorptivity of 0.937 over 300–4000 nm, suggesting that the proposed design provides an effective and spectrally balanced approach for broadband solar energy absorption.

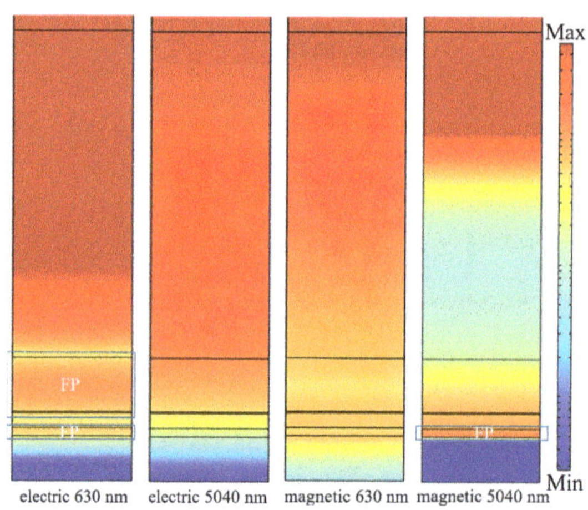

Figure 8. Simulation results of the electric and magnetic field intensity patterns of the designed ultra-wideband absorber when illuminated at normal incidence under various TE-polarized wavelengths.

The ultra-wideband optical absorber presented in this study exhibits strong light absorption over a broad spectral range spanning from 360 nm (near-ultraviolet) to 6300 nm (mid-infrared). This device has been meticulously designed and optimized to maintain consistently high absorption efficiency throughout this extensive wavelength range. To

https://doi.org/10.3390/photonics13010086

elucidate the physical mechanisms underlying the observed absorption behavior, Figure 8 presents the simulated electromagnetic field distributions at two representative wavelengths, 630 nm and 5040 nm, as shown in Figure 6b. The placement of the WSe_2 and Mo layers is critical to the ultra-broadband absorption behavior of the proposed structure. The ultrathin WSe_2 layer is positioned adjacent to the reflective W substrate to enhance interfacial optical loss and near-field coupling, which predominantly contributes to absorption in the short-wavelength region. In contrast, the thick Mo layer functions as a lossy plasmonic medium that dominates absorption at longer wavelengths. Notably, the absorption bandwidth is not governed by the Mo layer alone, but by its coupled interaction with the top SiO_2 layer. This thickness-dependent interplay facilitates relative effective impedance matching and Fabry–Pérot–assisted field confinement, thereby jointly defining the spectral range and bandwidth over which the absorptivity exceeds 0.900. As a result, systematic tuning of the Mo and top SiO_2 layer thicknesses enables flexible extension of both the short- and long-wavelength cutoff edges.

At the shorter wavelength of 630 nm, the electric field distribution indicates that the region of high absorptivity extends from the top SiO_2 layer down through nearly the entire Mo layer. In this spectral region, the *h*4 Ni layer and the *h*6 Mo layer, separated by the *h*5 SiO_2 layer, act as reflective interfaces that form a Fabry–Pérot–type cavity, leading to resonance-assisted absorption. However, this absorption is not extremely strong, as it is mainly governed by moderate Fabry–Pérot resonance arising from constructive interference within the dielectric cavity. In addition to these primary mechanisms, a moderate Fabry–Pérot–type cavity effect is also observed within the *h*3 SiO_2 layer. As evidenced by the electric field distribution at 630 nm and the magnetic field distribution at 5040 nm, partial field enhancement occurs inside this dielectric layer, indicating auxiliary optical confinement rather than a dominant resonance. This cavity-assisted effect helps redistribute electromagnetic energy across different wavelength regions and contributes to stabilizing the broadband absorption response, while the main bandwidth modulation remains governed by the coupled interaction between the Mo layer and the top SiO_2 layer.

At a wavelength of 630 nm, the electric field distribution exhibits a gradual decay from the top *h*7 SiO_2 layer toward the underlying metallic layers, extending through most of the *h*6 Mo layer. No pronounced field localization at subwavelength metallic features or clear propagation along a single extended interface is observed, indicating that neither localized surface plasmon resonance nor propagating surface plasmon resonance (PSPR) plays a dominant role at this wavelength. Instead, a weak but noticeable field enhancement appears around the *h*4 Ni layer, where the Ni and Mo layers, separated by the *h*5 SiO_2 spacer, act as partially reflective interfaces. This configuration gives rise to a moderate Fabry–Pérot–type cavity effect, which contributes to resonance-assisted absorption but does not dominate the overall absorption mechanism. The magnetic field distribution at 630 nm shows a similar gradual attenuation from the top SiO_2 layer toward the bottom W substrate, while maintaining a non-negligible field intensity throughout the Mo layer. This behavior suggests distributed energy dissipation within the lossy metallic layers rather than strong magnetic confinement associated with a well-defined resonant cavity.

At the longer wavelength of 5040 nm, both electric and magnetic field distributions exhibit a different characteristic behavior. The electric field intensity again decays progressively from the top SiO_2 layer into the Mo layer, consistent with delocalized field penetration rather than localized confinement. Meanwhile, the magnetic field distribution reveals a distinct enhancement within the *h*3 SiO_2 layer, indicating the presence of a Fabry–Pérot–type cavity mode formed by reflections between adjacent metallic layers. In this wavelength regime, extended metal–dielectric interfaces, particularly those involving the Mo and W layers, support PSPR, which facilitates efficient coupling of incident radiation

https://doi.org/10.3390/photonics13010086

into lossy plasmonic modes. Overall, the field profiles indicate that absorption at shorter wavelengths is primarily governed by distributed loss and moderate Fabry–Pérot interference, while at longer wavelengths, PSPR-driven energy coupling, assisted by auxiliary Fabry–Pérot resonance in the dielectric layers, becomes more prominent. Localized surface plasmon resonance does not play a dominant role in either spectral region, as no strong nanoscale field localization is observed.

3.4. Comparison with Other Research Results and Potential Applications

To underscore the advantages of this structure, Table 1 provides a performance comparison with several recently reported ultra-wideband absorbers. As summarized in Table 1, although some previously reported optical absorbers can extend further into the deep-ultraviolet or achieve slightly higher average absorptivity across limited spectral ranges [28–32], attaining ultra-broadband performance often requires complex nanostructures, as noted in the table. Even with such intricate designs, their overall absorption bandwidths remain narrower than that of our proposed absorber. The proposed absorber, constructed from only seven fully planar thin-film layers, covers an unprecedented spectral range of 360–6300 nm, bridging the near-ultraviolet to the mid-infrared region. Such wide coverage is rarely attainable without resorting to highly complex nanostructures or multilayer stacks. Even when compared with absorbers specifically optimized for the infrared domain, such as one spanning 2000–8410 nm with an average absorptivity of 0.850 [7], our design demonstrates clear advantages.

Table 1. Comparison of the absorption characteristics of the investigated ultra-wideband absorber with recent literatures. The bandwidth corresponds to the spectral range with absorptivity exceeding 0.900.

	Bandwidth	Average Absorptivity	The Used Technology in Each Study
Ref. [28]	380–2150 nm	~0.962	In this design, surface plasmon modes are excited by placing a periodic Ti-SiO_2 composite cap, either circular or square in shape, on top of a metallic substrate. This study examines the visible-to-near-infrared absorption characteristics of a plasmonic metamaterial absorber based on a metal–dielectric–metal (MDM) architecture. Surface plasmon resonances are excited by introducing a periodic metal–dielectric cap above a metallic substrate. The absorption bandwidth is strongly influenced by both the cap geometry and the presence of a glass overlayer.
Ref. [29]	515–1945 nm	0.924	An ultra-wideband tungsten-based solar absorber is numerically investigated using the finite-difference time-domain method. The absorber features a cross-shaped resonator placed above a metallic back reflector, separated by a dielectric spacer. Numerical results show sustained high absorptivity exceeding 90% over the 514–1945 nm wavelength range, with an average absorptivity of 92.36%. This broadband absorption originates from the combined excitation of propagating and localized surface plasmon resonances, together with Fabry–Pérot cavity effects.

https://doi.org/10.3390/photonics13010086

Table 1. *Cont.*

	Bandwidth	Average Absorptivity	The Used Technology in Each Study
Ref. [30]	300–2400 nm	0.957	This study reports an ultra-broadband metamaterial absorber operating over the 250–4000 nm wavelength range. The proposed design integrates a Ti disk resonator with stacked TiO_2/Ti square resonators supported by TiO_2/Ti thin films, forming an effective metal–insulator–metal architecture that enhances optical absorption. The influence of key geometric parameters-including resonator thickness, disk radius, and square width-on the absorption response is systematically examined. To maximize the average absorptivity, particle swarm optimization is applied to identify the optimal structural configuration.
Ref. [31]	350–2670 nm 300–3600 nm	~0.900 ~0.980	Initially, the authors proposed a planar multilayer thin film absorber composed of Cu/Zr-ZrO_2/Zr-ZrO_2/Zr-ZrO_2/Al_2O_3. This work presents the design and experimental demonstration of a hybrid metamaterial absorber based on tapered multilayer cermet thin films. By integrating optical interference within the multi-cermet stack and strong electromagnetic field localization induced by nano-cone structures, the absorber achieves an average absorbance exceeding 98% over an ultra-broad spectral range from 300 to 3000 nm, in good agreement between simulation and experimental results.
Ref. [32]	180–2750 nm	0.981	This article presents a simple-structured metamaterial solar absorber capable of achieving ultra-broadband near-perfect absorption. The proposed design adopts a four-layer SiO_2-TiN-SiO_2-TiN configuration, consisting of a TiN substrate, an intermediate SiO_2 dielectric layer, and a top array of TiN double-ring resonators. The inner ring is capped with a SiO_2 layer and extends higher than the outer ring, enabling enhanced electromagnetic coupling. The near-perfect absorption performance originates from the synergistic interaction of surface plasmon resonance, cavity resonance, and magnetic resonance, as evidenced by the electric and magnetic field distributions.
Ref. [7]	2000–8410 nm	0.850	The authors designed and experimentally validated an ultra-broadband mid-infrared absorber based on a periodic Ge/Cr multilayer structure capped with MgF_2 and ZnS coatings. Both simulations and measurements confirm that the absorber exhibits robust absorption performance over a wide range of polarizations and incident angles. The observed ultra-broadband absorption originates from the combined effects of photonic topological transitions and enhanced anti-reflection behavior.
This study	360–6300 nm	0.925	The proposed structure consists of seven stacked planar layers arranged from bottom to top: W ($h1$, acting as a reflective substrate and transmission blocker), WSe_2 ($h2$), SiO_2 ($h3$), Ni ($h4$), SiO_2 ($h5$), Mo ($h6$), and SiO_2 ($h7$). These results indicate that coordinated tuning of the Mo and SiO_2 layers provides effective control over both the short- and long-wavelength cutoff positions as well as the overall absorption bandwidth. Rather than relying on complex nanostructuring or noble-metal components, the proposed seven-layer architecture realizes ultra-broadband absorption using a fully planar, scalable multilayer thin-film configuration.

https://doi.org/10.3390/photonics13010086

Beyond its absorption capability, a key characteristic of the proposed absorber is its tunability. By varying the thickness of selected layers, both the short-wavelength onset and long-wavelength cutoff can be precisely controlled, effectively tailoring the absorption bandwidth to meet different application requirements. As illustrated in Figure 3, this flexibility allows the absorption onset to be shifted to below 300 nm, offering a rare combination of ultra-broad coverage, high absorptivity, and design simplicity. At the same time, the broadband absorption performance is highly advantageous for solar energy harvesting, where maximizing energy conversion efficiency across a wide spectrum is critical. The unique combination of structural simplicity, tunable spectral characteristics, and material scalability makes this absorber a promising candidate for diverse optoelectronic and energy-related applications.

4. Conclusions

In this study, after fixing the thickness of the bottom W substrate ($h1$) at 150 nm, the optimal values for the intermediate layers were determined as follows: $h2$ WSe_2 layer at 5 nm, $h3$ SiO_2 layer at 50 nm, $h4$ Ni layer at 5 nm, and $h5$ SiO_2 layer at 170 nm. However, the thicknesses of the $h6$ Mo layer and the $h7$ SiO_2 layer were found to be strongly interdependent, jointly defining the spectral regions and bandwidths where absorptivity remains above 0.900. Specifically, when the Mo layer thickness was set to 700 nm, the optimal thickness of the top SiO_2 layer was 40 nm, yielding a high-absorptivity range from 300 nm to 4650 nm. Increasing the Mo layer thickness to 900 nm, 1100 nm, and 1300 nm, with corresponding optimal SiO_2 layer thicknesses of 50 nm, 50 nm, and 60 nm, and extended the high-absorptivity ranges to 360–5550 nm, 360–6300 nm, and 430–7050 nm, respectively. These findings clearly demonstrate that the synergistic control of the Mo and SiO_2 layers enables precise adjustment of both cutoff wavelengths and absorption bandwidth. Unlike many previously reported absorbers that rely on complex nanostructures or costly noble metals, the proposed seven-layer fully planar design achieves ultra-broadband absorption through a scalable multilayer thin-film structure. The absorber maintains an absorptivity above 0.900 over a broad spectral range, extending from the near-ultraviolet through the visible and into the mid-infrared, with tunable characteristics that allow adjustment for different application conditions. This wavelength- and bandwidth-tunable absorber not only provides a promising pathway for high-efficiency solar energy harvesting but also shows great potential for optoelectronic and photonic devices operating across diverse regions of the electromagnetic spectrum.

Author Contributions: Conceptualization, K.-P.M., Y.-T.G., C.-F.Y., W.W. and C.-T.H.; methodology, K.-P.M., Y.-T.G. and C.-F.Y.; validation, K.-P.M., Y.-T.G., C.-F.Y., W.W. and C.-T.H.; formal analysis, K.-P.M., Y.-T.G., C.-F.Y., W.W. and C.-T.H.; investigation, K.-P.M., Y.-T.G. and C.-F.Y.; data curation, K.-P.M., Y.-T.G., C.-F.Y., W.W. and C.-T.H.; writing—original draft preparation, K.-P.M., Y.-T.G., C.-F.Y., W.W. and C.-T.H.; writing—review and editing, K.-P.M. and C.-F.Y.; visualization, K.-P.M., Y.-T.G., C.-F.Y., W.W. and C.-T.H. All authors have read and agreed to the published version of the manuscript.

Funding: The works are supported by Summit-Tech Resource Corp., and by projects under Nos. NSTC 113-2622-E-390-001 and NSTC 113-2221-E-390-011.

Institutional Review Board Statement: Not applicable.

Informed Consent Statement: Not applicable.

Data Availability Statement: The original contributions presented in this study are included in the article. Further inquiries can be directed to the corresponding author.

Acknowledgments: We would like to thank Pitotech Co. LTD. for their help in teaching the use of COMSOL Multiphysics® software.

https://doi.org/10.3390/photonics13010086

Conflicts of Interest: The authors declare no conflicts of interest.

References

1. Mehrabi, S.; Rezaei, M.H.; Zarifkar, A. Ultra-broadband metamaterial absorber based on cross-shaped TiN resonators. *J. Opt. Soc. Am. A* **2020**, *37*, 697–704. [CrossRef]
2. Ahmad, M.; Alam, T.; Islam, M.T.; Hakim, M.L.; Rmili, H.; Alshammari, A.S.; Islam, M.S.; Soliman, M.S. Broadband plasmonic metamaterial optical absorber for the visible to near-infrared region. *Nanomaterials* **2023**, *13*, 626. [CrossRef]
3. Hung, C.C.; Lin, X.Y.; Wu, T.L.; Liao, S.H.; Chen, H.S.; Yang, C.F. Investigation of an ultra-wideband optical absorber with the bandwidth from ultraviolet C to middle infrared. *Photonics* **2025**, *12*, 83. [CrossRef]
4. Jiang, Y.; Deng, J.; Lu, Y.; Xie, Z.; Chen, Y. Large-scale, transferable and double-sided metasurface absorber for the visible to near-infrared spectrum. *Surf. Interfaces* **2025**, *56*, 105708. [CrossRef]
5. Lin, J.J.; Tseng, L.C.; Yang, C.F. Analysis and comparison of multilayer planar structure ultra–wideband absorbers for visible to mid–infrared light using different simulation software. *Appl. Funct. Mater.* **2021**, *3*, 1–8.
6. Sakotic, Z.; Raju, A.; Ware, A.; Estévez, F.A.H.; Brown, M.; Behar, Y.M.; Hungund, D.; Wasserman, D. Mid-infrared perfect absorption with planar and subwavelength-perforated ultrathin metal films. *Adv. Phys. Res.* **2024**, *3*, 2400012. [CrossRef]
7. Zhang, H.; Chen, X.; Dong, S.; Ye, B.; Yang, R.; Liao, Y.-L.; Chen, Z.; Zhao, Y. Ultra-broadband mid-infrared absorption based on photonic topological transition and anti-reflection effect. *Opt. Laser Technol.* **2025**, *185*, 112607. [CrossRef]
8. Acharyya, J.N.; Desai, N.R.; Gangineni, R.B.; Prakash, G.V. Effect of photonic cavity interactions on femtosecond multiphoton optical nonlinear absorptions from Bi_2O_3-based one-dimensional photonic crystal. *ACS Photonics* **2022**, *9*, 2092–2100. [CrossRef]
9. Wu, F.; Wu, X.; Xiao, S.; Liu, G.; Li, H. Broadband wide-angle multilayer absorber based on a broadband omnidirectional optical Tamm state. *Opt. Express* **2021**, *29*, 23976–23986. [CrossRef] [PubMed]
10. Mokhtari, A.; Rezaei, M.H.; Zarifkar, A. Near-perfect wide-band absorbers based on one-dimensional photonic crystal structures in 1–20 THz frequencies. *Appl. Opt.* **2023**, *62*, 3660–3671. [CrossRef]
11. Xue, C.-H.; Wu, F.; Jiang, H.-T.; Li, Y.; Zhang, Y.-W.; Chen, H. Wide-angle spectrally selective perfect absorber by utilizing dispersionless Tamm plasmon polaritons. *Sci. Rep.* **2016**, *6*, 39418. [CrossRef] [PubMed]
12. Huang, M.; Cheng, Y.; Cheng, Z.; Chen, H.; Mao, X.; Gong, R. Graphene-based tunable dual-band terahertz metamaterial absorber with wide-angle performance. *Opt. Commun.* **2018**, *415*, 194–201. [CrossRef]
13. You, X.; Upadhyay, A.; Cheng, Y.; Bhaskaran, M.; Sriram, S.; Fumeaux, C.; Withayachumnankul, W. Ultra-wideband far-infrared absorber based on anisotropically etched doped silicon. *Opt. Express* **2019**, *27*, 14841–14853. [CrossRef]
14. Luo, H.; Xiong, Y.; Cheng, Y.; Chen, F.; Li, X. Optical transparent metamaterial structure for microwave–infrared-compatible camouflage based on indium tin oxide. *Sci. China Technol. Sci.* **2023**, *66*, 2850–2861. [CrossRef]
15. Zhang, Y.; Ugeda, M.M.; Jin, C.; Shi, S.-F.; Bradley, A.J.; Martín-Recio, A.; Ryu, H.; Kim, J.; Tang, S.; Kim, Y.; et al. Electronic structure, surface doping, and optical response in epitaxial WSe_2 thin films. *Nano Lett.* **2016**, *16*, 2485–2491. [CrossRef]
16. Ma, Q.; Kyureghian, H.; Banninga, J.D.; Ianno, N.J. Thin film WSe_2 for use as a photovoltaic absorber material. *MRS Proc.* **2014**, *1670*, mrss14-1670-e01-02. [CrossRef]
17. Sierra-Castillo, A.; Haye, E.; Acosta, S.; Colomer, J.-F. Synthesis and characterization of highly crystalline vertically aligned WSe_2 nanosheet. *Appl. Sci.* **2020**, *10*, 874. [CrossRef]
18. Zheng, J.; Zong, M.; Ye, K.; Feng, X.; Liu, J.; Liu, J.; Liu, D.; Liu, Z. Monolayer WSe_2 film as saturable absorbers for mid-infrared passive Q-switching. *Microw. Opt. Technol. Lett.* **2025**, *67*, e70119. [CrossRef]
19. Makeswaran, N.; Orozco, C.; Battu, A.K.; Deemer, E.; Ramana, C.V. Structural, optical and mechanical properties of nanocrystalline molybdenum thin films deposited under variable substrate temperature. *Materials* **2022**, *15*, 754. [CrossRef]
20. Founta, V.; Soulié, J.-P.; Sankaran, K.; Vanstreels, K.; Opsomer, K.; Morin, P.; Lagrain, P.; Franquet, A.; Vanhaeren, D.; Conard, T.; et al. Properties of ultrathin molybdenum films for interconnect applications. *Materialia* **2022**, *24*, 101511. [CrossRef]
21. Oliveira, A.d.S.; Pamplona de Sousa, G.C.; Valença, A.K.A.; da Silva Neto, J.F.; Gomes, K.C. Effect of surface treatments on absorptance and morphology of molybdenum and silica absorbing thin films. *Sol. Energy* **2023**, *254*, 158–167. [CrossRef]
22. Pamplona de Sousa, G.C.; da Silva Oliveira, A.; Alves Queiroga, R.; da Silva, U.C.; da Costa, N.P.; de Moura Reis, A.F.; da Silva Neto, J.F.; Gomes, K.C. Mathematical modeling study to optimize the production of molybdenum and silica absorbing thin films. *Sol. Energy* **2024**, *282*, 112962. [CrossRef]
23. Akafzade, H.; Sharma, S.C. New metamaterial as a broadband absorber of sunlight with extremely high absorption efficiency. *AIP Adv.* **2020**, *10*, 035209. [CrossRef]
24. Wilson, D.P.; LaPierre, R.R. Simulation of optical absorption in conical nanowires. *Opt. Express* **2021**, *29*, 9544–9552. [CrossRef] [PubMed]
25. Yang, C.F.; Wang, C.H.; Ke, P.X.; Meen, T.H.; Lai, K.K. Design and manufacture of a multi-layer planar solar light absorber with high absorptivity and ultra-wideband from visible light to infrared. *Nanomaterials* **2024**, *14*, 930. [CrossRef]

 https://doi.org/10.3390/photonics13010086

26. Zhang, Y.; Hou, B.; Song, Q.; Yi, Z.; Zeng, Q. A THz Smart Switch Based on a VO_2 Metamaterial That Switches Between Wide-Angle Ultra-Wideband Absorption and Transmission. *Dalton Trans.* **2024**, *53*, 19264–19271. [CrossRef] [PubMed]
27. Zhang, Y.; Hou, B.; Yang, Y.; Song, Q.; Yi, Z.; Zhou, Z. An Active Controllable Wide-Angle and Ultrawideband Terahertz Absorber/Reflector Based on VO_2 Metamaterial. *Phys. Chem. Chem. Phys.* **2024**, *26*, 6091–6098. [CrossRef]
28. Tharwat, M.M.; Alsulami, A.R.; Mahros, A.M. Exploring the Absorption Spectra of an Ultra-Wideband Metamaterial Absorber in the Visible and Near-Infrared Regions. *Materials* **2022**, *15*, 7160. [CrossRef]
29. Wang, Y.; Liu, F.; Ni, B.; Chen, L.; Ji, K. Design of an ultra-wideband solar absorber based on tungsten. *AIP Adv.* **2023**, *13*, 095024. [CrossRef]
30. Rezaei, M.H.; Vatandoust, Y.; Afshari-Bavil, M.; Liu, D. Ultra-broadband, polarization-independent, and wide-angle met-amaterial absorber based on fabrication-friendly Ti and TiO_2 resonators. *Opt. Quantum Electron.* **2024**, *56*, 400. [CrossRef]
31. Yang, F.; Li, R.-H.; Tan, S.-L.; Dong, J.-W.; Jiang, S.-J. Visible–mid infrared ultra-broadband and wide-angle metamaterial perfect absorber based on cermet films with nano-cone structure. *Nanophotonics* **2023**, *12*, 2451–2460. [CrossRef] [PubMed]
32. Song, S.; Chen, Y.; Chen, S.; Hou, J.; Huang, X. An ultra-broadband high-performance solar energy perfect absorber from deep ultraviolet to mid-infrared. *Mater. Today Commun.* **2024**, *39*, 108712. [CrossRef]

https://doi.org/10.3390/photonics13010086

photonics

MDPI

Article

Analysis and Design of a Hybrid Graphene/Vanadium-Dioxide Terahertz Metasurface with Independently Reconfigurable Reflection Phase and Magnitude

Eric Amoateng [1,2], Ellis Mubarak Sani [2], Kingsford Sarkodie Obeng Kwakye [2] and Alexandros Pitilakis [3,*]

1 Department of Electrical and Electronics Engineering, Ghana Communication Technology University, Accra GA-167-2979, Ghana; eamoateng@gctu.edu.gh

2 Department of Telecommunication Engineering, Kwame Nkrumah University of Science and Technology, Kumasi AK-448-4944, Ghana; smellis.coe@knust.edu.gh (E.M.S.); ksobengkwakye@knust.edu.gh (K.S.O.K.)

3 School of Electrical and Computer Engineering, Aristotle University of Thessaloniki, 54124 Thessaloniki, Greece

* Correspondence: alexpiti@auth.gr

Abstract

A reconfigurable THz metasurface (MS) capable of independent reflection amplitude and phase modulation is designed and analyzed. The tunability is achieved in a simple few-layer structure by control over the chemical potential of a graphene monolayer patterned in square patches and over the bulk conductivity of an overlying vanadium dioxide (VO_2) patch array; these impart control over the reflection phase and magnitude, respectively. To design and analyze the MS unit cell, we employ intuitive equivalent circuit and transmission line modeling, which is validated against full-wave simulations, showing good agreement in the regime of interest, i.e., on the first resonance for normal plane wave incidence. The simulated phase modulation approaches 250°, enabling binary-encoded digital metasurface designs, while the magnitude modulation spans more than 20 dB, from −3 dB almost down to perfect absorption. The flexibility of dynamic phase and amplitude control can unlock the full potential of such THz MS hybrid designs for future wireless communications (6G and beyond) and for sensing applications. Finally, the analytical modeling can be extended to polarization-dependent, anisotropic, or non-local EM responses and/or to include aspects of the multiphysical control mechanisms.

Keywords: metasurface; reconfigurable; terahertz; graphene; vanadium dioxide; phase-change material; equivalent circuit model; transmission line model

1. Introduction

Metasurfaces (MS) are artificially engineered ultra-thin planar structures arranged in periodic configurations, enabling precise control of the scattered electromagnetic (EM) wavefront direction/shape, amplitude, and/or polarization [1–6]. Through tailored geometric design and material composition, MS can realize a wide range of functionalities, including wavefront shaping (e.g., beam steering or splitting), absorption, and transmission control [7,8]. Being the two-dimensional counterpart of bulk/3D metamaterials, MS offer distinct advantages such as low profile, reduced loss, ease of fabrication, and seamless integration with existing systems, while maintaining high efficiency and versatile EM wave manipulation capabilities [9,10]. By reconfigurability [5], we refer to the ability to non-negligibly adjust their EM response after deployment, in real time and without moving parts, which makes reconfigurable MS (RMS) particularly attractive for modern adaptive applications, as opposed to conventional passive metasurfaces.

The ongoing work related to RMS can be broadly categorized based on their capability to modulate the phase, amplitude, and/or polarization of scattered EM waves; theoretical and numerical techniques for modeling, analysis, and inverse design of such RMS are particularly interesting. Phase-modulating RMS shapes the phase of an incident wavefront and has been extensively investigated for applications such as dynamic beam steering [11], beam focusing [12], holography [13], and polarization control [6,14–16]. Amplitude-modulating RMS regulate the magnitude of the scattered EM wave within specific frequency bands and have been widely explored for functionalities including tunable perfect absorbers and transmission or backscatter control [3,4,7,8,17,18]. However, for more sophisticated and performance-critical applications—such as high-fidelity holography, multi-beamforming, high-resolution imaging, and sidelobe suppression—the ability of RMS to simultaneously control both the phase and amplitude of the reflected wavefront is highly desirable. The independent and continuous modulation of these two degrees of freedom enables precise wavefront shaping and significantly improves system performance in 'smart' EM environments. Nevertheless, most reported RMS capable of amplitude and phase tuning operate at microwave and/or millimeter-wave frequencies, employing PIN diodes or varactors [19]. Unfortunately, extending such designs to the THz band, even the sub-THz (e.g., 300 GHz), remains challenging due to increased material losses, fabrication constraints, and the reduced effectiveness of conventional tuning mechanisms at shorter wavelengths [1,2]. This gap highlights the need for simple, compact, low-loss, and efficient RMS platforms suitable for emerging THz or hybrid applications [20].

To address this demand, focus has shifted to advanced materials that exhibit strong THz light-matter interaction and offer additional degrees of freedom for realizing complex and dynamic EM functionalities. This has driven the development of smart materials and nanostructures capable of achieving efficient and versatile tunability in the THz band [10,21]. Atomically thin 2D materials, such as graphene [22], and phase-change materials, such as vanadium dioxide (VO_2) [5], are two emerging and promising material platforms for achieving dynamic and reconfigurable THz wavefront scattering. In the THz, graphene exhibits superior carrier mobility, which allows for efficient electrical control of its reactive surface conductivity, making it particularly attractive for dynamic phase control. VO_2, on the other hand, exhibits a reversible insulator-metal transition (I-M-T) near 340 K, during which its bulk electrical conductivity changes by several orders of magnitude; this enables strong and controllable amplitude modulation, which cannot be achieved with graphene alone or other phase-change materials [23,24], which require more complicated resonant structures, reducing fabrication feasibility. Therefore, using this hybrid MS design, independent control of the amplitude and phase is realized, which is difficult to achieve using a single material platform [25].

In this work, we employ a hybrid unit cell that integrates graphene and VO_2 in a single THz RMS design to exploit their combined strong points: Electrically tuning graphene's chemical potential (μ_c) primarily alters the reactive component of its surface impedance, leading to a smooth and broadband phase modulation; as a result, graphene is suitable for phase control but is intrinsically limited in amplitude modulation due to its relatively weak absorption in the THz range [26,27]. The I-M-T of VO_2 [28] largely modifies its resistive impedance through its bulk conductivity (σ_{VO_2}) which can be exploited to modulatethe reflection magnitude in resonant MS structures. We choose a hybrid but

https://doi.org/10.3390/photonics13020195

decoupled MS architecture where varying graphene μ_c minimally perturbs the absorption (governed by the VO_2 state) and, conversely, changing σ_{VO_2} has limited impact on the reflection phase. This architecture enables flexible and independent modulation of the complex (amplitude and phase) EM wave scattering, which is not readily achievable with either material alone. The graphene–VO_2 hybrid MS therefore provides a practical pathway toward multifunctional, reconfigurable THz metasurfaces capable of dynamic and independent reflection phase and magnitude modulation, which is highly desirable for emerging 6G communication and sensing applications [25]. Normal incidence is considered in all theoretical studies, for the sake of simplicity, but polarization-dependent performance under oblique illumination is also numerically evaluated.

The remainder of this paper is organized as follows: Section 2 presents a systematic design methodology of the proposed RMS, highlighting the unit cell architecture, material dispersion models, equivalent circuit/transmission line model (ECM/TLM), and full-wave EM simulations. Section 3 reports and compares the results obtained from three investigated configurations. Section 4 provides a detailed discussion and comparison of the results to the current state-of-the-art. Finally, Section 5 concludes the paper by summarizing the main findings.

2. Design of the Unit Cell

2.1. Unit Cell Architecture

A three-dimensional schematic of the proposed MS array and the unit cell structure are shown in Figure 1a and Figure 1b, respectively, with periodic extension along the x and y directions. This unit cell adopts a multilayer architecture consisting of five stacked layers arranged sequentially from top to bottom: The top layer is VO_2 which is investigated in two configurations, as a uniform bulk slab and as a patterned square patch occupying a large fraction of the unit cell area [29,30]. VO_2 (PCM) is toggled between transparent and metallic state. Thus, without being highly absorptive itself, metallic VO_2 can form an absorption resonance inside the unit cell (where the low losses of graphene can be exploited) which eventually controls the magnitude of the reflected wave; this control is effectuated by tuning the σ_{VO_2} via heating. Next, an oxide (SiO_2) dielectric layer is introduced beneath the VO_2 layer, to provide electrical isolation and reduce the coupling between the reconfigurable materials; in the THz, SiO_2 is described by a relative permittivity $\varepsilon_r = 3.88$ (refractive index is 1.97), assumed dispersionless. A graphene monolayer is incorporated below the dielectric oxide spacer; graphene is modeled either as an infinite sheet or in square-patterned patches, with zero thickness in both cases.High-quality graphene predominantly controls the phase of the impinging THz wave, owing to its largely reactive surface impedance. Graphene also has some losses (resistive impedance) which can be considered as parasitic for the phase-manipulation, i.e., they are not strong enough to create an absorption resonance by themselves.Consequently, by electrically tuning graphene's μ_c, we control only the phase of the reflected wave. The graphene layer is supported by a silicon substrate with a relative permittivity $\varepsilon_r = 11.8$ (dispersionless), which enhances field confinement and mechanical robustness in the multilayer structure [31]. Finally, the back of the cell is terminated with a continuous gold mirror, modeled as a perfect electric conductor, to suppress transmission and enable reflective mode operation [32,33]. The combination of these two qualities, brought by graphene and VO_2, highlight the potential of this hybrid unit cell design in achieving independent modulation in phase and magnitude.

https://doi.org/10.3390/photonics13020195

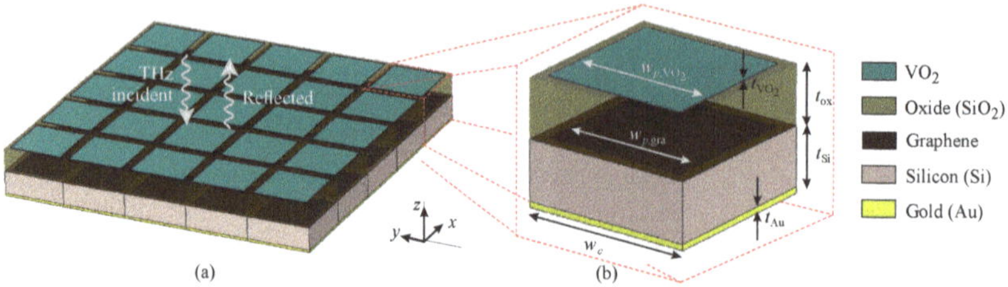

Figure 1. The reconfigurable hybrid MS architecture. (**a**) Example of a 5×5 cell array. (**b**) Unit cell schematic with material and dimension annotations.

2.2. Material Dispersion Models

Graphene is modeled as a zero-thickness transparent impedance sheet, e.g., with a custom "tabulated surface impedance" (TSI) material in CST Microwave Studio; its corresponding complex surface conductivity σ_s is given by Kubo formulas [34]. Graphene's surface conductivity generally has intraband and interband contributions, but the intraband term is dominant in the THz regime; so, the interband term (which is dominant in the near-infrared and visible spectrum) can be neglected [35,36]. According to the Kubo formulas, the complex σ_s depends on the interacting EM wave frequency (ω), lattice temperature (T), chemical potential (μ_c), and relaxation time (τ), i.e., $\sigma_s = \sigma_s(\omega, T, \mu_c, \tau)$. In this work, we exploit the dependence on μ_c by assuming it can be externally controlled, e.g., by electrical gating or biasing. The relaxation time is related to the carrier mobility (μ_{mob}), which quantifies the purity of the graphene sample [37], i.e., its losses. With these assumptions, the graphene THz surface conductivity is expressed as

$$\sigma_s \approx \sigma_{\text{intra}} = j\frac{2q^2 K_B T}{\pi\hbar^2(\omega + j\tau^{-1})} \ln\left[2\cosh\left(\frac{E_F}{2K_B T}\right)\right], \tag{1}$$

where $\hbar = h/(2\pi)$ is the reduced Planck constant, K_B is the Boltzmann constant, and $\tau = (\mu_c\mu_{\text{mob}})/(qv_f^2)$, i.e., it depends on the chemical potential. We consider a fixed Fermi velocity $v_f = 10^6$ m/s and a fixed $\mu_{\text{mob}} = 20{,}000\ \text{cm}^2\,\text{V}^{-1}\,\text{s}^{-1}$; q is the electron charge. For $\mu_c \gg K_B T$, i.e., for highly doped graphene at room temperature $T = 300$ K, it holds that $\mu_c \approx E_F$ and that the surface conductivity can be further simplified into the Drude-like spectra as

$$\sigma_s(\omega; \mu_c, \tau) = j\frac{q^2\mu_c}{\pi\hbar(\omega + j\tau^{-1})}, \tag{2}$$

which highlights its μ_c-dependent dispersive properties.

For the VO_2 THz dispersion, we used the Drude formula

$$\varepsilon_r(\omega; \sigma_{\text{VO}_2}, \gamma) = \varepsilon_\infty - \frac{\omega_p^2(\sigma_{\text{VO}_2})}{\omega^2 + j\gamma\omega}, \tag{3}$$

where ω is the EM wave frequency, ϵ_∞ is the high-frequency relative dielectric permittivity, ω_p is the plasma frequency, and $\gamma = 5.75 \times 10^{13}$ rad/s is the collision frequency that dictates the losses. As implied by Equation (3), the plasma frequency depends on its heat-tunable bulk conductivity and can be approximated as $\omega_p^2 = \omega_p^2(\sigma_{\text{VO}_2}) = (\sigma_{\text{VO}_2}/\sigma_{\text{ref}})\omega_{p,\text{ref}}^2$ with reference bulk conductivity $\sigma_{\text{ref}} = 3 \times 10^5$ S/m and reference plasma frequency $\omega_{p,\text{ref}} = 1.4 \times 10^{15}$ rad/s. Importantly, the bulk conductivity of VO_2 in the metallic state is four orders of magnitude higher than its conductivity in the insulator (dielectric) state. In this work, we set the two σ_{VO_2} extremes, for dielectric and metallic states, as 20 S/m and 200,000 S/m, respectively. Furthermore, we assume this bulk conductivity can be

https://doi.org/10.3390/photonics13020195

continuously varied between the insulator and metallic states, e.g., by temperature control, to consider its behavior in the intermediate states.

Note that the heating-mediated phase transition in VO_2 is non-volatile and is intended to be used at slow rates (not in fully dynamic scenarios), i.e., as a 'toggle' switch between absorptive and wavefront-shaping RMS functionalities; the former can be used in wireless sensing (e.g., if graphene patches are seen as planar integrated plasmonic antennas or photodiodes), while the latter is used in wireless communications (beam steering to extend non-line-of-sight coverage).

2.3. Transmission Line Modeling

A transmission line model (TLM) is be used to approximately quantify the reflection coefficient spectra when the MS cell of Figure 1b is illuminated by a plane wave. The unit cell cross-section side view and the corresponding TLM are shown in Figure 2 for two configurations of the VO_2 layer. The purpose is to calculate the input impedance Z_{in} at the top side, i.e., at the exposed surface where THz light impinges normally on the surface. Importantly, the model must use *only* the wave frequency, geometric dimensions, and material properties of the unit cell, in closed-form expressions, so that we can compute the spectra *without* conducting full-wave simulations. To compute Z_{in}, we can start from the bottom of the TLM, i.e., from the reflective mirror, and move to the top, using transmission line theory [37–39].

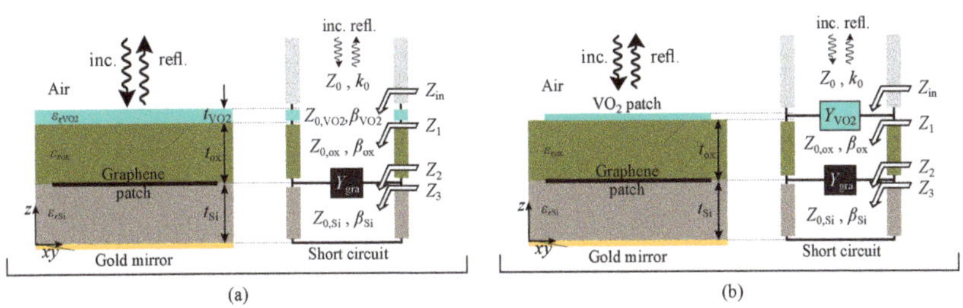

Figure 2. Unit–cell cross section and corresponding transmission line model (TLM) for the case where the VO_2 material is (**a**) unpatterned, i.e., a slab, or (**b**) patterned into wide patches. The shunt admittances Y_{gra} and Y_{VO_2} of the patch arrays can be computed by equivalent circuit models (ECM).

Starting from the bottom, the gold ground plane is considered a short circuit. Then, the silicon substrate slab with thickness t_{Si} is modeled as a transmission line (TL) segment of that same length, of characteristic impedance $Z_{0,Si}$ and of phase constant β_{Si}. For normal incidence, it holds that $Z_{0,Si} = \eta_0/\sqrt{\varepsilon_{r,Si}}$ and $\beta_{Si} = k_0\sqrt{\varepsilon_{r,Si}}$, where $\eta_0 \approx 377\ \Omega$ is the vacuum wave impedance and $k_0 = 2\pi/\lambda_0$ is the vacuum wavenumber. The input impedance of this TL segment, marked as Z_3 in Figure 2, is given by the grounded TL equation

$$Z_3 = jZ_{0,Si}\tan(\beta_{Si}t_{Si}). \tag{4}$$

Continuing, the impedance Z_2 is the parallel combination of the surface impedance of the graphene layer ($Z_{gra} = 1/Y_{gra}$) and the input impedance of the previous layer, Z_3, computed as

$$Z_2 = \frac{Z_{gra}Z_3}{Z_{gra} + Z_3}. \tag{5}$$

When the graphene sheet is infinite, then $Z_{gra} = 1/\sigma_s$, i.e., its surface conductivity suffices. When the sheet is patterned, e.g., in wide patches with narrow gaps, the surface impedance of this array has an extra a capacitive (reactive) term which can be computed by an equivalent circuit model (ECM), presented in the following subsection.

https://doi.org/10.3390/photonics13020195

Then, since the top SiO_2 substrate is also a uniform slab, it can be similarly modeled as a TL segment of length t_{SiO_2} 'loaded' by Z_2; the input impedance at the slab's top port, marked as Z_1 in Figure 2, is given by the impedance transformation equation (ITE)

$$Z_1 = Z_{0,ox} \frac{Z_2 + jZ_{0,ox} \tan(\beta_{ox} t_{ox})}{Z_{0,ox} + jZ_2 \tan(\beta_{ox} t_{ox})}. \tag{6}$$

Note that for a short-circuit load, i.e., $Z_2 = 0$, the ITE is equivalent to Equation (4).

Finally, on the top of the cell, the VO_2 slab of thickness t_{VO_2} depicted in Figure 2a is modeled as a thin TL segment, similar to the modeling of the two dielectric substrates; the difference here is that the characteristic impedance and phase constant of the transmission line are Drude-*dispersive* as the relative permittivity of the VO_2 slab depends on ω according to Equation (3). The total input impedance Z_{in} of the MS unit cell is computed by an ITE, now using Z_1 as the 'load',

$$Z_{in} = Z_{VO_2} \frac{Z_1 + jZ_{VO_2} \tan(\beta_{VO_2} t_{VO_2})}{Z_{VO_2} + jZ_1 \tan(\beta_{VO_2} t_{VO_2})}, \tag{7}$$

where $\tan(\beta_{VO_2} t_{VO_2}) \approx \beta_{VO_2} t_{VO_2}$ for very thin layers. Alternatively, if the VO_2 layer is patterned in deeply subwavelength-thickness square patches, i.e., similar to the graphene sheet, an ECM can be used to compute the equivalent surface impedance of this VO_2 patch array which shunts Z_1 to compute Z_{in}, instead of using the ITE; see the following subsection.

Having computed Z_{in} and assuming normal THz wave incidence from the air-exposed side of the unit cell ($Z_0 = \eta_0$), the co-polarized complex-valued reflection coefficient (Γ) at the input port where the THz light impinges is given by

$$\Gamma = \frac{Z_{in} - Z_0}{Z_{in} + Z_0}. \tag{8}$$

2.4. Equivalent Circuit Model for Surface Impedance of Capacitive Patch Arrays

The ultra-thin graphene wide-patch array sandwiched between two dielectric substrates with permittivities ε_{r1} and ε_{r2} can be approximated with a surface impedance Z_{gra} given by [29,30]. The surface impedance consists of two parts in series, $Z_{gra} = Z_{sc} + Z_{grid}$: The first term mainly depends on the surface conductivity of the graphene patch material, and the second term represents the equivalent surface impedance of the periodic capacitive patch array, which can be computed analytically as [30]

$$Z_{grid} = -j\frac{n_{eff}}{2\alpha}, \tag{9}$$

where α is the so-called 'grid parameter', which relates the tangential electric field at the MS plane to the induced surface current density on the periodic patch array. Physically, it quantifies the strength of electric coupling between adjacent metallic elements and therefore governs the capacitive behavior of the grid when the inter-element gaps are much smaller than the unit-cell period [30,40]. According to the classical Kontorovich–Tretyakov grid theory, for an electrically dense periodic array of metallic patches or strips, the grid parameter α is given in this case as [30,41],

$$\alpha = \frac{k_{eff} w_p}{\pi} \ln\left\{ \csc\left[\frac{\pi(w_c - w_p)}{2w_p} \right] \right\}. \tag{10}$$

https://doi.org/10.3390/photonics13020195

In this manner, the final ECM-computed surface impedance of the graphene capacitive patch array is

$$Z_{\text{gra}} = Z_{sc} + Z_{\text{grid}} = \frac{1}{\sigma_s}\frac{w_c}{w_p} - j\frac{\eta_{\text{eff}}}{2}\frac{\pi}{k_{\text{eff}} w_p} \ln^{-1}\left\{\csc\left[\frac{\pi(w_c - w_p)}{2w_p}\right]\right\}. \tag{11}$$

where σ_s is the dispersive surface conductivity according to Equation (2) above, η_{eff} is the wave impedance in the *equivalent* bulk medium in which the graphene patch is embedded [with relative effective permittivity $\varepsilon_{r,\text{eff}} = (\varepsilon_{r1} + \varepsilon_{r2})/2$], w_p is the width of the graphene patch, and w_c is the period of the unit cell. Furthermore, $k_{\text{eff}} = k_0\sqrt{\varepsilon_{r,\text{eff}}}$ is the wave number of the incident wave in the equivalent medium. In the case of the graphene patch array in the unit cell of Figure 1b, it holds $\varepsilon_{r,\text{eff}} = (\varepsilon_{r,\text{Si}} + \varepsilon_{r,\text{ox}})/2$.

Now, concerning VO_2 when patterned in wide patches: We can model it as a shunt load equivalent to how graphene is modeled in Equation (11). In this case, the complex surface complex conductivity is expressed according to [42,43] as

$$\sigma_{\text{VO}_2} = j[\varepsilon_r(\omega) - 1]\omega\varepsilon_0 t_{\text{VO}_2} \tag{12}$$

in units of Siemens. The surface impedance of the capacitive VO_2 patch array is given by Equation (11) after replacing $\sigma_s \to \sigma_{\text{VO}_2}$ and computing k_{eff} and η_{eff} with the corresponding $\varepsilon_{r,\text{eff}} = (1 + \varepsilon_{r,\text{ox}})/2$, as the VO_2 patches are between air and silicon dioxide. Obviously, the width of the VO_2 patches can be different from the width of the graphene patches, but both must be small in all cases. The cell side view and the corresponding TLM for this case are shown in Figure 2b, and the reflection coefficient at the input port at the top is determined in the same way as outlined earlier.

2.5. Full-Wave Simulation

We employ the CST Microwave Studio simulation software [44] to numerically investigate the EM response of the proposed MS unit cell, assuming infinite periodic repetition. The simulation is carried out using the frequency domain solver with tetrahedron mesh, which provides high accuracy for steady-state spectral, phase and absorption analyses [44,45]. The optimized geometric dimensions of the unit cell are shown in Table 1, which are the same throughout the following sections, unless otherwise stated. The unit cell periodic boundary conditions are applied along the x- and y-directions. We excite the structure at the upper Floquet port under normal plane wave incidence along the z-axis. An open boundary condition is assigned at the excitation top boundary (z_{max}), where the reflection is to be computed, and the bottom boundary (z_{min}) is modeled as a perfect electric conductor (PEC) to emulate the gold backplane [44,46]; use of realistic materials (such as gold) has a minimal impact on the response. The temperature-dependent phase transition behavior of VO_2 is incorporated in the simulation by varying its bulk electric conductivity σ from 20 S/m to 200,000 S/m, which transitions from the insulator to metallic state [5]; this change affects the plasma frequency, and the final dispersive permittivity used in the simulations is implemented by a Drude model. Also, the graphene layer is modeled with a TSI with surface conductivity values taken by the Kubo formula (intraband); such a zero-thickness impedance sheet properly captures its two-dimensional nature without introducing artificial thickness, which is negligible compared to THz wavelengths, that should be meshed, increasing computational burden. The graphene surface impedance, which is highly reactive at THz wavelengths (imaginary part of σ_s is non-negligible), is tuned simply by varying the chemical potential μ_c, typically between 0.1 (near-pristine graphene) and 1 eV. This approach reduces the computational cost while preserving the essential electromagnetic behavior of the graphene-wave interaction [22,47,48].

https://doi.org/10.3390/photonics13020195

Table 1. Geometric parameters of the metasurface unit cell (in µm).

w_c	t_{Au}	t_{Si}	t_{SiO_2}	t_{VO_2}	$w_{p,gra}$	w_{p,VO_2}
30	1.2	20	15	0.1	27	27

3. Results

In this section, we present and compare the unit cell response results by ECM/TLM modeling and by full-wave simulations. The analytical ECM/TLM provides a good starting tool for pre-selecting the geometric dimensions before an optimization with full-wave simulations at the unit cell is performed. We consider normal illumination; however, the ECM and TLM can both be extended to oblique incidence, according to [29,30], to study the polarization sensitivity of the device.

3.1. VO_2-Only Configuration for Reflection Amplitude Modulation

We first investigate the EM response of the MS when graphene is absent, considering the unit cell consisting of VO_2 on top of the oxide/Si substrate backed by the gold ground plane. We assume the bulk conductivity of VO_2 can be continuously tuned between the dielectric and metallic phases, e.g., by heating.

We start with the unpatterned VO_2 slab case, which can be modeled by a simple TLM. The slab thicknesses are varied to 0.1 µm, 1 µm, 2 µm, and 5 µm to investigate its influence on THz reflection [49]. The input impedance Z_{in} and the reflection coefficient Γ are subsequently computed as discussed in the methodology section by setting $\sigma_s = 0$. The equivalent circuit is shown in Figure 3a, where the unpatterned VO_2, oxide, and Si slabs are cascaded as TL segments, while the gold mirror corresponds to a short-circuit termination [29,30,43].

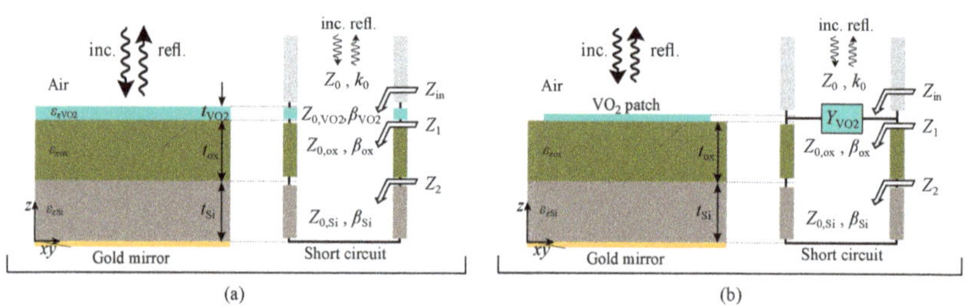

Figure 3. Unit–cell cross section and transmission line model (TLM) for the case where the VO_2 material is (**a**) unpatterned, i.e., a slab, or (**b**) patterned into wide patches. Graphene has been removed in both cases.

The corresponding reflection amplitude spectra are presented in Figure 4a, where we vary the VO_2 bulk conductivity. These spectra clearly show the Fabry–Pérot resonant behavior of the stratified medium backed by a PEC ground. For low σ_{VO_2}, below 200 S/m, the structure behaves as a loss-low dielectric slab, exhibiting minimal resonance with a reflection coefficient close to unity, indicating weak electromagnetic coupling. This is consistent with the insulator phase of VO_2, where the interaction with incident THz radiation is dominated by dielectric reflections [5,50]. As the conductivity increases towards the intermediate and metallic phases, deeper and sharper resonance dips appear at nearly 1 THz periodicity (free-spectral range, FSR), which is consistent with Fabry–Pérot standing-wave resonances [50,51]. Figure 4b presents a comparison between the analytical ECM/TLM and the full-wave CST simulation for the metallic VO_2 state; as expected, for the unpatterned slab unit cell configuration, the agreement is exact.

https://doi.org/10.3390/photonics13020195

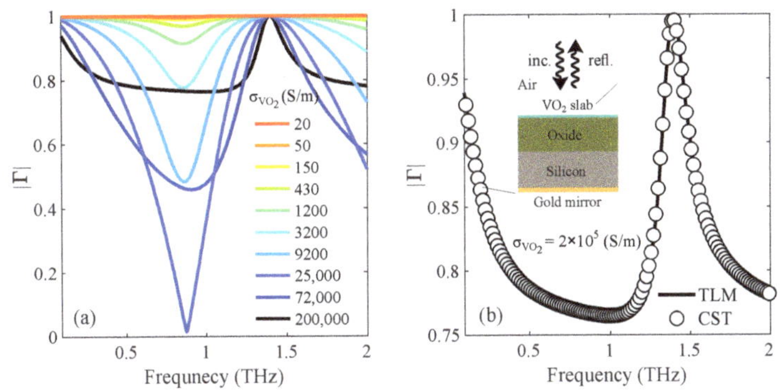

Figure 4. (**a**) Reflection amplitude tuning by changing the top VO_2 slab bulk conductivity, in S/m, in the unit cell. (**b**) Comparison between full-wave simulation and transmission line model, for the fully metallic case, σ_{VO_2} = 200,000 S/m.

Figure 5 studies the effect of the thickness of the unpatterned VO_2 slab on the reflection spectra as it is increased from 0.1 μm to 5 μm. For small thickness (0.1 μm), the structure supports discrete resonance frequencies, with relatively broad absorption minima within each frequency band. In this thin-film regime, the effective optical path length is short, leading to weak multiple reflections and reduced field confinement within the cavity. As a result, the supported resonances are fewer in number less sharply and weakly defined in frequency. As the VO_2 thickness increases progressively up to 5 μm, the optical thickness of the multilayer stack increases substantially, enhancing internal reflections between the VO_2 layer and the metallic ground plane. This enhances field confinement and the excitation of higher-order longitudinal cavity modes, leading to multiple sharp absorption dips and a decrease in the spacing between adjacent resonances. This behavior is typical of Fabry–Pérot cavity dynamics, where the resonance condition is governed by the round-trip phase condition imposed by the geometry of the cavity. Specifically, the free spectral range (FSR) of a Fabry–Pérot cavity is inversely proportional to the effective optical length, such that increasing the VO_2 thickness reduces the FSR. As a result, a larger number of resonant modes can be accommodated within the same spectral window. Moreover, the increased thickness enhances impedance modulation and loss accumulation within the VO_2 layer, further deepening the absorption minima. These combined effects explain the transition from sparse, broadband absorption features at small thicknesses to more dense and higher Q-factor resonances at larger thicknesses [52].

Next, we examine the unit cell configuration shown in Figure 3b, in which the VO_2 layer is now patterned in patches with a width equal to 90% of the unit-cell width (period). In this case, the VO_2 is represented by a parallel admittance, whose value is given by the ECM. The resulting reflection amplitude and phase spectra are presented in Figure 6a and Figure 6b, respectively [29,30,53,54]. Compared with the unpatterned slab configuration, the VO_2 patch configuration generally exhibits higher reflection levels, which are seen as shallower resonance dips in the reflection spectrum. This behavior arises due to the fact that patterning reduces the effective volume of the lossy material participating in the light-matter interaction. As a result, both ohmic dissipation and plasmonic losses associated with induced surface currents are reduced, which alters the surface impedance, so less incident power is absorbed and more is reflected. In addition, the edge capacitance from the patch gaps also modify the resonant condition compared with the uniform slab case. These effects shift the resonance condition and weaken absorption by reducing current continuity. This further contributes to the shallower reflection minima which is observed in the patch configuration. The corresponding reflection phase unveils that the

https://doi.org/10.3390/photonics13020195

VO_2 patch configuration produces minimal phase change when the conductivity is swept from dielectric to metallic state. This combination makes the patch geometry particularly suitable for applications where a high reflected amplitude is desired, with comparatively less emphasis on wide phase tunability. The decoupling of amplitude and phase responses makes the patch geometry particularly attractive for applications requiring large reflection amplitude combined with wide phase tunability, such as reconfigurable reflect-arrays, beam steering, and wavefront shaping metasurfaces [54–56].

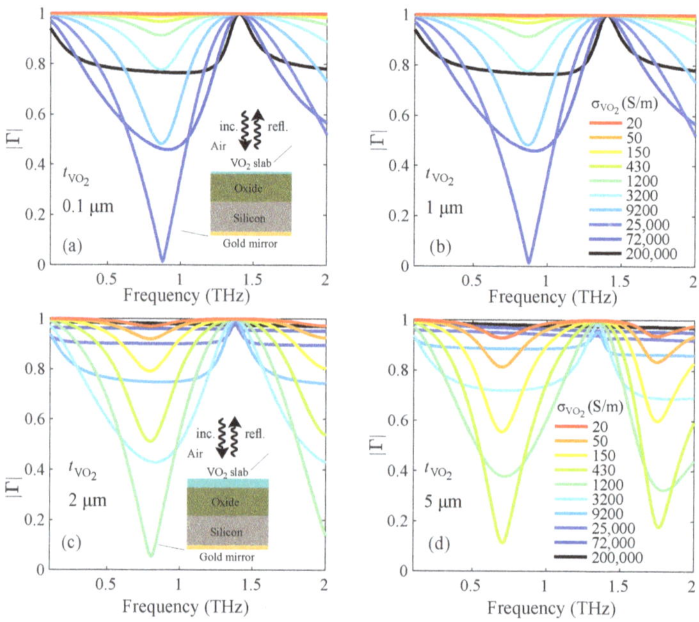

Figure 5. Analytical TLM reflection amplitude spectra for four different VO_2 slab thicknesses as the bulk conductivity is changed; in all four panels, the different colored curves correspond to different σ_{VO_2} values, given in the legends of panels b and d. (**a**) $t_{VO_2} = 0.1$ μm, (**b**) $t_{VO_2} = 1$ μm, (**c**) $t_{VO_2} = 2$ μm, (**d**) $t_{VO_2} = 5$ μm.

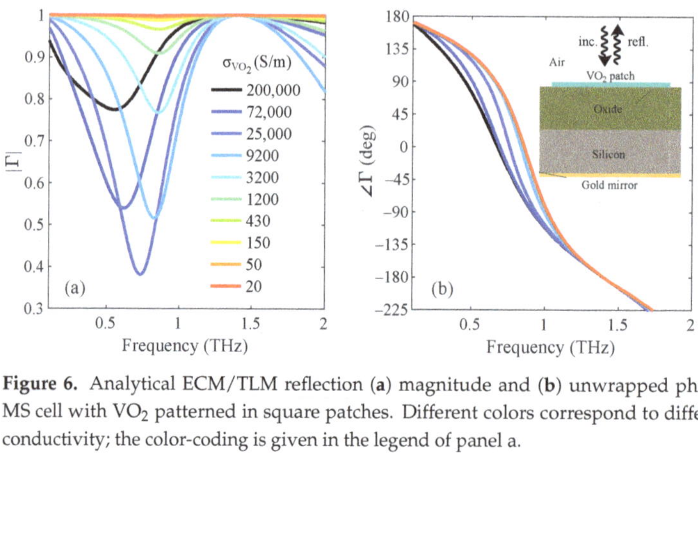

Figure 6. Analytical ECM/TLM reflection (**a**) magnitude and (**b**) unwrapped phase spectra for MS cell with VO_2 patterned in square patches. Different colors correspond to different VO_2 bulk conductivity; the color-coding is given in the legend of panel a.

https://doi.org/10.3390/photonics13020195

Comparison between the ECM/TLM and full-wave simulations also reveals excellent agreement as shown in Figure 7. The ECM/TLM accurately reproduces the resonance frequencies as well as the overall dispersion trends of both the reflection magnitude and phase across the considered spectral range. Only minor discrepancies are observed in the depth of the amplitude minima as frequency increases, which can be attributed to higher-order modes and edge-induced fringing fields that are inherently captured in the full-wave simulations but approximated in the analytical formulation. Nevertheless, the very good agreement near the first resonance validates the proposed ECM/TLM approach and confirms its robustness and reliability for accurately estimating the resonance frequency and Q-factor of the VO_2-patch unit cell.

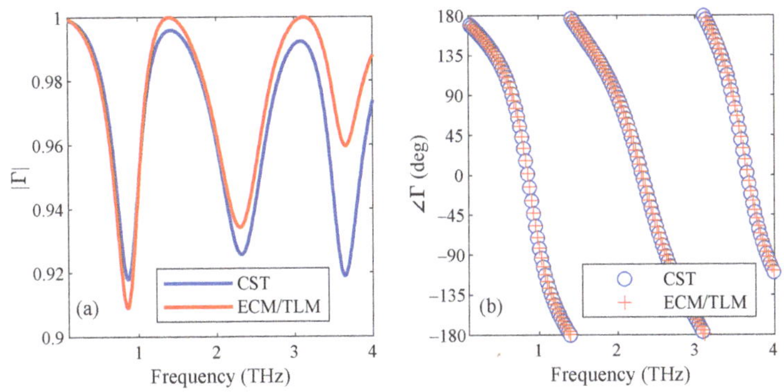

Figure 7. Comparison of reflection spectra, (**a**) magnitude and (**b**) phase, between full-wave CST simulation and the ECM/TLM, for the unit cell of Figure 3b, where VO_2 is patterned in patches and $\sigma_{VO_2} = 200$ S/m.

3.2. Graphene-Only Configuration for Reflection Phase Modulation

In this subsection we consider the unit cell configuration of Figure 8 in which the graphene is present, sandwiched between the Si and SiO_2 layers, while the top VO_2 layer is removed. For this more well-studied configuration, e.g., [37,54], we examine both the reflection magnitude and phase responses obtained from our analytical ECM/TLM and full-wave simulations.

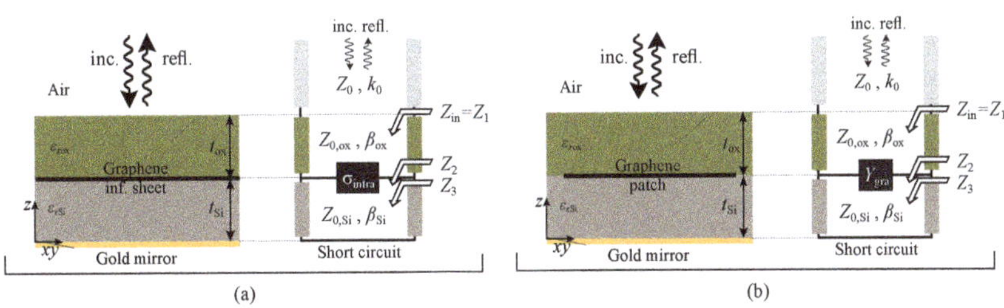

Figure 8. Unit-cell cross section and transmission line model for the case where the graphene monolayer is (**a**) unpatterned, i.e., an infinite sheet, or (**b**) patterned in square patches. VO_2 has been removed in both cases.

Starting from the unpatterned graphene sheet configuration, the results are presented in Figure 9. The reflection amplitude spectra obtained from the analytical ECM/TCM and the CST full-wave simulations show very good overall agreement near the first resonance, around 1 THz, confirming the validity of the analytical approach in this region. Both approaches capture the main resonance frequency and the high-frequency reflective behavior of the structure. The small discrepancies in the magnitude and the progressive

https://doi.org/10.3390/photonics13020195

mismatch of the phase can be attributed to material dispersion (difference in CST and our own version of the Kubo formula), and the more realistic treatment of graphene and dielectric interfaces, which are not fully captured by the simplified assumptions of the analytical model [9,35]. Overall, the comparison demonstrates that the analytical model reliably predicts the qualitative first-resonance electromagnetic response of the multilayer structure, while CST provides further accuracy. The reflection-phase results also show the same resonant dispersive behavior with phase flips at the resonance frequencies. The slight difference is expected because the CST includes realistic losses and full wave interactions.

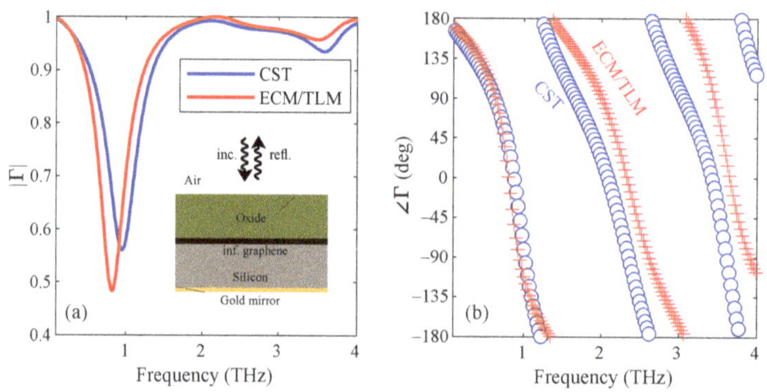

Figure 9. Reflection (**a**) magnitude and (**b**) phase spectra in the infinite graphene monolayer unit cell configuration; comparison between ECM/TLM and full-wave simulation for graphene chemical potential at 0.1 eV.

Next, the graphene layer is assumed to be patterned in square patches of width 90% that of the unit cell. The reflection amplitude spectra from the sheet and patch case are compared in Figure 10 and one can observe that the graphene patch gives higher reflection amplitude [36,57]. That is because reducing the graphene coverage decreases the interaction between the incident wave and the conductive surface. The continuous sheet supports stronger surface currents and higher ohmic and plasmonic dissipation, producing a deeper resonance and lower $|\Gamma|$. In contrast, the patch introduces current confinement and additional capacitive discontinuities, which weaken the effective coupling to the incident field and increase the surface-impedance mismatch. As a result, the patch absorbs less energy and reflects more, leading to a shallower reflection dip compared with the uniform sheet.

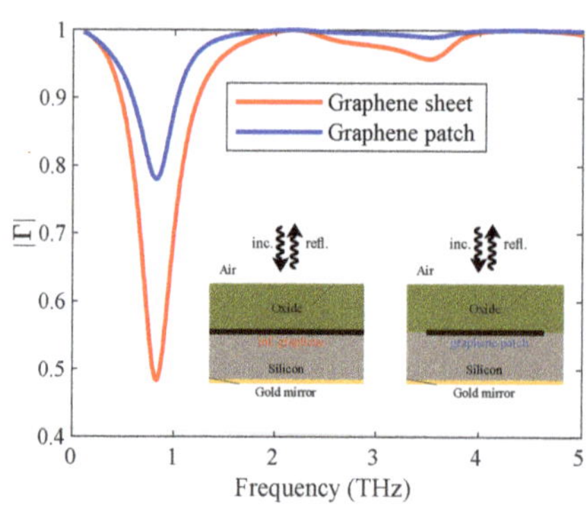

Figure 10. Reflection amplitude for graphene sheet and patch configurations, both assumed at low chemical potential, $\mu_c = 0.1$ eV.

https://doi.org/10.3390/photonics13020195

The reconfigurable reflection-phase results shown in Figure 11 demonstrate strong and broadband tunability as the graphene chemical potential is varied from 0.1 eV to 1.0 eV. The maximum phase tuning is achieved near 1.2 THz and exceeds 240°. Increasing μ_c shifts the resonance features to higher frequencies. This behavior arises from the change in graphene's complex-valued surface conductivity and its impact on the effective impedance of the multilayer structure can be understood through the ECM. The smooth phase evolution and relatively large reflection magnitude confirm the suitability of the design for THz wave phase-modulated metasurfaces, even without the contribution of VO_2. More meticulous optimization of the dimensions (dielectric material and/or thickness) and graphene patch shape and placement might further improve the performance.

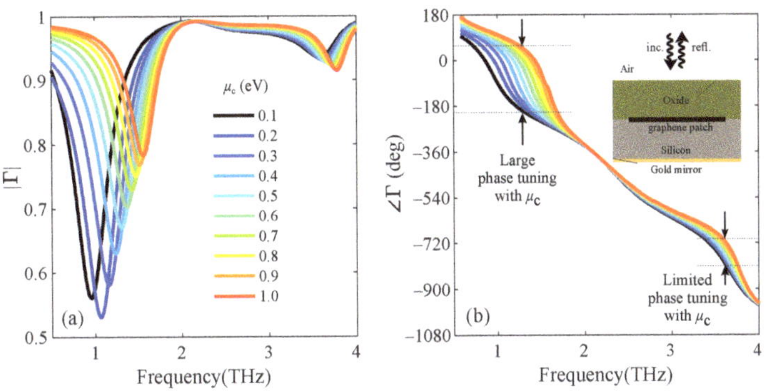

Figure 11. Full-wave simulated reflection (**a**) magnitude and (**b**) unwrapped phase spectra as the chemical potential is varied from 0.1 eV to 1 eV, in the graphene-patch configuration without VO_2.

3.3. *Hybrid Graphene/VO_2 Configuration for Independent Reflection Phase and Amplitude Modulation*

In this section, graphene and VO_2 patches are simultaneously present to investigate the potential of such hybrid configurations for the independent control of the reflection amplitude and phase, enabled by the combined tuning mechanisms: changing graphene μ_c (phase tunability) and/or VO_2 bulk conductivity (amplitude modulation).

In Figure 12a we present the reflection magnitude spectra when the VO_2 conductivity is varied from its insulating state to the metallic state, while the graphene chemical potential is fixed at 0.5 eV. We observe the reflection amplitude modulation particularly around the resonance near 1.2 THz. As the conductivity of VO_2 increases, both the depth and the spectral position of the reflection minima vary significantly, demonstrating strong amplitude tunability of the MS. At the resonance position, the reflection magnitude is almost -30 dB, indicating near-perfect impedance matching and high absorption induced by the metallic state of VO_2. In contrast, when VO_2 is in the insulating state, the reflection magnitude is considerably higher, corresponding to a weakly lossy and predominantly reflective regime. This wide dynamic range in $|\Gamma|$ highlights the effectiveness of VO_2 as an active amplitude-modulation element in the proposed metasurface design. Also, maintaining the graphene chemical potential at 0.5 eV results in only marginal changes in the reflection amplitude, confirming that graphene has a minimal impact on amplitude modulation under these conditions. One can observe that there exists a critical σ_{VO_2} at which full absorption is achieved, in this case 2.5×10^4 S/m. Further increasing σ_{VO_2} reduces the absorption and induces a shift of the resonance frequency. This behavior highlights an intrinsic limitation of the present design: that the tuning of graphene and VO_2 on the EM response—namely the amplitude and phase—are not fully decoupled for one frequency. In particular, variations of σ_{VO_2} could lead to more complete coverage

https://doi.org/10.3390/photonics13020195

of the magnitude and phase, as shown below, but possibly not at the same frequency. This behavior is consistent with the predominantly reactive response of graphene in the THz regime, where it primarily influences the phase rather than introducing significant dissipative losses.

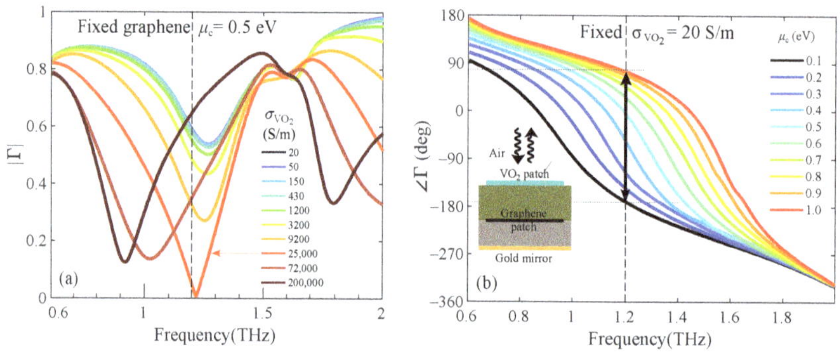

Figure 12. Hybrid graphene/VO_2 unit cell simulated reflection (**a**) magnitude spectra as σ_{VO_2} is varied with fixed $\mu_c = 0.5$ eV and (**b**) phase spectra as μ_c is varied with fixed $\sigma_{VO_2} = 20$ S/m.

In Figure 12b, we present the reflection phase spectra of the hybrid MS unit cell as graphene μ_c is varied from 0.1 eV to 1.0 eV while the VO_2 patch array (thickness 0.1 µm) is fixed in its insulating state, i.e., at $\sigma_{VO_2} = 20$ S/m. As μ_c increases, the resonance frequency blue-shifts, as indicated by the $\angle\Gamma = 0$ crossing. This behavior originates from the tunable surface conductivity of graphene, which alters the effective surface impedance of the metasurface and, consequently, the phase of the reflected wave. The wide and continuous phase evolution across the considered range of μ_c confirms the strong capability of graphene to enable dynamic phase modulation in the proposed unit cell. The maximal reflection phase coverage of 245° at 1.2 THz, when $\mu_c = 0.1 \to 1$ eV, is almost enough to create a 2-bit (4-state) digital holographic metasurface.

To showcase the full tunability map of the complex reflection coefficient at the hybrid unit–cell, in Figure 13 we present $|\Gamma|$ and $\angle\Gamma$ as a function of both μ_c tuning (horizontal axis) and σ_{VO_2} tuning (vertical axis) at a fixed operating frequency of 1.2 THz. The magnitude heatmap in panel Figure 13a shows that there exists one specific combination of control parameters μ_c and σ_{VO_2} that leads to perfect absorption ($|\Gamma| < -20$ dB). On the bright side, tuning *any* of the two control parameters can lead to absorption modulation of over 15 dB. For instance, for a $|\Gamma| = -20 \to -5$ dB modulation, one could either toggle $\mu_c = 0.1 \to 0.5$ eV (at fixed $\sigma_{VO_2} \approx 10^4$ S/m) or toggle $\sigma_{VO_2} = 10^4 \to 10^5$ S/m (at fixed $\mu_c = 0.5$ eV).

The phase heatmap in Figure 13b clearly reveals a monotonic and nearly uniform phase variation covering 245° along the μ_c axis for σ_{VO_2} up to 3500 S/m; low σ_{VO_2} values are preferable for holographic phase control as $|\Gamma|$ does not fluctuate much as μ_c is swept. Now, the absence of noticeable phase variation along the σ_{VO_2} axis indicates that, in its insulating state, VO_2 has a negligible influence on the phase response at this frequency. This demonstrates that the phase modulation mechanism is effectively decoupled from VO_2 and is predominantly governed by graphene. Consequently, graphene serves as the phase-tuning element, while VO_2 is reserved for independent amplitude (i.e., absorption) control in the hybrid metasurface designs. This decoupled behavior is highly advantageous as it enables independent and flexible control of phase and magnitude within a single MS.

https://doi.org/10.3390/photonics13020195

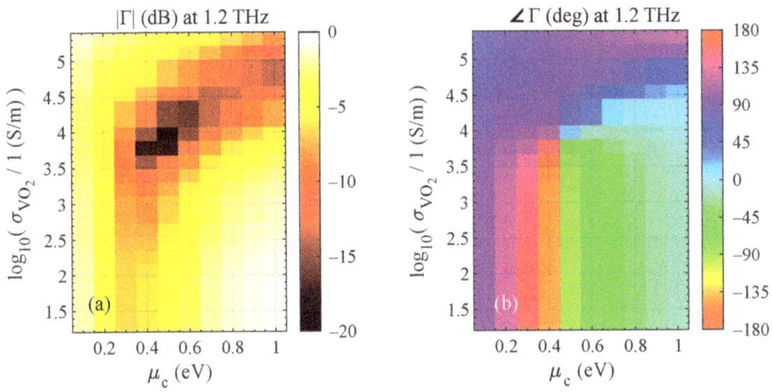

Figure 13. Hybrid graphene/VO_2 unit cell simulated reflection (**a**) magnitude and (**b**) phase heatmaps at 1.2 THz as both σ_{VO_2} and μ_c are varied.

Concluding the hybrid graphene/VO_2 unit cell performance analysis, in Figure 14 we evaluate the $\{\sigma_{VO_2}, \mu_c\}$-reconfigurable reflection at 1.2 THz under oblique 30° incidence, in both TE and TM planes. The CST simulated results show only slight degradation in performance, i.e., higher overall absorption (more evident in TM polarization) and a small reduction in $\angle\Gamma$-coverage. These indicate that the proposed RMS unit cell performance is robust under oblique illumination in both polarization planes.

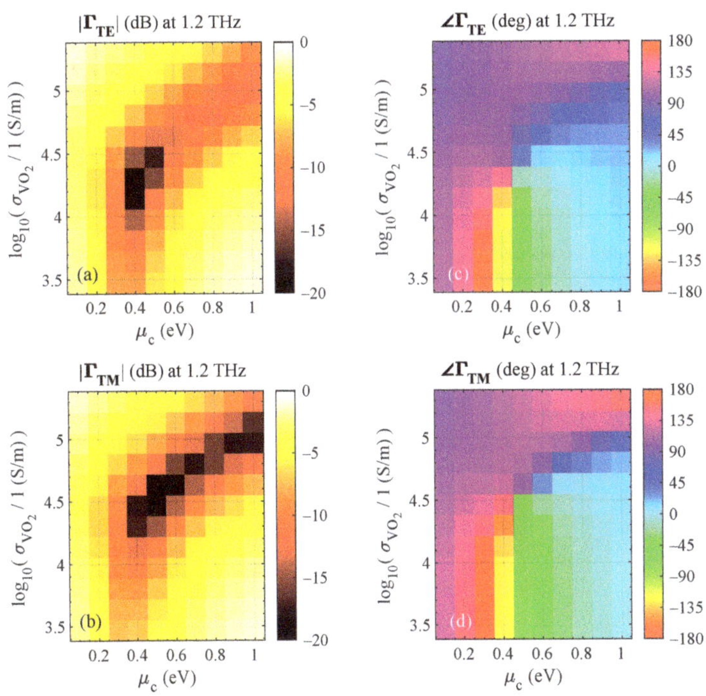

Figure 14. Hybrid graphene/VO_2 unit cell simulated reflection (**a**,**b**) magnitude and (**c**,**d**) phase heatmaps at 1.2 THz as both σ_{VO_2} and μ_c are varied, under oblique 30° incidence in both (**a**,**c**) TE and (**b**,**d**) TM planes.

https://doi.org/10.3390/photonics13020195

4. Discussion and Comparison

As presented in the Introduction, the independent control of the phase and the magnitude of the scattered THz wave is desired to implement high-performing multifunctional and reconfigurable metasurfaces (RMS), enabling wavefront shaping and sensing applications within a single platform. In this section, we present similar RMS designs and associated functionalities, with a special emphasis on the material platforms employed here, i.e., graphene and VO_2, and the involved technological complexity.

In [58], the authors used VO_2 to achieve near-perfect absorption and reflection by switching conductivity between metallic and dielectric states. In [59], VO_2 is used to exhibit dual functionalities of broadband absorption and reflective beam steering by switching from metallic state to insulator state. In [60], the authors used VO_2 together with patterned metallic resonators to achieve both broadband and triple narrowband absorption in a simple 3-layer configuration by switching between the metal state and insulator state, respectively [61]. In [62], graphene is used to achieve the same effect as in [60] by controlling the chemical potential of graphene in a 4-layer design consisting of metal-dielectric-graphene-dielectric architecture from bottom to top. In [37], graphene was employed to achieve nearly 100% reflection and absorption by reconfiguring the MS response via the chemical potential of graphene; the 3-layer structure of graphene and silicon over a mirror was analyzed using both equivalent circuits and simulation; the Authors also achieved both anomalous reflection and wavefront modulation by controlling a group of unit cells. Phase-modulated RMS was studied in [63], where a graphene-based MS gave rise to more than 330° phase modulation in the mid-infrared spectrum. Also, in [64], the authors achieved efficiently tunable reflective MS consisting of a dielectric substrate sandwiched between a hollow Z-shaped graphene structure and a ground plane. The phase tuning was achieved by adjusting the rotation angle of the graphene. In [65], the authors achieved phase shift of 180° by changing the Fermi level based on a graphene hybrid MS. In 2018 [66], the authors achieved an efficient control of both amplitude (>50 dB) and phase control (>90°) in the microwave frequencies at 11.8 GHz using a hybrid combination of graphene capacitively coupled with a split ring resonator (SRR) for radar absorbing materials. In [64], the authors used multi-layer varactor diodes loaded on metallic patches to achieve full control of phase and amplitude response. The metasurface operated at 6 GHz. In 2023 [67], the dual amplitude and phase tuning was achieved similarly by an arrangement of two varactor diodes on the top and a lumped resistor on the bottom layer at a frequency band of 4 to 9 GHz. In 2024 [43], the authors proposed an equivalent circuit approach for independent phase and amplitude control based on a Floquet modal expansion method [68,69] in the mmWave frequency band. The amplitude control layer consisted of two monolayer graphene and an electrolyte layer between them, and the phase control layer consisted of two metallic patches connected by an ideal varactor diode and a ground dielectric slab.

Both graphene and VO_2 are used in [70,71] for polarization conversion, triple band, and dual band absorption. Also, in [72], the authors used integration of a metal patch, graphene, and VO_2 to design a programmable coding MS in the THz band capable of switching between dynamic beam steering and dual-band absorber. In [25], a multifunctional absorber based on graphene and VO_2 with band selection capability was designed and analyzed. Broadband absorption exceeding 90% was achieved when VO_2 was set as metal and graphene chemical potential was tuned. Similarly, the authors achieved double narrowband absorption with VO_2 in insulator mode and setting graphene μ_c at specific values. Similarly, broadband and narrowband near-perfect absorption and transmission tuning using both graphene and VO_2 was studied in [56,73,74].

https://doi.org/10.3390/photonics13020195

Having reviewed the existing literature, it is evident that substantial potential remains in the use of hybrid graphene–VO_2 metasurfaces. Most reported studies employ graphene and VO_2 either individually or in hybrid forms, but primarily for amplitude-modulated applications. Thus, relatively limited investigation was devoted to phase modulation and independent control, which are essential for enabling more advanced smart radio environments [75,76]. In Table 2, we compare the unit cell design and methodology outlined in the present work to some of the similar papers found in the literature to showcase its comparative advantage in *independent* and wide control of reflection phase and magnitude.

Table 2. Performance comparison of this work with previously reported graphene- and/or VO_2-based MS unit cell designs.

Ref.	Material Used	Number of Layers	Operating Frequency	Modeling Method	Functionality (Amplitude/Phase)	Functionality Description
[60]	VO_2–Metal	3	THz	Simulation	Amplitude	Triple Band/ Broadband Absorption
[37]	Graphene	3	THz	Simulation/ECM	Amplitude/Phase	Reflection/Absorption Phase Gradient MS.
[64]	Metal Patches	5	GHz	Simulation/ECM	Amplitude/Phase	Simultaneous Phase/Amplitude at GHz
[65]	Graphene–Metal	3	THz	Simulation	Phase	Maximum phase shift 170°
[72]	Metal-Graphene-VO_2	3	THz	Simulation/ECM	Amplitude/Phase	Phase gradient Coded MS Absorption/Beam steering
[18]	VO_2-Photosensitive Silicon	7	THz	Simulation	Amplitude/Phase	Broadband Absorption/Polarization Conversion
This work	VO_2–Graphene	5	THz	Simulation/ECM	Amplitude/Phase	Reflection/Absorption Independent control −28 dB, 245° at 1.2 THz

5. Summary and Outlook

In this work, we incorporated graphene and VO_2 materials in a simple multi-layer hybrid metasurface unit cell to achieve independent (decoupled) reflection amplitude and phase modulation of incident THz beams. The tuning of the phase, spanning over 245°, was achieved by changing the chemical potential of graphene patches sandwiched between an oxide and a silicon layer; the tuning of the amplitude, from approximately −3 dB to near-perfect absorption, was achieved by changing the bulk conductivity of an overlying array of VO_2 patches.

The proposed configuration resulted in a hybrid unit cell with minimal technological complexity, which was initially designed using intuitive and efficient all-analytical methods. The designs were validated and subsequently optimized using full-wave numerical simulations at the unit-cell level, showing good agreement with the model in the target zone, around 1.2 THz. The tuning range and performance can possibly be enhanced, or translated in central frequency, by further optimization.

As future research steps, we envision the study of non-local response (accounting for inter-cell coupling in heterogeneous RMS configurations), finite-aperture diffraction, and the multiphysical aspects of the required electro-thermo-optic controls.

https://doi.org/10.3390/photonics13020195

Author Contributions: Conceptualization, E.A. and A.P.; methodology, A.P. and E.A.; software, E.A.; validation, E.A., A.P., E.M.S., and K.S.O.K.; formal analysis, A.P. and E.M.S.; investigation, E.A. and A.P.; resources, E.A.; data curation, E.A.; writing—original draft preparation, E.A.; writing—review and editing, A.P., E.M.S., and K.S.O.K.; visualization, E.A. and A.P.; supervision, A.P., E.M.S. and K.S.O.K. All authors have read and agreed to the published version of the manuscript.

Funding: This research received no external funding.

Institutional Review Board Statement: Not applicable.

Informed Consent Statement: Not applicable.

Data Availability Statement: Data supporting the findings of this study are available from the corresponding author upon reasonable request.

Conflicts of Interest: The authors declare no conflicts of interest.

Abbreviations

The following abbreviations are used in this manuscript:

MS	Metasurface
RMS	Reconfigurable Metasurface
TSI	Tabulated Surface Impedance
ECM	Equivalent Circuit Model
TL	Transmission Line
TLM	Transmission Line Model
ITE	Impedance Transformation Equation
FSR	Free Spectral Range

References

1. Liu, L.; Zhang, X.; Kenney, M.; Su, X.; Xu, N.; Ouyang, C.; Shi, Y.; Han, J.; Zhang, W.; Zhang, S. Broadband metasurfaces with simultaneous control of phase and amplitude. *Adv. Mater.* **2014**, *26*, 5031–5036. [CrossRef]
2. Lee, G.Y.; Yoon, G.; Lee, S.Y.; Yun, H.; Cho, J.; Lee, K.; Kim, H.; Rho, J.; Lee, B. Complete amplitude and phase control of light using broadband holographic metasurfaces. *Nanoscale* **2018**, *10*, 4237–4245. [CrossRef]
3. Ding, F.; Cui, Y.; Ge, X.; Jin, Y.; He, S. Ultra-broadband microwave metamaterial absorber. *Appl. Phys. Lett.* **2012**, *100*, 103506. [CrossRef]
4. Song, Q.; Zhang, W.; Wu, P.C.; Zhu, W.; Shen, Z.X.; Chong, P.H.J.; Liang, Q.X.; Yang, Z.C.; Hao, Y.L.; Cai, H.; et al. Water-resonator-based metasurface: An ultrabroadband and near-unity absorption. *Adv. Opt. Mater.* **2017**, *5*, 1601103. [CrossRef]
5. Zhang, Z.; Shi, H.; Wang, L.; Chen, J.; Chen, X.; Yi, J.; Zhang, A.; Liu, H. Recent advances in reconfigurable metasurfaces: Principle and applications. *Nanomaterials* **2023**, *13*, 534. [CrossRef]
6. Dong, T.; Zhang, Y.; Li, Y.; Tang, Y.; He, X. Development of a terahertz metamaterial polarization converter with multiple tuning modes based on graphene and vanadium dioxide. *Diam. Relat. Mater.* **2023**, *139*, 110373. [CrossRef]
7. Luo, H.; Liu, H.; Chen, C.; Feng, Y.; Gao, P.; Ren, Z.Y.; Qiao, Y.J. Dual-broadband terahertz absorber based on phase transition characteristics of VO_2. *Results Phys.* **2022**, *34*, 105270. [CrossRef]
8. Zhuang, L.; Zhang, W.; Liu, J.; Chao, M.; Liu, Q.; Cheng, B.; Xu, Y.; Song, G. Switchable trifunctional terahertz absorber for both broadband and narrowband operations. *Opt. Express* **2022**, *30*, 45848–45861. [CrossRef] [PubMed]
9. Holloway, C.L.; Kuester, E.F.; Gordon, J.A.; O'Hara, J.; Booth, J.; Smith, D.R. An overview of the theory and applications of metasurfaces: The two-dimensional equivalents of metamaterials. *IEEE Antennas Propag. Mag.* **2012**, *54*, 10–35. [CrossRef]
10. Chen, H.T.; Taylor, A.J.; Yu, N. A review of metasurfaces: Physics and applications. *Rep. Prog. Phys.* **2016**, *79*, 076401. [CrossRef]
11. Yu, S.; Kim, Y.; Shin, E.; Kwon, S.H. Dynamic beam steering and focusing graphene metasurface mirror based on fermi energy control. *Micromachines* **2023**, *14*, 715. [CrossRef]
12. Pors, A.; Nielsen, M.G.; Eriksen, R.L.; Bozhevolnyi, S.I. Broadband focusing flat mirrors based on plasmonic gradient metasurfaces. *Nano Lett.* **2013**, *13*, 829–834. [CrossRef]
13. Venkatesh, S.; Lu, X.; Saeidi, H.; Sengupta, K. A high-speed programmable and scalable terahertz holographic metasurface based on tiled CMOS chips. *Nat. Electron.* **2020**, *3*, 785–793. [CrossRef]
14. Ding, X.; Monticone, F.; Zhang, K.; Zhang, L.; Gao, D.; Burokur, S.N.; De Lustrac, A.; Wu, Q.; Qiu, C.W.; Alù, A. Ultrathin Pancharatnam–Berry metasurface with maximal cross-polarization efficiency. *Adv. Mater.* **2015**, *27*, 1195–1200. [CrossRef]

https://doi.org/10.3390/photonics13020195

15. He, X.; Wu, M.; Lu, G.; Zhang, Y.; Geng, Z. High-efficiency multi-channel focusing and imaging enabled by polarization-frequency multiplexing non-interleaved metasurfaces. *Photon. Res.* **2025**, *13*, 976–986. [CrossRef]
16. Wu, M.; He, X.; Lu, G.; Zhang, Y.; Geng, Z.; Liu, H.; Zhang, K. On-Chip Multi-Channel Wavefront Manipulation of Spoof Surface Waves with Structural Heterogeneous Metasurfaces. *Laser Photon. Rev.* **2025**, e01634. [CrossRef]
17. Qi, L.; Liu, C.; Shah, S.M.A. A broad dual-band switchable graphene-based terahertz metamaterial absorber. *Carbon* **2019**, *153*, 179–188. [CrossRef]
18. Dong, T.; Zhang, Y.; Li, Y.; Tang, Y.; He, X. Dual-function switchable terahertz metamaterial device with dynamic tuning characteristics. *Results Phys.* **2023**, *45*, 106246. [CrossRef]
19. Pitilakis, A.; Seckel, M.; Tasolamprou, A.; Liu, F.; Deltsidis, A.; Manessis, D.; Ostmann, A.; Kantartzis, N.; Liaskos, C.; Soukoulis, C.; et al. Multifunctional Metasurface Architecture for Amplitude, Polarization and Wave-Front Control. *Phys. Rev. Appl.* **2022**, *17*, 064060. [CrossRef]
20. Taghvaee, H.; Pitilakis, A.; Khodadadi, M.; Tsilipakos, O.; Boulogeorgos, A.A.A.; Khalily, M. Holographic mmWave Metasurface Integrating THz Sensing for 6G Wireless Networks. *IEEE Wirel. Commun.* **2025**, *32*, 54–62. [CrossRef]
21. Wang, J.; Yang, R.; Ma, R.; Tian, J.; Zhang, W. Reconfigurable multifunctional metasurface for broadband polarization conversion and perfect absorption. *IEEE Access* **2020**, *8*, 105815–105823. [CrossRef]
22. Balci, O.; Polat, E.O.; Kakenov, N.; Kocabas, C. Graphene-enabled electrically switchable radar-absorbing surfaces. *Nat. Commun.* **2015**, *6*, 6628. [CrossRef]
23. Zou, W.; Zhong, C.; Hong, L.; Lei, J.; Shen, Y.; Deng, X.; Chen, J.; Guo, T. Switchable Vanadium Dioxide Metasurface for Terahertz Applications. *Micromachines* **2024**, *15*, 967. [CrossRef]
24. Liu, M.; Hwang, H.Y.; Tao, H.; Strikwerda, A.C.; Fan, K.; Keiser, G.R.; Sternbach, A.J.; West, K.G.; Kittiwatanakul, S.; Lu, J.; et al. Terahertz-field-induced insulator-to-metal transition in vanadium dioxide metamaterial. *Nature* **2012**, *487*, 345–348. [CrossRef] [PubMed]
25. Liu, Y.; Hu, L.; Liu, M. Graphene and vanadium dioxide-based terahertz absorber with switchable multifunctionality for band selection applications. *Nanomaterials* **2024**, *14*, 1200. [CrossRef]
26. Low, T.; Chaves, A.; Caldwell, J.D.; Kumar, A.; Fang, N.X.; Avouris, P.; Heinz, T.F.; Guinea, F.; Martin-Moreno, L.; Koppens, F. Polaritons in layered two-dimensional materials. *Nat. Mater.* **2017**, *16*, 182–194. [CrossRef]
27. Farzin, P.; Soleimani, M. Graphene-based metasurface for real-time control of three electromagnetic wave modes and polarization state. *Diam. Relat. Mater.* **2023**, *139*, 110279. [CrossRef]
28. Wang, L.; Gao, F.; Teng, S.; Tan, Z.; Zhang, X.; Lou, J. Terahertz tunable vanadium dioxide metasurface for dynamic illusion and cloaking. *iScience* **2024**, *27*, 108609. [CrossRef]
29. Luukkonen, O.; Simovski, C.R.; Raisanen, A.V.; Tretyakov, S.A. An efficient and simple analytical model for analysis of propagation properties in impedance waveguides. *IEEE Trans. Microw. Theory Tech.* **2008**, *56*, 1624–1632. [CrossRef]
30. Luukkonen, O.; Simovski, C.; Granet, G.; Goussetis, G.; Lioubtchenko, D.; Raisanen, A.V.; Tretyakov, S.A. Simple and accurate analytical model of planar grids and high-impedance surfaces comprising metal strips or patches. *IEEE Trans. Antennas Propag.* **2008**, *56*, 1624–1632. [CrossRef]
31. Chen, J.; Hong, L.; Lei, J.; Shen, Y.; Deng, X.; Chen, J.; Guo, T. High-Performance Terahertz Coherent Perfect Absorption with Asymmetric Graphene Metasurface. *Photonics* **2024**, *11*, 544. [CrossRef]
32. Landy, N.I.; Sajuyigbe, S.; Mock, J.J.; Smith, D.R.; Padilla, W.J. Perfect metamaterial absorber. *Phys. Rev. Lett.* **2008**, *100*, 207402. [CrossRef]
33. Jiao, X.F.; Zhang, Z.H.; Li, T.; Xu, Y.; Song, G.F. Tunable dual broadband terahertz metamaterial absorber based on vanadium dioxide. *Appl. Sci.* **2020**, *10*, 7259. [CrossRef]
34. Tang, X.; Jia, H.; Liu, L.; Li, M.; Wu, D.; Zhou, K.; Li, P.; Tian, L.; Yang, D.; Wang, W. A tunable terahertz absorber based on double-layer patterned graphene metamaterials. *Materials* **2023**, *16*, 4166. [CrossRef]
35. Hanson, G.W. Dyadic Green's functions for an anisotropic, non-local model of biased graphene. *IEEE Trans. Antennas Propag.* **2008**, *56*, 747–757. [CrossRef]
36. Hanson, G.W. Dyadic Green's functions and guided surface waves for a surface conductivity model of graphene. *J. Appl. Phys.* **2008**, *103*. [CrossRef]
37. Dash, S.; Psomas, C.; Krikidis, I.; Akyildiz, I.F.; Pitsillides, A. Active control of THz waves in wireless environments using graphene-based RIS. *IEEE Trans. Antennas Propag.* **2022**, *70*, 8785–8797. [CrossRef]
38. Barzegar-Parizi, S.; Ebrahimi, A.; Ghorbani, K. Two bits dual-band switchable terahertz absorber enabled by composite graphene and vanadium dioxide metamaterials. *Sci. Rep.* **2024**, *14*, 5818. [CrossRef]
39. Wang, X.C.; Tretyakov, S.A. Graphene-based tunable metasurface for all-angle perfect absorption. In *2018 12th International Congress on Artificial Materials for Novel Wave Phenomena (Metamaterials)*; IEEE: Espoo, Finland, 2018; pp. 430–432.
40. Pasdari-Kia, M.; Memarian, M.; Khavasi, A. Generalized equivalent circuit model for analysis of graphene/metal-based plasmonic metasurfaces using floquet expansion. *Opt. Express* **2022**, *30*, 35486–35499. [CrossRef] [PubMed]

https://doi.org/10.3390/photonics13020195

41. Tretyakov, S. *Analytical Modeling in Applied Electromagnetics*; Artech House: Norwood, MA, USA, 2003.
42. Chatzidimitriou, D.; Pitilakis, A.; Kriezis, E.E. Rigorous calculation of nonlinear parameters in graphene-comprising waveguides. *J. Appl. Phys.* **2015**, *118*. [CrossRef]
43. You, X.; Kosmas, P. Equivalent Circuit Modeling of a Novel Reconfigurable Metasurface With Independent Control of Amplitude and Phase Based on Floquet Modal Expansion. *IEEE J. Microwaves* **2024**, *4*, 428–439. [CrossRef]
44. Dassault Systèmes. *CST Studio Suite—Electromagnetic Field Simulation Software*; Dassault Systèmes: Waltham, DC, USA, 2024.
45. Jin, J.M. *The Finite Element Method in Electromagnetics*, 3rd ed.; Wiley: New York, NY, USA, 2014.
46. Pozar, D.M. *Microwave Engineering*, 4th ed.; Wiley: Hoboken, NJ, USA, 2012.
47. Vakil, A.; Engheta, N. Transformation Optics Using Graphene. *Science* **2011**, *332*, 1291–1294. [CrossRef] [PubMed]
48. Gong, Y.; Liu, N. Advanced numerical methods for graphene simulation with equivalent boundary conditions: A review. *Photonics* **2023**, *10*, 712. [CrossRef]
49. Albooyeh, M.; Ra'Di, Y.; Adil, M.; Simovski, C. Revised transmission line model for electromagnetic characterization of metasurfaces. *Phys. Rev. B Condens. Matter Mater. Phys.* **2013**, *88*, 085435. [CrossRef]
50. Zhang, J.; Yan, H.; Yang, X.; Lyu, H. Dual-Function Metasurface for Tunable Selective Absorption. *Micromachines* **2022**, *13*, 2087. [CrossRef]
51. Yang, D.; Liu, X.; Wang, Y.; Zhang, Y. Programmable VO_2 Metasurface for Terahertz Wave Beam Manipulation. *Materials* **2022**, *15*, 104824.
52. Mou, N.; Tang, B.; Li, J.; Dong, H.; Zhang, L. Switchable ultra-broadband terahertz wave absorption with VO_2-based metasurface. *Sci. Rep.* **2022**, *12*, 2501. [CrossRef]
53. Simovski, C.R.; Tretyakov, S.A. Modeling and Analysis of Metasurfaces as Two-Dimensional Electromagnetic Structures. *IEEE Trans. Antennas Propag.* **2012**, *60*, 2587–2595.
54. Wang, X.; Tretyakov, S.A. Toward ultimate control of terahertz wave absorption in graphene. *IEEE Trans. Antennas Propag.* **2018**, *67*, 2452–2461. [CrossRef]
55. Yu, N.; Genevet, P.; Kats, M.A.; Aieta, F.; Tetienne, J.P.; Capasso, F.; Gaburro, Z. Light propagation with phase discontinuities: Generalized laws of reflection and refraction. *Science* **2011**, *334*, 333–337. [CrossRef]
56. Chen, K.; Wang, Z.; Guan, M.; Cheng, S.; Ma, H.; Yi, Z.; Li, B. Tunable ultra-wideband VO_2–graphene hybrid metasurface terahertz absorption devices based on dual regulation. *Photonics* **2025**, *12*, 987. [CrossRef]
57. Gómez-Díaz, J.S.; Perruisseau-Carrier, J. Graphene-Based Plasmonic Switches at Terahertz Frequencies. *Opt. Express* **2013**, *21*, 15490–15504. [CrossRef] [PubMed]
58. Afra, T.; Fuscaldo, W.; Zografopoulos, D.C.; Natale, T.; Dell'Olio, F. Tunable wide band near-perfect absorber for terahertz waves based on a vanadium dioxide metasurface. *Opt. Quantum Electron.* **2025**, *57*, 272. [CrossRef]
59. Zhang, Y.; Ma, S.; Fu, W.; Huang, S.; Zhu, Y.; Luo, X. Vanadium dioxide-based terahertz metasurface device with switchable broadband absorption and beam steering functions. *Opt. Commun.* **2024**, *560*, 130486. [CrossRef]
60. Zou, Y.; Lin, H.; Tian, G.; Zhou, H.; Zhu, H.; Xiong, H.; Wang, B.X. Triple-Band and Ultra-Broadband Switchable Terahertz Meta-Material Absorbers Based on the Hybrid Structures of Vanadium Dioxide and Metallic Patterned Resonators. *Materials* **2023**, *16*, 4719. [CrossRef]
61. Liu, C.; Xu, Y.; Huang, R.; Zha, S. Terahertz metamaterials for broadband, high modulation depth modulator, and tunable dual-band absorber based on metal-vanadium dioxide hybrid structure. *Aip Adv.* **2023**, *13*, 095204. [CrossRef]
62. Chen, Z.; Chen, J.; Tang, H.; Shen, T.; Zhang, H. Dynamically switchable broadband and triple-band terahertz absorber based on a metamaterial structure with graphene. *Opt. Express* **2022**, *30*, 6778–6785. [CrossRef]
63. Sun, Z.; Huang, F.; Fu, Y. Graphene-based active metasurface with more than 330° phase tunability operating at mid-infrared spectrum. *Carbon* **2021**, *173*, 512–520. [CrossRef]
64. Zheng, B.; Ren, H.; An, S.; Tang, H.; Li, H.; Haerinia, M.; Dong, Y.; Fowler, C.; Zhang, H. Tunable metasurface with dynamic amplitude and phase control . *IEEE Access* **2021**, *9*, 104522–104529. [CrossRef]
65. Zhou, S.; Hu, F.; Wang, H.; Su, A.; Jiang, M.; Zhang, L.; Gao, Z.; Luo, W. Large dynamic range terahertz phase shifter based on a graphene–metal hybrid metasurface. *Appl. Opt.* **2025**, *64*, 5188–5194. [CrossRef]
66. Balci, O.; Kakenov, N.; Karademir, E.; Balci, S.; Cakmakyapan, S.; Polat, E.O.; Caglayan, H.; Özbay, E.; Kocabas, C. Electrically switchable metadevices via graphene. *Sci. Adv.* **2018**, *4*, eaao1749. [CrossRef]
67. Phon, R.; Lee, M.; Lor, C.; Lim, S. Multifunctional reflective metasurface to independently and simultaneously control amplitude and phase with frequency tunability. *Adv. Opt. Mater.* **2023**, *11*, 2202943.
68. Bhattacharyya, A.K. *Phased Array Antennas: Floquet Analysis, Synthesis, BFNs and Active Array Systems*; John Wiley & Sons: Hoboken, NJ, USA, 2006.
69. Rodriguez-Berral, R.; Mesa, F.; Medina, F. Analytical multimodal network approach for 2-D arrays of planar patches/apertures embedded in a layered medium. *IEEE Trans. Antennas Propag.* **2015**, *63*, 1969–1984. [CrossRef]

https://doi.org/10.3390/photonics13020195

70. Mistri, H.; Ghosh, A.; Sardar, A.R.; Roy, P. Multifunctional voltage and temperature controlled metasurface using graphene and vanadium dioxide for terahertz applications. *Opt. Quantum Electron.* **2025**, *57*, 332. [CrossRef]
71. Fang, J.; Zhu, W.; Cao, L.; Huang, X.; Zhang, B.; He, D.; Wang, S. Hybrid dual-mode tunable polarization conversion metasurface based on graphene and vanadium dioxide. *Opt. Express* **2023**, *31*, 23095–23105. [CrossRef]
72. Naghshpour, N.; Pakizeh, T. Hybrid VO_2-graphene bi-functional digital metasurface for beam steering and dual-band absorption. *J. Appl. Phys.* **2025**, *138*, 203103. [CrossRef]
73. Li, J.; Liu, Y.; Chen, Y.; Chen, W.; Guo, H.; Wu, Q.; Li, M. Tunable broadband-narrowband and dual-broadband terahertz absorber based on a hybrid metamaterial vanadium dioxide and graphene. *Micromachines* **2023**, *14*, 201.
74. Zhuo, S. Switchable dual ultra-broadband perfect absorber based on graphene and vanadium dioxide hybrid metamaterials. *J. Opt. Soc. Am. B* **2024**, *41*, 1483–1489. [CrossRef]
75. Di Renzo, M.; Zappone, A.; Debbah, M.; Alouini, M.S.; Yuen, C.; De Rosny, J.; Tretyakov, S. Smart radio environments empowered by reconfigurable intelligent surfaces: How it works, state of research, and the road ahead. *IEEE J. Sel. Areas Commun.* **2020**, *38*, 2450–2525. [CrossRef]
76. Renzo, M.D.; Debbah, M.; Phan-Huy, D.T.; Zappone, A.; Alouini, M.S.; Yuen, C.; Sciancalepore, V.; Alexandropoulos, G.C.; Hoydis, J.; Gacanin, H.; et al. Smart radio environments empowered by reconfigurable AI meta-surfaces: An idea whose time has come. *Eurasip J. Wirel. Commun. Netw.* **2019**, *2019*, 1–20. [CrossRef]

https://doi.org/10.3390/photonics13020195

hv photonics

MDPI

Article

Quasi-Bound States in the Continuum in PDMS-Supported Silicon Metasurfaces

Sy Khiem Nguyen [1], Ba Thong Trinh [1], Dayeon Kim [1], Netrapal Singh [1], Young Kyu Hwang [2], Vu Dinh Lam [3] and Ilsun Yoon [1,*]

[1] Department of Chemistry, Chungnam National University, Daejeon 34134, Republic of Korea
[2] Chemical Process Technology Division, Korea Research Institute of Chemical Technology, Daejeon 34114, Republic of Korea
[3] Graduate University of Science and Technology, Vietnam Academy of Science and Technology, 18 Hoang Quoc Viet, Nghia Do, Hanoi 100000, Vietnam
* Correspondence: ilsunyoon@cnu.ac.kr

Abstract

Quasi-bound states in the continuum (quasi-BICs) in all-dielectric metasurfaces support high-Q resonances that are highly sensitive to structural symmetry and radiative coupling. Most previous studies have focused on static configurations on rigid substrates, whereas the behavior of quasi-BIC modes in the presence of low-index polymer supports remains less explored. In this work, we present a numerical investigation of quasi-BIC resonances in a silicon nanodimer metasurface on a polydimethylsiloxane (PDMS) substrate by systematically analyzing the effects of in-plane asymmetry, light incident angle and substrate thickness variation on their spectral position and quality factor. The results demonstrate pronounced tuning of the resonance wavelength and linewidth while preserving the characteristic high-Q behavior of quasi-BIC modes. This study establishes PDMS-supported silicon nanodimers as a viable platform for quasi-BIC metasurfaces and provides guidelines for future mechanically or chemically reconfigurable infrared devices based on polymer substrates.

Keywords: quasi-bound states in the continuum; dielectric metasurfaces; high-Q resonance; Si nanodimers; angle-dependent optical response

1. Introduction

Bound states in the continuum (BICs) and their leaky counterparts, quasi-bound states in the continuum (quasi-BICs), have emerged as a powerful route to trap and manipulate light in compact photonic structures [1–3]. In contrast to conventional cavity modes that are isolated from the radiation continuum by a bandgap or total internal reflection, BICs reside inside the continuum of propagating modes yet remain perfectly confined due to symmetry mismatch or destructive interference, ideally exhibiting infinite radiative lifetimes [1,2,4]. In realistic structures, fabrication imperfections or intentional perturbations relax this perfect confinement and give rise to quasi-BICs with finite but very large quality factors and ultra-narrow spectral linewidths, which are more relevant for experiments and devices [5,6]. The ability of BICs and quasi-BICs to support strong field enhancement in subwavelength volumes has motivated extensive work in photonic crystal slabs, resonant metasurfaces, and polaritonic nanoresonators [2,3,7,8].

Because of these properties, BIC-based platforms have been exploited across a broad range of applications. In passive systems, quasi-BIC resonances underpin narrow-band absorbers and thermal emitters [9], highly sensitive refractive index and biochemical sensors [10], and waveguides or beam-shaping elements that use the topological nature of BIC polarization vortices [2]. In plasmonic and phonon polaritonic systems, mirror-coupled and polariton-based BIC or quasi-BIC metasurfaces have enabled tunable perfect absorption and compact, narrow-band mid-infrared thermal emitters with significantly enhanced Q-factors compared to earlier metasurface designs [11,12]. More broadly, recent reviews have emphasized that BIC-enabled light confinement, sharp Fano resonances, and far-field polarization control are now core tools for lasing, nonlinear frequency conversion, sensing, and wavefront engineering [4,7,10].

Most of these demonstrations, however, rely on rigid substrates such as quartz, silicon-on-insulator, sapphire, or CaF_2 [13–16]. Such platforms are ideal for maintaining the delicate geometric and symmetry conditions required for high-Q quasi-BICs, but they inherently yield static devices that cannot easily conform to curved surfaces or undergo large mechanical deformations. Even in active or reconfigurable implementations—using electro-optic modulation in $BaTiO_3$ [5], thermo-optic tuning, or mirror-coupled plasmonic absorbers with engineered loss and gap sizes—the underlying substrates remain mechanically rigid. As a result, the tuning mechanisms are typically based on refractive index modulation, carrier injection, or local phase-change materials rather than large-amplitude elastic deformation.

In parallel, there has been rapidly growing interest in tunable metasurfaces, particularly for terahertz and infrared sensing. These platforms are often realized by patterning metallic resonators on reconfigurable polymer or elastomer substrates, taking advantage of their low refractive index, mechanical compliance, and low cost [17–22]. However, many of these flexible metasurfaces rely on conventional resonances with moderate Q-factors because radiation loss and material absorption limit the achievable linewidths. The recent demonstration of bending sensing based on quasi-BIC resonances in a flexible terahertz metallic metasurface shows that BIC physics can indeed be combined with mechanical deformation, but it also highlights that most BIC-enabled flexible sensors to date operate in the THz regime and employ lossy metallic elements [17,23]. At the same time, recent reviews stress that tunability and integration with novel materials remain major challenges and opportunities for next-generation BIC photonics [4,24].

Motivated by these developments, it is natural to ask how high-Q quasi-BIC metasurfaces behave when they are placed on such low-index polymer substrates, and to what extent the substrate properties themselves can shape their resonant response. Building on this concept, we numerically investigate a silicon double-ellipse metasurface on a PDMS substrate in the mid-infrared (3–6 μm) spectral range, where both silicon and PDMS exhibit relatively low absorption for thin structure. Our results show that PDMS-supported silicon nanodimer metasurfaces can sustain quasi-BICs with Q-factors on the order of 10^3 while exhibiting sensitivity to angle and substrate thickness. We hope these results can provide design guidelines for future mechanically or chemically reconfigurable infrared devices based on polymer substrate platforms suitable for next-generation optical sensing and wearable devices.

2. Design and Simulation Methods

The optical response of the quasi-BIC metasurface was investigated numerically using the Wave Optics Module of COMSOL Multiphysics 6.2. A single unit cell of a periodic array of silicon (Si) nanodimers supported by a substrate was modeled, as schematically shown in Figure 1. The lateral dimensions of the unit cell were set to $P_x = 4.2$ μm, $P_y = 2.9$ μm. The substrate is a polydimethylsiloxane (PDMS) slab with thickness $t_{sub} = 1$ μm, covered by an upper air region of thickness $t_{air} = 6$ μm. Two identical silicon (Si) elliptical cylinders

 https://doi.org/10.3390/photonics13030226

of height $h_{Si} = 0.2$ µm were placed on top of the PDMS substrate. Each ellipse had in-plane semi-axes $D_x = 1.2$ µm (along the x-direction) and $D_y = 2.2$ µm (along the y-direction) and was rotated by an angle $\alpha = 20°$ with respect to the vertical axis. The half center-to-center distance between the two ellipses was set to $d = 1.05$ µm along the x-direction. The refractive indices of Si and PDMS were taken from data in the mid-infrared region [25,26]. The refractive index of the air superstrate was set to $n = 1.0$.

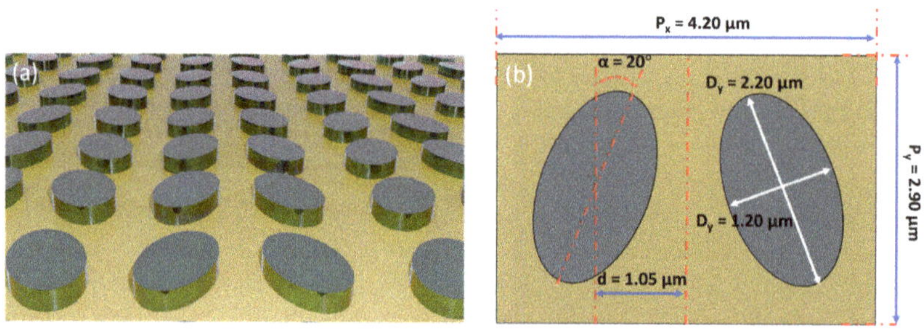

Figure 1. Schematic of the silicon nanodimer metasurface on a PDMS substrate. (**a**) Perspective view of the periodic array of two elliptical silicon resonators on a PDMS layer. (**b**) Top view of a single unit cell.

To represent an infinite metasurface, Floquet periodic boundary conditions were applied to the lateral boundaries of the unit cell along the x and y directions. A normally incident plane wave was launched from the top port, with the electric field polarized along the *x*-axis, while the bottom port absorbed the transmitted field. The reflection spectra were investigated by sweeping the free-space wavelength from 3000 to 6000 nm. A tetrahedral mesh was employed throughout the computational domain, with local refinement around the Si–air and Si–PDMS interfaces.

3. Results and Discussion

We first examine how the in-plane symmetry of the silicon nanodimer controls the appearance of the high-Q resonance at normal incidence. Figure 2a shows the symmetric configuration in which the two ellipses are parallel to each other and aligned with the vertical axis ($\alpha = 0°$) so that the unit cell possesses a mirror plane perpendicular to the incident electric field. In this case, the reflectance spectrum axis ($\alpha = 0°$ trace in Figure 2b) exhibited a resonance centered at $\lambda = 3.68$ µm. This feature originated from a bright mode lattice-guided resonance supported by the periodic Si metasurface and PDMS slab. The mode coupled efficiently to the external radiation field and was therefore observed in the reflectance spectrum. No additional narrow resonances were observed in this configuration, indicating that any BICs present in the structure are completely decoupled from the normally incident plane wave in this limit.

When the in-plane mirror symmetry was broken by rotating the two ellipses by a finite angle α with respect to the vertical axis, a new narrow spectral feature emerges in the 5.1–5.3 µm range, as seen in the reflectance map of Figure 2b. The lattice-guided resonance at $\lambda = 3.68$ µm remains almost dispersionless with respect to α. This behavior indicates that this lattice-guided resonance is governed by vertical waveguiding and lattice periodicity rather than by the dimer orientation. In contrast, the high-Q resonance is absent at $\alpha = 0°$, and appears for any nonzero rotation angle, with its linewidth increasing as α grows. This resonance is completely absent in the symmetric configuration and appears only after the in-plane symmetry is perturbed, strongly suggesting that it originates from a symmetry-protected BIC that is dark at $\alpha = 0°$ and becomes radiative when $\alpha \neq 0°$. In other words, the rotation of the ellipses lifts the symmetry mismatch between the bound mode

https://doi.org/10.3390/photonics13030226

and the outgoing plane wave, converting the ideal BIC into a high-Q quasi-BIC whose radiative coupling can be controlled by the degree of asymmetry.

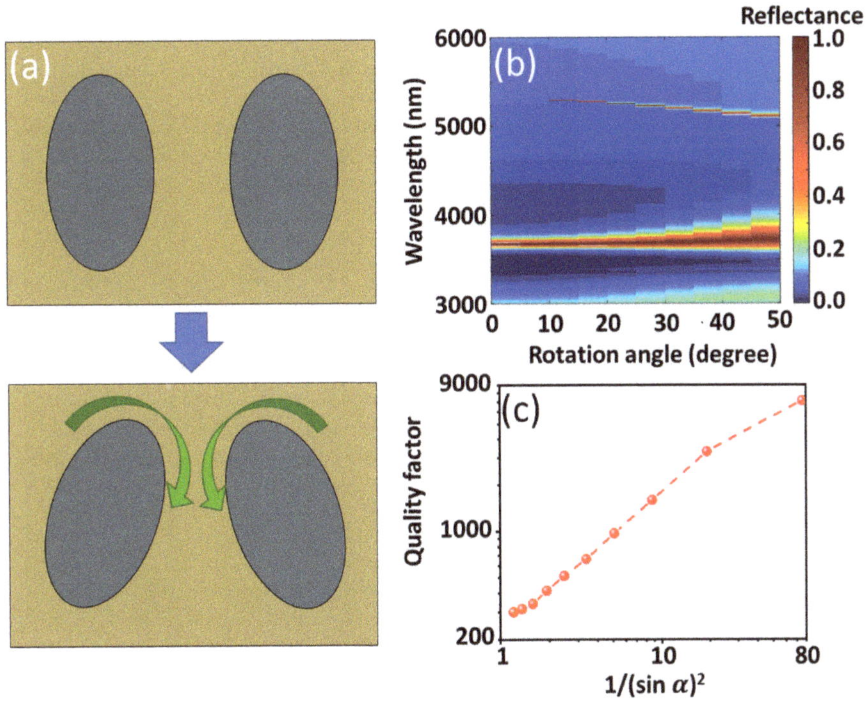

Figure 2. (**a**) Top-view schematics of the symmetric dimer configuration with two parallel ellipses ($\alpha = 0°$) and the rotated dimer in which the ellipses are tilted by an angle α with respect to the vertical axis, the green arrows indicate the rotation direction of the ellipses (**b**) Simulated reflectance as a function of wavelength and rotation angle α at normal incidence. (**c**) Extracted quality factors of the q-BIC as a function of $1/(\sin\alpha)^2$.

To quantify this dependence, we extracted the quality factor of the high-Q resonance from one-dimensional cuts of the reflectance spectrum at each rotation angle and plotted it as a function of $1/(\sin\alpha)^2$ in Figure 2c. The data points follow an approximately linear trend over the explored range, consistent with the theoretical scaling $Q \propto 1/(\sin\alpha)^2$ predicted for symmetry-protected quasi-BICs in perturbed periodic structures [27]. This scaling confirms that the narrow resonance indeed originates from a symmetry-protected BIC of the $\alpha = 0°$ metasurface whose radiative leakage is controlled by the dimer rotation angle. From now on, unless otherwise noted, we fix the rotation angle to $\alpha = 20°$ as a representative asymmetry value that yields a high-Q quasi-BIC with a less sensitive linewidth to fabrication imperfections.

To clarify the nature of the two resonances identified in Figure 2, we decomposed the multipole contributions of the Si dimer to analyze the physical mechanism of the two resonance modes. The induced electric current density in the Si dimer were calculated, from which the multipole moments were obtained following the formulation in Ref. [28]:

$$p_\alpha = -\frac{1}{i\omega}\{\int d^3r J^{\omega}_{\alpha} j_0(kr) + \frac{k^2}{2}\int d^3\mathbf{r}[3(\mathbf{r}\cdot\mathbf{J}_\omega)r_\alpha - r^2 J^{\omega}_{\alpha}]\frac{j_2(kr)}{(kr)^2}\} \quad (1)$$

$$m_\alpha = \frac{3}{2}\int d^3\mathbf{r}(\mathbf{r}\times\mathbf{J}_\omega)_\alpha \frac{j_1(kr)}{(kr)^2} \quad (2)$$

https://doi.org/10.3390/photonics13030226

$$Q^{e}_{\alpha\beta} = -\frac{3}{i\omega}\{\int d^{3}r[3\left(r_{\beta}J^{\omega}_{\alpha} + r_{\alpha}J^{\omega}_{\beta}\right) - 2(r\cdot J_{\omega})\delta_{\alpha\beta}]\frac{j_1\left(kr\right)}{\left(kr\right)^{2}} + 2k^{2}\int d^{3}r[5r_{\alpha}r_{\beta}(r\cdot J_{\omega}) - (r_{\alpha}J_{\beta} + r_{\beta}J_{\alpha})r^{2} - r^{2}(r\cdot J_{\omega})\delta_{\alpha\beta}]\frac{j_3\left(kr\right)}{\left(kr\right)^{3}} \tag{3}$$

$$Q^{m}_{\alpha\beta} = 15\int d^{3}\mathbf{r}\{r_{\alpha}(\mathbf{r}\times\mathbf{J}_{\omega})_{\beta} + r_{\beta}(\mathbf{r}\times\mathbf{J}_{\omega})_{\alpha}\}\frac{j_2(kr)}{(kr)^{2}} \tag{4}$$

where $p_{\alpha}, m_{\alpha}, Q^{e}_{\alpha\beta}, Q^{m}_{\alpha\beta}$ are the electric dipole, magnetic dipole, electric quadrupole, and magnetic quadrupole, respectively. $\alpha, \beta = x, y, z$ are the Cartesian component indices, $\mathbf{r}$ is position vector inside the dimer, $\mathbf{J}_{\omega}$ is the induced current density at angular frequency ω, $\delta_{\alpha\beta}$ is the Kronecker delta, k is the wave number in the surrounding medium, and $j_1(kr)$ is the spherical Bessel function with order 1. For the multipole decomposition, we integrate induced current density over the combined volume of two silicon ellipses. The origin of the coordinate system is placed at the geometric center of the dimer (midpoint between the ellipse centers and at the mid-height of the silicon layer); z is normal to the substrate, x lies along the dimer axis (center-to-center line), and y is the in-plane transverse direction.

The multipole spectra in Figure 3a show that, for the resonance at $\lambda = 3.68$ μm, the response is overwhelmingly dominated by the electric dipole (ED) contribution, while the magnetic dipole (MD), electric quadrupole (EQ), and magnetic quadrupole (MQ) terms remain an order of magnitude smaller across the entire linewidth. The ED dominance directly implies that the radiation into the far field is governed by a bright dipolar channel.

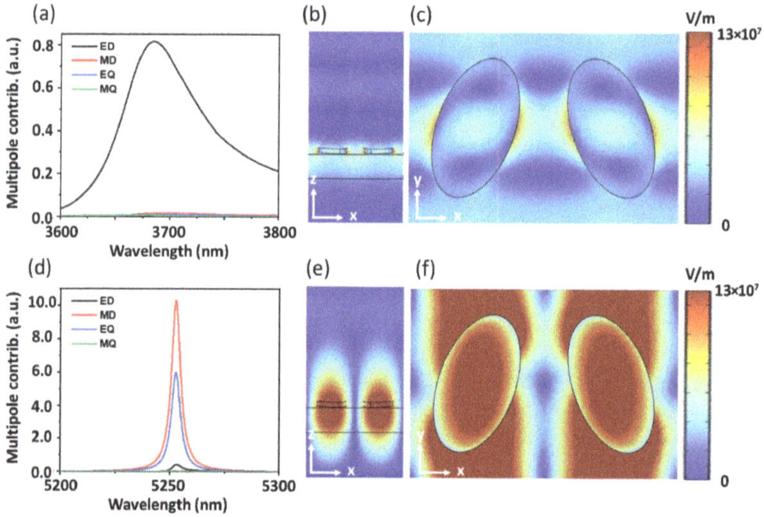

Figure 3. (**a**) Multipole contributions of the lattice-guided resonance at $\lambda = 3.68$ μm. (**b**) xz-plane and (**c**) yz-plane cross-sectional electric field distribution at the same wavelength. (**d**) Multipole contributions of the symmetry-protected quasi-BIC at $\lambda = 5.25$ μm. (**e**) xz-plane and (**f**) yz-plane cross-sectional electric field distribution at the same wavelength. In panels (**b**,**c**,**e**,**f**), the black outlines indicate the boundaries of the Si ellipses and the relevant material interfaces (air/Si/PDMS) shown for visual guidance.

The field maps in Figure 3b,c are consistent with this interpretation. In the *xz* plane (Figure 3b), the electric field intensity is only moderately enhanced near the dimer and decays gradually into the PDMS substrate, forming a standing-wave pattern characteristic of a leaky guided mode of the Si–PDMS slab. The field is not tightly confined to the resonators, which explains the relatively strong radiation leakage inferred from the dominant ED contribution. In the *yz* plane (Figure 3c), the in-plane electric field distribution shows that the electric field is

https://doi.org/10.3390/photonics13030226

concentrated near the edges of the Si ellipses and extends laterally along the x direction across the unit cell. Altogether, these features identify the $\lambda = 3.68$ μm resonance as a conventional electric-dipole-type lattice-guided resonance with relatively low Q.

In contrast, the high-Q resonance at $\lambda = 5.25$ μm exhibited a different multipolar signature, as shown in Figure 3d. Here the MD and EQ contributions are strongly enhanced and exhibit very sharp, high-amplitude peaks, whereas the ED contribution is strongly suppressed and remains comparatively small. This redistribution of scattering power from the ED channel into magnetic and higher-order multipoles indicates that the electric dipole radiation of the unit cell is largely canceled by destructive interference between different current distributions inside the dimer. As a result, the mode couples only weakly to the external radiation continuum and attains a much higher Q-factor.

The field profiles in Figure 3e,f further support this picture. In the *xz* plane cross-section (Figure 3e), the electric field is strongly localized around the Si ellipses, with intense lobes that decay rapidly away from the resonators, indicating that most of the energy is confined within a subwavelength volume near the high-index inclusions. In the *yz* plane (Figure 3f), there is a highly symmetric field distribution across the dimer geometry. While intense "hotspots" exceeding 13×10^7 V/m are localized within the high-index material, the field also extends into the surrounding medium as evanescent "tails".

To examine the angular dependence of the q-BIC supported by the metasurface, we investigated its evolution as a function of the incident angle. Figure 4a showed the reflectance as a function of wavelength and incident angle for the rotated dimer configuration ($\pm\alpha = 20°$) at fixed PDMS thickness $t_{sub} = 1$ μm. The bright, nearly horizontal band around $\lambda = 3.68$ μm corresponds to the lattice-guided resonance already discussed previously; its resonance position is almost insensitive to the incident angle because it is mainly governed by lattice periodicity and vertical waveguiding. In contrast, the high-Q features in the 4.7–5.4 μm range form a set of dispersive branches, indicating the presence of angularly dispersive modes whose coupling to free space is strongly angle dependent.

To analyze these high-Q modes more clearly, Figure 4b presents a magnified view of the circled region. Two distinct branches can be identified. The upper branch, labeled s-q-BIC, originates at normal incidence and shifts with the change in incident angle. This branch is the angular continuation of the symmetry-protected quasi-BIC discussed in Figures 2 and 3: at incident angle 0° the mode is weakly coupled to the radiation continuum due to the quasi-BIC symmetry, and as the angle increases it remains predominantly localized in the nanodimer. The lower branch, which we refer to as a high-Q resonance branch, emerges at a smaller wavelength and exhibits a strong angular dispersion. Its trajectory suggests hybridization between the nanodimer resonance and a guided mode of the Si–PDMS stack: as the incident angle increases, the in-plane wavevector component matches with the guided spectrum, leading to a pronounced variation in radiative loss.

The corresponding quality factors, extracted from one-dimensional cuts of the reflectance spectrum, are plotted in Figure 4c for both branches. For the symmetry-protected quasi-BIC (magenta curve), the Q factor starts at Q = 973 at normal incidence and gradually increases to Q = 2041 as the angle approaches 15°. This trend indicates that oblique incidence further reduces the mode overlap with the available radiation channels in the surrounding media, leading to slightly lower radiative leakage. In contrast, the high-Q resonance branch (blue curve) exhibits a non-monotonic behavior: the Q factor starts at Q = 4383 at the angle 1°, and then drops rapidly as the angle is increased further. Beyond the angle 6°, the mode becomes strongly leaky, and its Q factor falls to much lower values, Q = 801 at 6.5° and Q = 142 at 7°. We interpret this lower branch as a guided-resonance-assisted high-Q mode whose radiation loss is minimized only in a finite angular window.

https://doi.org/10.3390/photonics13030226

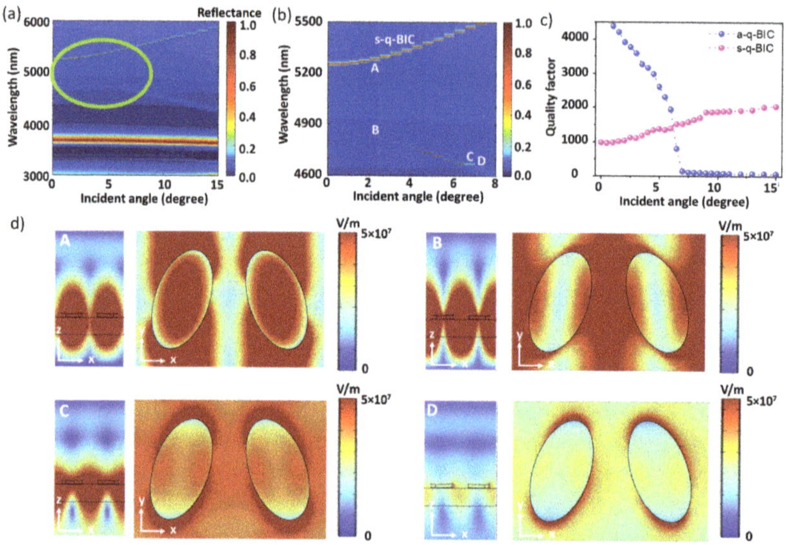

Figure 4. (**a**) Simulated reflectance as a function of wavelength and angle of incidence for the rotated dimer metasurface (α = 20°) at fixed PDMS thickness t_{sub} = 1 μm. (**b**) Enlarged view of the high-Q region circled in (**a**). (**c**) Extracted quality factor Q of the symmetry-protected-quasi-BIC (s-q-BIC) (magenta) and high Q-resonance (blue) as a function of incident angle. (**d**) Electric field distribution corresponding to four marked points in (**b**): A, B, C and D. In panels (**d**), the black outlines indicate the boundaries of the Si ellipses and the relevant material interfaces (air/Si/PDMS) shown for visual guidance.

The near-field distributions at representative points A–D along the two branches are shown in Figure 4d. Point A lies on the s-q-BIC branch at incident angle 2°. Both the xz- and yz-plane cross-sections reveal intense, nearly uniform fields around the Si ellipses, consistent with the dimer-localized mode we interpreted in Figure 3d–f. Point B is chosen on the high-Q resonance branch at incident angle 2°. Here the field remains strongly concentrated around the ellipses but develops additional lobes, but still maintains high destructive interference in the far field. At point C, taken at incident angle 6.5° along the same branch, the guided mode contribution in the resonance become more dominant: the field becomes more delocalized in the vertical direction, and the contrast between the ellipses and the background is reduced, reflecting the increased radiative leakage and reduced Q factor. Finally, point D (at incident angle 7°) samples the vicinity where the high-Q branch approaches a nearby guided mode; the field pattern shows strong energy in the background and only moderate enhancement in the ellipses, characteristic of a leaky guided resonance.

To gain further insight into how the polymer substrate influences both the spectral position and radiation loss of the resonances, we investigate the dependence on PDMS thickness t_{sub}, shown in Figure 5, in more detail. The reflectance map in Figure 5a contains two clearly distinguishable bands. The lower, tilted band around $\lambda \approx 3.4$–3.8 μm corresponds to a leaky guided-mode lattice resonance of the Si–PDMS slab. Its wavelength increases almost linearly with PDMS thickness. As the PDMS layer becomes thicker, the optical path length grows and the resonance condition is satisfied at longer wavelengths. The field of this mode, as seen in Figure 3b,c, spans the entire slab thickness and extends into the surrounding media; the Si metasurface provides the lateral periodicity required to couple this guided mode to free-space radiation.

https://doi.org/10.3390/photonics13030226

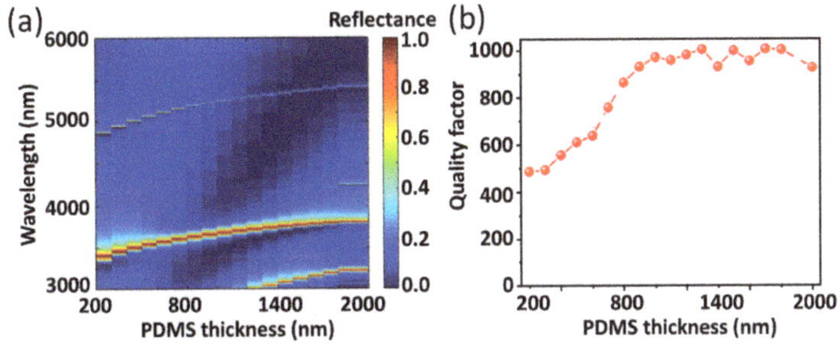

Figure 5. (**a**) Simulated reflectance as a function of wavelength and PDMS thickness at normal incidence for the rotated dimer metasurface. (**b**) Extracted quality factor Q of the symmetry-protected quasi-BIC as a function of PDMS thickness.

In contrast, the upper band in Figure 5a, located in the range 4.8–5.4 μm, is associated with the symmetry-protected quasi-BIC. In the absence of the PDMS slab (or for an infinitely thick, homogeneous background), the eigenfrequency of this mode would be governed almost exclusively by the in-plane geometry of the nanodimer and the incident angle. Introducing a finite-thickness PDMS layer, however, creates a vertically asymmetric environment consisting of air/Si/PDMS interfaces. As PDMS thickness varies, the vertical standing-wave pattern in the substrate shifts relative to the dimer array, which modifies both the effective index experienced by the quasi-BIC and its phase relation to the continuum of radiating slab modes. This leads to a slightly more pronounced dispersion of the quasi-BIC branch with thickness: the mode frequency tracks not only the intrinsic dimer resonance but also its hybridization with the continuum spectrum of slab-guided modes supported by the Si–PDMS stack.

The thickness dependence of the quality factor in Figure 5b provides a direct view of how this hybridization controls radiative loss. For very thin PDMS layers (200–500 nm), the field associated with the quasi-BIC has a substantial overlap with the lower interface and leaks efficiently into the continuum of substrate modes, resulting in moderate Q values of 487 to 610. As the PDMS thickness increases into the 800–1300 nm range, a larger fraction of the energy is concentrated around the high-index Si regions, while the field at the bottom interface is reduced. In this regime the coupling coefficient between the quasi-BIC and the radiative channels in the substrate decreases, and out-of-plane leakage is suppressed. Consequently, the quality factor increases and reaches values up to Q = 1003, indicating that the system approaches an optimal compromise between strong field confinement in the dimer and limited access to radiation continua. For even thicker substrates (≥1500 nm), the Q factor saturates and even exhibits a slight decline. Physically, further increasing PDMS thickness can introduce additional guided modes that approach the quasi-BIC frequency from below. Once these modes become nearly phase-matched, the quasi-BIC can hybridize with them and acquire additional leakage channels. The existence of a broad thickness window over which the Q factor is high therefore reflects a balance between vertical confinement (which improves with thickness) and avoided crossings or hybridization with leaky slab modes (which become more prominent at large PDMS thickness).

Taken together, these results show that the quasi-BIC is sensitive to the PDMS thickness change. The polymer layer does not simply act as a passive support; its thickness directly controls the vertical modal structure and, through that, the interference between quasi-BIC radiation and guided mode channels. This sensitivity suggests that modest changes in effective thickness—whether from fabrication tolerances, mechanical deformation, or index changes that alter the vertical mode profile—can be used to tune both the resonance wavelength and Q factor of quasi-BIC modes in dielectric metasurfaces.

https://doi.org/10.3390/photonics13030226

4. Conclusions

In this work, we numerically investigated quasi-bound states in the continuum in a silicon elliptical-dimer metasurface on a PDMS substrate in the mid-infrared. By using the relative rotation angle between the ellipses as a single asymmetry parameter, we confirmed the characteristic $Q \propto 1/(\sin \alpha)^2$ scaling of a symmetry-protected quasi-BIC and distinguished it from a lower-Q guided-mode lattice resonance that is mainly set by the Si–PDMS slab. Multipole analysis showed that the lattice resonance at $\lambda = 3.68$ μm is electric dipole dominated with fields extended through the slab, whereas the quasi-BIC at $\lambda = 5.25$ μm is governed by magnetic and higher-order multipoles with fields strongly confined in the Si dimers. Angle-resolved and thickness-dependent spectra revealed that the quasi-BIC maintains high Q-factors over a range of incidence angles and is sensitive to PDMS thickness, with an optimal thickness window where Q factor approaches 10^3. These results establish PDMS-supported silicon nanodimers as a viable platform for high-Q mid-infrared metasurfaces and provide guidelines for future flexible or reconfigurable photonic devices.

Author Contributions: Conceptualization, S.K.N. and I.Y.; data curation, S.K.N., B.T.T., D.K., N.S. and I.Y.; formal analysis, S.K.N. and I.Y.; investigation, S.K.N., V.D.L., Y.K.H. and I.Y.; methodology, S.K.N. and I.Y.; software, S.K.N., V.D.L. and I.Y., visualization, S.K.N., B.T.T., N.S. and I.Y.; writing—original draft, S.K.N., V.D.L., Y.K.H. and I.Y.; writing—review and editing, S.K.N., B.T.T., D.K., N.S., V.D.L., Y.K.H. and I.Y.; funding acquisition, I.Y.; supervision, I.Y. All authors have read and agreed to the published version of the manuscript.

Funding: This work was supported by Regional Innovation System & Education (RISE) program through the Daejeon RISE Center, funded by the Ministry of Education (MOE) and the Daejeon Metropolitan City, Republic of Korea (No. 2025-RISE-06-012).

Data Availability Statement: Data available on request.

Conflicts of Interest: The authors declare no conflicts of interest.

References

1. Hsu, C.W.; Zhen, B.; Stone, A.D.; Joannopoulos, J.D.; Soljačić, M. Bound states in the continuum. *Nat. Rev. Mater.* **2016**, *1*, 16048. [CrossRef]
2. Kang, M.; Liu, T.; Chan, C.; Xiao, M. Applications of bound states in the continuum in photonics. *Nat. Rev. Phys.* **2023**, *5*, 659–678. [CrossRef]
3. Chen, H.; Li, N.; Zhao, Y.; Ou, H.; Wang, Y.; Jing, X.; Zhang, N.; Su, Z.; Huang, L. Maximizing the chirality of bound states in the continuum by inverse design. *Photonics Res.* **2025**, *13*, 2054–2064. [CrossRef]
4. Wang, J.; Li, P.; Zhao, X.; Qian, Z.; Wang, X.; Wang, F.; Zhou, X.; Han, D.; Peng, C.; Shi, L.; et al. Optical bound states in the continuum in periodic structures: Mechanisms, effects, and applications. *Photonics Insights* **2024**, *3*, R01. [CrossRef]
5. Tian, H.; Li, J.; Wu, Y.; Wang, X. Dynamic switch between BIC and quasi-BIC supported by the electro-optic metasurface. *Opt. Commun.* **2025**, *574*, 131181. [CrossRef]
6. Kim, K.-H.; Kim, I.-P. Quasi-bound states in the continuum with high Q-factors in metasurfaces of lower-index dielectrics supported by metallic substrates. *RSC Adv.* **2022**, *12*, 1961–1967. [CrossRef] [PubMed]
7. Xu, G.; Xing, H.; Xue, Z.; Lu, D.; Fan, J.; Fan, J.; Shum, P.P.; Cong, L. Recent advances and perspective of photonic bound states in the continuum. *Ultrafast Sci.* **2023**, *3*, 0033. [CrossRef]
8. Wang, J.T.; Panoiu, N.C. Nonlinear optical metasurfaces empowered by bound-states in the continuum. *Rev. Phys.* **2025**, *13*, 100117. [CrossRef]
9. Zhu, G.; Hong, I.; Li, K.; Anyika, T.; Ugwu, M.T.; Nolen, J.R.; He, M.; Caldwell, J.D.; Ndukaife, J.C. Engineering Thermal Emission with Enhanced Emissivity and Quality Factor Using Bound States in the Continuum and Electromagnetically Induced Absorption. *Adv. Opt. Mater.* **2025**, *14*, e01257. [CrossRef]
10. Joseph, S.; Pandey, S.; Sarkar, S.; Joseph, J. Bound states in the continuum in resonant nanostructures: An overview of engineered materials for tailored applications. *Nanophotonics* **2021**, *10*, 4175–4207. [CrossRef]
11. Wang, J.; Weber, T.; Aigner, A.; Maier, S.A.; Tittl, A. Mirror-coupled plasmonic bound states in the continuum for tunable perfect absorption. *Laser Photonics Rev.* **2023**, *17*, 2300294. [CrossRef]

12. Yang, S.; He, M.; Hong, C.; Nordlander, J.; Maria, J.-P.; Caldwell, J.D.; Ndukaife, J.C. Single-peak and narrow-band mid-infrared thermal emitters driven by mirror-coupled plasmonic quasi-BIC metasurfaces. *Optica* **2024**, *11*, 305–314. [CrossRef]
13. Li, N.; Chen, H.; Zhao, Y.; Wang, Y.; Su, Z.; Liu, Y.; Huang, L. Ultrasensitive metasurface sensor based on quasi-bound states in the continuum. *Nanophotonics* **2025**, *14*, 485–494. [CrossRef] [PubMed]
14. Moretti, G.Q.; Weber, T.; Possmayer, T.; Cortés, E.; Menezes, L.d.S.; Bragas, A.V.; Maier, S.A.; Tittl, A.; Grinblat, G. Si metasurface supporting multiple quasi-BICs for degenerate four-wave mixing. *Nanophotonics* **2024**, *13*, 3421–3428. [CrossRef] [PubMed]
15. Huang, L.; Jin, R.; Zhou, C.; Li, G.; Xu, L.; Overvig, A.; Deng, F.; Chen, X.; Lu, W.; Alù, A.; et al. Ultrahigh-Q guided mode resonances in an All-dielectric metasurface. *Nat. Commun.* **2023**, *14*, 3433. [CrossRef]
16. Gupta, H.; Venturi, G.; Contino, T.; Janzen, E.; Edgar, J.H.; De Angelis, F.; Toma, A.; Ambrosio, A.; Tamagnone, M. Bound states in the continuum and long-range coupling of polaritons in hexagonal boron nitride nanoresonators. *ACS Photonics* **2024**, *11*, 4017–4026. [CrossRef] [PubMed]
17. Yang, F.; Li, J.; Wu, L.; Zheng, C.; Yue, Z.; Li, H.; Song, C.; Ding, X.; Zhang, Y.; Yao, J. Bending sensing based on quasi bound states in the continuum in flexible terahertz metasurface. *Adv. Opt. Mater.* **2023**, *11*, 2300909. [CrossRef]
18. Luo, Y.; Abidian, M.R.; Ahn, J.-H.; Akinwande, D.; Andrews, A.M.; Antonietti, M.; Bao, Z.; Berggren, M.; Berkey, C.A.; Bettinger, C.J.; et al. Technology roadmap for flexible sensors. *ACS Nano* **2023**, *17*, 5211–5295. [CrossRef]
19. Qu, J.; Cui, G.; Li, Z.; Fang, S.; Zhang, X.; Liu, A.; Han, M.; Liu, H.; Wang, X.; Wang, X. Advanced flexible sensing technologies for soft robots. *Adv. Funct. Mater.* **2024**, *34*, 2401311. [CrossRef]
20. Tian, J.; Cao, W. Reconfigurable flexible metasurfaces: From fundamentals towards biomedical applications. *PhotoniX* **2024**, *5*, 2. [CrossRef]
21. Zhu, X.; Xiao, S.; Shi, L.; Liu, X.; Zi, J.; Hansen, O.; Mortensen, N.A. A stretch-tunable plasmonic structure with a polarization-dependent response. *Opt. Express* **2012**, *20*, 5237–5242. [CrossRef]
22. Zhu, X.; Shi, L.; Liu, X.; Zi, J.; Wang, Z. A mechanically tunable plasmonic structure composed of a monolayer array of metal-capped colloidal spheres on an elastomeric substrate. *Nano Res.* **2010**, *3*, 807–812. [CrossRef]
23. Dai, Z.; Yan, C.; Ye, Y.; Chen, M.; Liang, Y.; Li, J. Quasi-BIC supported flexible terahertz metamaterial sensor for curvature measurement. *Opt. Express* **2024**, *32*, 35030–35038. [CrossRef]
24. Zeng, Y.; Zhang, X.; Ouyang, X.; Li, Y.; Qiu, C.W.; Song, Q.; Xiao, S. Manipulating light with bound states in the continuum: From passive to active systems. *Adv. Opt. Mater.* **2024**, *12*, 2400296. [CrossRef]
25. Li, H. Refractive index of silicon and germanium and its wavelength and temperature derivatives. *J. Phys. Chem. Ref. Data* **1980**, *9*, 561–658. [CrossRef]
26. Zhang, X.; Qiu, J.; Li, X.; Zhao, J.; Liu, L. Complex refractive indices measurements of polymers in visible and near-infrared bands. *Appl. Opt.* **2020**, *59*, 2337–2344. [CrossRef] [PubMed]
27. Koshelev, K.; Lepeshov, S.; Liu, M.; Bogdanov, A.; Kivshar, Y. Asymmetric metasurfaces with high-Q resonances governed by bound states in the continuum. *Phys. Rev. Lett.* **2018**, *121*, 193903. [CrossRef]
28. Alaee, R.; Rockstuhl, C.; Fernandez-Corbaton, I. An electromagnetic multipole expansion beyond the long-wavelength approximation. *Opt. Commun.* **2018**, *407*, 17–21. [CrossRef]

 https://doi.org/10.3390/photonics13030226

MDPI AG
Grosspeteranlage 5
4052 Basel
Switzerland
Tel.: +41 61 683 77 34

Photonics Editorial Office
E-mail: photonics@mdpi.com
www.mdpi.com/journal/photonics

www.ingramcontent.com/pod-product-compliance
Lightning Source LLC
Chambersburg PA
CBHW061603220926
51850CB00005B/414